# PROTEIN BIOSYNTHESIS
## and Problems of Heredity
## Development and Ageing

# PROTEIN BIOSYNTHESIS
## and Problems of Heredity, Development, and Ageing

by

## Zh. A. MEDVEDEV

Chief of the Laboratory of Molecular Radiobiology
Institute of Medical Radiology
Academy of Medical Sciences of the U.S.S.R., Obninsk

Translated by
ANN SYNGE

PLENUM PRESS

NEW YORK

1966

This is a translation of Биосинтез белков и проблемы онтогенеза
(*Biosintez belkov i problemy ontogeneza*) published by the State
Medical Publishing House, Moscow, in 1963, to which has been
added a supplementary chapter

Published in Great Britain by
Oliver and Boyd Ltd, Edinburgh and London, 1966

ENGLISH EDITION
First Published 1966

Published in the U.S.A. by
PLENUM PRESS
a division of PLENUM PUBLISHING CORPORATION
227 West 17th Street, New York, N.Y. 10011

Library of Congress Catalog Card Number 66-28598

Printed in Great Britain by
Robert Cunningham and Sons Ltd, Alva

# PREFACE TO THE RUSSIAN EDITION

This book is a study of various aspects of the problem of the biological synthesis of proteins and an analysis of the interaction between the mechanisms of synthesis of proteins and nucleic acids and the processes of individual development.

The need for an experimental and theoretical study of this problem is obvious, as almost all forms of biological growth and development are brought about, in the first place, by means of the synthesis of the fundamental components of living material, namely proteins and nucleic acids, which are the main material carriers of biological specificity. The reproduction of this specificity during ontogenesis is the basis of growth; its preservation through many generations forms the foundation for heredity; changes in it, whether determinate or accidental, direct the course of both ontogenesis and phylogenesis. The mechanisms of synthesis of proteins and nucleic acids are, in fact, the links which unite these universal biological phenomena and the elucidation of them provides the basis for directing various forms of biological development.

Biological chemistry as a whole, and especially the biochemistry of proteins and nucleic acids, is at present passing through a period of rapid development, characterized by very wide sweeps of experimental investigations and rapid changes in theoretical concepts. On the one hand this gives rise to demands for theoretical explanations of the new facts which are accumulating so quickly yet, on the other hand, the lifetime of these theories and hypotheses is far shorter than usual and their stability less sure. The theoretical concepts put forward in this book must be considered in the light of these circumstances.

As the problem of protein synthesis is closely associated with many far-flung branches of biology, such as genetics, virology, immunology, oncology, cytology etc., it is natural that, while we borrow facts from these branches of biology, we cannot give an exhaustive account of all the possible special aspects of protein synthesis. We have, therefore, tried to give the fullest possible treatment to the general laws governing protein synthesis, by means of a survey of all the work on the mechanism of the synthesis of proteins and the reproduction of their specificity. We have also tried to trace the connections between the mechanism of protein synthesis and the phenomena of heredity, morphogenesis and ageing.

In the past years a very large number of monographs and reviews has been published, dealing with particular aspects of the synthesis of proteins and nucleic acids and giving a fairly full account of the large quantity of factual material which had accumulated in this field. This

v

saves us from having to review the historical aspects of various problems and allows us to concentrate our attention mainly on an analysis of the most recent results and concepts which have not yet been adequately explained in the theoretical and review literature of biochemistry.

It must also be noted that a certain selectivity has been necessary in reviewing the factual and theoretical material. The most important and theoretically significant researches and hypotheses have been chosen for more detailed consideration from the general mass of results, which could not be considered in detail without unreasonably increasing the length of this book. In doing this we are striving for a theoretical and logical sequence and an objective account of the significance of the work, and not for the bibliographic completeness of a monograph.

The author would be very grateful to his colleagues for any critical observations or advice, which would certainly help with further work on generalizing from the material concerning the biological synthesis of proteins.

This book was written while the author was working in the Department of Agronomic and Biological Chemistry in the Timiryazev Agricultural Academy of Moscow. The author extends his thanks to all those in the Department and especially to its Head, Academician V. M. Klechkovskiĭ, for his constant help with the work. The author is also deeply grateful to his colleagues, Doctor of Biological Sciences V. G. Konarev and Professors I. B. Zbarskiĭ, S. Ya. Kaplanskiĭ, V. N. Nikitin, A. E. Braunshteĭn and V. S. Shapot, who have studied this work in manuscript and helped the author greatly by their comments and advice.

ZH. A. MEDVEDEV

*Institute of Medical Radiology,*
*Obninsk,*
*U.S.S.R.*

# PREFACE TO THE ENGLISH EDITION

This is a translation of *Biosintez belkov i problemy ontogeneza* which was published in Moscow (*Gosudarstvennoe Izdatel'stvo Meditsinskoĭ Literatury*) at the end of 1963. The Russian edition contained such material bearing on the problem as had been published up to the end of 1962. However, thanks to the kindness of Mrs A. Synge, who translated the book, and Dr R. L. M. Synge, who saw the English edition through the press, I have been able to write a special supplementary chapter. This surveys the main publications and the most important advances of 1963 and 1964. Certain minor alterations have also been made in some of the chapters where this was necessary.

Nowadays books on biochemistry and molecular biology become out of date very quickly. In view of this I have tried, in the preparation of this book, not just to present an exhaustive review of the material which had been assembled up to a given time, but to use the most important experimental and theoretical studies as a basis for tracing the history, over the preceding decade, of the working out of the problems under consideration, for discussing the lines which further studies may be expected to follow, and for showing the interdependence of different aspects of these problems.

Moreover, an attempt has been made in this book to deal in a comprehensive way with such very different problems as protein biosynthesis, nucleic acid biosynthesis, morphogenesis, differentiation and ageing on the basis of the molecular and genetic concepts which are common to them all.

In assembling the material I have naturally tried to give a fuller review of the work of Soviet authors than is usually given in monographs on biochemistry published in other countries.

The translation and publication of this work in English has only been made possible by the kindness and the tremendous amount of work expended on it by Mrs Ann Synge and Dr R. L. M. Synge, who showed great confidence in the author and agreed to undertake the translation of the book at a time when the Russian edition was still only being prepared for the press. For nearly four years the author has been discussing many questions concerning the content of the different chapters with the Synges. In the course of preparing the English edition for the press, Dr Synge has removed some mistakes from the bibliography and made it more systematic. Dr and Mrs Synge have also compiled an author index, which was not present in the Russian edition.

The author extends to Mrs Ann Synge and to Dr R. L. M. Synge his unbounded thanks for their invaluable and friendly help. He is also very grateful to Dr N. Cohen and to Dr N. A. Matheson for help in correcting the proofs of the English edition.

The author is also very grateful to the publishers of the English edition, Messrs Oliver & Boyd, for the honour they have done him by their decision to publish the book, and for the risk they have taken in publishing a book by a Russian author dealing with problems on which the main important advances have been made by the work of American and British scientists. The author will be glad if this risk should prove to be justified.

ZHORES ALEKSANDROVICH MEDVEDEV

*Laboratory of Molecular Radiobiology,*
*Institute of Medical Radiology of the Academy*
*of Medical Sciences of the U.S.S.R.*
*Obninsk,*
*U.S.S.R.*
*April 1965*

# TRANSLATOR'S NOTE

This translation has been read and approved by the author. In the earliest copies of the Russian edition to be distributed, part of Chapter XVII contained detailed and strongly worded attacks on T. D. Lysenko and his followers, which were replaced by more general discussion in the later copies. This translation follows the latter. I am depositing originals of both kinds in the Library of the British Museum.

One terminological point seems worth making. I have used the word "matrix" throughout, instead of the more usual word "template", because it is the word used in the Russian and it also seems to me to convey the idea of a three-dimensional mould (as used in printing newspapers), whereas "template" calls to mind a two-dimensional shape (as used in shipbuilding).

Dr Medvedev has already thanked those who have helped with the preparation of the English edition of the book. I must thank him for his patience and co-operativeness over the years which it has taken us to produce it.

The following illustrations are reproduced by courtesy of the authors and publishers cited (for brevity, where the reference is included in the lists at the end of chapters, the reference number only is given): 1: A. P. Ryle, F. Sanger, L. F. Smith & R. Kitai, *Biochem. J.* **60**, 541 (1955); 2: I, 30; 3: I, 24; 4: I, 41a; 5: I, 22; 7: I, 72; 8: I, 37; 9: I, 73; 10: P. Karlson, "Introduction to Modern Biochemistry" (Academic Press, New York, 1963), pp. 123 and 135; 12: II, 68; 13: II, 89; 14: II, 56; 15: IV, 122; 16: IV, 56; 17: IV, 61; 18 & 19: V, 73; 20: V, 213; 21: V, 253a; 22: V, 196; 23: V, 47: 24: V, 159; 25: V, 152; 26: V, 43; 27: VII, 70 & 157; 28: VII, 161; 29: VII, 46; 30: VII, 42; 34: XI, 63; 35: XIII, 16; 37: XIII, 178; 38-40: XIII, 171; 41: XIII, 125; 42: XIV, 65; 44: XIV, 14; 47 & 48: XIV, 116; 49: XV, 59; 50: XV, 57; 52: XV, 37; 54 & 55: XV, 18a; 58: XVIII, 85; 59 & 60: XVIII, 121; 61: XVIII, 64; 66: XX, 144; and 67: XX, 66.

ANN SYNGE

*Aberdeen,*
March 1966

# CONTENTS

## PART I
## GENERAL FEATURES OF THE BIOLOGICAL
## SYNTHESIS OF PROTEINS

# PART II

## THE SPECIAL FEATURES OF DIFFERENT FORMS OF BIOLOGICAL SYNTHESIS OF PROTEINS

CONTENTS

# SUPPLEMENT FOR THE ENGLISH EDITION

# ABBREVIATIONS

### Abbreviations Designating the Amino Acid Residues in Polypeptides and Proteins:

| | | | | | |
|---|---|---|---|---|---|
| Ala | = Alanine | Glu-NH$_2$ | = Glutamine | Phe | = Phenylalanine |
| Arg | = Arginine | Gly | = Glycine | Pro | = Proline |
| Asp | = Aspartic acid | His | = Histidine | Ser | = Serine |
| Asp-NH$_2$ | = Asparagine | Ileu | = Isoleucine | Thr | = Threonine |
| Cy-S- | = ½ Cystine | Leu | = Leucine | Try | = Tryptophan |
| CySH | = Cysteine | Lys | = Lysine | Tyr | = Tyrosine |
| Glu | = Glutamic acid | Met | = Methionine | Val | = Valine |

### Abbreviations Designating the Compounds Associated with the Metabolism of Nucleic Acids:

| | | | |
|---|---|---|---|
| ADP | = Adenosine diphosphate | PP | = Pyrophosphate |
| AMP | = Adenosine monophosphate (Adenylic acid) | RNA | = Ribonucleic acid |
| | | S-RNA | = "Soluble" or "transfer" RNA |
| ATP | = Adenosine triphosphate | | |
| CTP | = Cytidine triphosphate | RNP | = Ribonucleoprotein |
| DNA | = Desoxyribonucleic acid | TTP | = Thymidine triphosphate |
| GDP | = Guanosine diphosphate | UDP | = Uridine diphosphate |
| GMP | = Guanosine monophosphate | UMP | = Uridine monophosphate |
| GTP | = Guanosine triphosphate | UTP | = Uridine triphosphate |

### Miscellaneous Abbreviations

| | | |
|---|---|---|
| DEAE-cellulose | = | Diethylaminoethyl cellulose |
| MSH | = | Melanophore-stimulating hormone |
| S | = | Svedberg unit (sedimentation in ultracentrifuge) |
| TMV | = | Tobacco mosaic virus |

# PART I

# GENERAL FEATURES OF THE BIOLOGICAL SYNTHESIS OF PROTEINS

# INTRODUCTION

The biosynthesis of proteins is more involved and complicated than any of the syntheses which take place in living and non-living nature or any of the artificial syntheses which have ever been carried out by mankind.

The extraordinary complexity of the synthesis of many thousands of different proteins is determined by the necessity for the reproduction within the cell of a unique sequence of the twenty amino acids in long polypeptide chains and of a unique macromolecular structure which is directly linked with the ability of the protein to carry out its biological, structural and enzymic functions.

Biochemistry began to attack this problem several decades ago, but progress was very slow on this front and the advance lagged far behind that in other sectors of biochemistry, which were revealing the intimate mechanisms of metabolism of carbohydrates, fats, amino acids, phosphorus compounds, etc., and the mechanisms of biochemical energetics.

In the last decade, however, this front seems to have been opened up in two directions. The first of these concerns the mechanism for the reproduction of the specificity of a protein by means of a system of nucleic acids, which are capable of complementary autosynthesis and can act as matrices for protein synthesis and as carriers of a store of biochemical information. The other is concerned with the complicated biochemical enzymic system whereby amino acids are activated, thus accomplishing, in the course of their interaction with various forms of ribonucleic acid, the formation of peptide bonds between the amino acids, by means of which the sequence of the amino acids is determined in the polypeptide chain which is being formed.

Active research in these directions has had a decisive effect on the development of the biochemistry of protein and nucleic acid synthesis in recent years, so it is appropriate that we should concentrate our attention on these subjects in the first part of this book.

# CHAPTER I

# ESSENTIAL FEATURES OF THE STRUCTURE AND SPECIFICITY OF PROTEINS

## Introduction

The study of the structure of proteins is of tremendous theoretical and practical importance and is being actively pursued in dozens of laboratories. The work is at present progressing vigorously in a number of directions and is leading to the clear chemical delineation of those compounds which comprise the fundamental components of living material and which exercise functions connected with the essential manifestations of living processes, both simple and complicated. In this chapter we shall give a very short survey of the research and progress in this area.

In studying the structure of proteins a very extensive range of experimental and theoretical material has by now been accumulated, but a detailed analysis of all of it is not being attempted in this book. Quite full surveys are available in the current review journals and a number of monographs (5, 6, 8, 9, 26, 55, 56, 58, 62a, 64 and others).

We will only survey some problems of the structure of proteins which are logically connected with the analysis of possible ways in which they may be synthesized biologically and which are necessary for a proper understanding of the matter in later chapters.

## 1. The sequence of amino acid residues in the polypeptide chains of proteins

Proteins consist of polypeptide chains in which a single chemical group (the peptide bond) is exactly repeated at determinate intervals.

$$-CO-CH-NH-CO-CH-NH-$$
$$\phantom{-CO-}\underset{R}{|}\phantom{-NH-CO-}\underset{R}{|}$$

The various forms of specificity of the proteins depend on the characteristic sequence of the side-chains (R) of the twenty different amino acid residues.

A systematic study of the sequences of the amino acid residues in polypeptide chains started with the classical work of Sanger and his colleagues only ten years ago (61, 62). These studies led to the complete elucidation of the structure of the hormonal protein insulin which has a low molecular weight (Fig. 1).  In spite of the extreme laboriousness of the methods developed for the purpose, the structure of the protein ribonuclease has been fully worked out mainly as a result of work in two

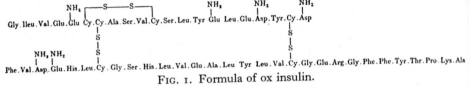

FIG. 1. Formula of ox insulin.

laboratories (18, 19, 21, 29, 30, 31, 47).  Its molecule is built up of 124 amino acid residues linked together to form a single unbranched polypeptide chain.  The complicated secondary structure of this protein depends on disulphide bridges and suggestions as to their probable nature were put forward by Smyth, Stein and Moore (30) in the form of the following scheme (Fig. 2).

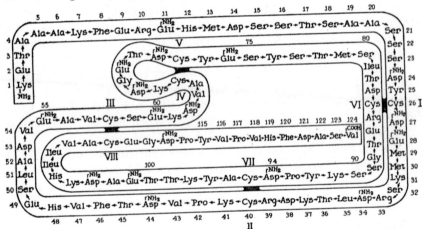

FIG. 2. The sequence of amino acid residues in bovine pancreatic ribonuclease A.

The recent work of Schramm and his colleagues (17) also represents a major achievement in biochemistry.  They have managed to deduce completely the sequence of amino acids in the polypeptide chain of the protein of the tobacco mosaic virus, which is composed of 157 amino acid residues.  The same structure (with some slight discrepancies) was also worked out by a group of American biochemists (68).  Work on the sequence of amino acid residues in lysozyme has been going on for many years and is now complete.  This is a protein with a polypeptide chain consisting of 126 residues (15, 67, 33, 34, 35).

In addition to these three proteins, workers in many laboratories have, in recent years, worked out the structures of many peptide and polypeptide hormones (vasopressin, oxytocin, angiotensin, α- and β-corticotrophins, glucagon, melanocyte-stimulating hormone) and a large number of peptide antibiotics. These studies have laid the basis for the artificial synthesis of series of hormones and antibiotics with a considerable range of variation in structure, like those which may be observed in nature.

As well as these studies, which are nearly complete, work is in progress on a very large number of proteins, the structures of which have fairly nearly been established. This usually takes several years even in large and specially equipped laboratories. As a result of intensive work over many years by a group of Czechoslovak biochemists, the nature of the majority of the bonds in the two enzymic proteins trypsin and chymotrypsinogen has been worked out. The investigations of this group of biochemists have been systematically published in a long series of papers and those which have already appeared have recently been collected into five articles (13, 39, 65, 66, 71). In the same laboratory a major piece of work on the constitution of haemoglobin is also being carried out (45, 46), as well as many other activities (27, 28, 32).

In the Soviet Union in the laboratory of Orekhovich work on the determination of the structure of proteins is also being successfully undertaken. The first studies on the problem carried out in this laboratory have led to the elucidation of the sequence of eighteen amino acids of the terminal polypeptide chain of pepsinogen (7).

The number of proteins of which the structure is in process of elucidation or of which a partial structure has been obtained included, by the end of 1961, more than 40 named proteins (chymotrypsin, carboxypeptidase, pepsin, pepsinogen, papain, elastase, ovalbumin, serum albumin, $\gamma$-globulin, myoglobin, cytochrome $c$, tropomyosin, actin, casein, $\beta$-lactoglobulin, salmine, collagen, clupeine, two fibroins, keratin, succinic dehydrogenase, fibrinogen and fibrin, protamines, histones, esterase, enolase, taka-amylase and some others) as well as those already mentioned.

If we add to these the proteins which have already been thoroughly studied and the eight hormones, we have more than 50 polypeptide formations which have been investigated. It must also be noted that many of the proteins and peptides studied have a number of specific varieties in which certain parts of the protein show definite specific variations (insulin from cattle, sheep, dogs, pigs, whales and man; corticotrophin from sheep, dogs and cattle; haemoglobin from dogs, fowls, rabbits, man, snakes etc.; cytochrome from horses, cattle, fowls etc.). If these varieties of proteins are regarded as independent substances, then the number of proteins and polypeptides undergoing investigation is increased to 80, not counting the antibiotic peptides (bacitracin, tyrocidines, gramicidins etc.) and the large number of proteins of which, as yet, only the terminal amino acids have been determined (zein, oryzenin, rhodopsin, lipoproteins, thrombin and prothrombin, edestin and many others), or modified forms of certain proteins arising under pathological conditions.

Thus, the structures of altogether more than 100 proteins and peptides have been examined. Therefore, although this work is still far from complete, we can already attempt to establish some regularities in the structure of different proteins. A demonstration of such regularities would enable us to solve many biological problems.

## 2. Some regularities in the distribution of identical and analogous peptide groups in different proteins

In establishing the structure of insulin Sanger has already drawn attention to the fact that this protein has three dipeptide groups (His. Leu.; CySH.Gly.; and Leu.Val.) which are repeated twice in the same phenylalanyl chain. He has suggested, in this connection, that it may be that ready-made peptide blocks can be used in the building up of these parts of the chain. It has since been found that one of these groups (CySH.Gly.) is very frequently met with in the crude proteins of *Escherichia coli* (14). The repetition of fairly large peptide chains has been established as a fundamental characteristic of the structure of silk fibroin (4).

Statistical studies with this in mind have been carried out in the laboratories of Gamow (23, 72), and of Šorm (12, 13, 65, 66). The work of the Czechoslovakian biochemists in this field has been particularly detailed and systematic. In their latest work they have studied all the variations in the sequences of amino acids which have been worked out up to 1961. As a result of this work it is possible to establish a number of interesting peculiarities in the structure of proteins which are undoubtedly connected with the nature of the system controlling their biosynthesis.

In the first place it was noted (65) that some dipeptide groups are specially frequently repeated in different proteins while the reverse combination of these residues is rare although, by the laws of probability, the rate of their occurrence in proteins should be the same.

Particularly suggestive, however, was the repetition of the same tripeptide structures in different proteins, the frequency of their occurrence being many times greater than that which could have resulted from a random distribution (chance) of peptide combinations in the different peptides.

Theoretically about 5000 different tripeptides and about 85000 different tetrapeptides could be constructed from the 17 most commonly occurring amino acids. It has been shown by the Czechoslovakian biochemists that three proteins which have different functions, namely chymotrypsinogen, trypsinogen and ribonuclease, have a large number of sequences in common. In one of his papers devoted to detailed statistics Šorm (65) gives the following calculations concerning combinations occurring in several compounds. Out of the 85 peptide sequences of ribonuclease which were known when he made his calculations in 1959, the 115 dipeptide combinations which had been ascertained to occur in chymotrypsinogen and the 98 similar combinations in trypsinogen, 17 dipeptide structures (20%) occurred in both ribonuclease and trypsinogen, 22 (25%) were common to ribonuclease and trypsinogen while 48 (50%) were present in both trypsinogen and chymotrypsinogen. Furthermore, it was found that there were 6 identical tripeptide sequences and 2 such tetrapeptide sequences. A large number of identical and similar tripeptide sequences were noticed by the authors on comparing the proteins just discussed with insulin and glucagon.

It is, however, important to note that all these proteins, even though they have different functions, are produced by the same organ, namely the pancreas.

A common site of origin is not, however, necessary for the occurrence of such coincidences. For example, ribonuclease and lysozyme contain a common tetrapeptide sequence (Ala. Lys. Phe. Glu.) and three identical tripeptides. The number of such examples could be extended considerably and these correspondences have been established and are being established by the Czechoslovakian biochemists with extreme thoroughness.

It is also particularly interesting that individual groupings are repeated along the polypeptide chain of a single protein. Šorm and Keil (13) have made a careful study of these repetitions in ribonuclease. They examined the di- and tripeptide sequences which occur in ribonuclease and noted that eleven dipeptides are repeated twice in the molecule of this protein and five dipeptides are found three times. Furthermore, there are three tripeptides, Ser. Thr. Ser., Ser. Ser. Asp., and Thr. Asp. Glu., which are repeated twice. The authors have also tried to establish the presence of other regularities in the structure of the molecule of this protein (for example a regular incidence of lysine residues, some selective predominance in the nature of the "neighbours" of lysine, the presence of cross-linkages). The existence of such regularities may only be confirmed by further repetitions in the structure of other proteins. An interesting and profound attempt to analyse the regularities in the distribution of amino acids in ribonuclease has recently been made by Lanni (42, 43). Some further regularities were described by Keil & Šorm (38a).

It is quite obvious that, even if the presence of two or three dipeptide combinations of a similar kind in two different proteins or along a single polypeptide chain could be a chance coincidence, the occurrence of a large number of dipeptides or several identical tripeptides common to several proteins or the repetition of a tetrapeptide sequence cannot be a chance coincidence but is a manifestation of some regularity of which we do not know the nature. In this connection the authors consider that the presence of identical sequences is the result of a phylogenic and ontogenic kinship between the different proteins. Just as new species of animals and plants, which arise in the course of evolution, retain many common peculiarities and systems depending on the closeness of their relationship to one another, so many proteins which now have different functions may have had common historical origins, represented materially in the form of similar sequences of amino acids. The clearest example of this is the common features in the three hormones of the hypophysis of pigs: cortico- trophin and the $\alpha$- and $\beta$-melanocyte-stimulating hormones. Cortico- trophin consists of 39 amino acid residues, $\alpha$-melanocyte-stimulating hormone of 13 and $\beta$-melanocyte-stimulating hormone of 18.

As may be seen by the comparison in Fig. 3 (taken from the work of Harris (24)), the first 13 amino acid residues of corticotrophin are, essenti-

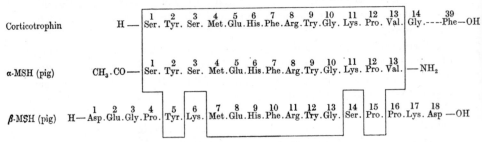

FIG. 3. Amino acid sequences of $\alpha$- and $\beta$ MSH, compared with the related $N$-terminal portion of corticotrophin.

ally, $\alpha$-melanocyte-stimulating hormone (MSH). The composition of $\beta$-melanocyte-stimulating hormone also shows clear indications of its relationship with the other two hormones.

It is interesting to note that corticotrophin also has a certain melano- cyte-stimulating activity, which is about 100-200 times less than that of the melanocyte-stimulating hormones.

The existence of common peptide groups in different proteins (es- pecially hormones) must certainly also reflect a functional relationship between these proteins, the existence of common links in the mechanisms of their enzymic actions. For example, one of the active centres of trypsin contains the group Asp. Ser. Gly. which is repeated in two other proteo- lytic enzymes—chymotrypsin and elastin (50). A study of the enzymic reaction brought about by trypsin (60) has shown that five or six different groups of the protein take part. It is quite evident that the specificity of the various proteolytic enzymes could be associated with one or two groups irrespective of the other parts of the molecule.

There can be no doubt as to the evolutionary nature of the phenomena of the identity and correspondence of particular parts of protein molecules. This also has a bearing on the problem of protein synthesis. The accom- plishment of the evolutionary alteration of the protein and the ontogenetic formation of the protein substances which gives effect to the course of evolution can be brought about in two ways. On the one hand one may suppose that, in the process of evolution, there arise gradually, alongside the matrices for the synthesis of protein or peptide $x$, groups of matrices forming the functionally new proteins $x.a.$, $x.a.b.$, $c.d.x.a.b$, etc. However, one may equally well suppose that in the course of evolution there arises what might be called a second, supplementary complex of matrices which complete the supplies of peptide or protein $x$ by adding to them the groups $a$, $a.b.$, $a.b.$ and $c.d.$ etc.

We shall return to an evaluation of these possibilities after we have surveyed the experimental evidence concerning the mechanisms of protein synthesis.

### 3. The possibility of a limitation to peptide combinations in the structure of proteins

The 20 amino acids of which ordinary proteins are composed can only be linked together to form 400 dipeptide combinations. However, although the total number of dipeptides identified from different proteins is already far more than 2000, some of the theoretically possible combinations have not been found. It is possible that nature avoids some dipeptide combinations. This idea was suggested in the papers of Yčas (72) and Šorm and his colleagues (66) and was confirmed by Morowitz in a number of papers in which he studied the statistical limitations of the sequence of amino acids in proteins, using whole cells of *Escherichia coli* which have a very heterogeneous composition so far as their proteins are concerned (48, 49).

The existence of such limitations will certainly become more apparent when, perhaps in the distant future, we can analyse the limitations to the occurrence of each of the possible 8000 tripeptide combinations. Such limitations are completely understandable. On the one hand they may be regarded as unused reserves, for the development of proteins, on the other, as a result of the elimination of peptide combinations which have no useful function. Each protein molecule is not merely a capricious selection of amino acids surrounding some active centre. Although the functional significance of the different parts of protein molecules are certainly not all of equal importance, nevertheless each part of the molecule must obviously play some part in carrying out its function and each has its own history. Natural selection has undoubtedly had its effect on the various enzymic processes, even at the stage of the combination of amino acids; the position of each amino acid residue in the polypeptide chain must be fixed by genetic factors.

### 4. The three-dimensional structure of proteins in connection with their structure and biological function

A unique sequence of amino acids in the polypeptide chains of proteins is a fundamental prerequisite for their specificity, and the explanation of the mechanism of reproduction of that specificity during protein synthesis is the essential basis for all the theoretical and experimental researches aimed at elucidating the mechanism of protein synthesis.

The building of a particular sequence of amino acid residues in the course of protein synthesis has a double significance. On the one hand, certain forms of combination of amino acids are the carriers of particular biochemical properties, for example the ability to catalyse some reaction or other. A series of investigations has already been carried out which shows that, when certain enzymes (pepsin, trypsin, chymotrypsin, myogen, ribonuclease) are broken down into peptides by proteases, i.e. when their secondary structure has been broken down, the peptides partially retain the specific enzymic activity of the original enzyme (2, 11, 38, 54, 59). On the other hand, the sequence of amino acid residues in the protein molecule as a whole, even in those parts which are far away from the so-called "active centres", is decisive in determining the form of folding of the polypeptide chain or for the individual character of the relationship between several polypeptide chains, i.e. for the formation of the three-dimensional spatial structure of the protein molecules which enables them to manifest their biological potential as fully as possible.

The study of the three-dimensional structure of proteins by means of X-ray structural analysis is beset by many technical difficulties which have not yet been overcome. At present, work in this new field of protein chemistry is only in its earliest stages but so far, owing to the brilliant work of Kendrew and his colleagues (40, 41, 41a) the first three-dimensional model of a protein, myoglobin, has been made. Fig. 4 is a photograph of this model.

The model of the molecule of myoglobin shown in Fig. 4 was obtained by three-dimensional X-ray analysis, which revealed the distribution of density in a series of parallel sections through the molecule of myoglobin, similar to the sections cut by a microtome through a tissue, but on a molecular scale.

Another group of workers in the same laboratory (57) has managed to reconstruct the three-dimensional structure of an even more complicated protein, haemoglobin. Myoglobin is a protein with a molecular weight of about 18,000 and takes the form of a single polypeptide chain of about 153 amino acid residues. The haemoglobin of vertebrates is a protein with a molecular weight of about 67,000 and is made up of four polypeptide chains, two each of two different kinds.

The formation of such structures depends on the sequence of amino acid residues as they interact with the reactive groups which are distributed in strictly determinate parts of the polypeptide chain. Disulphide bonds play an extremely important part in the formation of secondary structure, but other types of weak bonds (especially hydrogen and ionic bonds) can also play a determining part in the construction of the final "labyrinth" of the protein molecule.

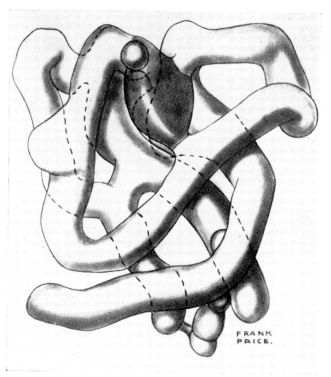

FRANK
PRICE.

FIG. 4. Molecule of the tertiary structure of the molecule of myoglobin.

The exact formation of the three-dimensional structure has a considerable bearing on the efficiency of the proteins in their biological functions. In some proteins, such as antibodies, it would seem that the variability of this structure is the essential basis for the manifestation of their physiological activity.

In many enzymes the "relief" or shape of their surfaces is a very important condition for the specificity of their reactions with their substrates. Finally, some forms of enzymic activity depend on particular congregations of amino acids, not merely along the polypeptide chain, but also on the cooperation of participating groups arranged in particular ways in various parts of the molecule.

It is clear, therefore, that in order to synthesize proteins biological systems must have the ability to reproduce the sequence of all the amino acids and polypeptide chains of the proteins.

It is also important to note that the macromolecules of many proteins are composed, not of one, but of several polypeptide chains. For example, the molecules of α-chymotrypsin and aldolase consist of three polypeptide chains of different lengths. The molecule of insulin is made up of four polypeptide chains in two pairs bound together by sulphide bonds. The molecule of haemoglobin consists of four chains. According to the latest information the molecule of the α-protein of elastin consists of 17 polypeptide chains, each of thirty-five amino acid residues. We have evidence to suggest that a number of other proteins have similar structures.

There can be no doubt that the forms of linkage between such fragments are specific. In other words, during the building up of a protein a selection is made from a more or less limited number of variants of the large number of theoretically possible forms of linkage between the polypeptide chains.

It is still hard to be sure whether this occurs "spontaneously", i.e. is determined by the arrangements within the chains of, for example, cysteine and cystine residues in a special form, or whether the orientation of the chains in relation to each other takes place on some sort of matrix. It is, however, certain that these polypeptide chains can develop independently on independent matrices and under the control of different genetic mechanisms. This has been made quite clear by study of diseases of the blood, some of which are associated with a disturbance of the balance between the production of the α- and β-chains which are the components of the haemoglobin molecule (36).

## 5. Species specificity of homologous proteins

In many of the studies which have been made of the sequences of amino acids in proteins and peptides, comparison has been made between the proteins of different species of animal with the object of finding out the basis of the species specificity of different homologous proteins, which had

PB C

been established long before by immunological methods (the immuno-logical species specificity of protein). So far, such comparisons have mainly been concerned with mammals; however, the facts which are already known in this field are of tremendous interest and lead to a number of interesting conclusions.

As long ago as 1950 Blagoveshchenskiĭ put forward the idea that the evolution of plants is brought about by the partial alteration of their proteins, resulting from non-equivalent substitutions in the amino acid sequences of their polypeptide chains (1).

In comparing the amino acid sequences in the polypeptide chains of homologous proteins which have the same function but are derived from animals of different species (insulins, albumins, fibrins, etc.) it has been found that these proteins often differ from one another only in the arrange-ment of one or two amino acids and it is these differences which confer the specific characteristics.

A typical example of species variability is provided by the small differences in the structures of the insulins of five different species of animals, established by Sanger and his colleagues (22, 25). As may be seen from the comparison given in Fig. 5, the specific characteristics of this protein are confined to a small part of the A polypeptide chain (in positions 8-10, between two cystine residues).

| 7 | 8 | 9 | 10 | 11 | |
|---|---|---|---|---|---|
| .... Cy | Thr | Ser | Ileu | Cy | ..... pig |
| .... Cy | Ala | Ser | Val | Cy | ..... ox |
| .... Cy | Ala | Gly | Val | Cy | ..... sheep |
| .... Cy | Thr | Gly | Ileu | Cy | ..... horse |
| .... Cy | Thr | Ser | Ileu | Cy | ..... whale |

FIG. 5. Species differences in the structure of insulins.

We see that the pig and the whale have insulins of identical structure while, in the other cases, one or two amino acids are replaced by others which are, as a rule, not very different from them structurally (the substi-tution of glycine for serine, threonine for alanine and isoleucine for valine —all of which are monoaminomonocarboxylic acids).

In their recent studies on the amino acid sequence of human insulin Nicol & Smith (51) have shown that, in the section 8-10, it has the same structure as that of the pig and the whale. At the same time, human insulin differs from the other insulins in that the terminal alanine of the B chain is replaced by threonine.

Another extremely interesting type of species variability of the sequence of amino acids is exemplified by the variability of the corticotrophins in the sections between the 24th and 33rd amino acids of three species of animals (Fig. 6). In the section 25-28 we find four different arrange-ments of the same four amino acids while in the section 31-32 there is both rearrangement and substitution (by structurally similar residues).

At positions 29-30 for the ox there is an inversion of sequence compared with that which occurs in the other species. These two examples in themselves, are enough to suggest the possible existence of some sort of regularity in the species interchange of amino acids.

| | 24 | | 25 | 26 | 27 | 28 | 29 | 30 | | 31 | 32 | | 33 |
|---|---|---|---|---|---|---|---|---|---|---|---|---|---|
| Sheep | | | Ala | Gly | Glu | Asp | | | | Ala | Ser | | |
| Pig (α) | Pro. | | Asp | Gly | Ala | Glu | Asp | Glu | | Leu | Ala | | |
| Pig (β) | | | Gly | Ala | Glu | Asp | | | | Leu | Ala | | Glu ... |
| Ox | | | Asp | Gly | Glu | Ala | Glu | Asp | | Ser | Ala | | |

FIG. 6.   Structural differences in the molecules of corticotrophins.

This suggestion was first put forward by Gamow et al. (23) and then made more precise by Yčas (72) who tried to draw up a generalization about the occurrence of interchange of position between amino acids in the proteins of different species in the form of a table (Fig. 7). The material in this table consisted of the species peculiarities, established by 1957, of the arrangements of the terminal amino acids and terminal peptides in four specimens of myoglobin (striped whale, sperm whale, horse and seal), two specimens of serum albumin (human and ox), two specimens of vasopressin (ox and pig), nine specimens of haemoglobin (horse, dog, pig, snake, rabbit etc.), two specimens of gliadin (rice and wheat), two specimens of fibrinogen (human and ox) and two strains of tobacco mosaic virus. In addition, the table includes data on the species differences in small sections from the middle of insulin, cytochrome c, corticotrophin and vasopressin.

| Type of substitution | No. of recorded instances |
|---|---|
| Val ↔ Ileu | 3 |
| Ala ↔ Thr | 2 |
| Ala ↔ Ser | 2 |
| Ala ↔ Gly | 2 |
| Ala ↔ Leu | 2 |
| Ser ↔ Gly | 2 |
| Ala ↔ Glu | 1 |
| Val ↔ Glu | 1 |
| Val ↔ Met | 1 |
| Phe ↔ Glu | 1 |
| Glu ↔ Asp | 1 |
| Arg ↔ Lys | 1 |

FIG. 7.   Frequency of substitution of amino acids associated with species variation of proteins (after Yčas (72)).

If we do not count the case in which phenylalanine is substituted for glutamine, which occurs in a plant, and analyse only the results from the proteins and peptides of animal origin, then we can clearly see that most of the exchanges involve the substitution of analogous amino acids (Glu ↔ Asp; Arg ↔ Lys; Ala ↔ Gly, etc.) and only three (Ala ↔ Glu;

Val ↔ Met and Val ↔ Glu) represent substitution by amino acids of a different type.

More data on the species variation of protein have now become available and this enables us to supplement the table given above.

The work of Tuppy and Paléus (52, 69, 70) shows that a segment of the cytochrome c of the silkworm (Bombyx mori), consisting of 11 amino acids, the sequence of which has been established, differs from the corresponding segment of the cytochrome c of mammals by the replacement of lysine by arginine. The cytochrome of the chicken differs from that of the salmon by the substitution of alanine by serine. Analogous segments of the cytochromes of yeast and Rhodospirillum rubrum differ from the cytochromes of mammals, birds and insects in the majority of their amino acids, but in any event the systematic differences between these species are so great that it is hardly possible to consider the proteins as being homologous.

It has been established that there is a species difference in the terminal amino acids of the fibrinogens of seven species of animals. (Fig. 8) (37)

|  | 1 | 2 | 3 | 4 | 5 | 6 |
|---|---|---|---|---|---|---|
| Ox | Tyr | Tyr | Glu | Glu . . . . . . . . . . . |  |  |
| Pig | Tyr | Tyr | Ala | Ala | Ala | Ala . . . . . |
| Man | Tyr | Tyr | Ala | Ala . . . . . . . . . . . |  |  |
| Sheep, goat | Tyr | Tyr | Ala | Ala | Gly | Gly . . . . . |
| Dog | Tyr | Tyr | Thr | Thr . . . . . . . . . . . |  |  |
| Horse | Tyr | Tyr | Thr | Thr . . . . . . . . . . . |  |  |

FIG. 8.  Species differences in the N-terminal sections
of fibrinogens (37).

Anfinsen and his colleagues (20) have recently compared the structures of the ribonucleases of sheep and cattle. They have found the following differences. In ox ribonuclease position 3 is occupied by threonine and position 37 by lysine, whereas in sheep ribonuclease these positions are occupied by serine and glutamic acid respectively. As the substitution of lysine for glutamic acid must lead to a marked change in the properties of that section of the chain, the authors suggest that the segment in the neighbourhood of the 37th residue cannot be essential for the manifestation of enzymic activity.

Of the other research on species differences in proteins, we should mention the work of Parcells and his colleagues (44, 53) which establishes the sequence of the three-residue N- and C-terminal segments of the growth hormones of man, the apes, whales, cattle and sheep; as well as the work of Alberti (16) in determining the sequence of the amino acid residues in the corticotrophin of the ox. A number of supplementary results concerning the species differences of several proteins (haemoglobins, growth hormone, albumins and others) have been published recently by Yčas in a new review. All these new data provide the following additions to Yčas' first table (Fig. 9).

| | |
|---|---|
| Arg ↔ Lys | 3 |
| Asp ↔ Glu | 1 |
| Phe ↔ Ala' | 1 |
| Ala ↔ Ser' | 3 |
| Ala ↔ Val | 1 |
| Ala ↔ Gly | 2 |
| Ileu ↔ Val | 1 |
| Ala ↔ Thr | 3 |
| Glu ↔ Ser | 1 |
| Ser ↔ Thr | 2 |
| Pro ↔ Ser | 2 |
| Ala ↔ Leu | 1 |
| Lys ↔ Glu | 1 |
| Lys ↔ Ser | 1 |

FIG. 9.   Additional cases of substitution of
amino acid residues (73).

Among the 21 additional cases there are only two cases (Lys ↔ Glu and Lys ↔ Ser) in which the substitution is between structurally dissimilar amino acids.

Of course, there is not sufficient material to permit any extensive generalization but, on the basis of the evidence we already have, we may draw the tentative conclusion that *the species differences of animal proteins having the same functions consist in alterations in the structures of small segments of their molecules involving (a) the substitution of one or two amino acids by ones with similar properties or (b) a simple interchange in the position of some amino acids e.g. 12 ↔ 13.* Such variations do not, as a rule, affect the functional properties but they do confer the properties of species specificity.

It is interesting to note, for example, that according to the evidence of X-ray structural analysis, the myoglobins of the cachalot and the seal have exactly the same molecular configurations in spite of the fact that their amino acid compositions are different (63).

But what is the purposive, evolutionary point of such variation? The sharp differences between species must, of course, also occur in the individual structures of particular proteins (the insulins of a whale, the cachalot, and the pig). Why should the evolution of a species and natural selection alter the positions of one or two amino acids in the composition of some protein or peptide hormone without altering its functional properties?

For the present the answer to this question can only be given in a very tentative form. The differentiation of a species must certainly be associated with a number of changes in the particular proteins which augment or change some of its functions and properties. This applies especially to those processes and peculiarities which undergo the greatest changes in connection with some particular evolutionary morphogenetic process (for example, the formation of milk proteins in the transition to the mammals, the formation of the keratin of horns, the proteins of wool, etc.). Some functions may, however, be retained, especially within the limits of a single class such as the mammals. One may imagine that changes in the effectors of such functions, e.g. changes in insulin, pepsin or corticotrophin, are progressive and that changes in the position of particular amino acids are also determined by evolution and are purposive and produce a better adaptation to the carrying out of some particular function under changed conditions. While not altering the qualitative nature of their functions these proteins, as it were, adapt themselves to altering internal conditions. A supposition of this kind is, of course, possible, but it has its weak side. There are, in fact, tens of thousands of species of vertebrates alone while the possibilities for alteration in, for example, insulin or ribonuclease, without destroying their important functions, are extremely limited and localized to small segments of their internal structures each consisting of only 3-4 amino acid residues. Very different species must therefore have identical proteins which are, nevertheless, completely adapted to very different conditions.

This fact suggests that there must be a somewhat different explanation for the phenomenon in question. It seems to us that although the possibilities for alteration are very small in proteins which have a function which is equally necessary to the different species, yet evolution still uses these very limited possibilities, simply in order to create the widest possible species specificity of proteins. It seems to us that the species specificity of proteins is in itself, quite apart from the functions of the proteins, a purposive, protective characteristic. It must be remembered that the reactions of immunity, which are vitally important to the organism, are based on this very species specificity of proteins.

In this way the evolution of parasite species is directed towards the surmounting of the immunity barrier while the evolution of animals and plants which are subject to infections and other forms of infestation is directed towards the erection of new barriers.

We find it very hard to agree with the view of Bresler (3) that only the so-called "functional centre" of the macromolecule of a protein is constructed in a determinate, purposive way and that the other parts of the molecule may not have any constant structure, being vestigial remains arising during the history of the species.

In this connection we may take note of the very interesting evidence obtained by Allinson (quoted by Pauling (10)). This author has shown that the formation of an anomalous haemoglobin S, which is formed in man in sickle-cell anaemia and differs from normal haemoglobin only in the position of one amino acid residue (32), could, under certain conditions, give rise to a resistance to malaria in the person affected. When this haemoglobin occurs in its dominant form (SS) the patient soon dies of sickle-cell anaemia. The person with normal haemoglobin (AA) is susceptible to malaria. In the mixed form (AS) in which the erythrocytes contain a mixture of haemoglobin A and haemoglobin S, the patient is protected against malaria. In some highly malarial regions of Africa hereditary mutations have occurred with the result that nearly 50% of the population has blood of the heterozygous type AS, and has an inherited immunity to malaria.

Thus, we may postulate (7a) that species variations in proteins can be of two types, morphogenic and protective. *The repetition of identical peptide groups in different proteins is a manifestation of the evolutionary, morphogenic type of variation while the species differences between proteins having the same function are apparently a result of "protective" variation*; it is the individual "escape device" of the organism from the attacks of the environment without transgressing the limits of a small and strictly circumscribed segment.

The same assumption was made recently by Efroimson (13a) in a theoretical paper on correlations between variations in human haemoglobin and resistance to malaria.

## *Conclusion*

In this first chapter we have simply made a very brief survey of some questions concerning the structure of proteins without touching on many of the aspects of the problem which have been studied intensively by protein chemists (the physical chemistry of proteins, their amino acid composition, their heterogeneity, etc.). However, even this meagre theoretical survey of a number of the recent achievements of protein chemistry is quite enough to enable us to see how complicated is the task which nature has set before the biological systems which are responsible for the continual synthesis and resynthesis of proteins and also how hard is the task before those research workers who are trying to carry out an artificial synthesis of these compounds.

The systems for the syntheses of proteins within the cell must be able to bring about the following processes: (1) the activation of amino acids so that they can surmount the energy barrier to their interaction; (2) the maintenance of the activated amino acids in an isolated state until they can interact with their neighbours in the polypeptide chain; (3) the orderly arrangement of amino acids in a way which corresponds with their sequence in the protein and at such a distance from one another that they can interact; (4) the synthesis of peptide bonds between the amino acids which are correctly orientated; (5) the fixation of the "stopping places" for the synthetic process when the polypeptide chain has reached the required length, and (6) the laying down of conditions for the necessary specific aggregation of several polypeptides or, in many cases, even of several proteins.

If we also take into account the fact that these same systems must ensure the reproduction within the cells, not of any one protein, but of hundreds of different proteins, including those enzymic proteins which determine the activity of the actual synthetic system itself, then the complexity of the problem of protein biosynthesis becomes quite clear. Nevertheless, a survey of the peculiarities of these systems, to which we shall turn in the next chapters, will show us that they are even more complicated than one might have thought, judging from what we know of the structure of protein molecules.

# REFERENCES

1. BLAGOVESHCHENSKIĬ, A. V. (1950). *Biokhimicheskie osnovy evolyutsionnogo protsessa u rastenii*. Moscow: Izd. Akad. Nauk S.S.S.R.
2. BRESLER, S. E. (1955). In *Belki, ikh spetsificheskie svoĭstva* (ed. M. F. Gulyĭ & V. A. Belitser), p. 73. Kiev: Izd. Akad. Nauk Ukr. S.S.R.
3. — (1959). *Proc. first internat. Sympos. on the origin of life on the Earth* (ed. A. I. Oparin *et al.*), p. 289. London: Pergamon.
4. IOFFE, K. G. (1954). *Biokhimiya*, 19, 495.
5. KAVERZNEVA, E. D. (1959). *Vestnik Akad. Nauk S.S.S.R.*, No. 8, 26.
6. LOKSHINA, L. A. & TROITSKAYA, O. V. (1958). *Uspekhi biol. Khim.* (ed. V. N. Orekhovich) 3, 3. Moscow: Izd. Akad. med. Nauk S.S.S.R.
7. LOKSHINA, L. A. & OREKHOVICH, V. N. (1960). *Doklady Akad. Nauk S.S.S.R.* 133, 472.
7a. MEDVEDEV, Zh. A. (1961). *Zhur. vsesoyuz. Mendeleev. Obshchestva*, 6, 268.
8. OREKHOVICH, A. N. (1959). In *Aktual'nye voprosy sovremennoĭ biokhimii* (ed. V. N. Orekhovich). *Trudy Konf. "Biokhimiya belkov"*, 1, 5. Moscow: Izd. Akad. med. Nauk S.S.S.R.
9. PASYNSKIĬ, A. G. (1958). *Izvest. Akad. Nauk S.S.S.R., Ser. biol.*, No. 6, 641.
10. PAULING. L. (1959). *Proc. first internat. Sympos. on the origin of life on the Earth* (ed. A. I. Oparin *et al.*), p. 215. London: Pergamon.
11. CHERNIKOV, M. P. (1956). In collective work *Voprosy meditsinskoĭ khimii* (ed. V. N. Orekhovich & S. E. Severin), Part 2, No. 1, p. 59. Moscow: Izd. Akad. med. Nauk S.S.S.R.
12. ŠORM, F. (1959). *Proc. first internat. Sympos. on the origin of life on the Earth* (ed. A. I. Oparin *et al.*), p. 231. London: Pergamon.
13. ŠORM, F. [SHORM] & KEIL, B. [KEĬL] (1959). In *Aktual'nye voprosy sovremennoĭ biokhimii* (ed. V. N. Orekhovich). *Trudy Konf. "Biokhimiya belkov"*, 1, 20. Moscow: Izd. Akad. med. Nauk S.S.S.R.
13a. EFROIMSON, V. P. (1961). In collective work *Problemy kibernetiki* (ed. A. A. Lyapunov), *no. 6*, p. 161.
14. ABELSON, P. H. (1952). *Science*, 115, 479.
15. ACHER, R., LAURILA, U. R., THAUREAUX, J. & FROMAGEOT, C. (1954). *Biochim. biophys. Acta*, 14, 151.
16. ALBERTI, C. (1958). *Farmaco (Pavia), Ed. sci.* 13, 602.
17. ANDERER, F. A., UHLIG, H., WEBER, E. & SCHRAMM, G. (1960). *Nature (Lond.)*, 186, 922.
18. ANFINSEN, C. B. (1957). *Fed. Proc.* 16, 783.
19. — (1959). *Ann. N.Y. Acad. Sci.* 81, 513.
20. ANFINSEN, C. B., ÅQVIST, S. E. G., COOKE, J. P. & JÖNSSON, R. (1959). *J. biol. Chem.* 234, 1118.

21. BAILEY, J. L., MOORE, S. & STEIN, W. H. (1956). *J. biol. Chem.* **221**, 143.
22. BROWN, H., SANGER, F. & KITAI, R. (1955). *Biochem. J.* **60**, 556.
23. GAMOW, G., RICH, A. & YČAS, M. (1956). *Advances in Biol. and med. Phys.* **4**, 23. New York: Academic Press.
24. HARRIS, J. I. (1959). *Biochem. J.* **71**, 451.
25. HARRIS, J. I., SANGER, F. & NAUGHTON, M. A. (1956). *Arch. Biochem. Biophys.* **65**, 427.
26. HAUROWITZ, F. (1950). *The chemistry and biology of proteins.* New York: Academic Press.
27. HILL, R. L. & SCHWARTZ, H. C. (1959). *Nature (Lond.),* **184**, 641.
28. HILSCHMANN, N. & BRAUNITZER, G. (1959). *Hoppe-Seyl. Z.* **317**, 285.
29. HIRS, C. H. W. (1960). *Ann. N.Y. Acad. Sci.* **88**, 611.
30. SMYTH, D. G., STEIN, W. H. & MOORE, S. (1963). *J. biol. Chem.* **238**, 227.
31. HIRS, C. H. W., STEIN, W. H., MOORE, S. & FALLON, B. M. (1956). *J. biol. Chem.* **221**, 151.
32. HUNT, J. A. & INGRAM, V. M. (1959). *Nature (Lond.),* **184**, 640.
33. JOLLÈS, J., BERNIER, I., JAUREQUI, J. & JOLLÈS, P. (1960). *Compt. rend. Acad. Sci., Paris,* **250**, 413.
34. JOLLÈS, P. & JOLLÈS, J. (1958). *Bull. Soc. Chim. biol.* **40**, 1933.
35. — (1961). *Proc. V int. Congr. Biochem., Moscow,* **9**, 90.
36. JONES, R. T., SCHROEDER, W. A., BALOG, J. E. & VINOGRAD, J. R. (1959). *J. Amer. chem. Soc.* **81**, 3161.
37. JORPES, J. E., BLOMBÄCK, G. E. B. & YAMASHINA, I. (1958). *Proc. intern. Symposium Enzyme Chem. Tokyo and Kyoto,* **2**, 400.
38. KALNITZKY, G. & ROGERS, W. I. (1956). *Biochim. biophys. Acta,* **20**, 378.
38a. KEIL, B. & ŠORM, F. (1962). *Coll. Czechoslov. chem. Communs.* **27**, 1310.
39. KEIL, B., ŠORM, F., HOLEYŠOVSKÝ, V., KOSTKA, V., MELOUN, B., MIKEŠ, O., TOMÁŠEK, V. & VANĚČEK, J. (1959). *Coll. Czechoslov. chem. Communs.* **24**, 3491.
40. KENDREW, J. C. (1959). *Fed. Proc.* **18**, 740.
41. KENDREW, J. C., BODO, G., DINTZIS, H. M., PARRISH, R. G., WYCKOFF, H. & PHILLIPS, D. C. (1958). *Nature (Lond.),* **181**, 662.
    DAVIES, D. R., PHILLIPS, D. C. & SHORE, V. C. (1960). *Nature (Lond.),* **185**, 422.
41a. KENDREW, J. C., DICKERSON, R. E., STRANDBERG, B. E., HART, R. G., DAVIES, D. R., PHILLIPS, D. C. & SHORE, V. C. (1960). *Nature (Lond.),* **185**, 422.
42. LANNI, F. (1960). *Proc. nat. Acad. Sci. U.S.* **46**, 1563.
43. — (1961). *Proc. nat. Acad. Sci. U.S.* **47**, 261.
44. LI, C.-H., PARCELLS, A. J. & PAPKOFF, H. (1958). *J. biol. Chem.* **233**, 1133.
45. MÄSIAR P., KEIL, B. & ŠORM, F. (1957). *Coll. Czechoslov. chem. Communs.* **22**, 1203.
46. — (1958). *Coll. Czechoslov. chem. Communs.* **23**, 734.
47. MOORE, S., HIRS, C. H. W. & STEIN, W. H. (1956). *Fed. Proc.* **15**, 840.
48. MOROWITZ, H. J. (1959). *Biochim. biophys. Acta,* **33**, 494.
49. MOROWITZ, H. J. & BARRA, R. V. (1959). *Biochim. biophys. Acta,* **33**, 505.
50. NAUGHTON, M. A., SANGER, F., HARTLEY, B. S. & SHAW, D. C. (1960). *Biochem. J.* **77**, 149.
51. NICOL, D. S. H. W. & SMITH, L. F. (1960). *Nature (Lond.),* **187**, 483.
52. PALÉUS, S. & TUPPY, H. (1959). *Acta chem. Scand.* **13**, 641.
53. PARCELLS, A. J. & LI, C.-H. (1958). *J. biol. Chem.* **233**, 1140.
54. PERLMANN, G. E. (1954). *Nature (Lond.),* **173**, 406.
55. PERLMANN, G. E. & DIRINGER, R. (1960). *Ann. Rev. Biochem.* **29**, 151.
56. PERUTZ, M. F. (1958). *Endeavour,* **17**, 190.
57. PERUTZ, M. F., ROSSMANN, M. G., CULLIS, A. F., MUIRHEAD, H., WILL, G. & NORTH, A. C. T. (1960). *Nature (Lond.),* **185**, 416.

58. NEURATH, H. & BAILEY, K. (Eds.) (1953-4). *The Proteins* (two vols.). New York: Academic Press.
59. RICHARDS, F. M. (1958). *Proc. nat. Acad. Sci. U.S.* **44**, 162.
60. RONWIN, E. (1959). *Biochim. biophys. Acta*, **33**, 326.
61. SANGER, F. & THOMPSON, E. O. P. (1953). *Biochem. J.* **53**, 353, 366.
62. SANGER, F. & TUPPY, H. (1951). *Biochem. J.* **49**, 463, 481.
62a. SCHERAGA, H. A. (1961). *Protein structure.* New York: Academic Press.
63. SCOULOUDI, H. (1959). *Nature (Lond.)*, **183**, 374.
64. SEGAL, J., BOLL-DORNBERGER, K. & KALAĬDZHIEV, A. T. (1960). *Globular protein molecules : their structure and dynamic properties.* Berlin: Verlag der Wissenschaften.
65. ŠORM, F. (1959). *Coll. Czechoslov. chem. Communs.* **24**, 3169.
66. ŠORM, F., KEIL, B., HOLEYŠOVSKÝ, V., KNESSLOVÁ, V., KOSTKA, V., MÄSIAR, P., MELOUN, B., MIKEŠ, O., TOMÁŠEK, V. & VANĚČEK, J. (1957). *Coll. Czechoslov. chem. Communs.* **22**, 1310.
67. THOMPSON, A. R. (1955). *Biochem. J.* **60**, 507.
68. TSUGITA, A., GISH, D. T., YOUNG, J., FRAENKEL-CONRAT, H., KNIGHT, C. A. & STANLEY, W. M. (1960). *Proc. nat. Acad. Sci. U.S.* **46**, 1463.
69. TUPPY, H. (1957). *Z. Naturforsch.* **12B**, 784.
70. TUPPY, H. & PALÉUS, S. (1955). *Acta chem. Scand.* **9**, 353.
71. VANĚČEK, J., MELOUN, B., KOSTKA, V., KEIL, B. & ŠORM, F. (1960). *Coll. Czechoslov. chem. Communs.* **25**, 2358.
72. YČAS, M. (1956). *Symposium Inform. Theory Biol., Gatlinburg, Tenn.* (pub. 1958), p. 70. Cf. *Chem. Abstr.* **53**, 18136*h* (1959).
73. YČAS, M. (1961). *J. theoret. Biol.* **1**, 244.

# CHAPTER II

# THE ESSENTIAL FEATURES OF THE STRUCTURE AND SPECIFICITY OF NUCLEIC ACIDS

## Introduction

The nucleic acids, ribonucleic acid (RNA) and desoxyribonucleic acid (DNA) are essential components of the self-reproducing biochemical system which specializes in synthesizing and reproducing the specificity of proteins and which thus exercises a determining influence upon many extremely important biological processes and phenomena.

There are many reasons to suppose that, in the cell, nucleic acids perform the peculiar function of "biochemical memory" and that, by means of a particular biochemical cycle, they can pass on such structural information as will regulate the arrangement of amino acids into a specific sequence during the syntheses of proteins. This function of nucleic acids has acquired the somewhat conventional name of "matrix function" by analogy with the purely typographical process of multiple reproduction of printers' type by the surface of typographic matrices.

The structure of nucleic acids must accordingly be such that they can carry out their functions as matrices. In this chapter we shall discuss, very briefly, recent ideas as to the structures of DNA and RNA without concerning ourselves for the moment with the mechanisms by which these polymers take part in the synthesis of proteins, the phenomena of heredity or the nature of their reproduction, localization, metabolism, etc. These are questions which will be considered in later chapters.

Progress in the study of the macromolecular structure of the nucleic acids, especially DNA, has been extremely fast and has focussed great interest on these compounds, not only among biochemists and biologists, but also among chemists, physicists, physical chemists and mathematicians. The extensive literature in the field has been pretty well reviewed in many reviews and monographs (2, 2a, 3, 13, 21, 8, 9, 10, 22, 23, 19, 24, 25, 28, 29, 47, 48, 46, 59a, 67, 73, 82, 83, 84, 90) and this saves us from having to discuss the history of these studies and from repeating many generally known concepts. We must, however, dwell on some questions of principle in this matter, as they will be used in our analysis of various hypotheses and theoretical concepts.

# 1. General features of the structures of DNA and RNA

It seems appropriate here to give a very short account of some factual evidence as to the general features of the structure of the nucleic

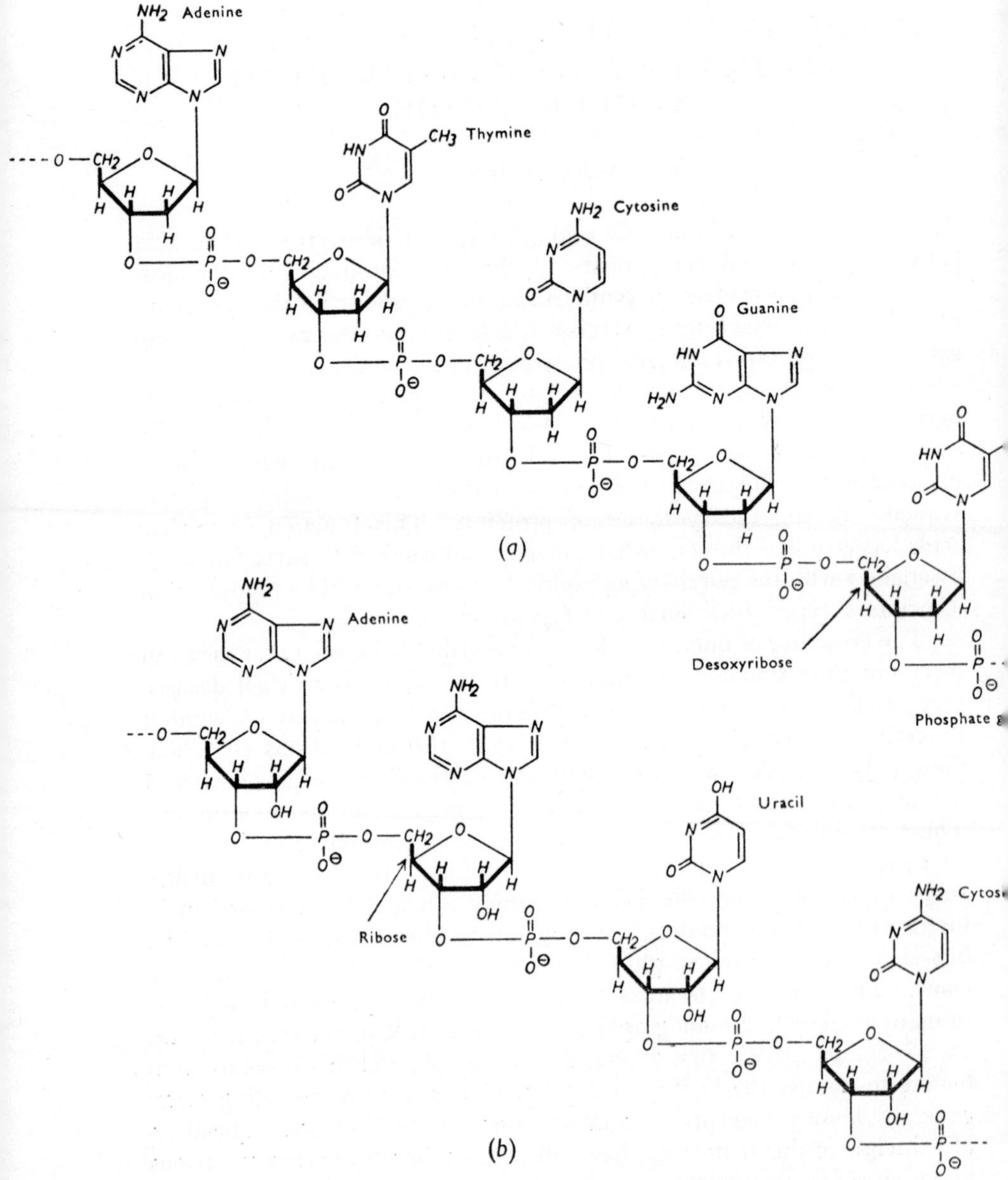

FIG. 10. Structure of fragments of molecules of (a) desoxyribonucleic acid and (b) ribonucleic acid.

acids, even though this may be elementary and already available in the textbooks. It must not be forgotten, however, that this seemingly elementary evidence was mainly obtained by organic chemists as a result of several decades of intensive work involving numerous experiments and comparisons with artificially synthesized compounds. Without this colossal amount of work by the organic chemists the present progress of our ideas in the field of biochemistry would have been quite impossible.

The nucleic acids, RNA and DNA, are high polymers built up of nucleotides (nucleoside phosphates). The nucleosides, in their turn, are complex compounds formed by the combination of purine and pyrimidine bases with a pentose. The difference between DNA and RNA is primarily that in DNA the carbohydrate component of the nucleosides is desoxyribose whereas in RNA it is ribose. The polymers of RNA consist of nucleosides arranged in a specific sequence and made up of four bases: uracil, cytosine, guanine and adenine. In some specimens of RNA minor amounts of other bases are present as well. In DNA the uracil is replaced by another pyrimidime base—thymine. Small amounts of other bases are also sometimes to be found among the nitrogen-containing bases of DNA.

In both types of nucleic acid the nucleosides are linked together by phosphodiester bonds. Fig. 10 is a diagram of the structure of RNA and DNA. It is the absence of a hydroxyl group attached to carbon atom 2 of desoxyribose which accounts for the resistance of DNA to alkaline hydrolysis, as it makes ring formation by the carbohydrate group impossible (32). According to the X-ray structural evidence at present available both RNA and DNA have an unbranched linear structure.

## 2. Peculiarities of the molecular structure of DNA

We must first point out that the macromolecular structure of DNA has been considerably more thoroughly studied than that of RNA. By means of X-ray structural analysis and by other methods it has been shown definitely that the long polymeric chain of DNA consists of two polynucleotide chains oriented in opposite directions, linked with one another and twisted into a characteristic helix. The individual polynucleotide chains are linked together in this helix by means of hydrogen bonds and treatment with the various reagents which are capable of breaking hydrogen bonds will usually halve the molecular weight of DNA.

There are many schemes for representing the way in which the polynucleotide chains of DNA are arranged in relation to one another but the one which is most generally accepted as corresponding to the factual evidence is that put forward by Watson & Crick in 1953 (86, 87). In suggesting this scheme the authors tried to solve two problems. On the one hand they had to take account of the experimental data concerning the structure of DNA obtained by X-ray structural and chemical analysis, while, on the other, they were trying to explain the mechanism for the reproduction of genetic material, that is to say, the replication of the chains of DNA.

According to the scheme worked out by Watson & Crick, the structure of DNA consists of two polynucleotide chains curled in a helix around a common axis and joined to one another by hydrogen bonds from one nucleotide to another (Fig. 11). The main linkage of the nucleotides within each polynucleotide chain is a phosphoester bond between the $C_5'$ and $C_3'$ of neighbouring nucleotides. Both chains form right-handed helices but the arrangements of the atoms in their phospho-carbohydrate skeletons are opposite and the nitrogen-containing bases are linked by secondary valencies lying perpendicular to the axis of the helix. The phosphate and carbohydrate residues are arranged on the periphery of the helix while the bases are in the middle. The way in which the bases are linked together is an essential feature of the structure. The bases lie in planes perpendicular to the axis and are joined together in pairs. Their combination into pairs is very specific and only certain pairs of bases will correspond with a particular structure. One of the members of the pair must always be a purine, the other a pyrimidine base in order to make a bridge between the two chains. If the pair consisted of two purines there would not be enough space for them, while if it consisted of two pyrimidines the space between them would be too great for the formation of a hydrogen bond.

The making of stereochemical models representing various means of linking the bases in pairs as carried out by the authors of the scheme and later, in greater detail, by Pauling & Corey (68) showed that adenine can only form a pair with thymine and guanine only with cytosine (or 5-methylcytosine or 5-hydroxymethylcytosine). Such combination in pairs is shown in Fig. 12. If adenine is brought into contact with cytosine, hydrogen bonds are not formed between them because, in the sites which would have permitted it, there are either two hydrogen atoms or none.

The members of any pair can change places. Adenine, for example, may be in either of the two chains but its partner must always be thymine.

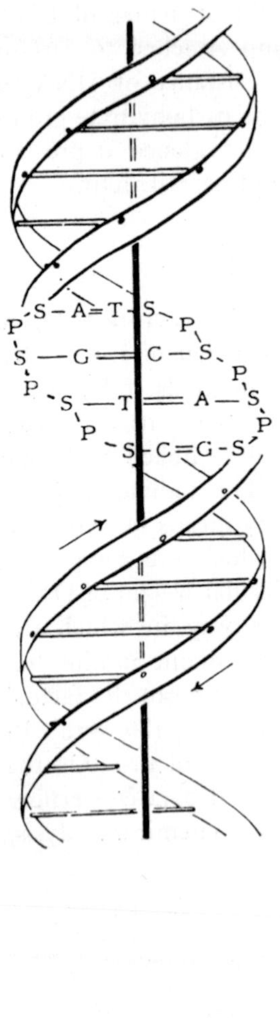

FIG. 11

Diagram illustrating the four types of sites present in the groove of the DNA molecule. S represents desoxyribose sugar; P, the phosphate group; A, adenine; T, thymine (5-methyluracil); G, guanine; and C, cytosine.

It must be emphasized that, so long as each base can form hydrogen bonds at different places it can unite to form a pair with isolated nucleotides by the most varied means. The specific combination of bases into pairs can only occur when there are certain limitations present and in the case in question these are a direct consequence of the existence of regularity in the phosphate-carbohydrate skeleton.

It should further be emphasized that, no matter what is the nature of

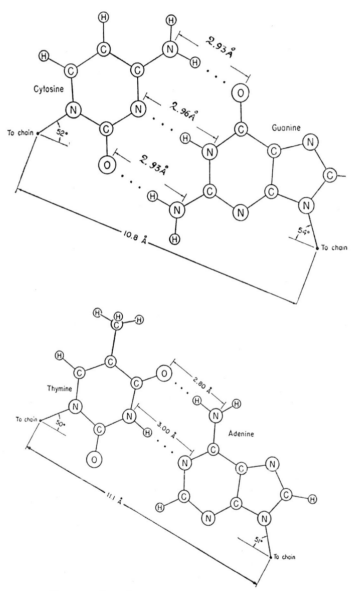

FIG. 12, Complementary pairs of nitrogenous bases.

the pair of bases at any particular point in the structure of DNA, this does not affect the neighbouring pairs and any sequence of pairs is possible.

Although this structure permits the occurrence of any sequence of bases, the necessity for specific combination in pairs requires a correspondence between the sequences of the nucleotides in the two chains. This means that, if we know the actual sequence of the bases in one chain, we can automatically deduce their sequence in the other. The structure thus consists of two chains, each of which is complementary to the other. At the same time, the vertical sequence of nucleotides in any isolated chain can be very varied and this variation of sequences gives rise to an enormous variety of forms of nucleic acid and is, according to theoretical ideas, the biochemical means of recording hereditary information.

Watson & Crick have adduced a great deal of physical and physico-chemical evidence confirming the agreement between the fundamental features of their model and the real structure of DNA. However, the most convincing evidence in favour of the complementary structure of DNA is that derived from numerous analytical studies of its nucleotide composition. In spite of the fairly wide species differences and quite large changes in nucleotide composition, the extensive material derived from the literature and from Chargaff's own researches has enabled Chargaff (23, 41, 42) to show that there is a consistent balance between the individual components of all known types of DNA corresponding to the complementary scheme of Watson & Crick. This balance is shown by the following features of the composition of DNA: The molar quantity of adenine is always the same as that of thymine while the molar amount of guanine is the same as that of (cytosine + methylcytosine); the sum of the purine nucleotides is the same as that of the pyrimidine nucleotides. Chargaff also noted that in almost all specimens of DNA and RNA there is an equality between the molar sum of the nucleotides having 6-amino groups (adenylic and cytidylic acids) and that of the nucleotides having 6-keto groups (guanylic and thymidylic or uridylic acids). At the same time the ratio of adenine to guanine and of thymine to cytosine may vary fairly widely depending on the source of the DNA.

Watson & Crick made use of the work of many authors in building their model of the structure of DNA but they indicated that the work of Wilkins and Franklin and other fellow workers in the physical laboratory of King's College, London (50, 90, 91) played a specially important part in the formation of their hypotheses. In these researches as well as other later investigations (49) this group has revealed the helical configuration of the polymer and measured many intramolecular distances, which has made it possible to calculate the length of the period of the helix and has provided the basis necessary for the construction of an accurate stereo-chemical model of this exceptional molecule. The attempt of Wilkins (89) to give a theoretical review of the work which has confirmed the fundamental idea of Watson & Crick is of special interest in this connection.

FIG. 13. Model of a helix of DNA. In the lower part of the model a polyarginyl chain lies in the groove of the molecule.

New and more accurate X-ray studies of DNA have enabled Wilkins to make certain corrections in the scheme of the structure of DNA (decreasing the distance between the nitrogen-containing bases and the axis of the helix and decreasing the diameter of the helix) and also to build a detailed stereochemical model of the molecule of DNA. This is of great interest, especially when considering the possibility of the stereochemical specificity of the surface of DNA, because the function of DNA as a matrix is bound up with the structure of its surface.

Fig. 13 shows a stereochemical model of the structure of one of the turns of a helix of DNA and is taken from Wilkins. A real molecule of DNA has about 1000-6000 such turns (molecular weight $5 \times 10^6$-$40 \times 10^6$.)

The complementary theory of the structure of DNA, based on a solid foundation of physical and physicochemical research, is undoubtedly one of the greatest advances in chemistry and biochemistry.


## 3. Characteristics of the structure of RNA

The molecular weight of RNA derived from organelles and viruses varies between $6 \times 10^5$ and $2 \times 10^6$. The molecular weight of the soluble RNA found in the cell sap is considerably lower (between 1 and $5 \times 10^4$); that of messenger RNA is in the range $3 \times 10^5$-$1 \times 10^6$. The main difference between RNA and DNA is, however, not just the lower molecular weight and the replacement of thymine by uracil and of desoxyribose by ribose, but the way in which the actual macromolecule of RNA is constructed. The polynucleotides of viral RNA, unlike those of DNA, are found in the cell as single chains, not double ones (18, 51, 53, 55). This would seem to be the reason why the exact molar ratios between the different bases as found in DNA are not usually found in chemically pure preparations of viral RNA. However, analyses of the total RNA isolated from many sources, which have been carried out in the laboratories of A. N. Belozerskiĭ and N. M. Sisakyan, show, fairly clearly, many of the correspondences demonstrated by Chargaff (2, 3, 4, 6, 12, 20, 26) and this may suggest that the synthesis of new molecules of RNA also occurs according to the principle of complementarity. Jehle (56) has proposed a model for the configuration of RNA.

In the next sections we shall discuss evidence from a number of laboratories which suggests that the single chain of RNA found in the organelles and in solution may also undergo secondary folding with the formation of complementary pairs of nucleotides and a twin helix, many of the physical properties of which are comparable to those of the twin helix of DNA.

## 4. The sequence of nucleotides in the polynucleotide chains of DNA and RNA

According to present-day genetical and biochemical ideas it is the actual sequence of nucleotides in the polynucleotide chain which constitutes the "alphabet" in which information is "written" in the polynucleotide chains, determining the specificity of the proteins which are synthesized on the surface of these chains. Naturally, alongside the studies of the order in which amino acids are arranged in proteins, attempts have been made in recent years and are being made to study experimentally the sequence of nucleotides in DNA and RNA. There are, however, considerable technical difficulties in such studies, associated with the high molecular weight of RNA and DNA, with the difficulty of fractionating them into individual groups and with the difficulty of fractionating oligonucleotides. Nevertheless we can see systematic progress in this field even though it is slow, and already some important regularities in the structure of RNA and DNA have been discovered which are of great theoretical significance.

As has been noted above, we already know the exact sequence of the amino acids in a very large number of peptide, polypeptide and protein structures whereas the study of the sequence of nucleotides is only in its earliest stages. Nevertheless it seems appropriate to compare the findings for DNA and RNA with what has been found out about the structure of proteins. Although we cannot yet compare a particular protein with the polynucleotide on which it may have been formed yet, even at the present stage of our investigations, we can draw certain conclusions as to the possibility or impossibility of such a method of recording information. For example, if a strictly regular sequence of nucleotides was observed in sections of the chains of RNA and DNA which have been studied and this sequence was found to continue along the whole chain, then such an arrangement of the units of RNA and DNA could not correspond completely with the sequence of amino acids in a protein, because we know that the sequence of amino acids in proteins does not follow such a rule. Thus, although we cannot yet compare the structure of an individual molecule of DNA or RNA with that of a molecule of the protein which was formed on its surface and therefore has a structure corresponding to its own, yet we can consider in an exploratory way whether there are not some characteristics of the known modes of arrangement of the nucleotides in RNA and DNA, which would rule out the possibility that they play the determining part in the reproduction of the specificity of proteins and which do not correspond with the theory of the direct transfer of information from the polynucleotide chain to the protein.

Our review of the work on the sequence of nucleotides in the polynucleotide chains should begin with several papers by Shapiro and Chargaff in which they describe investigations of the structure of DNA obtained from calf's thymus by studying the oligonucleotides produced by its partial breakdown (22, 76, 77). These authors found that 70% of the pyrimidine residues of DNA were arranged in the form of uninterrupted oligonucleotide chains containing three or more pyrimidine bases in succession. This means, according to the laws already given, that the same should be true of purine bases. Thus, only a part of the total length of the polynucleotide chain can have a random (without any simple regularity) distribution of nucleotides.

Studies of the sequence of the nucleotides in RNA and DNA now occupy the energies of workers in many laboratories but the collection of detailed information in this field is only just beginning. Several different methods have been proposed for this purpose (31, 57, 88). In the most recent of these researches, as in those of Chargaff and his colleagues, it has been found that there are parts of the molecule of DNA in which pyrimidine bases follow, one after another. Similar observations were made by Kent and his colleagues (61) while studying the distribution of pyrimidine bases in the DNA of herring roe by an original method. Jones, Stacey & Watson (59) have recently studied some of the di- and tri-nucleotide components of the DNA isolated from *Mycobacterium phlei* after its alkaline degradation. They found by statistical treatment that the combinations in the compounds which they had studied were very different from those which would have been predicted according to the laws of chance distribution.

However, the distribution of purines and pyrimidines in uniform groups is not the only way in which they are distributed in DNA. A detailed study of the DNA of the thymus carried out by Cohn & Volkin (44) has shown that almost all possible combinations of purines and pyrimidines in di- and trinucleotides are to be found in the products of the incomplete breakdown of DNA. The existence in DNA of varied groupings of nucleotides as well as sections with a more or less uniform composition has also been demonstrated by other workers (33, 34, 58).

The interesting work of Burton & Petersen (36) is particularly suggestive in this connection. They have worked out a new method for hydrolysing DNA which they believe to be more favourable for the study of the oligonucleotides which are released. In their analysis of the DNA of the thymus these authors did not fully confirm the findings of Shapiro & Chargaff as to the high percentage of polypyrimidine and polypurine blocks in the composition of this DNA. Burton & Petersen found that 39% of the pyrimidine residues in DNA occur in the form of mononucleotides or dinucleotides and 61% in trinucleotides or longer oligonucleotides. If the distribution had been random each of these forms should have contained 50% of the total amount of the pyrimidines. Thus, in this respect, there is a rather slight deviation from random distribution in the composition of DNA. This sort of distribution of nucleotides is perfectly compatible with a mathematical correlation between the composition of proteins and that of nucleic acids. In proteins, even if we use the total protein of a tissue or organism, as we do now in the case of nucleic acids, the distribution of amino acids is also not absolutely random, while yet not showing any strict regularity in combination. The common occurrence of isomeric series (such as...Ser.Arg...and...Arg.Ser...), which is characteristic of proteins, is also found in DNA. According to Burton & Petersen the ratio between the amounts of cytosyl-p-thymyl-p and thymyl-p-cytosyl-p found in DNA are markedly different from what would be expected theoretically on the basis of a random distribution (there is a preponderance of the former combination).

Similar results have been obtained from the same laboratory (34a, 35, 35a) on preparations of DNA from eight different sources (four species of bacteria and four species of animals). These results are completely reproducible, and are characteristic for each of these DNA's. The results from materials from different sources were different, even when the DNA's in question were similar in base composition (e.g. those from *Pseudomonas aeruginosa* and from *Alkaligenes faecalis*). Although the reaction products were often significantly different in composition from that which would be expected for random distribution of the nucleotides along the polynucleotide chain, these deviations were not in any constant sense. Neither purines nor pyrimidines showed any tendency to be aligned in any particular sequence. Among various products which were identified was not only "tetrathymidine triphosphate" (TpTpTpT) but also a substance which, judging from its ultraviolet absorption and chromatographic behaviour, was "pentathymidine tetraphosphate". Both these substances were found in about such yields as might be expected for the case of completely random distribution of the nucleotide residues along the polynucleotide chain.

Burton considers that the products containing both cytosine and thymine occur as a mixture of all the possible sequence isomers. The relative quantities of these isomers were different in his eight different DNA preparations, obtained from different sources, and showed no correlation with the overall base composition of the preparations. Species differences in nucleotide sequences in DNA were also observed by Swartz and colleagues (84a).

Some recent work of Shapiro & Chargaff (78, 79), mainly devoted to the character of the distribution of methylcytosine in DNA prepared from rice germ, is extremely interesting in connection with the study of the distribution of nucleotides in DNA. Methylcytosine is a fifth base commonly found in DNA. In DNA of animal origin it is usually only present in very small amounts but in plant DNA the amounts of methylcytosine are quite considerable, especially in Gramineae (about 6 mole % of the total nucleotides). Therefore, as the sum of the cytosine and 5-methylcytosine (on a molar basis) is equal to the amount of guanine in the DNA, so it is natural to suppose that, in the double helical structure of this polymer, guanine can form a pair with cytosine or with its derivative 5-methylcytosine. Thus, if the resynthesis of new polynucleotide chains were determined entirely on the basis of complementarity, then the replacement of cytosine by 5-methylcytosine would be random, i.e. uniform. In fact Shapiro and Chargaff obtained a quite different result. On separating the DNA obtained from rice germ by fractional precipitation into 7 fractions they found that the ratio of cytosine to methylcytosine (c/m) in these fractions was different, varying between 2·50 and 3·36. The ratios between the sum of the adenine and thymine on the one hand and that of guanine and the two cytosines on the other were also different in different fractions, varying from 1·09 to 1·47. The other ratios corresponded exactly with the rules of complementarity referred to above.

By using the method of partial breakdown of preparations of DNA in which the most labile, interpurine bonds are broken first while the more resistant purely pyrimidine blocks remain unaffected in the form of so-called apurinic acid (about $\frac{2}{3}$ of all the pyrimidine residues remain in such blocks) the authors determined the molar proportion of those pyrimidines which were isolated in the polynucleotide chain, i.e. had a purine on each side of them. It was curious that, in all seven fractions, the distributions of cytosine and methylcytosine were unequal. While from 12-22% of the total amount of methylcytosine occurred in the form of isolated inclusions in purine sections only 7-12% of the cytosine was found in this situation (the corresponding figures for thymine were 14-18%). Thus we have evidence that cytosine and methylcytosine cannot be substituted for one another in the chain and that they obey individual laws in their distribution, although each of them would seem to form a pair with a guanine in the complementary chain.

It is interesting to note that the proportion of pyrimidine (and purine) bases occurring in the mixed and uniform sections was not the same in the different fractions, indicating that the nuclei of cells contain an assortment of DNA's having different sequences of bases. In connection with these interesting findings the authors suggest that the specificity of the sequence of the nucleotides in DNA is determined, not only by the complementarity of the bases in pairs, but also by the nature of their neighbours along the line of the chain if both the competitors (the actual nucleotide and its analogue) satisfy the requirements of complementarity.

A new direct electron-microscopic approach to the determination of base sequence in DNA was developed recently by Beer & Moudrianakis (26a).

The sequence of nucleotides in RNA has been even less fully studied, especially as regards the RNA of tissues. The only exception is the RNA of tobacco mosaic virus, which has formed the subject of some recent investigations. Reddi (70, 72) published the first results concerning the distribution of purine and pyrimidine bases in the RNA of this virus.

In a study of the products obtained by breaking down the RNA of tobacco mosaic virus (TMV) with ribonuclease and phosphodiesterase, Reddi has confirmed the presence in the molecules of this RNA of poly-purine and polypyrimidine blocks which, however, represented in all less than 50% of the total length of the polynucleotide of the TMV.

According to these figures the polypurine blocks contain about 20% of the total number of adenyl and guanyl residues in the RNA of TMV, while the polypyrimidine blocks contain about 50% of the total number of uridylic and cytidylic residues. The data covered 55·2, 40·8, 77·9 and 72·3% of the total amounts of adenylic, guanylic, cytidylic and uridylic residues respectively in the RNA of the virus.

The author reaches the conclusion that, as well as the polypurine and polypyrimidine sections of the viral RNA, there must be some regions in which the individual nucleotides simply follow one another in turn. Reddi puts forward the idea that the order of arrangement in these zones has a certain regularity such that the number of adenylic and guanylic residues and the number of cytidylic and uridylic residues in the uniform blocks correspond in pairs.

It must be pointed out that, in his ribonuclease and phosphodiesterase digests, Reddi found hardly any dinucleotides. However, he does not think that this means that they do not exist, as they could be identified when other nucleases were used.

Miura & Egami (66) have shown recently that the distribution of the nucleotides in the RNA of TMV and yeast is mainly in groups.

An interesting technique for determining the base sequence in RNA was developed recently by Rushizky & Sober (73a). This uses ribo-nuclease $T_1$, which specifically hydrolyses phosphodiester bonds between guanosine 3'-phosphate and other nucleotides. See also (83a).

Thus, despite the manifest limitation of the evidence, it is still clear that the sequence of nucleotides in DNA and RNA, like that of amino acid residues in proteins, does not conform to any simple rule. Apart from the more orderly small zones there are sections of mixed distribution, the length of which is still unknown. The overall length of the poly-nucleotides of the nucleus and of the intracellular organelles is usually tens and sometimes hundreds of times that of the polypeptide chains of known proteins. (A protein molecule with a molecular weight of the order of 100,000-600,000 has an aggregated structure and usually consists of sub-units of considerably smaller size.) The arrangement of zones of different types of sequence in polynucleotides could, therefore, represent an alternation of zones of synthetic activity with intermediate sections separating them. It would be well, however, to leave consideration of such hypotheses until a later chapter, after a review of the main features of protein synthesis.

## 5. The specificity of nucleic acids

Summarizing work on the specificity and heterogeneity of nucleic acids Belozerskiĭ (1) points out that one may put forward the following theoretical possibilities for the form of their specificity: it may refer to a species, age, organ, tissue, organelle or molecule while a single organelle may contain a collection of different molecules of nucleic acids.

We shall consider the possible existence of age specificity of nucleic acids later.

One must point out that if nucleic acids actually are compounds which accumulate a particular stock of biochemical information within cells, regulating the reproduction of the specificity of proteins, then one might expect that nucleic acids would, above all, possess species and tissue specificity and molecular heterogeneity, corresponding with the possible forms of interspecies and intracellular variation of protein substance.

The factual material which has so far been collected by biochemists fully confirms this expectation. The existence of chemical differences (nucleotide composition) between the RNA's and DNA's collected from sources which differ widely in their evolutionary origins is a solidly grounded fact and there is no need to illustrate it by a large number of examples, the more so as this question is discussed in detail in the reviews referred to at the beginning of the chapter. In recent years detailed studies have been carried out, showing that the overall composition of DNA is not the same, not merely in specimens derived from sources differing widely in their evolutionary origins, but also in specimens derived from fairly closely related sources, e.g. in different species of bacteria or even different bacterial populations brought into being by experimental mutations (3, 4, 5, 7, 14, 15, 16, 17, 26, 11).

The composition of the RNA derived from different species of bacteria (1, 2, 4) is considerably less variable, but this does not exclude wide variations in the sequences of nucleotides, which would be of fundamental importance in determining the specificity of proteins.

In recent detailed studies (6, 20, 11) the existence of species specificity in the composition of RNA and DNA has been demonstrated for both lower and higher plants by comparison of material derived from a very large number of different species. The variations in the composition of RNA and DNA in the plants were found to be considerably less than in bacteria but still there can be no doubt that species differences do exist.

The study of the species specificity of nucleic acids has been carried out extensively, for the most part in two laboratories, that of Chargaff in the U.S.A. and that of Belozerskiĭ in the U.S.S.R. Already all the important groups of animals, microorganisms and plants have been investigated, and the question of the existence of this specificity may be said to be solved. The results of these researches, which have extended over more than ten years, has increased interest in the nucleic acids and has provided the factual basis for the hypothesis that nucleic acids are compounds which can accumulate specific biochemical information.

The facts which indicate the existence of intracellular as well as interspecies heterogeneity of RNA and DNA are also of very great interest, as they show that the cell has a sufficient chemical basis for "recording" the extensive information which is needed for the regulation of the reproduction of the specificity of the proteins which are being synthesized. Chargaff and his colleagues (43, 45) have obtained some very interesting results, showing that DNA from different sources can be fractionated on the basis of ability to form complexes with histones; the fractions thus obtained have different nucleotide compositions. Arising out of this the authors put forward the hypothesis that the DNA of cells consists of a large number of differently constructed individual molecules of DNA, constituting a continuous "spectrum" of molecular gradations. Many other authors have also managed to separate the DNA of nuclei of a particular type into a number of fractions by a variety of methods of fractionation such as adsorption and ultracentrifugation (27, 38, 65, 69, 74, 75).

Detailed studies of the physico-chemical heterogeneity of DNA have been carried out by Butler and his colleagues (37, 38, 80). They did not confine themselves to observing the fact of the heterogeneity of DNA but studied several possible subsidiary causes for this heterogeneity, having no direct relationship to nucleotide composition. For example, they found that DNA can undergo partial degradation during its extraction and that this degradation leads to a certain polydispersity of the molecular weight. They also found that native DNA, isolated from different tissues, usually contains an admixture of proteins or large peptides of specific composition (more similar to cytoplasmic proteins than to nuclear protamines or histones). Before the separation of these impurities from the DNA, the preparations of DNA showed considerable heterogeneity on ultracentrifugation. After the separation and hydrolysis of these protein components with chymotrypsin the proportion of material in the fractions with a high sedimentation coefficient fell sharply. This effect gives rise to the idea that the admixture of a protein-peptide component with DNA may facilitate the aggregation of molecules by the formation of cross-linkages. These results provide clear evidence that the facts of physico-chemical heterogeneity must be approached very cautiously and need to be supplemented by analysis of the chemical composition of the isolated fractions.

There is also much evidence in favour of the heterogeneity of RNA. It has been found possible to separate highly purified specimens of RNA from yeast into a number of fractions with different molecular weights and electrophoretic properties (52, 64). Organ specificity of RNA has been demonstrated by means of a precipitation reaction. Two fractions of RNA have been isolated from the nuclei of organisms, differing from one another in the rate of renewal of the phosphoric groups in their molecules (60, 85, 63).

Some interesting work on the species specificity of RNA in plants and the presence of fractions having different electrophoretic properties in plant tissues has been done by Lindner and his colleagues (62). They found fairly well-marked species differences in RNA and a clear-cut electrophoretic heterogeneity of the RNA, both free and combined with protein. The relationships between these fractions were different in different species.

It must be said that there are, as yet, only few and contradictory studies of the organ and tissue specificity of nucleic acids. Theoretically, the presence of such differences could be a result of the differentiation of the tissues and organs during ontogenesis. What evidence we have in this connection will be dealt with in the chapter devoted to a survey of the biochemical aspects of the morphogenic processes of ontogenesis.

The interesting researches of Reddi (71) gave a new direction to the study of the species specificity of nucleic acids. He found that there is a definite difference in the nucleotide sequence in preparations of RNA isolated from different strains of tobacco mosaic virus. The data obtained are, as yet, the only ones which point to a species difference in the sequence of nucleotides in RNA. It is certain, however, that further investigations will be made in this field, as any new discoveries would be of great interest. The essential genetic characteristic in which nucleic acids should differ from one another, even if many of their chemical and physical properties are the same, is the sequence of the nucleotides in their polynucleotide chains, which is the primary expression of the unlimited variety and universality of the structure of proteins.

Great interest also attaches to the studies of the species specificity of the physico-chemical and molecular structure of nucleic acids, which have still only just begun.

A large group of scientists from several laboratories (54) recently carried out jointly a detailed X-ray structural investigation of the configuration of the molecule of DNA isolated from a very large number of varied sources. In fact, all the preparations had very similar structures, corresponding to the model of Watson & Crick.

There are, however, some interesting exceptions to this rule. We have in mind, on the one hand, the unusual single-chain DNA found in the small phage $\phi \times 174$ (81) and, on the other, the DNA having four or more chains which has been isolated from some bacteria in the stage of logarithmic growth (39, 40).

It must be noted that, in view of the enormous molecular weight of DNA and RNA, the complete elucidation of their nucleotide sequence is an extremely difficult matter. Chargaff (22) made a characteristic remark in this connection when he said that although the mutual arrangements of 4 nucleotides were limited, yet, if a chain of 25,000 such fragments were constructed, it could store an extremely large amount of information. Chargaff thinks, however that: "Even if we possessed a strictly uniform (homogeneous) nucleic acid and had at our disposal methods for degrading it in strictly sequential fashion, it would still be hopeless to try to decipher its nucleotide sequence. For, by the time the cryptographer had fulfilled his assignment and written out the complete sequence, evolution, in all probability, would have obliterated it from nature and he would have to start again."

It is quite clear that the detailed comparison, between the sequence of nucleotides in the matrices and that of the amino acids in the proteins which have been synthesized on them, will only be achieved in the more or less distant future. An exception may be the case of tobacco mosaic virus, for which a solution to the problem may not be too distant. Nevertheless we have, as yet, no reason for Chargaff's scepticism. This scepticism would be justified if future generations of biochemists used the same cumbersome methods for the solution of problems as are in use nowadays. There is, however, no reason to expect such stagnation in the improvement and automation of laboratory methods and it is quite possible that, in the future, special electronic machines will be constructed to elucidate the sequences of amino acids and nucleotides.

## Conclusion

In this chapter we have surveyed only a very few of the questions which affect contemporary ideas on the structure of DNA and RNA and the connection between these structures and the biological functions of these remarkable polymers in living nature. In later chapters we shall repeatedly return to and shall examine in detail the peculiarities of the structures of DNA and RNA associated with their ability to reproduce themselves, with their linkage to proteins in the organelles, with their ability to act as matrices, in connection with the peculiarities of the multiplication of viruses etc.

However, even as a result of the general information discussed in this chapter, we can point to certain common principles in the structure of proteins and nucleic acids, although their chemical composition is quite different. Nucleic acids and proteins show a certain parallelism in the principles on which they are constructed; this concerns the linearity of the structure of polynucleotides and polypeptides, the equality of the distances between the points at which the monomers combine with their neighbours, their intracellular heterogeneity and the presence of specificity conferred by particular forms of arrangement of nucleotide and amino acid residues. At the same time the nucleic acids are themselves able to arrange their own nucleotides in order and to preserve their helical-linear structure (by complementarity) while the polypeptide chains, after they have been formed, acquire a complicated secondary and tertiary structure which is responsible for their ability to carry out their particular biological functions but makes autosynthesis of the molecules impossible. Obviously these purely structural parallels cannot be due entirely to chance and their existence has given rise to speculation as to the possibility of nucleic acids functioning as matrices determining the specificity of protein synthesis. It has long been supposed that some sort of specific matrices might exist in cells. In 1927 Kol'tsov put this idea forward in explanation of the function of the genetic apparatus. In 1930 the idea of complementary matrices was put forward by Breinl & Haurowitz (30) to explain the dependence of the synthesis of antibodies on the structure of the corresponding antigens. The existence of such a relationship between nucleic acids and proteins, however, met with many objections, mostly based on the structural differences between these compounds and the lack of evidence for specificity or heterogeneity among nucleic acids. These objections have now been overcome and there are no longer any serious "chemical" grounds for denying the theoretical possibility that nucleic acids may function as matrices for protein synthesis. However, if we take a look at the actual data concerning the mechanism of the biological synthesis of proteins, we shall find that the way in which nucleic acids carry out this function is far more complicated than might have been expected from the purely chemical parallelism.

## REFERENCES

1. BELOZERSKIĬ, A. N. (1959). *Proc. first internat. Sympos. on the origin of life* (ed. A. I. Oparin *et al.*), p. 322. London: Pergamon.
2. — (1959). *Nukleoproteidy i nukleinovye kisloty rastenii i ikh biologicheskoe znachenie.* (*Bakhovskaya lektsiya*). Moscow: Izd. Akad. Nauk S.S.S.R.
2a. BELOZERSKIĬ, A. N. (1961). *Nukleinovye kisloty i ikh biologicheskoe znachenie.* Moscow: Izd. "Znanie".
3. BELOZERSKIĬ, A. N. & SPIRIN, A. S. (1956). *Uspekhi sovremennoĭ Biol.* **41**, 144.
4. — (1960). *Izvest. Akad. Nauk S.S.S.R., Ser. biol.* **25**, No. 1, 64.
5. BELOZERSKIĬ, A. N., SHUGAEVA, N. V. & SPIRIN, A. S. (1958). *Doklady Akad. Nauk S.S.S.R.* **119**, 330.
6. VANYUSHIN, B. F. & BELOZERSKIĬ, A. N. (1959). *Doklady Akad. Nauk S.S.S.R.* **127**, 455.
7. GUMILEVSKAYA, N. A. & SISAKYAN, N. M. (1961). *Doklady Akad. Nauk S.S.S.R.* **137**, 206.
8. DUBININ, N. P. (1956). *Biofizika* **1**, 677.
9. — (1957). *Byul. Mosk. Obshchestva Ispytateleĭ Prirody, Otd. biol.* **62**, No. 2, 5.
10. MEDVEDEV, ZH. A. (1960). *Izvest. Timiryazev. sel'skokhoz. Akad.* No. 1, 103.
11. SERENKOV, G. P. & PAKHOMOVA, M. V. (1959). *Nauch. Doklady Vyssheĭ Shkoly, Biol. Nauki*, No. 4, 156.
12. SISAKYAN, N. M., ODINTSOVA, M. S. & CHERKASHINA, N. A. (1960). *Biokhimiya*, **25**, 160.
13. SPIRIN, A. S. (1961). *Zhur. Vsesoyuz. khim. Obshchestva im. D. I. Mendeleeva*, **6**, 260.
14. SPIRIN, A. S. & BELOZERSKIĬ, A. N. (1956). *Biokhimiya*, **21**, 768.
15. — (1957). *Doklady Akad. Nauk S.S.S.R.* **113**, 650.
16. SPIRIN, A. S., BELOZERSKIĬ, A. N., KUDLAĬ, D. G., SKAVRONSKAYA, A. G. & MITEREVA, V. G. (1958). *Biokhimiya*, **23**, 154.

17. SPIRIN, A. S., BELOZERSKIĬ, A. N., SHUGAEVA, N. V. & VANYUSHIN, B. F. (1957). *Biokhimiya*, **22**, 744.
18. SPIRIN, A. S., GAVRILOVA, L. P. & BELOZERSKIĬ, A. N. (1959). *Doklady Akad. Nauk S.S.S.R.* **125**, 658.
19. SPITKOVSKIĬ, D. M., TONGUR, V. S. & DISKINA, B. S. (1958). *Biofizika*, **3**, 129.
20. URYSON, S. O. & BELOZERSKIĬ, A. N. (1959). *Doklady Akad. Nauk S.S.S.R.* **125**, 1144.
21. KHESIN, R. B. (1960). *Biokhimiya tsitoplazmy*. Moscow: Izd. Akad. Nauk S.S.S.R.
22. CHARGAFF, E. (1958). *Izv. Akad. Nauk S.S.S.R.*, *Ser. biol.*, No. 2, 144.
23. — (1959). *Proc. first internat. Sympos. on the origin of life on the Earth* (ed. A. I. Oparin *et al.*), p. 297. London: Pergamon.
24. CHEPINOGA, O. P. (1956). *Nukleinovye kisloty i ikh biologicheskaya rol'*. Kiev: Izd. Akad. Nauk Ukr. S.S.R.
25. ENGEL'GARDT, V. A. (1959). *Uspekhi Khim.* **28**, 1011.
26. BELOZERSKY [BELOZERSKIĬ], A. N. & SPIRIN, A. S. (1960). In *The nucleic acids —chemistry and biology* (ed. E. Chargaff & J. N. Davidson), Vol. **3**, p. 147. New York: Academic Press.
26a. BEER, M. & MOUDRIANAKIS, E. N. (1962). *Proc. nat. Acad. Sci. U.S.* **48**, 409.
27. BENDICH, A., PAHL, H. B., KORNGOLD, G. C., ROSENKRANZ, H. S. & FRESCO, J. R. (1958). *J. Amer. chem. Soc.* **80**, 3949.
28. BRACKET, J. (1957). *Biochemical cytology*. New York: Academic Press.
29. — (1960). *The biological role of ribonucleic acids*. Amsterdam: Elsevier.
30. BREINL, F. & HAUROWITZ, F. (1930). *Hoppe-Seyl. Z.* **192**, 45.
31. BROWN, D. M., FRIED, M. & TODD, A. R. (1953). *Chem. & Ind. (Lond.)*, p. 352.
32. BROWN, D. M. & TODD, A. R. (1952). *J. chem. Soc.*, p. 52.
33. — (1955). *Ann. Rev. Biochem.* **24**, 311.
34. — (1955). In *The nucleic acids—chemistry and biology* (ed. E. Chargaff & J. N. Davidson), Vol. **1**, p. 409. New York: Academic Press.
34a. BURTON, K. (1960). *Biochem. J.* **77**, 547.
35. — (1960). *Biochem. J.* **74**, 35P.
35a. — (1961). *Proc. V int. Congr. Biochem.*, *Moscow*, **1**, 61.
36. BURTON, K. & PETERSEN, G. B. (1960). *Biochem. J.* **75**, 17.
37. BUTLER, J. A. V., PHILLIPS, D. M. & SHOOTER, K. V. (1957). *Arch. Biochem. Biophys.* **71**, 423.
38. BUTLER, J. A. V. & SHOOTER, K. V. (1957). *Johns Hopkins Univ., McCollum-Pratt Inst., Contrib.* No. 153, 540.
39. CAVALIERI, L. F., FINSTON, R. & ROSENBERG, B. H. (1961). *Fed. Proc.* **20**, 352.
40. — (1960). *Nature (Lond.)*, **189**, 833.
41. CHARGAFF, E. (1950). *Experientia*, **6**, 201.
42. — (1952). *2-ème Congr. intern. Biochimie, Chim. biol. V, Symposium sur le Métabolisme microbien* (Paris), p. 41.
43. CHARGAFF, E., CRAMPTON, C. F. & LIPSHITZ, R. (1953). *Nature (Lond.)*, **172**, 289.
44. COHN, W. E. & VOLKIN, E. (1957). *Biochim. biophys. Acta* **24**, 359.
45. CRAMPTON, C. F., LIPSHITZ, R. & CHARGAFF, E. (1954). *J. biol. Chem.* **211**, 125.
46. CROOK, E. M. (ed.) (1957). *Biochem. Soc. Symposia*, No. 14.
47. DAVIDSON, J. N. (1960). *The biochemistry of the nucleic acids* (4th edn.). London: Methuen.
48. DOTY, P. (1957). *J. cell. comp. Physiol.* **49**, Suppl. 1, 27.
49. FEUGHELMAN, M., LANGRIDGE, R., SEEDS, W. E., STOKES, A. R., WILSON, H. R., HOOPER, C. W., WILKINS, M. H. F., BARCLAY, R. K. & HAMILTON, L. D. (1955). *Nature (Lond.)*, **175**, 834.
50. FRANKLIN, R. E. & GOSLING, R. G. (1953). *Nature (Lond.)*, **171**, 740.

51. GIERER, A. (1958). *Z. Naturforsch.* **13B**, 788.
52. HAKIM, A. A. (1957). *J. biol. Chem.* **225**, 689.
53. HALL, B. D. & DOTY, P. (1959). *J. mol. Biol.* **1**, 111.
54. HAMILTON, L. D., BARCLAY, R. K., WILKINS, M. H. F., BROWN, G. L., WILSON H. R., MARVIN, D. A., TAYLOR, H. E. & SIMMONS, N. S. (1959). *J. biophys. biochem. Cytol.* **5**, 397.
55. HART, R. G. (1958). *Biochim. biophys. Acta,* **28**, 457.
56. JEHLE, H. (1959). *Proc. nat. Acad. Sci. U.S.* **45**, 1360.
57. JONES, A. S. & LETHAM, D. S. (1954). *Biochim. biophys. Acta,* **14**, 438.
58. JONES, A. S., LETHAM, D. S. & STACEY, M. (1956). *J. chem. Soc.* p. 2579.
59. JONES, A. S., STACEY, M. & WATSON, B. E. (1957). *J. chem. Soc.* p. 2454.
59a. JORDAN, D. O. (1960). *The chemistry of nucleic acids.* London: Butterworths.
60. KAY, E. R. M., SMELLIE, R. M. S., HUMPHREY, G. F. & DAVIDSON, J. N. (1956). *Biochem. J.* **62**, 160.
61. KENT, P. W., LUCY, J. A. & WARD, P. F. V. (1955). *Biochem. J.* **61**, 529.
62. LINDNER, R. C., KIRKPATRICK, H. C. & WEEKS, T. E. (1956). *Plant Physiol.* **31**, 1.
63. LOGAN, R. & DAVIDSON, J. N. (1957). *Biochim. biophys. Acta,* **24**, 196.
64. MALLETTE, M. F. & LAMANNA, C. (1953). *Arch. Biochem. Biophys.* **47**, 174.
65. MARMUR, J. & DOTY, P. (1959). *Nature (Lond.),* **183**, 1427.
66. MIURA, K.-I. & EGAMI, F. (1960). *Biochim. biophys. Acta,* **44**, 379.
67. CHARGAFF, E. & DAVIDSON, J. N. (Eds.) (1955-60). *The nucleic acids— chemistry and biology.* Vols. **1, 2 & 3**. New York: Academic Press.
68. PAULING, L. & COREY, R. B. (1956). *Arch. Biochem. Biophys.* **65**, 164.
69. POUYET, J. & WEILL, G. (1957). *J. polymer Sci.* **23**, 739.
70. REDDI, K. K. (1959). *Proc. nat. Acad. Sci. U.S.* **45**, 293.
71. — (1959). *Biochim. biophys. Acta,* **32**, 386.
72. — (1960). *Nature (Lond.),* **188**, 60.
73. RICH, A. (1959). *Revs. modern Phys.* **31**, 191.
73a. RUSHIZKY, G. W. & SOBER, H. A. (1962). *J. biol. Chem.* **237**, 834, 2883.
74. SCHACHMAN, H. K. (1957). *J. cell comp. Physiol.* **49**, Suppl. 1, 71.
75. SEMENZA, G. (1956). *Bull. Soc. ital. Biol. sper.* **32**, 1298.
76. SHAPIRO, H. S. & CHARGAFF, E. (1957). *Biochim. biophys. Acta,* **23**, 451.
77. — (1957). *Biochim. biophys. Acta,* **26**, 596.
78. — (1960). *Biochim. biophys. Acta,* **39**, 62, 68.
79. — (1960). *Nature (Lond.),* **188**, 62.
80. SHOOTER, K. V. & BUTLER, J. A. V. (1955). *Nature (Lond.),* **175**, 500.
81. SINSHEIMER, R. L. (1959). *J. mol. Biol.* **1**, 43.
82. — (1959). *Brookhaven Symposia in Biol.,* No. *12*, 27.
83. — (1959). *Symposium Mol. Biol. Univ. Chicago,* p. 16.
83a. STAEHELIN, M. (1961). *Biochim. biophys. Acta,* **49**, 11, 20, 27.
84. STEINER, R. F. & BEERS, R. F., Jr. (1961). *Polynucleotides: natural and synthetic nucleic acids.* Amsterdam: Elsevier.
84a. SWARTZ, M., TRAUTNER, T. & JOSSE, J. (1961). *Fed. Proc.* **20**, 354.
85. VINTER, V. (1959). *Nature (Lond.),* **183**, 998.
86. WATSON, J. D. & CRICK, F. H. C. (1953). *Cold Spring Harbor Symposia quant. Biol.* **18**, 123.
87. — (1953). *Nature (Lond.),* **171**, 737.
88. WHITFELD, P. R. (1954). *Biochem. J.* **58**, 390.
89. WILKINS, M. H. F. (1956). *Cold Spring Harbor Symposia quant. Biol.* **21**, 75.
90. WILKINS, M. H. F., SEEDS, W. E., STOKES, A. R. & WILSON, H. R. (1953). *Nature (Lond.),* **172**, 759.
91. WILKINS, M. H. F., STOKES, A. R. & WILSON, H. R. (1953). *Nature (Lond.),* **171**, 738.

# CHAPTER III

# THE BIOCHEMICAL MECHANISM OF ACTIVATION OF AMINO ACIDS IN PROTEIN SYNTHESIS

## Introduction

The formation of peptide bonds is known to require the expenditure of a certain amount of energy, on the average, about 2-3000 cal./mole of substance in the formation of dipeptides. According to the calculations of Tarver (56) the combination of higher peptides (polymerization of peptides) can occur with the expenditure of rather less energy, that is to say, under more favourable energetic conditions. Thus, whichever way it happens, the synthesis of protein within the cell requires the occurrence of some parallel reactions to provide the energy needed for some sort of energetic activation of the amino acids. Although this now seems a perfectly straightforward idea it only arose as a result of researches carried out over quite a long period, in the course of which hundreds of different works on the subject were published.

There is no need, now, to go into a detailed account of all the stages of these investigations, as the material has already been reviewed and discussed in the many reviews of protein synthesis (1, 2, 3, 9, 25, 6, 7, 8, 10, 18, 32, 33, 35, 56).

The only fact which needs to be mentioned is that, as a result of all these investigations, it was found that there is some relationship between the synthesis of peptides and proteins and processes of phosphorylation. In the last analysis what was established by all this extensive series of researches, covering practically all the main groups of living things, was that the presence of adenosine triphosphate (ATP) was essential for the synthesis of proteins and peptides in the cell.

In the years 1954-55 this "energetic stage" in the study of the synthesis of peptide bonds was virtually finished and a new period began in which attention was mainly focussed on the mechanism whereby ATP takes part in the activation of amino acids. This period was considerably shorter and in the next two years this mechanism was elucidated. Naturally, until the reactions taking part in the process were worked out, the various hypotheses, which were always being put forward to explain the general way in which proteins are synthesized, were extremely tentative and conditional in character.

In the present chapter we shall look at the essential data mainly obtained in the course of this second stage in the study of the energetic activation of amino acids. As the fundamental lines of this activation may be held to have been fairly thoroughly established it will serve our purpose to present them as concisely as possible.

## 1. The enzymic system of activation of amino acids by the formation of aminoacyl adenylates

In 1954-56, as a result of the brilliant work of Hoagland, Zamecnik and their colleagues (28, 29, 30, 64, 65, 66) on the supernatant fraction from homogenates of animal tissues, a group of new enzymes was discovered, which bring about the activation of amino acids in the presence of ATP and magnesium ions. In the first series of these researches the authors used the ordinary method for the incorporation of amino acids into the proteins of cytoplasmic fractions *in vivo* and *in vitro*. The only unusual feature of these experiments was that the authors introduced into the organisms small amounts of [14C]amino acids of very high specific activity and that they could therefore observe the incorporation of these amino acids into proteins over very short intervals of time (2-10 min.). In order to exclude contamination by adsorbed material they made a systematic check on the occurrence of genuine synthetic inclusion of 14C by calculating the rate at which [14C]amino acids were split off when the isolated proteins were submitted to slow acid hydrolysis.

By carrying out a careful fractionation of the protoplasm of the livers or ascitic tumours of rats a few minutes after the animals had received injections of labelled amino acids ([14C]leucine or [14C]valine) Hoagland, Zamecnik and their fellow workers found that in these experiments the greatest activity (about 7-10 times that of the total protein) occurred in the proteins of the microsomes. Microsomal material also carried out active protein synthesis *in vitro*, but this required the presence of a particular biochemical system. This biochemical system comprised five essential components: (1) the microsomal fraction; (2) an enzymic complex precipitated at pH 5 from the supernatant liquid after removal of the microsomes by centrifugation at 105,000 g; (3) ATP or a system for regenerating ATP; (4) guanosine triphosphate (GTP) or guanosine diphosphate (GDP) and (5) a collection of amino acids. If one of these components was absent the rate of incorporation became far slower.

Among these components, the one which was of greatest interest was the enzymic complex and many further experiments were later devoted to studying it. These experiments showed that the enzymic complex brings about the carboxyl activation of the amino acids by means of a reaction in which the amino acid is linked to ATP with the elimination of pyrophosphate (PP) according to the following scheme:

$$\text{enzyme} + \text{ATP} + [^{14}\text{C}]\text{amino acid} \leftrightharpoons$$
$$[^{14}\text{C}]\text{aminoacyl-AMP-enzyme} + \text{PP}$$

In these experiments the part played by GTP and GDP remained unclear. They could not act as substitutes for ATP yet seemed to be necessary and very specific cofactors in the process. However, it has now been shown that GTP does not participate directly in the activation of amino acids but in later stages of protein synthesis.

The existence of a specific enzymic system for the activation of amino acids was soon confirmed by work in many other laboratories, and it was clearly demonstrated that the system has several components and is widely distributed in different organs and tissues.

For example, Work (20) and his colleagues showed that each component of a mixture of eleven different amino acids is activated by its own specific enzyme, so that amino acids do not compete against one another in this respect. In this work the enzymic complex was studied in a model system in which it catalysed the formation of hydroxamates of amino acids by incubation with amino acids, ATP and hydroxylamine. The enzymic system with the greatest activity was found in the pancreas.

The heterogeneity of the activating enzyme system was also clearly revealed by the work of Schweet and his colleagues (50-52), who separated the enzymes activating tyrosine from those activating tryptophan by fractionating the proteins of the supernatant liquid from a homogenate of the pancreas. Berg (15) isolated from yeast an enzyme which specifically activates methionine. Rendi et al. (48) identified an analogous enzyme for carboxyl activation of methionine in the liver of rats and in doing so they not only activated free methionine, but also activated some peptides containing this amino acid (alanylmethionine and glycylmethionine). A method has been worked out in Lipmann's laboratory (34) for obtaining a high degree of purification of the enzymic protein which specifically activates tryptophan and the activation of many other amino acids in the supernatant liquid of a homogenate of the pancreas has been studied. Detailed investigations of the activation of amino acids in the tissues of animals and in the cells of a number of microorganisms have been undertaken by Novelli (44, 45).

The activation of amino acids has also been observed in the supernatant liquid from homogenates of plant tissues (16, 19, 36, 24, 60, 61). This liquid activates practically the whole assortment of amino acids found in proteins and exhibits an even greater activity in this respect than the supernatant liquid from animal tissues.

The study of the activation of amino acids by a specific enzymic system is still going on intensively, as many details of the process are still not properly understood. The presence of these enzymes in the organelles of cells is being investigated (5, 37, 58) as are the details of the chemical mechanism by which the reaction proceeds and the need for a number of cofactors (22, 41, 43, 46, 54, 55), their ubiquity in various biological materials (47, 49) and preparative work is also being undertaken on the isolation and purification of individual activating enzymes for the further study of their chemical and biochemical properties (26, 39, 40, 42, 50, 61a).

## 2. Selectivity of the enzymic systems activating amino acids

The formation of adenylates is not a process peculiar to the system of protein synthesis. In recent years it has been found that very many processes of activation (the activation of acetate, fatty acids, pantothenic acid etc.) are brought about by the formation of intermediate compounds with adenylic acid, formed when the substrate reacts with ATP. Intermediate activated compounds of this kind have been given the collective name of acyl adenylates. The excellent reviews of Habermann (4) and Strominger (53) discuss in detail the biochemical properties of such compounds and it is therefore hardly worth while for us to enter into any lengthy discussion of the subject. It is, however, interesting for us to spend some time on another problem, that of the nature of the selectivity of activating enzymes.

Each such enzyme has at least a double selectivity, in respect of a nucleotide and in respect of an amino acid (we shall see later that the properties of activating enzymes are even more complicated). The molecules of these enzymes can only react with ATP; other nucleoside triphosphates (GTP, CTP, UTP) cannot replace this compound. On the other hand, each molecule of the enzyme is also specific in respect of an amino acid, that is to say it can form a triple complex (aminoacyl $\sim$ AMP-enzyme) only with one particular amino acid, tryptophan for example, while the activation of other amino acids requires other analogous enzymes. The specificity of the activating enzymes is also manifested in relation to analogues of the amino acids concerned, though it is not always very sharp. For example, purified tryptophan-activating enzyme did not activate 5-methyltryptophan or 6-methyltryptophan but did activate some other analogues (7-azatryptophan, tryptazan, etc.) though more slowly than tryptophan itself (23).

Connell and his colleagues have recently shown (21) that DL-ethionine and $\beta$-2-thienylalanine completely suppress the activation of alanine in the cells of *Azotobacter*.

The inhibitory action of thienylalanine on the activation of alanine has been clarified by the recent work of Wolfe & Hahn (62) who found that, although this substance is activated by the enzymic system of *E. coli*, the activated complex formed does not take part in the further reactions of protein synthesis.

Interaction between the activating enzymes and D-amino acids also inhibits the synthesis of proteins (41). However, in those rare cases in which the D-amino acids are themselves normal metabolites, they seem to be activated by specific enzymes. For example, a specific enzyme activating D-alanine has been found in *Lactobacillus arabinosus* and in a number of other microorganisms in which this amino acid enters into the composition of a special component of the cell wall (11). In algae an enzyme has been found which activates αε-diaminopimelic acid, which is a component of the cell wall, though not of the protein (18a).

The existence of a certain selectivity in respect of nucleotide and amino acid components in activating enzymes is at present explained hypothetically in terms of steric factors. It is suggested that the molecules of these enzymes have, on their surfaces, what might be called pits which are complementary in configuration and chemical properties to the side chains of the compounds which react with them (Fig. 14).

The existence of such specificity, even at the earliest stage of protein synthesis, is very important, biologically speaking. In synthesizing a protein it is not only important to arrange amino acids on a matrix in some particular order but also to supply in advance some strictly determinate assortment of reacting components in order that the task of the matrix may, as far as possible, be limited and thus the accuracy of the reproduction of the specificity of the proteins being synthesized may be increased to the required degree.

However, the selectivity of the activating enzymes is, as we shall see,

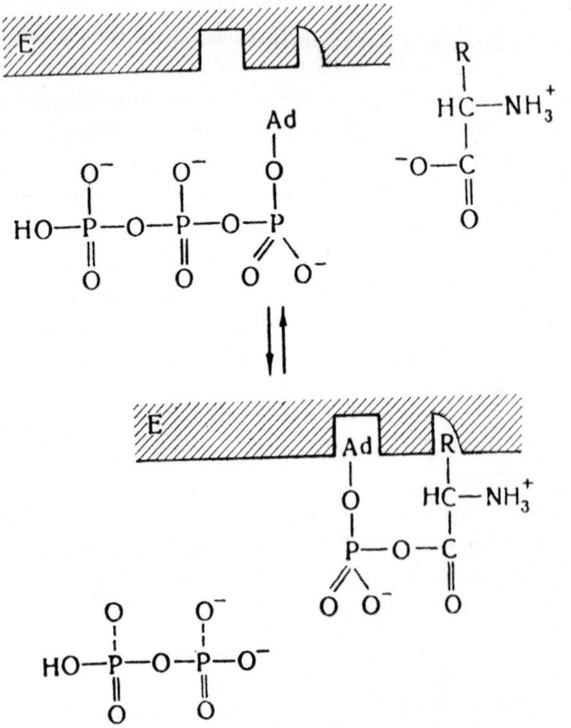

FIG. 14. Activation of amino acids.

E = molecule of enzyme; Ad = adenosine; R = side chain of amino acid.

not absolute and "mistakes" may happen, especially by reaction with analogous amino acids. This inaccuracy in the reaction of activating enzymes with analogues and normal metabolites has also been studied recently in connection with the reverse reaction in which the enzyme catalyses the formation of ATP from pyrophosphate and aminoacyl adenylate (31). Some analogues of normal aminoacyl adenylates catalysed this reaction, while others did not react with the enzyme.

An interesting case is that of the activation of peptides, which has been studied in a number of works (48, 49a, 57). There can be no doubt that, in this case, only the amino acid of the peptide grouping which has the free carboxyl group reacts with the enzyme.

## 3. Chemical synthesis of aminoacyl adenylates and the study of their role in the synthesis of proteins

The system of activating enzymes plays a double part in protein synthesis. On the one hand it activates amino acids by combining the adenyl component of ATP with their carboxyl groups. On the other hand it selects, from the general assortment of metabolites, just the collection of amino acids to be used in the synthesis of proteins. However, having formed the aminoacyl adenylates of the amino acids, the activating enzymes cannot release the activated products into the common stock of metabolism, as these activated compounds would begin to react in a disorderly way with all the substrates in the neighbourhood. Moldave, Castelfranco & Meister (17, 38) showed clearly that this is what happens to aminoacyl adenylates. These authors also carried out the chemical synthesis of the aminoacyl adenylates of many amino acids labelled with $^{14}$C and introduced these compounds into incubation mixtures of the supernatant liquid and organelles from homogenates of cells. It was found that the amino acids of the aminoacyl adenyates combined quickly but non-specifically with proteins and RNA, reacting with various active groups. In such cases systems which had been heated reacted with the aminoacyl adenylates even more rapidly than unheated ones because denatured proteins have a considerably larger number of free active groups than natural proteins.

In later works Moldave, Meister and their colleagues (31, 63) found that the chemically synthesized aminoacyl adenylates (e.g. L-tryptophyl adenylate) in the presence of purified tryptophan-activating enzyme can react specifically with it, producing the enzyme-aminoacyl adenylate complex. The aminoacyl adenylates are more stable when in this complex than in the free state. Thus, chemically synthesized L-tryptophyl adenylate can replace L-tryptophan and ATP in the reaction:

$$\text{enzyme} + \text{L-tryptophan} + \text{ATP} \xrightarrow{\text{Mg}^{++}}$$

$$\text{enzyme} + \text{L-tryptophyl adenylate} + \text{pyrophosphate}.$$

Here, one of the stages of the reaction of protein synthesis is being re-placed by a chemically synthesized product and the normal synthesis proceeds without making use of the synthetic action of the activating enzyme. It has been found that, in the presence of organelles in the supernatant fluid of cellular homogenates, such chemically synthesized complexes are, in fact, used as ordinary intermediate products in protein synthesis.

The non-enzymic nature of the incorporation into proteins of chemi-cally synthesized aminoacyl adenylates and adenylates of peptides has been demonstrated by Zioudrou & Fruton (67).

## Conclusion

In recent years there have appeared in the literature several papers indi-cating that in certain microorganisms, as mentioned above, it is possible to substitute for the activating systems certain other auxiliary systems which also bring about the incorporation of amino acids into the protein being synthesized (12, 13, 14, 59). These systems have, however, only a limited distribution, while the complex of activating enzymes would appear to be ubiquitous and brings about the first stage of the synthesis of proteins. The activation of the amino acids is, however, by no means the whole explanation of their reactions with one another in a definite sequence, the more so as the ability of the activating complex which is present to activate the individual amino acids is by no means always proportional to the quantitative relationships between the amino acids in the total protein being synthesized in the cell. This phenomenon was discovered by studying the activation of amino acids in the silk-secreting glands of the mulberry silkworm which mainly synthesize fibroin (27).

These facts make it evident that there must be some additional step between the system of the activating enzymes and the synthesized proteins. It must be some sort of biochemical system regulating the process of use of the activated amino acids (aminoacyl adenylates) and determining the sequence in which they are used in the reactions of formation of peptide structures.

Until 1957 it was supposed that these functions were fulfilled by stable matrices in the form of high-polymeric molecules of ribonucleic acid localized in the organelles.

In 1957-58, however, another specific biochemical system was dis-covered in cells, fulfilling the special function of transferring the activated amino acid residues from the enzyme-aminoacyl adenylate complexes to the stable matrices. Although the first researches revealing the functions of this new system have only been published comparatively recently, they very quickly attracted a lot of attention from biochemists. Quite a large amount of material in this field has been collected in a very short time and we should look at it in a separate chapter.

# REFERENCES

1. BRAUNSHTEĬN, A. E. (1955). *Biokhimiya*, **20**, 392.
2. BRESLER, S. E. (1950). *Uspekhi sovremennoĭ Biol.* **30**, 90.
3. — (1954). *Uspekhi biol. Khim.* **2**, 66.
4. GABERMANN [HABERMANN], V. (1959). *Uspekhi sovremennoĭ Biol.* **47**, 19.
5. GVOZDEV, V. A. (1960). *Biokhimiya*, **25**, 920.
6. LESTROVAYA, N. N. (1958). *Uspekhi biol. Khim.* **3**, 97.
7. MEDVEDEV, ZH. A. (1955). *Uspekhi sovremennoĭ Biol.* **40**, 159.
8. SISAKYAN, N. M. (1959). In *Proceedings of the first international Symposium on the origin of life on the Earth* (ed. A. I. Oparin *et al.*), p. 400. London: Pergamon.
9. KHESIN, R. B. (1960). *Biokhimiya tsitoplazmy*. Moscow: Izd. Akad. Nauk S.S.S.R.
10. SHAPOT, V. S. (1954). *Uspekhi sovremennoĭ Biol.* **37**, 37.
11. BADDILEY, J. & NEUHAUS, F. C. (1959). *Biochim. biophys. Acta*, **33**, 277.
12. BELJANSKI, M. (1960). *Biochim. biophys. Acta*, **41**, 104.
13. — (1960). *Biochim. biophys. Acta*, **41**, 111.
14. — (1960). *Compt. rend. Acad. Sci.*, Paris, **250**, 624 (1960).
15. BERG, P. (1956). *J. biol. Chem.* **222**, 1025.
16. BERNLOHR, R. W. & WEBSTER, G. C. (1958). *Arch. Biochem. Biophys.* **73**, 276.
17. CASTELFRANCO, P., MEISTER, A. & MOLDAVE, K. (1958). *Microsomal particles and protein synthesis, Papers Symposium Cambridge, Mass.*, p. 123.
18. CHANTRENNE, H. (1960). In *Comparative biochemistry* (ed. M. Florkin & H. S. Mason). Vol. 2, p. 139. New York: Academic Press.
18a. CIFERRI, O., GOROLAMO, M. DI, & GOROLAMO BENDICENTI, A. DI. (1961). *Biochim. biophys. Acta*, **50**, 405.
19. CLARK, J. M., Jr. (1958). *J. biol. Chem.* **233**, 421.
20. COLE, R. D., COOTE, J. & WORK, T. S. (1957). *Nature (Lond.)*, **179**, 199.
21. CONNELL, G. E., LENGYEL, P. & WARNER, R. C. (1959). *Biochim. biophys. Acta*, **31**, 391.
22. CORMIER, M. J., STULBERG, M. P. & NOVELLI, G. D. (1959). *Biochim. biophys. Acta*, **33**, 261.
23. DAVIE, E. W., KONINGSBERGER, V. V. & LIPMANN, F. (1956). *Arch. Biochem. Biophys.* **65**, 21.
24. DAVIS, J. W. & NOVELLI, G. D. (1958). *Arch. Biochem. Biophys.* **75**, 299.
25. HAUROWITZ, F. (1950). *The chemistry and biology of proteins*. New York: Academic Press.
26. HELE, P. & FINCH, L. R. (1960). *Biochem. J.* **75**, 352.
27. HELLER, J., SZAFRAŃSKI, P. & SUŁKOWSKI, E. (1959). *Nature (Lond.)*, **183**, 397.
28. HOAGLAND, M. B. (1955). *Biochim. biophys. Acta*, **16**, 288.
29. HOAGLAND, M. B., KELLER, E. B. & ZAMECNIK, P. C. (1956). *J. biol. Chem.* **218**, 345.
30. KELLER, E. B. & ZAMECNIK, P. C. (1956). *J. biol. Chem.* **221**, 45.
31. KRISHNASWAMY, P. R. & MEISTER, A. (1960). *J. biol. Chem.* **235**, 408.
32. LINDERSTRØM-LANG, K. (1949). *Exptl. Cell Research*, Suppl. **1**, 1.
33. LIPMANN, F. (1949). *Fed. Proc.* **8**, 597.
34. — (1958). *Proc. nat. Acad. Sci. U.S.* **44**, 67.
35. LOFTFIELD, R. B. (1957). *Progress in Biophysics*, **8**, 347.
36. MARCUS, A. (1959). *J. biol. Chem.* **234**, 1238.
37. McCORQUODALE, D. J. & ZILLIG, W. (1959). *Hoppe-Seyl. Z.* **315**, 86.
38. MOLDAVE, K., CASTELFRANCO, P. & MEISTER, A. (1959). *J. biol. Chem.* **234**, 841.

39. MUDD, S. H. & CANTONI, G. L.  (1958).  *J. biol. Chem.* **231**, 481.
40. NEIDHART, F. C. & GROS, F.  (1957).  *Biochim. biophys. Acta*, **25**, 513.
41. NISMAN, B.  (1959).  *Biochim. biophys. Acta*, **32**, 18.
42. NISMAN, B. & FUKUHARA, H.  (1959).  *Compt. rend. Acad. Sci., Paris*, **248**, 1438.
43. NISMAN, B., HIRSCH, M. L. & BERNARD, A. M.  (1958).  *Ann. Inst. Pasteur.* **95**, 615.
44. NOVELLI, G. D.  (1958).  *Proc. nat. Acad. Sci. U.S.* **44**, 86.
45. NOVELLI, G. D. & DeMoss, J. A.  (1957).  *J. cell. comp. Physiol.* **50**, Suppl. 1, 173.
46. OGATA, K., NOHARA, H. & MIYAZAKI, S.  (1959).  *Biochim. biophys. Acta*, **32**, 287.
47. PENNINGTON, R. J.  (1960).  *Biochem. J.* **77**, 205.
48. RENDI, R., DI MILIA, A. & FRONTICELLI, C.  (1958).  *Biochem. J.* **70**, 62.
49. VAN ROOD, J., BOOT, J., BRUNING, J. W. & KASSENAAR, A.  (1960).  *Biochim. biophys. Acta*, **39**, 232.
49a. SCHUURS, A. H. W. M., KLOET, S. R. DE & KONINGSBERGER, V. V.  (1960). *Biochem. biophys. Res. Communs.*, **3**, 300.
50. SCHWEET, R. S. & ALLEN, E. H.  (1958).  *J. biol. Chem.* **233**, 1104.
51. SCHWEET, R. S., BOVARD, F. C., ALLEN, E. & GLASSMAN, E.  (1958).  *Proc. nat. Acad. Sci. U.S.* **44**, 173.
52. SCHWEET, R. S., HOLLEY, R. W. & ALLEN, E. H.  (1957).  *Arch. Biochem. Biophys.* **71**, 311.
53. STROMINGER, J. L.  (1960).  *Physiol. Revs.* **40**, 55.
54. SZAFRAŃSKI, P., BAGDASARIAN, M. & TOMASZEWSKI, L.  (1960).  *Acta biochim. Polon.* **7**, 3.
55. SZAFRAŃSKI, P. & SUŁKOWSKI, E.  (1959).  *Acta biochim. Polon.* **6**, 133.
56. TARVER, H.  (1954).  In *The Proteins* (ed. H. Neurath & K. Bailey), Vol. **2B**, 1199.  New York: Academic Press.
57. TUBOI, S. & HUZINO, A.  (1960).  *Arch. Biochem. Biophys.* **86**, 309.
58. WACHSMANN, J. T., FUKUHARA, H. & NISMAN, B.  (1960).  *Biochim. biophys. Acta*, **42**, 388.
59. WAGLE, S. R., MEHTA, R. & JOHNSON, B. C.  (1960).  *Biochim. biophys. Acta*, **39**, 500.
60. WEBSTER, G. C.  (1956).  *Plant Physiol.* **31**, 482.
61. — (1959).  *Arch. Biochem. Biophys.* **82**, 125.
61a. — (1961).  *Biochim. biophys. Acta*, **49**, 141.
62. WOLFE, O. D. & HAHN, F. E.  (1960).  *Biochim. biophys. Acta*, **41**, 545.
63. WONG, K. K. & MOLDAVE, K.  (1960).  *J. biol. Chem.* **235**, 694.
64. ZAMECNIK, P. C. & KELLER, E. B.  (1954).  *J. biol. Chem.* **209**, 337.
65. ZAMECNIK, P. C., KELLER, E. B., LITTLEFIELD, J. W., HOAGLAND, M. B. & LOFTFIELD, R. B.  (1956).  *J. cell comp. Physiol.* **47**, Suppl. 1, 81.
66. ZAMECNIK, P. C., KELLER, E. B., HOAGLAND, M. B., LITTLEFIELD, J. W. & LOFTFIELD, R. B.  (1956).  *Ciba Foundation Symposium, Ionizing Radiations and Cell Metabolism*, p. 161.
67. ZIOUDROU, C. & FRUTON, J. S.  (1959).  *J. biol. Chem.* **234**, 583.

# CHAPTER IV

# SELECTIVE TRANSFER OF ACTIVATED AMINO ACIDS TO THE MATRICES OF PROTEIN SYNTHESIS BY MEANS OF MOLECULES OF "SOLUBLE" RIBONUCLEIC ACID (S-RNA) OF LOW-MOLECULAR WEIGHT

## Introduction

The important part played by ribonucleic acid in the biosynthesis of proteins is already generally known but, until comparatively recently, the function of RNA in this process was regarded as solely that of a polymeric matrix on the surface of which polypeptide chains were assembled selectively. Although many authors had found in cells fractions of RNA of low molecular weight, these forms of RNA have usually been considered as products of the degradation of the fractions of high molecular weight, formed as a result of the analytical procedure.

At that time it was widely held by biochemists that all the essential ingredients of cytoplasm were well known and that the cell could scarcely provide research workers with any new "surprise". However, this turned out not to be so. The use of gentler methods for isolating RNA showed that the low-molecular forms of RNA are normal constituents of the cell, while study of the process of activation led to the discovery, in 1957, of a close connection between the activating enzymes of the cell sap and the soluble RNA of low molecular weight which is to be found in it and which is necessary for the incorporation of labelled (activated) amino acids during protein synthesis. Further experiments showed that the rather small molecules of this RNA play the specific part of go-between between the activating enzymes of amino acids and the cellular organelles in which the polypeptide chains of proteins are formed. The rather small molecules of this RNA accept the amino acids which have been activated by the enzymes and transfer them to the cellular organelles where they are used directly for the synthesis of the polypeptide chains of proteins.

The discovery of soluble RNA and of a number of specific peculiarities of its structure and metabolism directed the theoretical and experimental studies of protein synthesis along new paths and had a great effect on the working out of the general problem of the synthesis of proteins. Although we have surveyed the material on the functions of this RNA in a special review (3), we still think we should analyse the question again as this is necessary for a proper understanding of the material in later chapters.

It must also be pointed out that excellent surveys of the literature on the functions of soluble RNA in the synthesis of proteins, published up to the end of 1959, have also been made in 1960 in the reviews of Hoagland (60) and Cohen & Gros (30) as well as in the monographs of Khesin (8), Brachet (20) and Chantrenne (29), considerable parts of which are devoted to the problem of the synthesis of proteins. In this connection we must, of course, give a short account of the history of the research, and concentrate the greater part of our attention on an evaluation of the latest work which has been published in 1960-2 and considerably extends our ideas about the mechanisms of the intermediate reactions of protein synthesis.

## 1. The discovery of soluble RNA and its connection with the enzymic systems for the activation of amino acids

In the last chapter we have already seen that there are at least 20 specific activating enzymes in the cytoplasm, each of which will react with its own particular amino acid. On the surface of each such enzyme one particular amino acid combines with a molecule of adenosine triphosphate with the formation of aminoacyl adenylate (AMP $\sim$ a) which remains attached to the surface of the enzyme. This reaction is accompanied by the liberation of pyrophosphate. The complex of activating enzymes is usually found in the supernatant fraction of a homogenate of cells, that is to say, it is localized in the cell sap.

In the period when the study of the enzymic complex activating amino acids was the object of the work of many biochemical laboratories, it was found to contain a peculiar fraction consisting of soluble RNA of low molecular weight, the hydrolysis of which with ribonuclease put a stop to the incorporation of the activated amino acids into proteins which were being synthesized (12, 64, 70, 91, 92, 119). The function of this soluble low-molecular-weight RNA in bringing about the incorporation of amino acids in protein synthesis became the central problem of a new series of experiments carried out in the laboratory of Zamecnik and Hoagland which revealed the existence of a supplementary specific stage following the process of activation and comprising the intermediate combination of the activated amino acids with the molecules of the peculiar so-called "soluble" RNA and their subsequent transfer to the microsomes (55, 63, 65, 121, 122).

These experiments showed that the labelled amino acids which had been activated by the enzymic system of the plasma juice are taken up almost instantaneously by the soluble RNA, which binds them quite firmly and only later transfers them to the microsomes, where they are used for the synthesis of protein in the ribonucleoprotein complexes of the microsomes. It was found possible to "bind" [$^{14}$C]leucine to the RNA of the supernatant liquid (by incubation without microsomes) and then, after ridding the system of activating (pH 5) enzyme and ATP, to bring it into contact with microsomes. This led to a rapid transfer of the leucine and RNA to the protein of the microsomes. Guanosine triphosphate (GTP) and a nucleoside-triphosphate-regenerating system were necessary co-factors for this process. Fig. 15 is a graphic representation of one of these experiments.

In later experiments it was established (9, 13, 59, 67) that the transfer of amino acid residues from the aminoacyl adenylate complex to the molecule of soluble RNA is brought about on the surface of the activating enzymes while the adenyl residue of this complex is split off into the cell sap.

It should be noted that the widely used term soluble RNA or S-RNA has recently begun to be replaced by the term transfer RNA, which is functionally more accurate. There are other terms besides these which are sometimes used as synonyms for them, such as acceptor RNA, adaptor RNA and carrier RNA.

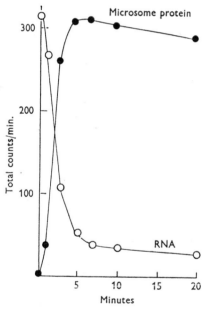

Fig. 15. Time curve of transfer of [$^{14}$C]leucine from pre-labelled pH 5 enzyme fraction to microsome protein. The pH 5 enzyme fraction containing [$^{14}$C]leucine was incubated with rat liver microsomes at 37°C. in the presence of guanosine triphosphate, phosphoenolpyruvate and pyruvic kinase. Samples were taken at various time intervals and centrifuged at 105,000 $g$ for 60 min. The radioactivity of the amino acid attached to the pH 5 RNA in the supernatant and incorporated into the microsomal protein was determined.

## 2. Methods of isolation of soluble RNA

The molecular weight of soluble RNA isolated from various sources, according to the results of many measurements, ranges between 20,000 and 40,000 while the molecular weight of RNA isolated by gentle methods from cellular structures varies between 600,000 and 2,000,000.

Quantitatively the soluble RNA represents 10-20% of the total RNA of the cell.

Soluble RNA can easily be obtained from the supernatant fraction of a cellular homogenate after it has been centrifuged at 100,000 $g$ in a refrigerated ultracentrifuge. If this supernatant fraction is deproteinized with phenol by the method described by Kirby (76) it is possible to isolate a soluble RNA of low molecular weight which is fully active as a donor and acceptor of amino acids (63).

The supernatant fraction of a cellular homogenate contains more RNA than the amount of RNA which is responsible for the intermediate acceptance of amino acids. This is because it contains products of degradation of the RNA of the microsomes and some other forms of RNA of unknown nature. In this connection the simplest method of purifying S-RNA is to centrifuge it at 100,000 $g$ for the longest possible time. The experiments of Hoagland and his colleagues (63) have shown that centrifugation at 100,000 $g$ for three hours instead of one diminished the amount of RNA in the supernatant fraction by almost a half without any loss of the power to accept amino acids. According to the findings of Medvedev, Zabolotskiĭ et al. (5) soluble RNA can be isolated from plant tissues without homogenizing them, by phenol treatment of cell sap which has been expressed from frozen leaves under pressure and represents an ultrafiltrate of the cytoplasm. In this case the cellulose envelopes of the plant cells and the numerous intracellular membranes themselves serve as the ultrafilter.

As the ergastoplasm is not homogenized by freezing and is only partly disrupted, the microsomes are largely bound to the system of internal membranes and are not expressed from the frozen leaves by pressure. Only the vacuolar juice can be expressed from freshly gathered leaves by pressure, but, when the sheets of pressed leaves have been frozen the cell sap can be removed from them very quickly and easily with a press

TABLE I

Content of RNA in $\mu$g./g. wet weight of leaves
(average of findings from 3 samples)

| Lower, mature leaves | | Upper, young leaves | |
|---|---|---|---|
| Cell sap | Organelles | Cell sap | Organelles |
| 14·8 | 326 | 39·5 | 372 |

(85). We have found that this sap also contains a soluble RNA with properties similar to those of the adaptor RNA isolated from the supernatant liquid of centrifugates. This similarity was established by studying the peculiarities of the metabolism of the RNA of the cell sap.

These experiments were done with the leaves of haricot beans. The content of RNA in various fractions was determined (Table 1).

Monier, Stephenson & Zamecnik (88) have recently described the possibility of isolating soluble RNA from yeast by the direct treatment of the yeast cells with phenol. When the cells are treated in this way their walls become permeable to the small molecules of soluble RNA and impermeable to the organelles of the cytoplasm.

Another interesting method for simplifying the isolation of soluble RNA has been published recently by Smith (111). He found that, if a tissue homogenate is treated with a molar solution of NaCl in the cold, the polymeric RNA is precipitated while the low-molecular fraction, having the properties of S-RNA, remains in solution. A similar method of separating the high- and low-molecular forms of RNA was used in the work of Mil'man (6).

Another method for speeding up the isolation of S-RNA without using prolonged high-speed centrifugation has been described recently by Rosenbaum & Brown (101). These authors first extracted the tissues directly with phenol and then precipitated the polymeric RNA with a molar solution of NaCl. The S-RNA which was dissolved in the salt solution was then precipitated with 70% alcohol.

By no means all the molecules of the S-RNA isolated from the supernatant fraction from a tissue homogenate after the removal of all the structures and high-polymeric forms of RNA possess the same molecular weight and acceptor functions. This was very clearly shown in the recent

TABLE 2

Fractionation of S-RNA containing [$^{14}$C]amino acids (117)

| Fraction | S-RNA content mg. | Content of [$^{14}$C]amino acids in the S-RNA | |
|---|---|---|---|
| | | Counts/min./mg. | $\mu$g./mg. |
| Unfractionated S-RNA | 80 | 470 | 0·21 |
| RNA precipitated with 33% alcohol | 65 | 70 | — |
| RNA precipitated with 50% alcohol | 11 | 1430 | 0·65 |
| RNA precipitated with 67% alcohol | 2 | 5950 | 2·70 |
| RNA precipitated with 80% alcohol | 2 | 2300 | 1·05 |

work of Webster (117). Having isolated S-RNA from a homogenate of pea shoots and incubated it with a mixture of amino acids labelled with [14C], Webster then carried out a fractional precipitation of the RNA with alcohol on the grounds that the concentration of alcohol required to precipitate the RNA is inversely proportional to its molecular weight.

The results of this experiment are given in Table 2 and show that the fraction precipitated at a concentration of 67% and comprising 2·5% of the total amount of S-RNA has the greatest power of accepting amino acids, 12 times that of the S-RNA as a whole.

## 3. The acceptance of individual amino acids by individual fractions of soluble RNA

The occurrence within cells of differential activation of amino acids by activating enzymes and the isolation of tyrosine-, valine-, tryptophan- etc. activating enzymes naturally gave rise to the idea that the acceptor S-RNA might also be heterogeneous in that each amino acid was accepted and transferred to the matrix by a particular specific fraction of the S-RNA. This idea has been indirectly confirmed by the absence of competition between the various amino acids when they react with unfractionated S-RNA (12, 63) and this has stimulated a series of successful attempts to fractionate S-RNA into individual amino acid-polynucleotide complexes.

Even in their first experiments Hoagland, Zamecnik and colleagues (63) managed to fractionate S-RNA into three components. One of these, comprising 36% of the total S-RNA, carried 68% of the total activity of labelled leucine associated with the S-RNA.

Schweet and his colleagues (79, 105, 106, 112) have made a number of further successful attempts to fractionate S-RNA derived from liver. These authors succeeded in showing that [14C]threonine, [14C]leucine and [14C]tyrosine were each associated with different fractions of this RNA. Different attempts to fractionate S-RNA may be based on differences between either the nucleotide or the amino acid components. For example, Brown et al. (21) carried out separation and partial purification of complexes of histidine-S-RNA and tyrosine-S-RNA by means of the ability of the amino acid components of these complexes to form diazonium compounds with polyaminoazostyrene resin at neutral pH. They used counter-current distribution very successfully for the separation of amino acid-specific forms of S-RNA. Holley and his colleages (68, 66) gave a detailed account of a simple method of separating the tyrosine-, alanine- and valine-accepting fractions of S-RNA from yeast. The polynucleotide chains of these forms of S-RNA differed from one another to such an extent that six transfers of counter-current distribution were enough for their substantial separation.

In recent publications (38, 39) other workers in the same laboratory have described the separation of 11 fractions from the S-RNA of yeast (Ala, Val, Pro, Thr, His, Leu, Try, Ser, Ileu, Lys and Tyr-accepting RNA's). The nucleotide compositions of these forms of S-RNA were different as well as their functions. Having separated by counter-current distribution fractions of S-RNA accepting alanine and tyrosine respectively, Marini (84) also established differences in their nucleotide compositions (Table 3).

TABLE 3

Nucleotide composition of S-RNA (84)
(moles %)

|                | C     | A     | U     | G     |
|----------------|-------|-------|-------|-------|
| S-RNA-alanine  | 28·3  | 16·2  | 19·3  | 36·2  |
| S-RNA-tyrosine | 28·1  | 22·5  | 22·0  | 30·5  |

In another series of experiments Holley and his colleagues (45, 46) used partition chromatography for fractionation of S-RNA into 5 fractions (Tyr, Try, His, Val and Ala-containing S-RNA's). Cantoni (27, 28) used the different solubilities of the complexes of different forms of S-RNA with spermine as a basis for fractionating it and separated a valine-containing fraction from a proline-containing one. In recent years many other methods for and attempts at fractionation of S-RNA have been described (16, 19, 50, 51, 52, 54a, 69, 102, 103, 104, 123) and, altogether, specific polynucleotide acceptors for fourteen different amino acids have been identified. As the total S-RNA of cells usually contains an almost complete assortment of amino acids in a bound state (9, 116, 120) it may be supposed that each of these amino acids has its own type of molecule of S-RNA and that the activating enzyme in some way "recognizes" the molecule of S-RNA with which it must combine any particular amino acid.

Finally, we must mention an unusually interesting fact described by Bates (10) in the Proceedings of the Fifth International Congress of Biochemistry (cf. 10a). He found that a special form of S-RNA with a molecular weight 35,000 was necessary for the synthesis of the specific tripeptide glutathione in the cells of *Neurospora*. This RNA accepted the activated dipeptide γ-glutamylcysteine and the author was able to separate it in a pure form, the rates between the dipeptide and the polynucleotide being 1:1. Lane & Lipmann (78a) were unable to repeat this work.

## 4. Peculiarities of the chemical and physical-chemical properties of soluble RNA

We have already noticed that the molecular weight of soluble RNA derived from various sources ranges between 20,000 and 40,000. This corresponds with about 70-130 nucleotide residues to the polynucleotide chain. According to the findings of many authors, the nucleotide composition of total S-RNA is not substantially different from that of the RNA of the microsomes as concerns the ordinary nucleotides but soluble RNA's from various sources differ, as a rule, from insoluble RNA by having a larger amount of unusual nucleotides which are hardly ever encountered in the high-polymeric fractions. Comparatively large proportions of methylcytosine, pseudouridylic acid, 6-methylaminopurine etc. have been found in S-RNA (33, 40, 43, 96, 94).

Dunn, Smith & Spahr (44) have given the following composition for S-RNA isolated from *Escherichia coli.*

TABLE 4

Base composition of S-RNA (44)

|  | Moles/100 moles nucleotide |
|---|---|
| Adenine | 20·3 |
| Guanine | 32·1 |
| Cytosine | 28·9 |
| Uracil | 15·0 |
| Pseudouracil | 2·1 |
| Thymine ribonucleotide | 1·1 |
| 2-Methyladenylic acid | 0·3 |
| 6-Methylaminopurine ribonucleotide | 0·1 |
| 1-Methylguanylic acid | 0·1 |

Dunn has recently (41, 42) found a new base, 1-methyladenine, in S-RNA. There is reason to suppose that such "impurities" are not accidental but are connected with the functions of S-RNA. Thus, for example, Otaka & Osawa have obtained five fractions from yeast S-RNA and the one which contained the greatest proportion of 5-ribosyluridine (pseudouridine) (5·57%) also had the greatest power to accept [14C]leucine (93, 95, 97). This result caused them to suggest that the presence of pseudouridine in S-RNA is, in some way, connected with its acceptor function, at least so far as concerns [14C]leucine. Osawa (93, 94) has also found pseudouridine in S-RNA from other sources.

In this work Japanese biochemists succeeded in isolating from yeast a soluble RNA of low molecular weight and of a high degree of homogeneity on ultracentrifugation. The molecular weight of this RNA was 27,000.

A detailed analysis of the nucleotide composition of low-molecular S-RNA from the liver of the rabbit has recently been made by Singer & Cantoni (28, 109). The molar ratios of the nucleotides were found to be as follows: AMP 1·0; UMP 1·0; GMP 1·8; CMP 1·7 and 5-ribosyluracil monophosphate 0·2. In these figures attention should be paid to the equality of the molar relationships in accordance with Chargaff's rules. In a recent paper on the same subject Herbert & Canellakis (57) established the same relationship for three fractions of S-RNA isolated from the livers of rats and having different molecular weights (15,000, 12,700 and 22,100). In this connection both groups of authors suggest that soluble RNA may have a secondary structure involving the formation of pairs of complementary bases. This suggestion received indirect confirmation from another piece of work by Singer, Cantoni and their colleagues (110) in which they found that, while phosphorolysis of polymeric RNA by polynucleotide phosphorylase proceeds to completion, the low-molecular RNA from the liver of the rabbit is only broken down to the extent of about 20-30%, 70-80% of its polynucleotide chain being resistant to the action of this enzyme. Such resistance is also characteristic of multistranded synthetic polynucleotides.

Direct proof of the "folding" of the molecules of low-molecular RNA derived from the cells of E. coli has recently been obtained by Cox & Littauer (32) who determined the viscosity of solutions of RNA at various temperatures and in various concentrations of NaCl. Using methods involving X-ray diffraction and hypochromasia on heating, Brown & Zubay (22) recently found out that this secondary structure of S-RNA is a double helical structure but without any very regular arrangement of pairs of bases.

These authors think that the molecule of soluble RNA consists of a chain of 70-80 nucleotides reacting with one another to form a double helix. They find that heating and rapid cooling of concentrated solutions of S-RNA produces a marked decrease in this secondary structure without at the same time diminishing the ability of the S-RNA to accept amino acids. In this connection Brown & Zubay consider it improbable that combination of activated amino acids with one end of the molecule of S-RNA requires cooperation by the other end. According to them this indicates that the particular sequence of nucleotides which determines the specificity of the combination of the S-RNA with amino acid is situated very close to an end group. It is, however, not impossible that the secondary structure of the "transport RNA" may be essential for the specificity of the final stages of its interaction with the transferring enzymes and with the matrices.

A large amount of supplementary material concerning the connection between the secondary structure of S-RNA and its ability to bind amino acids has recently been presented by Monier (87). It is also interesting to notice that there are many facts which indicate that, in the cell, soluble RNA may exist in the form of labile nucleoproteins (25).

## 5. The special role of the terminal group of soluble RNA in the activation and transfer of amino acids

The exact nature of the bond between the amino acid and the polynucleotide chain of the molecule of S-RNA has now been established with fair certainty. One molecule of S-RNA can only combine with one molecule of amino acid, which joins itself on to the terminal nucleotide at one end of the polynucleotide chain. The amino acid forms a bond between its carboxyl group and the 2' or 3' hydroxyl of the ribose of the terminal nucleotide (9, 56, 98, 120).

It is also of theoretical importance and has now been established with some certainty that the terminal trinucleotide fragment, to the last segment of which the amino acid becomes attached, is identical in all molecules of S-RNA regardless of the nature of the amino acid which combines with it and also, it would appear, of the source of the preparation. The

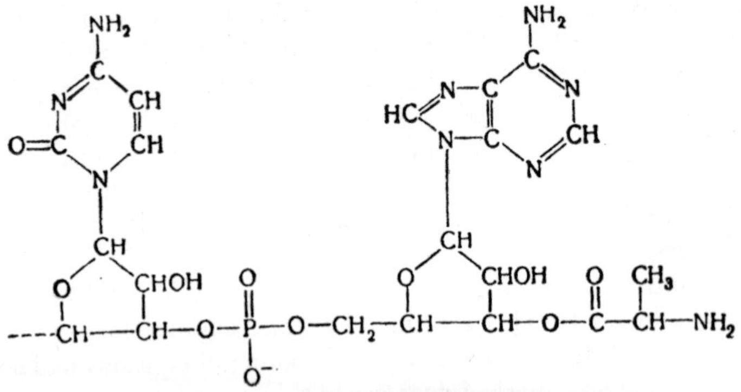

FIG. 16. Terminal group of alanyl-S-RNA.

nature of the terminal fragment of the molecule was finally established by Hecht, Stephenson & Zamecnik (56). They found that the polynucleotide chains of all molecules of S-RNA end with the group RNA—3'—5' —cytidyl—3'—5'—cytidyl—3'—5'—adenyl—3'—amino acid (Fig. 16). Only molecules of S-RNA which have this terminal group can accept amino acids. It must be mentioned that these interesting data were obtained by extremely refined investigations using nucleosides and triphosphates labelled with $^{14}$C and $^{32}$P at different particular points in their molecules.

Owing to the identity of the trinucleotide group the "charged" molecule of S-RNA consists of three parts: a polymeric polynucleotide part, the terminal trinucleotide (ACC-) and the amino acid residue attached to the hydroxyl of the ribose and retaining the store of energy which it received from ATP. The macroergic nature of the combination between the amino acid moiety and the molecule of S-RNA was clearly demonstrated by the work of Schweet and his colleagues (49, 80) who discovered the possibility of regenerating ATP from AMP and pyrophosphate by splitting the compound of soluble RNA with an amino acid.

The terminal trinucleotide is, as Hecht, Stephenson & Zamecnik showed in the paper to which we have already referred, a supplementary addition to the molecule of S-RNA. It combines with the ready-formed polynucleotide by means of a special enzyme. Moldave (86) has recently separated the enzyme which inserts the adenyl component into the polynucleotide from that which is responsible for the addition of the terminal adenyl group. Only the triphosphates of the nucleotides in question will act as substrate for carrying out the addition of this appendix. Pyrophosphate is eliminated when the triphosphates are added on to the main chain.

In recent researches Furth *et al.* (48a) have isolated from the cells of *E. coli* and purified an enzyme which carries out the reaction of terminal addition of adenyl and cytidyl nucleotides to S-RNA.

In reviewing his own results and those of others, Hoagland (60) suggests that the formation of the terminal trinucleotide group is carried out in three stages. At first two molecules of CMP derived from two molecules of CTP add themselves on in tandem to the 3'-hydroxyl group of the ribose of the terminal nucleotide of the chain. After this AMP attaches itself in the same way to the CMP at the end of the chain. This combination completes the reaction and gives the molecule its ability to act as an acceptor in relation to aminoacyl adenylate combined with the activating enzyme. When the aminoacyl adenylate combines with the terminal trinucleotide group the adenylic group of the adenylate is set free. Furthermore, it has been shown that synthetic aminoacyl adenylates can only serve as sources of amino acids for transfer to S-RNA after they have been combined with an activating enzyme.

In this connection one may assume that any particular enzyme has a specific area on its surface which can bind a particular amino acid and also a place for the adenyl component. The molecule of S-RNA is brought into approximation with the aminoacyl adenylate here, on the surface of the enzyme, in such a way that the end of its main polynucleotide chain lies in its own specific predestined place, just the length of the trinucleotide link away from the amino acid residue. The sequence of these reactions has been sketched out by M. Hoagland in the form of the following hypothetical diagram (61) (Fig. 17).

A number of other investigations by various authors have also established the fact that nucleoside triphosphates are necessary for the synthesis of S-RNA, although the difference in this respect between the terminal group and the rest of the S-RNA molecule has not always been brought out in such cases (53, 83).

Hoagland considers (60) that we can take it as firmly established that the reaction of the terminal addition of nucleotides to the molecule of S-RNA is different from the reactions catalysed by polynucleotide phosphorylase.

The peculiarity of these reactions is not only that nucleotide triphosphates act as the substrates for them, but also that only ATP and CTP and, to a slight extent, UTP take part in them, so that the nature of the terminal group does not change in the presence of other nucleotide precursors. A number of features of the reaction indicate that both AMP and CMP are attached to the end of the chain of S-RNA by means of the same enzyme, which may easily be separated from the enzymes activating amino acids. This peculiarity of the reaction suggests that the AMP group of the terminal trinucleotide of S-RNA is derived directly from ATP and not from the AMP which forms part of the aminoacyl adenylates.

The formation of the terminal trinucleotide group is a specific feature of the molecule of S-RNA. The polymeric RNA's of the nucleus and ribosomes cannot replace acceptor RNA in the system responsible for

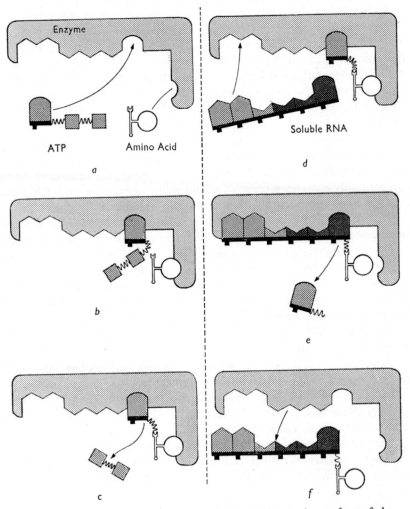

FIG. 17. Stages in the activation of amino acids on the surface of the activating enzyme and the acceptance of the activated amino acid by the terminal group of S-RNA.

this process. However, specimens of S-RNA derived from various widely differing sources (yeasts, bacteria, mammals) can easily be provided with their terminal nucleotides in the presence of the appropriate enzymes of ascites tumours of rats (60).

Intensive studies have been going on in several other laboratories (24, 31, 90b, 98a) on the enzymic system responsible for the arrangement of the terminal trinucleotide group of S-RNA. These workers have carried out a purification of this enzymic system isolated from the cells of the liver and have established the existence of a similar enzyme in bacterial cells. It is interesting to note that the terminal position at the other end of S-RNA in its various forms is occupied by a guanyl residue (109, 124).

## 6. Some peculiarities of the synthesis and metabolism of soluble RNA

It is important to note that the part played by S-RNA in the synthesis of proteins is active, not merely that of a passive "carrier". In performing this function the molecules of this RNA undergo extremely intensive rebuilding.

Even before the discovery of this RNA as a biochemically specific substance it had been established by many authors that the RNA of the supernatant fraction of a cellular homogenate (the plasma juice) has a very high rate of metabolism of phosphoric groups, while the RNA of the microsomes, where protein synthesis is mainly carried on, shows only very slight renewal (7, 11, 74, 75, 107). The nucleus of the cell was found to be another centre of active synthesis of RNA.

In connection with the discovery of the special acceptor function of S-RNA Shigeura & Chargaff (107) performed a very interesting piece of work on the rate of renewal of the phosphorus of the individual nucleotides to be found in various fractions of RNA. They injected radioactive sodium phosphate into rats and isolated for analysis RNA from various fractions of the liver, removed after short intervals (5, 20, 120 min.).

In all cases the $^{32}$P taken up in the first 120 min. after the injection was mainly found as part of adenyl nucleotides, in which phosphorus metabolism goes on 6-10 times as fast as in the other nucleotides of RNA.

Table 5 shows a selection from the corresponding data concerning S-RNA but even in RNA from different fractions a similar distribution of specific activity was found.

The numerical material obtained from these experiments demonstrates clearly the special part played by the adenylic component in carrying out the functions of this RNA. The cytidylic component takes second place in this respect and behind it come the uridylic and guanylic components. Sachs (103) reports similar findings.

From this it is clear that the nucleotides entering into the terminal group have the greatest rate of renewal and that among these the adenyl

TABLE 5

Specific activity of $(2' + 3')$-nucleotides of acceptor RNA
(in counts/min. for 10 $\mu$moles of nucleotide) (115)

| Nucleotides | Time after injection in minutes | | |
|---|---|---|---|
| | 5 | 20 | 120 |
| Adenyl | 580 | 1200 | 4100 |
| Guanyl | 50 | 120 | 1750 |
| Cytidyl | 90 | 380 | 3950 |
| Uridyl | 76 | 250 | 2000 |

nucleotide, occurring at the very end of the polynucleotide chain, is out-standing. Such inequality of metabolism can only be explained on the assumption that the terminal group of the RNA can be renewed independently of the breakdown and resynthesis of the whole molecule.

In evaluating this fact, however, one must not discount recently published evidence (99) which shows that adenyl nucleotides are also turned over very rapidly in fractions of acid-soluble nucleotides which may, in practice, accelerate the increase of specific activity of the corresponding component of RNA, even when all the components are being incorporated at the same rate.

In another work Shigeura & Chargaff (108) have carried out a further fractionation on S-RNA from the supernatant fraction which, in the experiments described above, had the highest rate of phosphorus metabolism. They divided it into four fractions (A, B, C, D) each of which had its own particular rate of phosphorus metabolism. The greatest specific activity was nearly 100 times that of the RNA of the microsomes and was found 3 minutes after the injection of $^{32}$P in fraction D which consisted of pure RNA (fraction C was labile nucleoprotein).

The independent metabolism of the terminal group of S-RNA has been demonstrated recently by Canellakis (23, 26, 58).

In our own experiments (5) we have also studied some features of the synthesis and metabolism of the soluble RNA of the plasma juice of plants. We discovered the presence of RNA in plasma juice as early as 1957 (1) but only began to study the properties of this fraction of RNA in 1959.

In a series of experiments [$^{32}$P]phosphate was introduced into the region of the root system of haricot beans and oats and it was also infiltrated into the leaves of haricots. Measurements were then made of the rate of inclusion of $^{32}$P into the RNA of the structures and that of the plasma juice over short periods of time. In all the experiments it was found that the rate of inclusion of $^{32}$P was considerably greater (sometimes tens of times greater) into soluble RNA.

## 7. Possible nature of the selectivity of the reactions of soluble RNA

The metabolism of soluble, low-molecular RNA involves many stages in which the phenomenon of molecular specificity may or may not occur. It is very important to classify them so that inconsistencies may not arise in our later assessment of the problem of the specificity of soluble RNA. At present we can point to the following reactions which are involved and in which it is theoretically possible that some sort of specificity may exist: (a) interaction between molecules of soluble RNA and individual enzymes activating carboxyl groups, (b) interaction between molecules of RNA and amino acid residues on the surfaces of activating enzymes (the power to discriminate between one amino acid and another analogous one) and (c) the interaction between molecules of soluble RNA and matrices in organelles (the transfer of amino acids to particular regions of the matrices).

We shall consider the problem of the interaction of molecules of soluble RNA with matrices concerned with synthesis in a later section; here we are only concerned with the first two cases in which specificity may manifest itself.

The mechanism whereby activating enzymes differentiate accurately between molecules of RNA is still not clear. Hoagland (60) has suggested that the polynucleotide components of soluble RNA bound to the terminal trinucleotide group have different sequences of nucleotides. According to Hoagland's hypothesis the activating enzymes "recognize" the sequence of nucleotides in the molecules of soluble RNA and combine the 20 amino acids with the corresponding 20 types of sequence. However, when there is a greater variety of sequences this hypothetical means of arrangement immediately ceases to apply. In this connection we suggested that the activating enzymes may only "recognize" the dimensions of the accepting molecule of soluble RNA and that there are 20 types of dimensions of soluble RNA corresponding with the 20 amino acids (2, 3). This suggestion, however, was not confirmed by the work of Klee & Cantoni (77) who showed that molecules of RNA combined with different amino acids can have the same molecular weight and not be separable with the ultracentrifuge.

Crick (cf. 60) has put forward the suggestion that the polynucleotide chain of soluble RNA may be twisted up into some more or less complicated shape owing to the formation of hydrogen bonds between pairs of bases. There are further facts, as has been noted above, which are suggestive of the existence of a secondary structure in soluble RNA. If these suggestions are confirmed one may suppose that the activating enzymes "recognize" the shape rather than the length of the molecules of soluble RNA.

There is not, as yet, enough evidence as to the presence or absence of species specificity in molecules of soluble RNA for any definite conclusions to be drawn, but there are some facts which support the idea that it may exist.  For example, Berg & Ofengand (12) found a high degree of specificity in soluble RNA isolated from *Escherichia coli*.  They tried to substitute preparations of RNA isolated by the same method from *Azotobacter* and from liver for that from *E. coli* but these preparations had practically no activity in the systems of *E. coli* (only 5% of the activity of its own RNA).  Synthetic polynucleotides were also inactive.

In another piece of work from the same laboratory (13) it was found that of a fraction of soluble RNA from *E. coli*, capable of accepting methionine, only 40% could accept methionine which had been activated by the enzymic system of yeast.  This fact also shows that there may be heterogeneity among the different sub-fractions of S-RNA separated on the basis of the amino acids with which they are combined.

In an interesting recent piece of work by Benzer & Weisblum (11a) the species specificity of the interaction between "soluble" RNA and the enzymes activating the amino acids has been studied in detail.  These authors were trying to compare the enzymes activating particular amino acids as isolated from different sources with the different forms of S-RNA (in respect of the amino acid accepted and species-specific characteristics). Inter-species differences were observed but they were very capricious. For example, the arginine-accepting S-RNA of *E. coli* is different from that of yeast.  The arginine-activating enzymes are also different.  Arginine can be combined with the S-RNA of *E. coli* only by means of the activating enzyme of *E. coli*, not by means of that of yeast.  A similar picture was also found in respect of the acceptance of tyrosine and leucine. Nevertheless, the lysine-accepting S-RNA of *E. coli* fulfilled its function just as well in the presence of the appropriate enzyme from yeast as with that from *E. coli*.  Similar variations were found in experiments with other amino acids.  When three species were used as sources for the materials (e.g. yeast, *E. coli* and rabbit liver) the picture became even more complicated but the results all tended to show that the possibilities for species-specific variation of S-RNA are very restricted.

We must also take note of the short communication of Bernlohr & Webster (14) who studied the activation of amino acids by the cells of *Azotobacter*.  In this case the intermediate acceptors of the activated amino acids were polynucleotides which were not precipitated by ethyl alcohol, that is to say, they have an even lower molecular weight than has been found in work with other materials.

The differences between the molecular weight of preparations of S-RNA from various sources have also been noted in the works of other authors.

At the same time we must mention the contradictory evidence which indicates that if the activating enzymes can "recognize" which molecule of RNA should combine with any particular amino acid, then it would seem that they could not "recognize" with certainty the species specificity of the molecules. We have already referred to the experiments of Hecht, Stephenson & Zamecnik (56) in which they showed that the activating enzymes of the cells of ascites tumours transferred labelled amino acids to soluble RNA from different sources (liver, yeast) at almost the same rate. This does not, however, exclude the possibility that the species specificity conferred by the sequence of bases and the properties of the molecule "recognized" by the activating enzyme may not be the same qualitatively. While it seems possible that there is a secondary form of selectivity in the reactions of S-RNA, we must emphasize that even the enzymes activating amino acids, which have substrate specificity, bring about a certain selection of those amino acids which take part in the synthesis of proteins. This selection is, however, totally inadequate. The controlling mechanism of the activating systems cannot always distinguish between amino acids and their derivatives and, also, they do not by any means put amino acids through the process of activation in the proportions in which they are required for the synthesis of protein in the cell in question. It is important to notice, in this connection, that, according to the preliminary findings of several authors, a biochemical complex of soluble RNA carries out a supplementary sorting of the intermediate products.

Loftfield and colleagues have shown recently (81, 82) that soluble RNA may itself be the discriminator which prevents the inclusion of certain analogues of amino acids in the synthesis of proteins. They have called attention to the fact that *allo*isoleucine is utilized 2000 times less well in the synthesis of proteins than is normal isoleucine, while the carboxyl activation of these substances proceeds at practically the same rate. In the second stage of the reaction, however, when the amino acid combines with the acceptor RNA, *allo*isoleucine only competes very weakly with the normal amino acid, for this reaction seems to be more selective. However, this stage of the reaction does not account for 100% of the discrimination against *allo*isoleucine. At later stages of the reactions there are further (but still not absolute) barriers to the inclusion of such unnatural substances in protein synthesis.

The interesting results of Fraser and his colleagues (47, 48) also indicate the great importance of S-RNA in connection with the selection and transfer of amino acids in the proportions required for the synthesis of proteins. These authors showed that the six amino acids which they studied were accepted and transported by soluble RNA isolated from the mammary gland roughly in the same proportions in which they occur in the proteins of milk.

Bergmann, Berg & Dieckmann (13a) have recently published an interesting paper indicating that the individual activating enzymes of amino acids (the authors call them enzymes synthesizing amino acid-ribonucleic acid complexes) are not always absolutely specific. For example, valyl-

RNA synthetase, having no activity in respect of D-valine, L-isoleucine, L-leucine or L-methionine nevertheless brings about the acceptance of L-threonine by S-RNA at about 25-28% of the rate of the acceptance of valine, but this only happened when no valine was present in the medium. Isoleucyl-RNA synthetase acts as a weak catalyst for the synthesis of L-valyl-RNA and L-methionyl-RNA and other examples are known.

A beginning has been made in the direct as well as the indirect study of the molecular basis for the specificity and selectivity of the various fractions of soluble RNA. In a recent short communication Lagerkvist, Berg and their colleagues (78) described work on the sequence of nucleotides in the part of the chain of soluble RNA adjacent to the terminal group. The authors determined the nature of the fourth and fifth groups from the end, i.e. those immediately adjacent to the terminal ACC trinucleotide. The nature of the fourth nucleotide varied. In 65% of molecules of acceptor RNA it was an adenyl nucleotide but in some cases this position was occupied by a uridyl or guanyl nucleotide or occasionally by a pseudouridyl nucleotide. The variation in the nature of the fifth nucleotide was found to be even greater. According to Asano & Egami (9a) the terminal sequence of the S-RNA of yeast is GpApCpCpA-amino acid.

In recent years it has been found possible to isolate S-RNA from the organelles of cells, especially the nucleus, as well as from the cell sap. In this connection it is interesting to note that although the S-RNA of the nucleus also contains adenylic acid as the receptor terminal nucleotide (71), it still, according to Webster (117a), cannot replace cytoplasmic S-RNA in the reactions of the transfer of amino acids to the cytoplasmic ribosomes. Khesin (8a), on the other hand, has come to the conclusion that in liver cells the acceptor form of RNA occurring in the nucleus is identical with that in the cytoplasm.

## 8. Transfer of the complex of S-RNA with an amino acid to the microsomes and incorporation of amino acids into the protein undergoing synthesis

When they are released from the aminoacyl-adenyl complex on the surface of the activating enzyme and form their new bonds with the terminal fragments of S-RNA, amino acid residues still retain some of the extra supply of energy which they received when the complex was formed (59). This extra energy is later used for the formation of peptide bonds when the amino acid, now combined with S-RNA, gets the opportunity to take part in this process which is going on in the organelles of the cell.

Amino acids which have been accepted by S-RNA cannot combine directly with one another and, for the truly specific synthesis of proteins to occur, a further stage is required which has hardly been studied at all. This is the interaction between the molecules of acceptor S-RNA and the matrices of protein synthesis, namely the high-polymeric molecules of RNA.

In his review at the Fourth International Congress of Biochemistry, in 1958 Hoagland (59) was the first to report the existence of evidence that, at a certain moment in the synthetic process, molecules of S-RNA associate with microsomal RNA, the association depending on guanosine triphosphate (GTP) and occurring only when the molecules of S-RNA were "charged" with amino acids. The incorporation of amino acids into proteins was, in fact, found to be connected with this process of association of the two types of RNA.

The mechanism of this interaction and its significance in the reproduction of the specificity of the protein molecule has still scarcely been studied, but the activities of several well-known biochemical laboratories are at present concentrated on this narrow field and it is certain that this important part of the problem of protein synthesis will be solved in the very near future.

The interaction between S-RNA and the RNA of the organelles is undoubtedly a very complicated process and we are still only groping after methods of studying it. The findings of Hoagland alluded to above, concerning the interaction between S-RNA and RNA from the organelles, have recently been confirmed by a number of authors (17, 18, 34, 35, 36, 73, 100). This process is accomplished very quickly and, according to Bosch and his colleagues, equilibrium between S-RNA labelled with $^{32}P$ and RNA from the microsomes has already been attained after 5 minutes.

According to Schweet et al. (104a), during the synthesis of haemoglobin by the reticulocytes of the rabbit, each molecule of S-RNA transfers about 20-40 amino acid residues per minute. The molecules of S-RNA react with the ribosomes even if they are not carrying amino acids (19a, 16a, 17a). Webster (116, 117) observed the same process in artificial systems made from pea shoots. In a second piece of work he succeeded in observing the incorporation of labelled amino acids combined with S-RNA and brought about an increase of the amount of protein in a cell-free system in vitro and thus confirmed the evidence in this matter derived from materials of animal origin. The proteins which were synthesized in Webster's experiments were specific and had an enzymic activity. In our own work (4) we also made some observations on the interaction between the RNA and the cellular structures of plants in vivo.

For this purpose we used vacuum infiltration of labelled S-RNA into the leaves of plants. Owing to the smallness of its molecules S-RNA can easily be infiltrated into tissues and passes through cell walls. Sufficient amounts of labelled S-RNA were isolated from organelle-free plasma juice from frozen leaves. Molecules of S-RNA may be labelled either by inclusion of $^{32}P$ or by means of [$^{14}C$]- or [$^{35}S$]amino acids which are firmly accepted by the terminal groups of the molecules and are not removed from the RNA during its deproteinization. To prepare our labelled S-

RNA we used young plants of beans and sunflowers which had been grown in water into which $^{32}$P had been introduced in the form of sodium phosphate and $^{35}$S in the form of sulphate with a very high specific activity. The S-RNA isolated from the plasma juice of leaves by the phenol method was doubly labelled. Solutions of this RNA were also used for the infiltration. In order to make observations on the possible occurrence of species specificity in the processes, the infiltration was "crossed", that is to say, RNA derived from beans was infiltrated into the leaves of both beans and sunflowers and the same for RNA from sunflowers. The results of one such experiment are set out in Table 6.

No definite manifestations of species specificity in the distribution of labelled S-RNA were observed in this case. There are only species peculiarities—the leaves of the sunflower bind S-RNA (both their own and foreign) more actively and compounds labelled with $^{35}$S derived from beans go almost completely into the structures in leaves, whether these are their own or foreign.

In the study of this final stage of protein synthesis great attention is now being paid, not so much to the occurrence of interaction between the molecules of S-RNA and the matrices of protein synthesis, but rather to its mechanism and to such features of this as might enable one to find out the significance of this interaction with regard to determination of the sequence of the amino acids in the polypeptide chain undergoing synthesis.

In our current theoretical ideas concerning the mechanism of reproduction of the specificity of proteins, it is assumed that the acceptor molecules of S-RNA interact with strictly determinate parts of the matrices and that the position of the amino acids in the polypeptide chain is determined in accordance with the nature of this interaction. A theoretical analysis of the different variants of this hypothesis will be given in detail in Chapter XV, which deals with the theoretical problems of protein synthesis. Here we are concerned with a number of phenomena which will gradually build up a biochemical picture of how the final reactions of soluble RNA actually take place.

In his review Hoagland (60) quotes some interesting findings of Acs which point to some features of the synthesis of proteins during the interaction of S-RNA with the matrices. Soluble RNA, as has already been mentioned, carries a stock of activated amino acids "ready" for incorporation into proteins. Acs "cut off" these amino acids with a weak alkali and then labelled the RNA with [$^{14}$C]threonine only. When he put this RNA into a synthesizing system he found that it was less effective as a donor of [$^{14}$C]threonine to proteins than it had been before treatment. This result can be explained by the necessity for the presence of all amino acids if any individual one is to be incorporated in a protein which is being synthesized.

According to the evidence put forward in Hoagland's review (60), when S-RNA labelled with $^{32}$P on the terminal adenosine group and

TABLE 6

Distribution of radioactivity in fractions of leaves 25 minutes after infiltration of S-RNA into the leaves (4)

| | Overall activity of the components (whole sample) counts/min. | | | | | |
|---|---|---|---|---|---|---|
| Variants of experiment | Labelled with $^{32}P$ | | | Labelled with $^{35}S$ | | |
| | Plasma juice | Structures | % activity in structures | Plasma juice | Structures | % activity in structures |
| 1. Leaves of bean, RNA of bean | | | | | | |
| a. protein + RNA | 627 | 250 | 25·1 | 210 | 1507 | 85·6 |
| b. soluble compounds | 120 | | | 45 | | |
| 2. Leaves of bean, RNA of sunflower | | | | | | |
| a. protein + RNA | 2318 | 593 | 18·8 | 3060 | 3140 | 49·1 |
| b. soluble compounds | 230 | | | 192 | | |
| 3. Leaves of sunflower, RNA of sunflower | | | | | | |
| a. protein + RNA | 1640 | 1470 | 42·1 | 1626 | 1420 | 44·2 |
| b. soluble compounds | 370 | | | 160 | | |
| 4. Leaves of sunflower, RNA of bean | | | | | | |
| a. protein + RNA | 470 | 426 | 42·2 | 270 | 1320 | 81·5 |
| b. soluble compounds | 115 | | | 32 | | |

The activity of the soluble compounds in the structures was not determined because homogenization of the structures was carried out in an isotonic solution of KCl in which it is known that there is always a small admixture of radioactive potassium ($^{40}K$).

combined with [$^{14}$C]valine reacts with the RNA of the microsomes, both labelled components combine with it.

Until recently only one necessary co-factor for the reaction between adaptor RNA and the organelles was known, and that was GTP. At the end of 1960, however, a more complicated biochemical picture of this phase of protein synthesis began to unfold. It was discovered almost simultaneously in three laboratories that some non-diffusible protein component of the supernatant fraction was essential to the occurrence of this reaction (54, 37, 72, 89, 90). It was further found that the reaction of transfer of amino acids was accelerated by glutathione and other compounds containing sulphydryl groups (15, 15a).

These findings formed the basis for the idea that the protein component might be a special transfer enzyme which would accept molecules of RNA and "plant" them on the matrices of protein synthesis. Takanami & Okamoto (114, 113) have recently carried out a successful piece of work on the isolation and purification of this enzyme by chromatography on diethylaminoethylcellulose (DEAE-cellulose). Bishop & Schweet (15) have also made a partial purification of the transfer enzyme of S-RNA. If it is to fulfil its function the following co-factors must also be present: magnesium ions, glutathione and GTP. Webster (118) has isolated a similar enzyme from homogenates of shoots of peas. In its purified form it was only active in the presence of these same co-factors. The formation of peptide linkages and polypeptide chains is accompanied by a degradation of GTP in strictly equivalent amounts. According to Webster the liberation of the completed protein also requires the participation of a special protein fraction, ATP and Mg$^{++}$.

Many important features of this final reaction in which S-RNA takes part have been studied in the recent work of Hoagland & Comly (62). These authors carried out a series of extremely elegant experiments designed to reveal the nature of the reversible transfer of molecules of S-RNA on to the matrices, which is connected with incorporation of amino acids into the proteins being synthesized. Having combined ribosomal RNA with S-RNA labelled with $^{32}$P or with $^{14}$C (in the amino acid fragment), these authors then studied the "displacement" of this S-RNA by fractions of unlabelled S-RNA. As a result of these experiments they showed that molecules of S-RNA participate reversibly in the reactions by which amino acids are quickly transferred to the matrices, without themselves being fragmented. The cycles of transfer are not accompanied either by breakdown of the polynucleotide chains or by stripping off of the terminal carrier group. The molecules of S-RNA which have been combined with the matrix are then displayed by new portions of S-RNA in such a way that the process of displacement is similar to the original incorporation. Molecules of S-RNA which are not carrying amino acids can also react with the matrices. The transfer enzyme thus appears not to be heterogeneous. Nathans & Lipmann (90a) have recently shown that, after purification in various ways, a preparation of the enzyme could accomplish transfer of S-RNA carrying different amino acids. At the same time the enzyme has marked species specificity. The material obtained recently by Ehrenstein & Lipmann (44a) is extraordinarily interesting. These authors found that the aminoacyl-S-RNA complexes when made had no species specificity. Any molecule of S-RNA carrying an amino acid (e.g. one isolated from *E. coli*) could react with the ribosomes of the reticulocytes of the rabbit and incorporate the amino acid component into the haemoglobin being synthesized. This may indicate that the coding sequence in RNA's which determines their interaction with enzymes and matrices may differ. If the interaction is with the enzyme the code may be species-specific, if with the matrix it will be universal.

## Conclusion

A survey of the material given in this chapter shows clearly that the cells of animals, plants and microorganisms contain, in addition to the complex system of enzymes activating amino acids, a no less complicated system of comparatively small polynucleotide structures which carry out an intermediate acceptance of the activated amino acids and their transfer to the organelles (microsomes, mitochondria and plastids) where the actual synthesis of protein structures takes place.

The systems which have been considered above carry out important preliminary work without which the structural fragments of the cell could not fulfil their functions. They not only get the amino acids into an active state, but they also carry out two stages in their qualitative and quantitative sorting, ensuring the unhesitating and purposive working of the synthetic system. The significance of the complex of soluble RNA and of the complex of enzymes for activating and transferring amino acids is not, however, confined to their performance of these tasks. There are many reasons to believe that the molecules of S-RNA also play an important part in the actual process of polymerization and that their reaction with the matrices of protein synthesis is specific and determines the positions of the amino acids in the polypeptide chains. Consideration is given to possible mechanisms whereby this reaction may occur in theoretical schemes of protein synthesis. A large number of such theoretical schemes are based on the assumption that there is a complementary relation between the molecules of S-RNA and some parts of the matrices, which have the appropriate information encoded in them. It is believed that the transfer of amino acids to the matrices is not haphazard but strictly determinate.

Such beliefs have a logical foundation but we cannot as yet point to any facts which would confirm the view that this is actually the way in which the matrices react with the molecules of S-RNA. It is only to be expected that, if we are to make a thorough theoretical analysis of this decisive stage of protein synthesis, we must take account, not only of material derived from the field of the activation and transfer of amino acids, but also of material derived from a study of the biochemistry of the matrices. We must take a more careful look at the processes and phenomena which take place in the cellular organelles which are the sites of the polymeric forms of RNA which act as the specific matrices for protein synthesis.

## REFERENCES

1. MEDVEDEV, ZH. A. (1957). *Doklady Moskov. sel'skokhoz. Akad. Im. K. A. Timiryazeva, Nauch. Konf.* 29, 55.
2. — (1959). *Doklady Moskov. sel'skokhoz. Akad. Im. K. A. Timiryazeva,* No. 47, 77.
3. — (1960). *Uspekhi sovremennoĭ Biol.* 50, 121.
4. — (1961). *Biofizika,* 6, 279.
5. MEDVEDEV, ZH. A., ZABOLOTSKIĬ, N. N., SHEN', TS.-S., MO, S.-M., DAVIDOVA, E. G. & DAVIDOV, E. R. (1960). *Biokhimiya,* 25, 1001.
6. MIL'MAN, L. S. (1960). *Biokhimiya,* 25, 796.
7. KHESIN, R. B. (1952). *Biokhimiya,* 17, 664.
8. — (1960). *Biokhimiya tsitoplazmy.* Moscow: Izd. Akad. Nauk S.S.S.R.
8a. — (1961). *Proc. V int. Congr. Biochem., Moscow,* 2, 257.
9. ACS, G., HARTMANN, G., BOMAN, H. G. & LIPMANN, F. (1959). *Fed. Proc.* 18, 178.
9a. ASANO, K. & EGAMI, F. (1961). *J. Biochem. (Tokyo),* 50, 467.
10. BATES, H. M. (1961). *Proc. V int. Congr. Biochem., Moscow,* 9, 81.
10a. BATES, H. M. & LIPMANN, F. (1960). *J. biol. Chem.* 235, PC 22.
11. BENNETT, E. L. (1953). *Biochim. biophys. Acta,* 11, 487.
11a. BENZER, S. & WEISBLUM, B. (1961). *Proc. nat. Acad. Sci. U.S.* 47, 1149.
12. BERG, P. & OFENGAND, E. J. (1958). *Proc. nat. Acad. Sci. U.S.* 44, 78.
13. BERGMANN, F. H., BERG, P., PREISS, J., OFENGAND, E. J. & DIECKMANN, M. (1959). *Fed. Proc.* 18, 191.
13a. BERGMANN, F. H., BERG, P. & DIECKMANN, M. (1961). *J. biol. Chem.* 236, 1735.
14. BERNLOHR, R. W. & WEBSTER, G. C. (1958). *Nature (Lond.),* 182, 531.
15. BISHOP, J. & SCHWEET, R. (1961). *Fed. Proc.* 20, 389.
15a. BISHOP, J. & SCHWEET, R. (1961). *Biochim. biophys. Acta,* 49, 235.
16. BLOEMENDAL, H. & BOSCH, L. (1959). *Biochim. biophys. Acta,* 35, 244.
16a. BLOEMENDAL, H., LITTAUER, U. Z. & DANIEL, V. (1961). *Biochim. biophys. Acta,* 51, 66.
17. BOSCH, L. (1961). *Proc. V int. Congr. Biochem., Moscow,* 9, 122.
17a. BOSCH, L. & BLOEMENDAL, H. (1961). *Biochim. biophys. Acta,* 51, 613.
18. BOSCH, L., BLOEMENDAL, H. & SLUYSER, M. (1959). *Biochim. biophys. Acta,* 34, 272.
19. — (1960). *Biochim. biophys. Acta,* 41, 444.
19a. BOSCH, L., BLOEMENDAL, H., SLUYSER, M. & POUWELS, P. H. (1961). *Protein Biosynthesis* (Symposium—ed. R. J. C. Harris), p. 133. London: Academic Press.

20. BRACHET, J. (1960). *The biological role of ribonucleic acids.* Amsterdam: Elsevier.
21. BROWN, G. L., BROWN, A. V. W. & GORDON, J. (1959). *Brookhaven Symposia in Biology,* **12**, 47.
22. BROWN, G. L. & ZUBAY, G. (1960). *J. mol. Biol.* **2**, 287.
23. CANELLAKIS, E. S. (1960). *Biochem. J.* **77**, 15P.
24. CANELLAKIS, E. S. & HERBERT, E. (1960). *Proc. nat. Acad. Sci. U.S.* **46**, 170.
25. — (1961). *Biochim. biophys. Acta,* **45**, 133.
26. — (1961). *Biochim. biophys. Acta,* **47**, 78.
27. CANTONI, G. L. (1960). *Nature (Lond.),* **188**, 300.
28. — (1961). *Proc. V int. Congr. Biochem., Moscow,* **9**, 129.
29. CHANTRENNE, H. (1961). *The biosynthesis of proteins.* Oxford: Pergamon.
30. COHEN, N. G. & GROS, F. (1960). *Ann. Rev. Biochem.* **29**, 525.
31. COUTSOGEORGOPOULOS, C. (1960). *Biochim. biophys. Acta,* **44**, 189.
32. COX, R. A. & LITTAUER, U. Z. (1960). *J. mol. Biol.* **2**, 166.
33. DAVIS, F. F., CARLUCCI, A. F. & ROUBEIN, I. F. (1959). *J. biol. Chem.* **234**, 1525.
34. DECKEN, A. VON DER & HULTIN, T. (1958). *Exptl. Cell Research,* **14**, 88.
35. — (1958). *Exptl. Cell Research,* **15**, 254.
36. — (1959). *Exptl. Cell Research,* **17**, 188.
37. — (1960). *Biochim. biophys. Acta,* **40**, 189.
38. DOCTOR, B. P., APGAR, J. & HOLLEY, R. W. (1961). *J. biol. Chem.* **236**, 1117.
39. DOCTOR, B. P. & APGAR, J. (1961). *Fed. Proc.* **20**, 390.
40. DUNN, D. B. (1959). *Biochim. biophys. Acta,* **34**, 286.
41. — (1961). *Biochim. biophys. Acta,* **46**, 198.
42. — (1961). *Proc. V int. Congr. Biochem., Moscow,* **9**, 126.
43. DUNN, D. B., SMITH, J. D. & SIMPSON, M. V. (1960). *Biochem. J.* **76**, 24P.
44. DUNN, D. B., SMITH, J. D. & SPAHR, P. F. (1960). *J. mol. Biol.* **2**, 113.
44a. EHRENSTEIN, G. VON & LIPMANN, F. (1961). *Proc. nat. Acad. Sci. U.S.* **47**, 941.
45. EVERETT, G. A., MERRILL, S. H. & HOLLEY, R. W. (1960). *J. Amer. chem. Soc.* **82**, 5757.
46. — (1961). *Fed. Proc.* **20**, 388.
47. FRASER, M. J. & GUTFREUND, H. (1958). *Proc. roy. Soc. (Lond.)* **B149**, 392.
48. FRASER, M. J., SHIMIZU, H. & GUTFREUND, H. (1959). *Biochem. J.* **72**, 141.
48a. FURTH, J. J., HURWITZ, J., KRUG, R. & ALEXANDER, M. (1961). *J. biol. Chem.* **236**, 3317.
49. GLASSMAN, E., ALLEN, E. H. & SCHWEET, R. S. (1958). *J. Amer. chem. Soc.* **80**, 4427.
50. GOLDTHWAIT, D. A. (1958). *Biochim. biophys. Acta,* **30**, 643.
51. — (1959). *J. biol. Chem.* **234**, 3245, 3251.
52. GOLDTHWAIT, D. A. & STARR, J. L. (1960). *J. biol. Chem.* **235**, 2025.
53. GREGORY, L. & MOLDAVE, K. (1959). *Fed. Proc.* **18**, 238.
54. GROSSI, L. G. & MOLDAVE, K. (1960). *J. biol. Chem.* **235**, 2370.
54a. HARTMANN, G. & COY, U. (1961). *Biochim. biophys. Acta,* **47**, 612.
55. HECHT, L. I., STEPHENSON, M. L. & ZAMECNIK, P. C. (1958). *Biochim. biophys. Acta,* **29**, 460.
56. — (1959). *Proc. nat. Acad. Sci. U.S.* **45**, 505.
57. HERBERT, E. & CANELLAKIS, E. S. (1960). *Biochim. biophys. Acta,* **42**, 363.
58. CANELLAKIS, E. S. & HERBERT, E. (1961). *Biochim. biophys. Acta,* **47**, 78.
59. HOAGLAND, M. B. (1960). *Proc. IV int. Congr. Biochem., Vienna, 1958,* Vol. **8**, p. 199. London: Pergamon.
60. — (1960). In *Nucleic Acids* (ed. E. Chargaff & J. N. Davidson), Vol. **3**, p. 349. New York: Academic Press.

61. — (1959). *Scient. American*, **201**, *no. 6*, p. 55.
62. HOAGLAND, M. B. & COMLY, L. T. (1960). *Proc. nat. Acad. Sci. U.S.* **46**, 1554.
63. HOAGLAND, M. B., STEPHENSON, M. L., SCOTT J. F., HECHT, L. I. & ZAMECNIK, P. C. (1958). *J. biol. Chem.* **231**, 241.
64. HOAGLAND, M. B., ZAMECNIK, P. C. & STEPHENSON, M. L. (1957). *Biochim. biophys. Acta*, **24**, 215.
65. — (1959). *Symposium Mol. Biol., Univ. Chicago*, p. 105.
66. HOLLEY, R. W., APGAR, J., DOCTOR, B. P., FARROW, J., MARINI, M. A. & MERRILL, S. H. (1961). *J. biol. Chem.* **236**, 200.
67. HOLLEY, R. W., BRUNNGRABER, E. F., SAAD, F. & WILLIAMS, H. H. (1961). *J. biol. Chem.* **236**, 197.
68. HOLLEY, R. W., DOCTOR, B. P., MERRILL, S. H. & SAAD, F. M. (1959). *Biochim. biophys. Acta*, **35**, 272.
69. HOLLEY, R. W. & MERRILL, S. H. (1959). *Fed. Proc.* **18**, 249.
70. HOLLEY, R. W. & PROCK, P. (1958). *Fed. Proc.* **17**, 244.
71. HOPKINS, J. W., ALLFREY, V. G. & MIRSKY, A. E. (1961). *Biochim. biophys. Acta*, **47**, 194.
72. HÜLSMANN, W. C. & LIPMANN, F. (1960). *Biochim. biophys. Acta*, **43**, 123.
73. HULTIN, T. & DECKEN, A. VON DER (1959). *Exptl. Cell Research*, **16**, 441.
74. JARDETZKY, C. D. & BARNUM, C. P. (1957). *Arch. Biochem. Biophys.* **67**, 350.
75. JEENER, R. & SZAFARZ, D. (1950). *Arch. Biochem.* **26**, 54.
76. KIRBY, K. S. (1956). *Biochem. J.* **64**, 405.
77. KLEE, W. A. & CANTONI, G. L. (1960). *Proc. nat. Acad. Sci. U.S.* **46**, 322.
78. LAGERKVIST, U., BERG, P., DIECKMANN, M. & PLATT, F. W. (1961). *Fed. Proc.* **20**, 363.
78a. LANE, B. G. & LIPMANN, F. (1961). *J. biol. Chem.* **236**, PC 80.
79. LEAHY, J., ALLEN, E. & SCHWEET, R. (1959). *Fed. Proc.* **18**, 270.
80. LEAHY, J., GLASSMAN, E. & SCHWEET, R. S. (1960). *J. biol. Chem.* **235**, 3209.
81. LOFTFIELD, R. B., EIGNER, E. A. & HECHT, L. I. (1960). *Proc. IV int. Congr. Biochem., Vienna, 1958*, Vol. **8**, p. 222. London: Pergamon.
82. LOFTFIELD, R. B., HECHT, L. I. & EIGNER, E. A. (1959). *Fed. Proc.* **18**, 276.
83. MANDEL, P., WEILL, J.-D., LEDIG, M. & BUSCH, S. (1959). *Nature (Lond.)*, **183**, 1114.
84. MARINI, M. A. (1961). *Fed. Proc.* **20**, 355.
85. MASON, T. G. & PHILLIS, E. (1939). *Ann. Botany (N.S.)*, **3**, 531.
86. MOLDAVE, K. (1960). *Biochim. biophys. Acta*, **43**, 188.
87. MONIER, R. (1961). *Proc. V int. Congr. Biochem., Moscow*, **9**, 135.
88. MONIER, R., STEPHENSON, M. L. & ZAMECNIK P. C. (1960). *Biochim. biophys. Acta*, **43**, 1.
89. NATHANS, D. (1960). *Ann. N.Y. Acad. Sci.* **88**, 718 (1960).
90. NATHANS, D. & LIPMANN, F. (1960). *Biochim. biophys. Acta*, **43**, 126.
90a. — (1961). *Proc. nat. Acad. Sci. U.S.* **47**, 497.
90b. OFENGAND, E. J., DIECKMANN, M. & BERG, P. (1961). *J. biol. Chem.* **236**, 1741.
91. OGATA, K. & NOHARA, H. (1957). *Biochim. biophys. Acta*, **25**, 659.
92. OGATA, K., NOHARA, H. & MORITA, T. (1957). *Biochim. biophys. Acta*, **26**, 656.
93. OSAWA, S. (1960). *Biochim. biophys. Acta*, **42**, 244.
94. — (1960). *Biochim. biophys. Acta*, **43**, 110.
95. OSAWA, S. & OTAKA, E. (1959). *Biochim. biophys. Acta*, **36**, 549.
96. OTAKA, E., HOTTA, Y. & OSAWA, S. (1959). *Biochim. biophys. Acta*, **35**, 266.
97. OTAKA, E. & OSAWA, S. (1960). *Nature (Lond.)*, **185**, 921.

98. PREISS, J., BERG, P., OFENGAND, E. J., BERGMANN, F. H. & DIECKMANN, M. (1959). *Proc. nat. Acad. Sci. U.S.* **45**, 319.
98a. PREISS, J., DIECKMANN, M. & BERG, P. (1961). *J. biol. Chem.* **236**, 1748.
99. PRICE, T. D., TSUBOI, K. K., HINDS, H. A. & HUDSON, P. B. (1960). *Nature (Lond.)*, **186**, 158.
100. REID, E. & SMITH, C. J. (1961). *Proc. V int. Congr. Biochem.*, *Moscow*, **9**, 138.
101. ROSENBAUM, M. & BROWN, R. A. (1961). *Anal. Biochem.* **2**, 15.
102. SACHS, H. (1958). *J. biol. Chem.* **233**, 643.
103. — (1958). *J. biol. Chem.* **233**, 650.
104. SAPONARA, A. & BOCK, R. M. (1961). *Fed. Proc.* **20**, 356.
104a. SCHWEET, R. S., BISHOP, J. & MORRIS, A. (1961). *Proc. V int. Congr. Biochem.*, *Moscow*, **9**, 204.
105. SCHWEET, R. S., BOVARD, F. C., ALLEN, E. & GLASSMAN, E. (1958). *Proc. nat. Acad. Sci. U.S.* **44**, 173.
106. SCHWEET, R. S., GLASSMAN, E. & ALLEN, E. (1958). *Fed. Proc.* **17**, 307.
107. SHIGEURA, H. T. & CHARGAFF, E. (1957). *Biochim. biophys. Acta*, **24**, 450.
108. — (1958). *Biochim. biophys. Acta*, **30**, 434.
109. SINGER, M. F. & CANTONI, G. L. (1960). *Biochim. biophys. Acta*, **39**, 182.
110. SINGER, M. F., LUBORSKY, S., MORRISON, R. A. & CANTONI, G. L. (1960). *Biochim. biophys. Acta*, **38**, 568.
111. SMITH, J. M. (1959). *Nature (Lond.)*, **184**, 956.
112. SMITH, K. C., CORDES, E. & SCHWEET, R. S. (1959). *Biochim. biophys. Acta*, **33**, 286.
113. TAKANAMI, M. (1961). *Biochim. biophys. Acta*, **51**, 85.
114. TAKANAMI, M. & OKAMOTO, T. (1960). *Biochim. biophys. Acta*, **44**, 379.
115. TARVER, H. & KORNER, A. (1957). *Fed. Proc.* **16**, 260.
116. WEBSTER, G. C. (1959). *Arch. Biochem. Biophys.* **82**, 125.
117. — (1960). *Arch. Biochem. Biophys.* **89**, 53.
117a. — (1960). *Biochem. biophys. Res. Commun.* **2**, 56.
118. — (1961). *Fed. Proc.* **20**, 387.
119. WEISS, S. B., ACS, G. & LIPMANN, F. (1958). *Proc. nat. Acad. Sci. U.S.* **44**, 189.
120. ZACHAU, H. G., ACS, G. & LIPMANN, F. (1958). *Proc. nat. Acad. Sci. U.S.* **44**, 885.
121. ZAMECNIK, P. C. & STEPHENSON, M. L. (1960). *Ann. N.Y. Acad. Sci.* **88**, 708.
122. ZAMECNIK, P. C., STEPHENSON, M. L. & HECHT, L. I. (1958). *Proc. nat. Acad. Sci. U.S.* **44**, 73.
123. ZAMECNIK, P. C., STEPHENSON, M. L. & SCOTT, J. F. (1960). *Proc. nat. Acad. Sci. U.S.* **46**, 811.
124. ZILLIG, W., SCHACHTSCHABEL, D. & KRONE, W. (1960). *Z. physiol. Chem.* **318**, 100.

# CHAPTER V

# THE ROLE OF THE CYTOPLASMIC
# ORGANELLES IN PROTEIN SYNTHESIS

## *Introduction*

It is generally accepted that the polymeric ribonucleic acid of the cellular organelles plays an important part in the accomplishment of the final stages of the synthesis of proteins.

Such a large number of theoretical reviews and sections of monographs dealing with the part played by RNA as matrices in protein synthesis have been published in recent years (1, 2, 8, 9, 23, 30, 35, 38, 40, 41, 45, 58, 90, 91, 106, 129, 135, 250, 59, 72, 67, 170, etc.) that any new treatment of this problem might seem redundant. This is, however, not altogether so. Some questions in this connection, which are of theoretical importance, have only arisen very recently. The problem of the connection between RNA and the synthesis of proteins, like any other major biological problem, requires periodical analysis and generalization from the extensive factual material which has been gathered.

Of course, there is no need for us to repeat in this chapter the exposition of many facts concerning the role of RNA in the synthesis of proteins which are already generally known and which have repeatedly been set out extremely fully in the review literature. These well-known theories concerning the role of RNA in the synthesis of proteins, which have been established as a result of thousands of experimental researches and which are by now unassailable and are continually being confirmed may be summarized as follows:

1. RNA is found in all biological structures which can synthesize proteins.
2. There is a definite correlation between the rate of synthesis of proteins and the content of RNA in organs, tissues, cells and fragments of cells.
3. Washing out RNA from the cell, its splitting with ribonuclease and the inhibition of its synthesis by antimetabolites or irradiation all interfere with the synthesis of proteins.
4. It has been established that there is a definite connection between the metabolism of the phosphorus in RNA and the synthesis of proteins which is especially marked during the synthesis of adaptive enzymes.

5. In many cases RNA can determine completely the specificity of the synthesis of a protein without the intervention of DNA in the process (in viruses, in fragments of cells not containing nuclei, when organelles are incubated *in vitro* etc.).

6. RNA is able to form complex compounds with proteins with different degrees of stability.

7. RNA shows species specificity in respect of its nucleotide composition, molecular weight, biological activity and other properties.

Nevertheless, the mechanism whereby RNA takes part in the synthesis and the reproduction of the specificity of proteins remains obscure in many ways. We shall therefore concentrate most of our attention on this question and shall not try to review the recent investigations which have merely confirmed facts which were already known, but will deal with the new and very new experimental material which will carry us a step or two forward in our understanding of the detailed biochemical mechanism by which RNA is involved in the synthesis of proteins.

We cannot consider the part played by polymeric RNA in protein synthesis apart from the part played by the cellular organelles themselves (ribosomes, mitochondria, and plastids) as the sites in which protein synthesis takes place.

The processes of activation of amino acids and their transfer to S-RNA occur in the cell sap. The actual final synthesis of the proteins occurs within the cellular organelles which not only form proteins for their own growth and multiplication, but also secrete them into the intracellular and extracellular fluids.

Under conditions of active growth of cells and tissues there is intensive synthesis of proteins in all the organelles of the cell and therefore the question of the predominance of any particular type of intracellular structure in protein synthesis is usually only discussed with reference to various functional syntheses the products of which are often secreted by the cell into the external medium (the synthesis of serum proteins, the synthesis of amylase, pepsinogen, trypsinogen and other enzymes, the synthesis of insulin, collagen, fibroin etc.). The question of the localization of synthesis may also be concerned with the discovery of the structures and surfaces within the organelles on which the new protein molecules are actually built up from the activated products which react with them from the intergranular medium.

In our investigation of the localization of protein synthesis we are mainly concerned with four types of structure: microsomes (and recently ribosomes), mitochondria, the plastids of plants and nuclei. In the present chapter we shall not touch upon the synthesis of the nuclear proteins—it is more suitable to do this when we are making an extensive survey of the role of the nucleus itself. Although, in developing tissues, the syntheses taking place in the cytoplasm are under the specific control of the nucleus, yet, under certain conditions, the cytoplasmic structures are capable of accomplishing the synthesis of proteins autonomously for a certain time. In this chapter we shall consider this cytoplasmic synthesis without yet referring to the mechanisms of biochemical and physiological regulation which are external to the cytoplasmic structures.

It must be noted that only a little of the evidence concerning the nature of protein synthesis in cellular structures has been obtained by calculating the actual increase in the amount of protein under a particular set of conditions. Most of the relevant material has been obtained in short-term experiments in which the synthesis of proteins is determined by the so-called "incorporation" of labelled amino acids into the proteins being synthesized. As we shall see from the material in Chapter X, such "incorporation" is sometimes non-specific and is associated with purely chemical combination or with the firm adsorption of amino acids. In this connection the occurrence of specific synthetic incorporation of amino acids into proteins must always be checked systematically by many control experiments.

According to Hoagland (129) the identification of processes of incorporation of amino acids into proteins *in vitro* with specific protein synthesis demands the establishment of the following criteria:

1. The irreversibility of incorporation: i.e. the impossibility of the later displacement of labelled amino acids by excess of unlabelled ones.
2. The dependence of the incorporation on energy-yielding systems (ATP or ATP-generating systems).
3. The combination of the amino acids in the proteins by α-peptide bonds, which may be established by a study of the kinetics of complete and partial acid hydrolysis of the proteins.
4. Combination of the amino acids with internal groups of the chain and not with terminal and superficial active groups.
5. The predominant incorporation of the amino acid into one particular protein (e.g. albumin) which is the main one being synthesized by the group of cells in question.
6. The dependence of the incorporation of one amino acid on the presence of the other amino acids of which the particular protein is composed.

The last two types of determination are the most critical but cannot be realized in *in vitro* experiments.

It must be mentioned that very detailed reviews of the evidence (especially cytological) concerning the localization of protein synthesis in the different cellular structures are to be found in the monographs of R. B. Khesin (35), B. V. Kedrovskiĭ (9) and Brachet (58) and in two specialized symposia (162, 163). This problem has also been surveyed in a number of recent reviews (4, 12, 23, 24, 34, 39, 67). This allows us not to spend long on the history of this work but to give a succinct account of the most important biochemical aspects of the subject, concentrating our main attention on certain very recent studies which have not yet received enough discussion in the theoretical and review literature.

## 1. Chemical, physical-chemical and biochemical properties of the polymeric RNA of the ribosomes

In Chapter II we have already met with some of the general evidence concerning the structure of RNA. We shall deal here with certain features of the RNA of the intracellular structures. We must first note that the RNA of the cellular structures and especially that of the ribosomes, which are the smallest and simplest of the cytoplasmic organelles, is very highly polymeric. The molecular weight of this RNA is sometimes as great as $1.5 \times 10^6$. Until recently it has not been possible to isolate RNA with a molecular weight of over $10^6$ from many organs because it undergoes degradation during isolation. In recent years, however, more gentle methods of extraction of RNA have been worked out and the presence of these compounds with molecular weights of $10^6$ and more has been established, not only in viruses, but also in the cytoplasmic structures of animals, plants and bacteria (94, 95, 123, 124, 154, 150, 146, 143, 232).

In the following survey of the material concerning the structure of the ribosomes we shall see that this RNA is arranged round the periphery. It has been shown by Ts'o, Bonner & Vinograd (243) that magnesium ions play an essential part in the combination of RNA and proteins. The molecular weight of the ribosomes isolated from the microsomes lies between $2 \times 10^6$ and $5 \times 10^6$. In this connection Hoagland (129) has put forward the suggestion that ribosomes often contain only one polymeric molecule of RNA which plays the part of a matrix. The RNA of microsomes after isolation consisted of single polynucleotide chains (95, 96, 124, 150).

The first of these papers and also certain other publications from the laboratory of P. Doty (95) describe an interesting attempt to determine the configuration of the polynucleotide of the ribosomes. For this purpose the authors studied the alterations in the structures of the polymers when the hydrogen bonds are weakened by heating. Under these circumstances they observed an increase in the optical density of the solution, which may indicate that native RNA has a special configuration. This conclusion could be strengthened by a study of changes in the optical rotation, which might reveal whether we have to deal with a continuous helical structure along the whole length of the molecules or with simple pairing of bases occurring at random. The results of such studies suggest that the particle of RNA may have hydrogen bonds repeated at regular intervals which lead to the formation of a helix. It has been found that, in the RNA of the ribosomes, only a part of the structure has helical elements. From their results the authors suggest that the RNA of the cellular particles may consist of zones with a continuous helical structure which alternate with non-helical parts which are readily available for reaction with proteins and other molecules.

Hall & Doty (124) found that RNA, isolated from the ribosomes of liver by careful phenol treatment, consisted of two components with molecular weights of $1\cdot3 \times 10^6$ and $0\cdot6 \times 10^6$. The molecular weight of the ribosomes themselves was $3\cdot7 \times 10^6$.

According to the preliminary results of these authors, these two types of molecule of RNA each consisted of sub-particles having molecular weights of $0\cdot12 \times 10^6$ and consisting of continuous nucleotide chains of uniform weight.

However, these results, concerning the presence of sub-particles of lower molecular weight in the molecules of ribosomal RNA, have not been confirmed by other authors and seem to be artefacts. A further contribution to this question has been made by A. S. Spirin et al. (3, 28, 29). For their study of this problem, these authors used the method which they had originally used to find out the nature of the loss of infectivity by viral RNA with time. The first observations of A. S. Spirin and L. Gavrilova (3, 28) showed that, when a solution of viral RNA is heated (in water or a urea solution) over the range 50-70°C, one may observe a sharp increase in the viscosity of the solution, associated with the unfolding of the convoluted molecules of RNA and their stretching out into long threads owing to the breaking of hydrogen bonds.

This effect was absent in the RNA of the tobacco mosaic virus which had lost its infectivity, because there were breaks in its polynucleotide chain which led to its breaking down into smaller fragments when it became unfolded. A similar phenomenon was also observed by Hall & Doty in their study of the RNA of the ribosomes, and also formed a basis for their conclusion that it is composed of sub-units having a molecular weight of about $0\cdot1 \times 10^6$.

A. S. Spirin & L. Mil'man (29) suggested that, in this case, the fragmentation of the RNA of the ribosomes might have a secondary origin in connection with the length of the procedure involved in the isolation of the ribosomes. They have confirmed this view by using the same methods for the study of RNA isolated directly from whole homogenates of liver by the phenol method. In this case the polymeric RNA did not have fractures and, on heating, it behaved like a complete molecule giving a well-marked rise in viscosity. The authors also established the existence of two types of polymeric RNA in tissues, having molecular weights of $0\cdot6 \times 10^6$ and $1\cdot5 \times 10^6$. In this connection Spirin & Mil'man believe that the breakdown of the RNA of the ribosomes on heating into the smaller fragments discovered by Hall & Doty is a result of the presence of fractures in the chain of RNA, which arise under the influence of ribonuclease during the isolation of the ribosomes. It is well known that the ribosomes contain latent ribonuclease which is closely associated with a ribonucleo-protein, the activity of which is increased during the purification of the ribosomes and also when they are treated with a urea solution or $0\cdot3$-$0\cdot5$ M-NaCl, and on heating (97).

The occurrence in ribosomes of two types of polymeric molecules of RNA with molecular weights of $0\cdot7\times10^6$ and $1\cdot7\times10^6$ has also been demonstrated in the recent work of Cheng Ping-Yao (74). It is quite clear that these polymeric molecules of RNA which have been discovered recently do constitute real units (166, 27).

Kurland (143) has made a further interesting observation on the state of the polymeric RNA in the ribosomes. This author isolated RNA (by the phenol method and by using detergents) from ribosomes of a particular type. Ribosomes which sedimented constantly at 70 S disintegrated into particles at 50 S and 30 S. When these particles were analysed it was found that the finer ribosomes (30 S) also contained the shorter polynucleotides (16 S, which corresponds to a molecular weight of $0\cdot56\times10^6$) while the larger molecules of RNA (23 S or mol. wt. $=1\cdot1\times10^6$) were only to be found in the coarser ribosomes (50 S and 70 S). Each of the smaller (30 S) ribosomes contained only one molecule of RNA (63% RNA). The coarser ribosomes (50 S) each contained either one molecule of RNA with a molecular weight of $1\cdot1\times10^6$ or two molecules of half that weight ($5\cdot6\times10^5$).

A. S. Spirin (27) has found that, depending on the object from which they are isolated, the size of the RNA of the organelles may vary, but when the size of the smaller component increases that of the larger component decreases. It is, however, interesting to note that Hall, Storck & Spiegelman (125) have found, quite recently, that in ribosomes of 70 S and 30 S there is, in addition to the large molecules of RNA (25 S and 16 S) a fraction with a comparatively low molecular weight (4 S) which is metabolized quickly. This same RNA of low molecular weight was also found in ribosomes by Rendi & Warner (195) who supposed that it was the S-RNA which combines with the matrices during the processes of transfer of amino acids.

Maeda (154a) determined the molecular weight of the RNA of yeast ribosomes by tentative methods in the presence of various concentrations of magnesium; he, like others, found two components with molecular weights $0\cdot7\times10^6$ and $1\cdot4\times10^6$. In this connection he postulated, and to some extent demonstrated, that the former are fundamental units while the latter are dimers formed from pairs of such units.

Green & Hall (117a) have obtained further evidence of the biochemical heterogeneity of ribosomes and their component RNA. They isolated from cells of E. coli various types of ribosomes. Part of their study related to the so-called "native" 30 S and 50 S ribosomes, i.e. those that exist as

such in the cytoplasm, not associated into 70 S aggregates. They compared these with the corresponding "derived" 30 S and 50 S ribosomes
obtained by the disruption of 70 S particles. In the course of this they
observed that 23 S RNA (mol. wt. $1 \cdot 1$-$1 \cdot 5 \times 10^6$) is only present in the
50 S moiety of the 70 S ribosomes. The native 50 S and 30 S particles (as
well as the 30 S derived ribosomes) contained only 16 S RNA, corresponding to mol. wt. $0 \cdot 6 \times 10^6$.

Zubay et al. (134, 258) have recently done some very interesting work
on the configuration of the RNA of the organelles. These workers showed
that, although the molecule of RNA consists of only a single polynucleotide
chain, yet in the organelles and in solution this chain may twist itself up
with the formation of a double helix similar to the helix of DNA, and that
about $\frac{3}{4}$ of the length of the ribosomal RNA is in the form of such helices.
These observations were confirmed by Schlessinger (204) in a study of the
hypochromicity of RNA in solution and in the ribosomes, and also by
Klug, Holmes & Finch (138) in a study on the diffraction of X-rays by
ribosomes from various sources (liver, yeast, E. coli). This double helix
was not, however, so regular as that of DNA. New results of the electron-
microscopic study of RNA from various sources have recently been
published by Danon, Marikovsky & Littauer (89). When isolated,
purified and prepared for electron microscopy the RNA of organelles in
their experiments usually took the form of a single polynucleotide with a
diameter of the order of 10 Å and a length of between 1500 and 4000 Å.
However, these polynucleotides had a tendency to curl up, forming pairs
of bases. Soluble RNA from E. coli had a pronounced secondary structure
and, in electron micrographs, it looked like ants' eggs. Similar results
were obtained in the electron-microscopic studies of high-polymeric RNA
by Kiselev, Gavrilova & Spirin (9a).

Doty and colleagues (101) used this evidence for the presence of a
secondary structure in ribosomal RNA to work out a tentative scheme of
the formation of pairs of complementary bases in the folding of the polynucleotide chain of RNA. In their view such spontaneous folding must
inevitably lead to the formation of many loops and lumps in the double
helix of the RNA on account of the nucleotide residues for which there
were no opposite numbers on the other side. The theoretical value of
these models will be considered later.

The RNA of the ribosomes acts as a matrix for the synthesis of protein.
If these matrices determine in some way the specificity of the structure
of the proteins, then one would naturally expect that the RNA would be
heterogeneous in its biochemical properties. The existence of such
heterogeneity is now receiving more and more confirmation. For a comparatively long time we have had evidence for the metabolic heterogeneity
of this RNA and we shall deal with this material in later chapters. Here
we must take note of the ever-accumulating material concerning the presence of fractions of varying composition in this RNA (137, 44, 145, 55,

149, 225). In their painstaking calculations Roberts, Aronson, Bolton and a group of other fellow-workers of the Carnegie Institute, Washington (44) have brought forward some very interesting material concerning differences between the RNA from the coarse (50 S) and fine (30 S) ribosomes in respect of their nucleotide composition and other properties. In a review article on the ribosomes Roberts (196) has given the following table showing the composition of the RNA of ribosomes belonging to the different groups (Table 7). It is quite clear that the 50 S ribosomes have a marked excess of purines. The RNA of the ribosomes is also heterogeneous in respect of the tenacity with which it is attached to the protein part of the organelles (15, 57). Aronson (43a) obtained evidence for differences in the nucleotide sequences of RNA isolated from 50 S and from 30 S ribosomes.

TABLE 7

Composition of RNA from various fractions of E. coli* (196)

| | Soluble | | From ribosomes | | | | |
|---|---|---|---|---|---|---|---|
| | S | CA-B | 30(n) | 30(70) | 50(n) | 50(70) | CA-D |
| C | 29·1 | 27·2 | 22·2 | 23·6 | 21·0 | 20·5 | 20·2 |
| A | 19·7 | 19·9 | 24·2 | 24·2 | 26·4 | 26·4 | 28·6 |
| G | 34·2 | 35·6 | 30·4 | 31·6 | 34·1 | 34·8 | 32·2 |
| U | 17·2 | 17·7 | (23·1) | 20·5 | 18·5 | 18·3 | 19·0 |

* The figures are the means of a series of determinations. The possible error is ± 3%. 30(n) and 50(n) refer to RNA from particles 30 (S) and 50 (S) taken on their own. 30(70) and 50(70) indicate ribosomes formed from particles 70( S). CA-B and CA-D indicate soluble and ribosomal RNA which had been formed in the presence of chloramphenicol.

## 2. Historical review of facts and ideas about the matrix function of RNA in protein synthesis before the discovery of "messenger" RNA

There is no longer any doubt that the RNA of the cellular organelles plays an important role in protein biosynthesis. The main change of emphasis in recent years has been as to precisely which kind of RNA serves directly as matrix. Before 1956-7 this function was attributed to the total RNA of the cell, without subdividing it into categories. After discovery of the transferring function of S-RNA, the matrix function was ascribed to the high-polymeric RNA of the various organelles, particularly of the ribosomes. Then in 1961 the search for the matrices narrowed still further and evidence was obtained for a special fraction of RNA in bacterial cells, formed in the nucleus and "settling" in the ribosomes ("information-transferring" or "messenger" RNA). The idea was developed that it is this form of RNA, having molecular weight approx.

300,000 and amounting only to a few percent of the total RNA of the cell, which serves directly as the matrix for protein synthesis, in the sense that it determines directly the sequence of the amino acids.  We shall discuss these studies in Chapter XV.  Here we shall confine ourselves to giving shortly the main facts ascertained by biochemists during the last decade which pointed to the direct involvement of RNA in that stage of protein synthesis at which the sequence of the amino acid residues in the polypeptide chains is determined.

The clearest demonstration of this is certainly to be found in the experiments in which complete virus particles are reproduced by the use of protein-free viral RNA.  Because of their theoretical importance, these studies soon became widely known and therefore we shall only discuss their interpretation briefly, as this question will be reviewed in more detail in the second part of the book, in the chapter devoted specially to the peculiarities of the synthesis of the proteins of viral particles.

Research directed towards the discovery of the specific part played by viral RNA in the reproduction of the protein part of the molecule began with the classical work of Fraenkel-Conrat & Williams (100) who managed to separate the RNA from the protein in the particles of tobacco mosaic virus by gentle preparative methods.  The viral protein, without its RNA, had no power of reproduction.  When the two components reacted with one another particles were formed which resembled the original ones and the infectivity was partly restored.  These results were confirmed in other laboratories (79, 118).  In later studies (31, 39) the authors obtained separate protein and RNA components from each of several different strains of tobacco mosaic virus (TMV) and cross-combined them.  In this way they were able to produce a series of "hybrid" strains which were infective.  It was shown that the descendants of these hybrid strains were always similar to those from which the RNA had been taken.  This similarity manifested itself both in the symptoms caused by the virus, and in its chemical composition.  Extending these investigations of Fraenkel-Conrat (99), and independently of them, Gierer & Schramm (115) found that the RNA of TMV, if used in concentrations 20-500 times as great as in the original virus (10 $\mu$g./ml. RNA or 0·02-0·5 $\mu$g./ml. TMV) has an infectivity similar to that of the virus, i.e., when introduced into the cells of a host it caused the production of specific viral protein. It was found that these properties were associated with the high-molecular fraction of the RNA.  No protein could be found in such fractions either by chemical or by serological means.  According to the authors' calculations (reckoning that the virus contains 5% of RNA) the infectivity of the isolated RNA was 0·1% of that of the virus itself.  In later experiments, however, this infectivity was considerably increased by the use of improved methods for the isolation of the viral RNA (cf. Chapter VII). Ramachandran & Fraenkel-Conrat (184) have recently shown, in a paper on preparative methods, that the infective RNA of TMV can really be completely freed of protein without loss of activity, but there still remains in this RNA an insignificant amount of amino acid-peptide material (about 0·3%) in which only serine and aspartic and glutamic acids have been identified.

The presence in the infective RNA of a small amount of protein which could not be removed in the earliest experiments in the laboratories of Schramm and Fraenkel-Conrat provided grounds on which a number of authors doubted the genuineness of the activity of pure RNA and tried to save the "reputation" of the protein by stressing the importance of even this admixture of protein. However, later experiments, the results of which were communicated to the Fourth International Congress of Biochemistry in Vienna (September 1958) and were discussed by the virus symposium of that congress, as well as the experiments of Ramachandran & Fraenkel-Conrat mentioned above, make it impossible to doubt that deproteinized RNA can reproduce a viral particle. As a result of these experiments it has become possible to rid viral RNA almost entirely of peptide material while markedly increasing its infectivity. It must be mentioned that the possibility of the reproduction of the protein component of the virus within the host by pure, native RNA is also supported by the results of a number of studies of the nature of the infective process when plants are infected with pure RNA and also by many data from other laboratories in which experiments demonstrating the infectivity of viral RNA have been performed, not only by studying the reproduction of TMV, but also during the multiplication of a number of animal viruses. The details of these experiments will be discussed by us in a separate chapter.

Experiments with viruses prompted the development of attempts to transfer the ability to synthesize a particular protein by transferring RNA and these also pointed to the part played by this RNA in the determination of the specificity of proteins.

Evidence which is strongly suggestive in this matter has been obtained by Kramer & Straub (141, 142). These authors, like Groth (121), also succeeded in showing that when the cells of a strain of *B. cereus* have been treated with ribonuclease the phenomenon of induction of the enzyme is preceded by resynthesis of a new RNA. Thus, if cells which are capable of induction are treated with ribonuclease and thus lose their ability to synthesize penicillinase in the presence of its inductor, penicillin, its ability to perform this synthesis can be restored by means of an extract of cells of another strain of *B. cereus* (NRRR-B-569/H) in which penicillinase is a constitutive enzyme, i.e. it is formed continually without any inductor. Investigations showed that the active factor in this extract is a specific RNA which carries the ability to synthesize a new protein from the one strain to the other. The effect of this RNA only lasted for 20 minutes and then fell off, presumably owing to its breakdown.

In 1953-57 yet another series of interesting experiments was carried out on the mechanism of synthesis of adaptive enzymes, which made a great advance in the biochemistry of the part played by RNA and DNA in the synthesis of protein and produced findings which, in some respects, fill out the picture sketched in as a result of the work discussed above. We have in mind the studies of Gale & Folkes on the synthesis of proteins by *Staphylococcus aureus* (107-113). As a result of the investigations of these authors over many years, extensive material has been gathered regarding many aspects of the problem of the synthesis and self-renewal of proteins. This evidence has been set out in detail by Gale in several reviews (103-106). Fragments of cells of the microorganism *Staph. aureus* were chosen as the subjects of these investigations. They were obtained by disrupting a culture of these bacteria with ultrasonic vibrations. After this disruption the cells lost their viability but many biochemical processes, among them protein synthesis, continued to occur in the fragments for some time under favourable conditions. This enabled the authors to wash RNA and DNA out from the fragments of the cells, as the preliminary disruption of the envelopes of the cells did away with the barrier which prevents the extraction of high-molecular compounds from undamaged objects. The fragments of the cells from which RNA and DNA had been removed selectively lost, to a considerable extent, their ability to synthesize proteins and there was a slowing down in them of processes such as the simple growth of structural proteins from the available amino acids as well as the formation of constitutive (catalase, glucozymase) and adaptive ($\beta$-galactosidase) enzymes. When the authors introduced into the culture some RNA isolated from the cells of staphylococci the synthesis of constitutive enzymes (catalase and glucozymase) returned to its previous level. The addition of such material did not, on the other hand, have any restorative effect on the synthesis of $\beta$-galactosidase even in the presence of galactose which is an inductor of this synthesis. The action of the RNA was species-specific; RNA obtained from other sources was completely inactive. The synthesis of $\beta$-galactosidase in the presence of galactose was only observed in cases in which the culture of damaged cells was supplied, not with RNA, but with degradation products of it from which the RNA was synthesized directly in the culture. Thus it was shown that the formation of an adaptive enzyme is associated with the formation of a new RNA. This was also confirmed by experiments in which the formation of $\beta$-galactosidase was suspended if the synthesis of RNA in staphylococcal cells from which the RNA had not been removed was inhibited by the action of azaguanine or other antimetabolites.

The results of these experiments have also been confirmed in other laboratories in the course of studying the formation of adaptive enzymes. The general significance and interest attaching to the mechanism of synthesis of adaptive enzymes in connection with the processes of variation of proteins and the biochemistry of morphogenetic processes are so great that we must review all these findings concerning adaptive synthesis later on in a special chapter.  What we must take note of here is that the difference between the conditions required for the synthesis of the two groups of proteins shows that the part played by RNA in the synthesis of a protein is specific and that the synthesis of different proteins in cells demands the presence of different forms of RNA.  If the RNA plays the part of a matrix then, naturally, the structure required for the formation of a protein which is new to the cell (an adaptive enzyme) may not exist among the collection of molecules of RNA and it will be necessary to synthesize this structure (under the influence of the inductor) if the new protein is to start being formed.

The work of Creaser (84-86) also gives a clear demonstration of the fact that the synthesis of adaptive enzymes requires the presence of some specific RNA, the formation of which precedes the beginning of formation of the enzyme.

In this connection we may also refer to the work of Brawerman & Chargaff (60, 62, 64) who have obtained a lot of results which indicate that, when chloroplasts are formed in etiolated plants which have been brought into the light, one may observe the formation of a specific RNA in the chloroplasts which is somewhat different from the RNA of the ribosomes of the cytoplasm (it has an increased content of adenylic acid and a decreased content of cytidylic acid).

All these results show plainly that the RNA of the cellular organelles affects the specificity of the proteins being synthesized and that its function as a set of matrices would appear to lead to the determination of the sequences of amino acids in the polypeptide chains.

In the biochemical literature there has, in recent years, been discussion of the problem of whether the RNA of the organelles is necessary as such for the synthesis of proteins or whether the important thing for this purpose is not the actual process of formation of this RNA.  This discussion was based on a series of observations which showed that, in a number of cases, the synthesis of proteins depends, not merely on the presence in the cells of polymeric RNA, but on the synthesis of new portions of this RNA.

In the course of this discussion the idea was put forward that the programming of the process of synthesis of a protein, i.e. the determination of its specificity, is not predetermined by the RNA itself but by the process of its synthesis.  In the course of the discussion some authors put forward the idea that there might be a combined synthesis of nucleoproteins from common nucleotide-amino acid precursors (cf. Chapter XV). These results must, of course, be briefly surveyed as they are connected with the ability of RNA to carry out its functions as a set of matrices.  In doing so we must decide whether the process of synthesis of RNA is associated with the synthesis of proteins or whether the synthesis of RNA is important in this matter as providing for replacement of "worn out" matrices and ensuring the synthesis of new (e.g. adaptive) proteins by means of new types of matrices.

Sabinin & Polozova (18) were some of the first authors to observe the stimulation of the growth of plants under the influence of components of nucleic acids, purines and pyrimidines, although their work was not published until 1957—long after it had been completed. In later years these observations were confirmed many times over on different experimental materials (16, 127, 172, 247, 248, 254). In itself, an increase in the synthesis of proteins under the influence of stimulation of the synthesis of RNA does not yet indicate that it is the actual synthesis of the RNA which is important here; it might equally be the effect of the RNA after it had been formed. More definite conclusions in this matter can, therefore, only be drawn on the basis of experiments to study the effect on growth when the synthesis of RNA is inhibited by specific antimetabolites. Although the synthesis of new molecules of RNA is largely suppressed under these circumstances and the antimetabolites do not have any noteworthy effect on the amount of RNA which has previously been made in the cells, this sort of inhibition usually brings the formation of protein to a standstill as well.

Detailed studies of this phenomenon in plants have been made by Webster et al. (248, 254). These authors made an in vitro study of the incorporation of labelled amino acids into granular particles isolated from the cytoplasm of pea shoots. This incorporation was markedly stimulated by RNA derived from shoots of the same sort. However, unlike the species-specific effect of RNA on the synthetic and metabolic incorporation of amino acids by bacteria, discovered by Gale (104-106), the effect of RNA in these experiments was not species-specific. RNA from various sources (yeast, liver, etc.) had a similar stimulant effect on the synthesis of proteins by the granules. Later experiments showed that the effect of the RNA was not associated with its native structure but with products of its decomposition. A similar effect was obtained by the addition of nucleotides, nucleosides, purines and pyrimidines to the medium. Antagonists to the nucleotides and other components of RNA, which slowed down its synthesis, also decreased the stimulant effect of the addition of normal components. All this led the author to the conclusion that the incorporation of amino acids into proteins is associated with the synthesis of RNA or with the renewal of some parts of its molecule.

It has already been mentioned that, according to Gale and colleagues, in Staph. aureus the direct synthesis of RNA from purine and pyrimidine products was only necessary for the production of one adaptive enzyme, β-galactosidase, while the constitutive enzymes only needed ready-made RNA for their formation. This rule, however, does not seem to be generally valid. In another classical subject of microbiological experiments, Escherichia coli (a mutant which does not form thymine), when the synthesis of RNA is impeded by antimetabolites, the synthesis of the proteins of the cellular structures as a whole is suspended (49). The same is true of the growth and synthesis of the total protein of *B. cereus* as Mandel observed (155-157) while studying the influence of 8-azaguanine on the metabolism of such a culture. Apparently the synthesis of many proteins depends on the synthesis of RNA, but in varying degrees. It was shown in a paper by Roodyn & Mandel (197) that the synthesis of the proteins of the cell-wall of *B. cereus*, unlike that of the cytoplasmic proteins of this bacillus, is not impeded by azaguanine.

Spiegelman *et al.* (226) have shown that when proteins and enzymic systems are being formed there is competition between them for "structural materials" and other things required for synthesis. It would therefore seem that the synthesis of new "adaptive" proteins must, at least, be capable of competing. For example, the synthesis of adaptive β-glucosidase by *E. coli* is impeded by the action of ultraviolet radiation which, in small doses, has a greater effect on the metabolism of RNA than on the increase in the total amount of protein. It is a characteristic feature that doses of ultraviolet irradiation, which halve the rate of metabolism of $^{32}P$ in RNA, completely suspend the synthesis of protein in the cells. When antagonists to the metabolism of RNA act on the cell, the synthesis of protein is also brought to a stop but it is said that this effect is at first confined to the formation of adaptive enzymes, which the authors believe to be weak competitors for the precursors of RNA. In the course of studying the adaptive synthesis of β-galactosidase in *E. coli* it has been found that 5-hydroxyuridine, which is an inhibitor of RNA synthesis, also prevents the formation of the adaptive enzyme. Stoppage of synthesis of the enzyme may be achieved even when the inhibitor is added to the medium after the inductor, during the period when maximal rate of formation of the enzyme has been attained and when the cell already contains ready-made molecules of RNA which were formed in association with the synthesis of the first portions of the enzyme and are able to function as ready-made matrices. A characteristic feature of this effect is that the 5-hydroxyuridine can lead to complete inhibition of the formation of β-galactosidase in concentrations which do not affect the formation of the proteins of the cell generally, suggesting that this adaptive biochemical system is particularly sensitive.

Similar results were obtained by Pardee (177) who found that mutant cultures of *E. coli*, which would not grow in the absence of uracil, stop synthesizing adaptive enzymes immediately the supply of uracil in the medium becomes exhausted. The importance of the actual synthesis of RNA for the synthesis of adaptive enzymes and the ultimate dependence of this synthesis on the presence of a supply of low-molecular products of the metabolism of nucleic acids (oligonucleotides, nucleotides and nucleosides, purines and pyrimidines) has been demonstrated many times (70, 71, 127, 169, 185).

However, when we are considering the nature of the interaction between the synthesis of RNA and the synthesis of proteins we must regard it from two different points of view. In the first place we may assume that the actual process of interaction of nucleotides with one another, which occurs during the formation of nucleic acids, is associated in some specific way with the formation of proteins from amino acids and peptides. That is to say, the synthesis of proteins and nucleic acids is a single integral process. On the other hand it is not impossible that the syntheses of molecules of RNA and proteins are not involved with one another at the low-molecular level and that the synthesis of proteins depends mainly on some active fractions of highly-polymeric RNA. The rapid breakdown of these fractions as a result of renewal requires the continual recreation of active RNA. There may, at the same time, be fractions of RNA within the cell which are inert in respect of protein synthesis and which are renewed comparatively slowly and this may give certain authors the impression that the pre-existing RNA, which has already been formed, is inert and that it is not the RNA itself but the process of its formation which is necessary for protein synthesis.

Although the experiments described above give a clear picture of the association between the synthesis of protein and that of RNA, they still do not give an exhaustive proof that what is important in this matter is the synthesis itself and not its final, high-molecular, specific product, which must always be replaced when it breaks down. In this connection we should like to draw attention to a number of facts of a somewhat different kind, which constitute more definite but still not absolutely convincing evidence for the existence of an interdependence between the synthesis of protein and that of RNA.

It has been shown in a number of papers (48, 76, 122, 167, 178) that, in the presence of certain inhibitors of protein synthesis (e.g. chloramphenicol), the cells of E. coli continue to form RNA and DNA actively without a corresponding accumulation of protein. When the inhibitor is removed and the cells suspended under favourable conditions, the synthesis of proteins does not begin again at once but only after the breakdown of the RNA and DNA which have just been formed (this was shown by labelling them with $^{14}$C and $^{32}$P in the period of unbalanced synthesis). Thus the ready-made RNA and DNA were not in a state to stimulate the synthesis of new proteins, and both processes had to go on in parallel. It is, however, not out of the question that the fractions of RNA and DNA formed under these conditions were different in their specific biochemical properties and just because of this difference they were functionally inert.

In this connection we must refer to an interesting communication (53) setting out the results of investigations of the biochemical properties of the RNA which accumulates in the cells of E. coli in the absence of a corresponding synthesis of proteins. This RNA does not differ in overall nucleotide composition from the rest of the RNA but the rate of incorporation of $^{32}$P into it in the course of its renewal was 20 times slower than in ordinary RNA.

Ben-Ishai & Volcani (49) obtained some very interesting material on the association between the synthesis of RNA and that of protein. In their paper these authors point out that, according to many workers, DNA is not directly associated with the formation of the cytoplasmic proteins and these can be formed in the absence of any new synthesis of DNA. They hold, however, that the synthesis of RNA is directly associated with protein synthesis. In order to study the interaction between RNA and proteins without the participation of DNA these authors carried out a series of experiments with mutant cultures of *E. coli* which cannot synthesize thymine. In the absence of thymine in the medium the synthesis of DNA is completely stopped while the cells retain, to a considerable extent, their ability to form RNA and proteins. Under these conditions it was found that stimulation of the synthesis of proteins was accompanied by a corresponding increase in the rate of synthesis of nucleic acids and *vice versa* in such a way that the ratio RNA:protein was maintained at a constant value of 1:4. By testing different inhibitors of the synthesis of proteins and RNA it was found that inhibition of the process of protein synthesis does not necessarily lead to cessation of the synthesis of RNA though inhibition of the new synthesis of RNA by means of analogues of purine and pyrimidine bases is always accompanied by a simultaneous suspension of protein synthesis, though the percentage of RNA in the cell does not fall. On the basis of these results the authors also come to the conclusion that the actual synthesis of RNA is necessary to ensure the simultaneous synthesis of protein.

The establishment of a correlation between the synthesis of proteins and that of nucleic acids is, however, only the start of the study of this problem. The essential part of the task is the establishment of the exact way in which these processes are associated with one another and whether the processes of synthesis of RNA are responsible for the reproduction of the specificity of the proteins, that is, whether they play any part in the transfer of functional "current" and hereditary information determining the sequence of amino acid residues in the polypeptide chains.

All the experiments described above still do not give enough proof of the importance of the actual synthesis of RNA for the determination of the sequence of the amino acid residues in the protein chains. The greater part of these experiments are concerned with the total RNA of the cell and not just with the RNA of the cytoplasmic organelles. However, all these results show quite clearly that the dependence of protein synthesis on RNA is complicated and many-sided. In fulfilling its function in the synthesis of protein the RNA of both the organelles and the soluble fraction undergoes metabolism and this metabolism has a direct relationship with the synthesis of proteins.

These facts, however, still do not prove the idea of an integral complex

PB H

synthesis of nucleoproteins. In most of these experiments the ready-made RNA failed to bring about protein synthesis only in cases in which the RNA had been formed under abnormal conditions. On the other hand, the synthesis of protein without an accompanying synthesis of RNA only failed to occur when the synthesis in question was that of a "new" protein such as an adaptive enzyme for which there might not be a sufficient supply of matrices within the cell, or when it was the overall synthesis of protein under conditions in which growth was being stimulated when the matrices which were present might not have had sufficient "power" to fulfil the requirements of the cell. It is apparently but not certainly the case that, for the actual synthesis of a particular protein it is not merely the synthesis of RNA which is important but the synthesis of a particular, specific RNA. This has been demonstrated most convincingly in the interesting work of Chantrenne's laboratory (73). The authors studied the relationship between the synthesis of proteins, RNA and DNA in the cells of

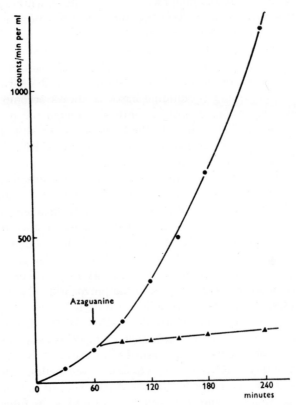

FIG. 18. Inhibition of the incorporation of [14][C]-L-phenylalanine into protein.

●—●—● without the inhibitor.
▲—▲—▲ with the addition of 35µg azaguanine to 1 ml. for 60 mins. during growth of culture.

*B. cereus* in a medium in which 8-azaguanine was present instead of ordinary guanine. In some organisms this analogue of guanine inhibits the synthesis of RNA. There are, however, some biological objects, including *B. cereus*, in which 8-azaguanine is incorporated in the synthesis of RNA in place of guanine. In such cases the quantitative synthesis of RNA is not suppressed, it is even stimulated, but in this case a somewhat abnormal RNA is formed which cannot, as the experiments show, bring about normal protein synthesis.

The authors succeeded in obtaining RNA in which nearly 40% of the guanine was replaced by 8-azaguanine. When this RNA was present the synthesis of various proteins was completely stopped. Figs. 18 and 19 give a graphic representation of one such experiment.

These authors put forward the somewhat tentative suggestion that there is some single mechanism within the cell for the synthesis of both RNA and proteins but they make no attempt to suggest what this mechanism may be in biochemical terms.

In assessing this material it must, however, be pointed out that it

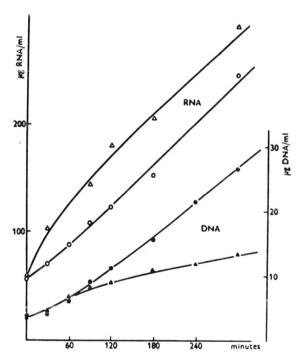

FIG. 19. The effect of azaguanine on the synthesis of nucleic acids.

RNA ○ — ○ — ○ without inhibitor.
    △ — △ — △ with the addition of 36 μg of azaguanine to 1 ml. before
              the beginning of growth (t = o).
DNA ● — ● — ● without inhibitor.
    ▲ — ▲ — ▲ with the addition of 36 μg azaguanine (t = o).

only indicates that RNA is necessary for the synthesis of proteins and that, being necessary, it must be formed and the new synthesis of it must make up for the inevitable metabolic losses. Not one of the experiments discussed proves that there is a common mechanism for the reproduction of proteins and of nucleic acids. In a number of recent papers concerned with the synthesis of proteins and RNA in very different objects (muscle (256), *E. coli* (179) and *Pseudomonas azotogena* (117)) it has been clearly shown that the formation of a particular fraction of RNA is important for the synthesis of new protein. It is not of decisive importance whether the synthesis of the RNA is simultaneous with or precedes the synthesis of the protein.

All these results, those concerned with viral proteins and the proteins of the adaptive enzymes and also many others, show, in most cases, that the synthesis of proteins requires the presence of particular specific forms of RNA. This circumstance would most plausibly suggest that the sequence of amino acids in the proteins being synthesized, which determines their specificity, is determined by the matrices on which they are synthesized, which are localized in the cellular organelles.

To solve the problem of protein synthesis, however, it is important, not merely to establish the connection and interdependence between the synthesis of specific proteins and the existence of specific matrices, but also to discover the mechanism of this connection. One of the most important approaches to this problem is the factual study of the synthesis of proteins by the various cytoplasmic organelles.

## 3. The synthesis of proteins in the microsomes and ribosomes

### *a) Biochemical characterization of the microsomes and ribosomes*

The microsomes are the finest and most "elementary" cytoplasmic inclusions and have many other names in the biochemical and cytological literature. These reflect, to some extent, the development of people's ideas about the structure of the cytoplasm (cytoplasmic granules, ultranetwork, ultramicrosomes, endoplasmic reticulum, ergastoplasm). Recent studies have shown, however, that the microsomes, which biochemists precipitate by differential centrifugation from cellular homogenates, constitute a mixed fraction consisting of rounded ribonucleoprotein particles and a proteinaceous membranous material which may be separated from the particles by treatment in various ways. These ribonucleoprotein particles have recently acquired a new name, that of ribosomes.

In what follows we shall use the term "microsomes" when the fraction is being studied as a whole and the term "ribosomes" when it is the ribonucleoprotein particles, freed of membranous material, which are being studied.

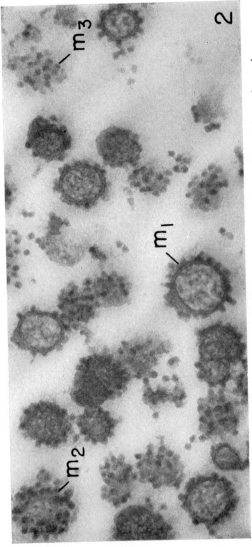

Fig. 20. Electron micrograph of part of cell with microsomal particles. According to the thickness of the sections, one may compare the structures of the microsomes ($m_1$, $m_2$, $m_3$) and observe the microsomal membrane. The small dark spheres present in the microsomes are ribosomes ($\times$ 116,000).

Although it is certainly true that the various organelles of the cells can synthesize proteins locally, yet, in most of the comparative studies of the synthetic activities of intracellular structures, special attention is given to the synthesis of proteins by the microsomes. The important and specific role of the microsomes and ribosomes is seen most clearly in the study *in vivo* of the synthesis of secreted proteins which are not used for the growth of the tissues in which they are made (digestive enzymes, proteins of blood and milk, hormones, etc.). This would seem to show that the ribosomes have a specific function. As the functions of the nucleus and chromosomes are determined to a significant extent by the part they play in the phenomena of heredity and as the functions of the plastids are determined by their part in the phenomena of photosynthesis and the functions of the mitochondria by theirs in the carrying out of glycolysis, oxidation and phosphorylation, so the granular ribosomal apparatus must supply the cell with the proteins which function in solution in the intergranular cell sap, in the lymph, plasma, secretions of the digestive glands, in the vacuoles of protoplasm and other such media in which the autonomous power of reproduction is slight or absent owing to their homogeneity. On this account there often falls on the microsomes a twofold, threefold or sometimes even tenfold burden of protein synthesis, as they not only have to reproduce themselves in the course of self-renewal and growth, but must also reproduce the whole mass of soluble protein which is formed by the organ, including that which circulates in the tissues.

We may get an idea of the scale of this synthesis from the rather rough calculation that if the whole mass of the protein secreted by, for example, the liver or the pancreas were used for the growth of the organ, then its weight would be doubled every few days. The specialization of the cytoplasmic ribosomes of certain organs for bringing about a very active protein synthesis makes these organelles the most interesting subjects in which to study the mechanisms and conditions of this synthesis without their being masked by the other complicated functions performed by, say, the nucleus, mitochondria or plastids.

In the first place, however, we must spend a little time on a number of facts concerning the nature of the microsomal fraction of cytoplasm. The dimensions of the microsomes vary between 50 and 150 m$\mu$. In biochemical studies, particles which are precipitated by centrifugation at about 100,000 $g$ after preliminary removal of the plastids and mitochondria are regarded as microsomes.

In order to obtain a more uniform fraction of ribosomes from this precipitate, it is treated with various detergents and then centrifuged again, the insoluble ribonucleoprotein particles, or ribosomes, being precipitated. The structural relationship between the microsomes and the ribosomes is to be seen very clearly in the electron micrographs made by Siekevitz & Palade (213) (Fig. 20). This picture shows rounded aggregates of fine, round particles. These are the microsomes. Under the influence of detergents they disintegrate and the ribosomes (the fine, round particles) enter the "free" state.

There is fairly extensive evidence as to the morphology, localization and internal structure of the microsomes and ribosomes and, judging from electron-microscopic observations, this group of particles is heterogeneous in many ways (7, 50, 114, 126, 176, 219, 241).

Very comprehensive surveys of the literature concerning the structure and synthetic activity of the microsomes and ribosomes have recently been published by Palade (175), Campbell (67), Hoagland (129) and Aronson et al. (44), and this allows us to make a concise summary of the essential features of this problem taking several new investigations into consideration.

Investigations in many laboratories have shown that, in the cytoplasm of cells of the most varied types, there is a heterogeneous population of ribosomes in the form of particles of various sizes. The relationship between the different groups of ribosomes is very dynamic and in different physiological states one may observe both dissociation and association of ribosomes of different types. Dissociation of ribosomes depends, to a considerable extent, on magnesium ions. When the concentration of $Mg^{++}$ is adequate, ribosomes from different sources have a sedimentation constant within the range 70-80 S which corresponds with a molecular weight of $3-4 \times 10^6$. A lowering of the concentration of $Mg^{++}$ from 0·005 M to less than 0·001 M leads to dissociation of the particles. According to Ts'o, Bonner & Vinograd (243) the content of bivalent ions in the intact particles is very high, being nearly 4 moles of ion per mole of nitrogenous base.

In E. coli, particles with a sedimentation constant of 70 S would seem to be the most common synthesizing structures. If the concentration of $Mg^{++}$ is decreased they break down into particles with sedimentation constants of 30 S and 50 S (the ratio of the molecular weights is 1:2·5) (143, 168, 239).

On further dissociation of the particles fragments with a sedimentation constant of 15 S are formed. These are pure RNA.

From one particle with a sedimentation constant of 70 S there may be produced one particle with a sedimentation constant of 50 S and one with a sedimentation constant of 30 S, the reaction being reversible.

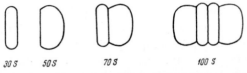

30 S    50 S    70 S    100 S

FIG. 21. Approximate form of the "dimer" and of the sub-units formed from the ribosomes of E. coli by changing the concentration of $Mg^{++}$ ions.

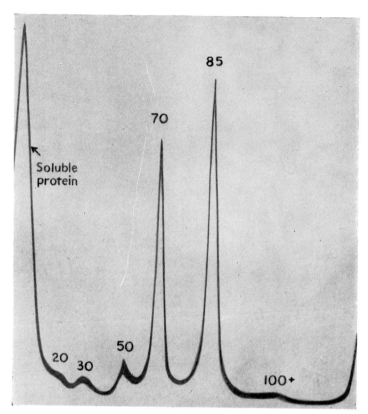

Fig. 22. Ribosome pattern of *E. coli* growing exponentially in broth medium as shown in the analytical ultracentrifuge.

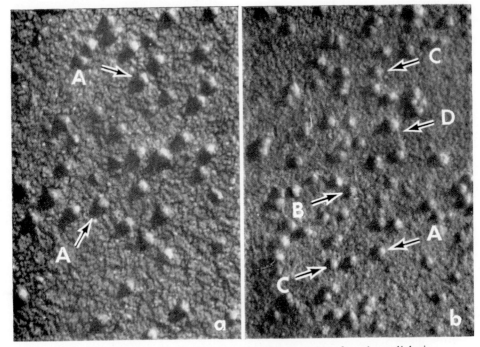

FIG. 23. (a). Electron micrograph of RNP particles after short dialysis from buffer into distilled water. Most of the particles are the 70 S doublets. The components of the doublets are readily seen at A. Magnification × 120,000

(b) Electron micrograph of RNP particles after long dialysis from buffer into distilled water. 50 S particles are visible at A and 30 S particles at B. Fibrils being released by the 50 S components can be seen at C. The 50 S and 30 S particles slightly separated and joined by a fibril are shown at D. Magnification × 120,000

Webster (253a) has put forward the following scheme of the structural relationships between the ribosomes of E. coli which have different sizes and different dissociation constants (Fig. 21). Particles with sedimentation constants of 30 S contain molecules of RNA with a lower molecular weight than particles with a sedimentation constant of 50 S (143).

Abdul-Nour & Webster (42) have recently obtained interesting results which suggest that the aggregated state of the nucleic acid particles is necessary to their synthetic activities. From homogenates of shoots of oats these authors isolated 80 S ribonucleoprotein particles which could synthesize proteins in vitro. When these particles had been broken down into 60 S and 40 S components the incorporation of [14C]glycine into the protein of these particles ceased, but regeneration of the particles by means of $Mg^{++}$ or cobalt restored the process of incorporation of amino acids to its earlier level.

The ratio between ribosomes of various sorts is a very important indicator of the state of the cytoplasm. Fig. 22 shows the character of the sedimentation diagram of the ribosomes in growing cells of E. coli as found in Roberts' laboratory (196). According to Roberts, 70 S particles are the most active in synthesizing proteins though finer particles incorporate isotopes more quickly in experiments with labelled amino acids. McCarthy (159) has recently shown that, in resting cells of E. coli, coarse (100 S and 85 S) components predominate. Components of this type are also found to accumulate when cells stop growing owing to a shortage of sources of energy. However, when glucose is added and the cells begin to grow quickly, it is seen that the coarse ribosomes are quickly replaced by smaller ones with sedimentation constants of 70, 50 and 32 S.

The aggregated structure of the coarse ribosomes of E. coli is clearly demonstrated in electron micrographs. The electron micrograph shown in Fig. 23 was obtained by Beer, Highton & McCarthy (47). It gives a striking picture of the ribosomal material of the cytoplasm of these cells.

Some very interesting material concerning the relationship between the structure of the microsomes and ribosomes and their biochemical functions has been put forward in a number of papers by Tashiro et al. (234-238). These authors propose a number of interesting schemes for the structure of microsomes according to which the microsomes, when in the cell, are double lipoprotein membranes on which there are, in various places, small, denser inclusions of a ribonucleoprotein nature which constitute the ribosomes. Studies of the separation of the RNA and protein in the ribosomes, when they are incubated with ribonuclease or trypsin or treated in other ways, have shown clearly that in the ribosomes the RNA is on the surface of the particles, which is fully consistent with the function of this RNA as a matrix for protein synthesis. In these particles the RNA is combined with the protein by hydrogen and electrostatic bonds. The authors note that in young embryonic tissues the ribosomes often occur in the cells partly in the free state.

Similar results have also been obtained recently by Slautterback & Fawcett (quoted in 129). Working on the development of the haematocyst of *Hydra* they found in undifferentiated cells only "naked" ribosomes distributed at random in the cytoplasm while, during the process of differentiation of the cells, they observed the formation of lipoprotein membranes and the organization of the ribosomes into their characteristic reticular system. It is interesting to note that ribosomes were also found within the nucleus (cf. Chapter XIII).

Palade & Siekevitz (209-213) have recently obtained interesting evidence that the membrane of the microsomes is composed of active points for the concentration and accumulation of the specific protein products synthesized in the ribosomes.

According to Webster (248) and Cohn (77) the ribosomes of plant and animal cells, freed from membranes by detergents, retain a residuum of the general amino acid-activating activity. In the microsomes they also found a certain amount of acceptor, low-molecular RNA (about 10%) (224) which would seem to give them a certain autonomy in synthesizing proteins.

The close similarity of the amino acid compositions of ribosomal proteins isolated from different tissues has also attracted attention (83, 242, 245). In this respect the protein of ribosomes differs from other cytoplasmic proteins by its high content of the basic amino acids lysine and arginine, being thus reminiscent of the nuclear proteins the histones (65a, 83, 254a). One may imagine an interaction of RNA and protein in the ribosomes similar to that of DNA and histone. At the same time there are definite differences of amino acid composition between the proteins forming the 50 S and the 30 S ribosomal particles of *E. coli* (228a). The significance of the membranous material in the actual synthesis of protein still remains unclear. Hoagland (129) suggests that the membranous part of the microsome may be necessary for the final stages of the specification of the protein being synthesized and for the release of the finished molecules from the points at which they were originally synthesized. The disruption of the cell may disturb these relationships. Hoagland considers that this idea is supported by many facts, the obligatory presence of lipids for protein synthesis *in vivo* (128), the dependence of the transport of proteins across endoplasmic membranes on energy donors (130) and also the results obtained by Sachs (200) and Campbell *et al.* (69) which can be interpreted as an indication that the proteins which have been synthesized by the ribosomes find it difficult to pass into the dissolved state after the particles of the ribosomes have been fractionated.

Bock & Yin (54) showed recently that 82 S ribosomes have a molecular weight of the order of $4 \cdot 7 \times 10^6$. According to the results of these authors the protein part of these ribosomes consists of 42-50 sub-units forming a regular polyhedron.

The observations of Elson & Tal (98) are also very interesting. They show that when 70 S ribosomes are broken down to give particles with constants of 50 S and 30 S the latent ribonuclease only remains associated with the 50 S particles.

Spirin and colleagues (28a) have made interesting observations on reversible unfolding of ribosomal particles into flexible ribonucleoprotein strands.

### b) The synthesis of proteins in the microsomes and ribosomes of animal tissues

As concerns present-day material about the synthesis of proteins in the microsomes and ribosomes it should be noted that the actual study of the biochemical systems by which amino acids are incorporated into the proteins of the microsomes, which was started by Zamecnik, Hoagland and their colleagues in 1953, later led to the discovery of the activating enzymes and soluble RNA in the cell sap.

In 1953-54 it was already known, from very many series of studies on the comparative rates of incorporation of labelled amino acids into the proteins of different fractions of cells, that, in tissues with intensive protein metabolism, this incorporation takes place most actively into the microsomal fraction. Zamecnik and colleagues again confirmed this phenomenon by a detailed study, and thus discovered a number of important features of it (136, 152). In these experiments they studied the incorporation of labelled amino acids into the proteins of the liver of rats over short

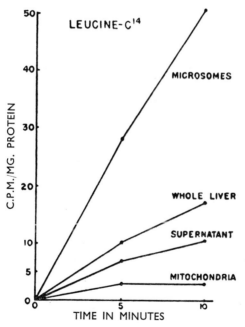

FIG. 24. Incorporation of $^{14}$[C] leucine into the protein fractions of the liver.

periods (1-10 min.) after the administration of the amino acids into the blood-stream. Analysis of the fractionated material after such short periods of time enabled the authors to detect the place in which synthesis first took place although, in experiments which involve more time, one cannot rule out the possibility that the radioactive proteins, which were originally synthesized in one place may later be found in another. [$^{14}$C]-Valine and [$^{14}$C]leucine were used for the injections because these amino acids are not converted into others but are incorporated into proteins as such. The results of one of these experiments are shown in Fig. 24.

Calculation showed that 7 minutes after the injection 70% of the radioactivity of the injected amino acids was concentrated in the microsomes, which contain only about 25% of the protein of the cell. Such a rate of synthesis corresponds quantitatively to the formation of 13 g. of new protein for each 100 g. of total protein of the liver in 24 hours. The bulk of this synthesis represents the plasma proteins secreted by the liver.

Having established that protein synthesis is predominantly carried out by the microsomal fraction Zamecnik and his colleagues tried to find out how the amino acids which had been "incorporated" into the proteins were distributed in the various biochemical fractions of the microsomal particles. Preliminary preparative studies by the authors had shown that treatment of microsomal material with sodium desoxycholate at a particular concentration leads to the separation of the material of the microsomes into two fractions, one of which is soluble and the other insoluble in desoxycholate. The soluble fraction accounts mainly for the proteins of the membranes (about four sixths of the total protein of the microsomes) while almost all the RNA and about one sixth of the total protein of the microsomes remain in the insoluble fraction. A study of the insoluble

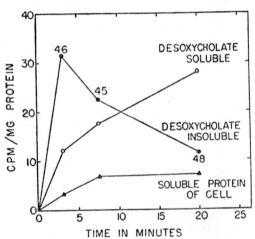

FIG. 25. Incorporation *in vivo* of small amounts of $^{14}$[C]leucine into two components of the microsomes and into the soluble proteins of the cell.

fraction by means of the electron microscope showed that it was composed of fairly uniform particles having a mean diameter of 240 Å. These particles are now called ribosomes. The RNA content of the particles varied between 42-50%. Analysis of the incorporation of labelled amino acids into each of these fractions of the microsomes of the liver, in experiments carried out *in vivo*, showed that in the first minutes after injection the incorporation of isotopes occurs mainly into the ribonucleoprotein particles which are insoluble in desoxycholate. The curve representing this process rises to a sharp maximum in a very short time and then begins to fall while there is a simultaneous increase in the activity of the other fractions of the microsomes and of the soluble proteins of the tissues (see Fig. 25). Incorporation of this type was also found after injection of large doses of labelled amino acids which ensured the maintenance of a constant intracellular specific activity over a long period (injection of 100 $\mu$moles of [$^{14}$C]leucine). In this case the sharp peak and rapid maximum of incorporation into the insoluble fraction was not associated with a subsequent fall of its specific activity but the general character of the curves does not differ in principle. The maximal incorporation of labels into this fraction after short intervals of time was also observed *in vitro*.

On the basis of these findings the authors built up the idea that the cytoplasmic ribonucleoprotein particles are the points at which free amino acids are first incorporated during protein synthesis. If this is correct the very rapidly occurring "ceiling" of specific radioactivity of the protein suggests that only a small part of the protein of the particles (about 1% according to the authors' calculations) takes part in the rapid reactions of synthesis and metabolism.

Similar results were obtained by the same group of authors (151) in studies of the synthesis of proteins in neoplastic tissues. In these it was noticed that the incorporation of amino acids into proteins in the first minutes is not an easily reversible reaction. The isolation of "labelled" particles and their incubation in complete mixtures containing unlabelled amino acids did not lead to a decrease in the specific activity of the proteins of the particles. It would therefore seem clear that what was observed in the experiments of Zamecnik's group was not "metabolism" but a true synthesis of new molecules of protein.

The active role of the microsomes and ribosomes in the synthesis and secretion of proteins has also been demonstrated by the work of many other laboratories.

An interesting mechanism stimulating the synthesis of protein by the microsomes of slices of liver has been found by Ziegler & Melchior (257). By incubating the liver slices in various solutions in the presence of labelled methionine the authors found that maximal incorporation of methionine into the proteins occurred in the microsomal fraction. The rate of this incorporation was regulated by the concentration of proteins in the surrounding solution. When this concentration was lower the rate of protein synthesis increased. This seems to be the mechanism which underlies the regulation of the concentration of the plasma proteins of the blood.

Many other pieces of research carried out both *in vitro* and *in vivo* may be adduced in which it has been shown in recent years that the microsomes are capable of active synthesis and secretion of soluble proteins (68, 191, 201, 214, 215). Now, however, it is important, not merely to demonstrate this ability of the microsomes actively to synthesize protein, but to try to find out the mechanism of this synthesis and its connection with the biochemical system of activation of amino acids in the intergranular medium. The study of this mechanism and the intramicrosomal localization of the synthesis of proteins was begun in the experiments of Zamecnik and his colleagues which have been discussed above and, like their experiments on activation, was soon extended and expanded in many biochemical laboratories. Various methods of fractionation of the microsomes of various experimental objects were used for this purpose. For example, Cohn & Butler (78) tried to fractionate the microsomes biochemically with several solvents other than desoxycholate which had been used by Zamecnik and colleagues. Good results were obtained by the use of non-ionic detergents, especially Lubrol, which is a condensate of polyoxyethylene. Treatment of the microsomes with solutions of this substance also led to the solution of about half of the protein, while the insoluble fraction retained all the RNA of the microsomes. In the first 6-8 minutes after injection of [¹⁴C]phenylalanine in *in vivo* experiments, the amino acid was incorporated into the insoluble fraction at maximal speed, while in later periods the soluble fraction of the protein began to rise towards a maximal level of activity. In other experiments the authors carried out further fractionation of the particles which were not soluble in Lubrol, ridding them of a significant amount of their proteins. The protein which remained in the insoluble precipitate was associated with RNA and, in this case it had a very high specific activity a short while after the injection of the isotopically labelled amino acid.

Theoretically similar conclusions as to the localization of synthesis in the microsomes were obtained by the work of Simkin & Work (216, 217) who used other methods of fractionation of the microsomal material of the liver. They used solutions of varying ionic strength and pH for this purpose.

Different methods of fractionation of the microsomes of animals, plants and microorganisms, with subsequent separation of the protein fractions having the greatest activity, has also been carried out in a number of other investigations (88, 132, 165, 182, 202, 207, 248, 140, 171, 233) and their results confirmed the main conclusions formed in Zamecnik's laboratory to the effect that this synthesis took place fastest in the fraction of the microsomes which was most closely bound to high-molecular RNA (ribonucleoprotein particles or ribosomes).

Extremely interesting information as to the nature of the incorporation of amino acids into the proteins of the microsomes has been obtained recently by Schweet *et al.* (206). These authors made simultaneous studies of the processes of activation of several amino acids in the precipitated fraction from the reticulocytes of rabbits and the incorporation of activated amino acids in the synthesis of haemoglobin. The formation of this protein is the main function of the microsomes of the reticulocytes, and after its formation it quickly becomes soluble. It is interesting that although valine, isoleucine and leucine were activated and accepted by the soluble RNA at the same rate, they became incorporated into the proteins of the microsomes at different rates, which were proportional to the amounts of each occurring in the haemoglobin (Table 8).

TABLE 8

Incorporation of amino acids into haemoglobin and composition of the haemoglobin in rabbit reticulocytes (206)

| Amino acids | "Incorporation" | | Composition | |
|---|---|---|---|---|
| | counts/min./ mg. | Proportion | Amino acid content (% by weight) | Proportion |
| [14C]Leucine | 748 | 1·00 | 11·7 | 1·00 |
| [14C]Isoleucine | 90 | 0·12 | 1·3 | 0·11 |
| [14C]Valine | 538 | 0·72 | 8·1 | 0·77 |

The composition of the total protein of the microsomes of the reticulocytes is different (leucine 8·7, isoleucine 5·7, valine 7·2%) (242) and microsomes isolated from the most varied sources (liver, plants, etc.) were also found to have the same amino acid composition. This indicates that the "incorporation" which occurred in the experiments under discussion was actually into the haemoglobin and was determined by the specificity of the synthesizing systems localized in the microsomes.

This result thus supports the suggestion that the determination of the sequence of the amino acids in protein synthesis occurs in the ribosomes under the influence of their RNA and that systems of this synthesis are, as it were, centres of attraction for the activated products which are formed in the intergranular medium. It would be possible to give an account of many other interesting recent experiments in which the process of incorporation of amino acids in the synthesis of protein in the microsomes and ribosomes has been studied using several other methods of fractionation and on preparations isolated from various tissues and organs (65, 92, 188, 192, 193, 231, 212, 213).

Detailed work on the process of incorporation of labelled amino acids *in vitro* into ribonucleoprotein particles isolated by means of detergents from various tissues has recently been carried out by Rendi & Hultin (194).

These authors made a detailed study of the conditions of the medium which are necessary to ensure maximal synthetic activity on incubation of the ribosomes. It is interesting to note that the absence of S-RNA did not stop the process of incorporation of amino acids in the synthesis of proteins, but only slowed it down. The same feature of the process was also revealed by Korner's experiments (139). The fact that S-RNA is not absolutely necessary for the incorporation of amino acids into proteins may be the result of two factors, in the first place, the presence of molecules of S-RNA in the ribosome, as mentioned above. In the second, it is possible that there are several ways in which amino acids are incorporated during the synthetic process. This second possibility will be discussed in Chapter XII.

Ts'o and colleagues (244a, 245a) made some interesting observations on the manner of synthesis of protein by the ribosomes of rabbit reticulocytes. Here the chief role in synthesis of haemoglobin is played by "complex" 78 S ribosomes. *In vitro*, at low concentrations of magnesium, these dissociate into 60 S and 40 S fragments. On *in vitro* incubation of [$^{14}$C]valine with 78 S ribosomes in the presence of the necessary co-factors, the labelled amino acid was rapidly incorporated into the protein of both components of the 78 S ribosomes. However, if, shortly after incubation with [$^{14}$C]valine these ribosomes were dissociated into 60 S and 40 S fragments and then re-associated, about 20% of the radioactivity appeared in the form of peptides containing valine having specific radioactivity 4-7 times that of the valine of the ribosomal protein. The authors consider that these peptides are "uncompleted blocks" for protein synthesis, formed in the groove between the 40 S and 60 S ribosomes.

Also of interest for the study of *in vitro* synthesis of proteins by microsomes and ribosomes is some work of Campbell & Kernot (69a) and Korner (140a). These authors found that the distribution of labelled amino acids in the peptide chains of serum albumin was the same whether it had been synthesized in the liver *in vivo* or by liver microsomes or ribosomes *in vitro*.

After completion of protein synthesis, the protein molecules have to be set free from the ribosomes, if the process is to continue. Studies of this stage are still very inadequate. It has been established that there must be present an energy source (ATP), $Mg^{++}$ and cell sap or some protein-rich fractions thereof (132a, 215a). It has been postulated that these last contain a special "liberation" enzyme (253a).

*c) The synthesis of proteins in the ribosomes of microorganisms*

Ribosomes, as structural elements of the microsomal fraction, have also been found in microorganisms (161) but, to judge from some of their properties, they are more closely associated with membranous structures in microorganisms than in other organisms. The intensive incorporation of amino acids into these structures has been established in a number of laboratories (80, 164, 173, 203, 240, 255). Extremely active incorporation has been observed in the membranous material in *Bacillus megaterium* (66, 133) but this material was evidently closely associated with ribosomal structures.

An extremely detailed investigation of the characteristics of the synthesis of proteins and RNA by the ribosomes in *E. coli* has recently been carried out in Roberts' laboratory (44, 196). Tissières, Schlessinger & Gros (240) have also made an intensive study of the synthetic activity of the ribosomes of *E. coli*. In their study of ribosomes of different sizes in this organism the authors found that among all the types of ribosome there was one special type of particle with a sedimentation constant of 70 S which they called the *active* 70 S particle.

Labelled amino acids were incorporated into these *active* 70 S particles 15-40 times as fast as into the 50 S and 30 S ribosomes. The *active* 70 S ribosomes did not constitute more than 10% of the total number of ribosomes. In the artificial system used in these experiments these ribosomes were neither formed nor broken down into their 50 S and 30 S components. After 40-50 minutes these ribosomes became "saturated" and could neither incorporate amino acids nor set free proteins which had already been made. The authors suggest that *in vivo* there is some factor present which brings about the liberation of the labelled products of synthesis and, in some way, makes the ribosomes active again.

The authors consider it possible that the amino acids which have been incorporated by the 70 S particles sink into a groove between the 30 S and the 50 S components of this ribosome and that, to liberate the polypeptide chain, both parts of the 70 S ribosome must move away in different directions. After this they may reunite to form a new active 70 S particle under the influence of some activator.

These interesting observations and theoretical considerations are drawing attention to the biochemical concept of the phenomena of the aggregation and disintegration of the ribosomes. Such processes would appear not to be artefacts but to be essential links in the mechanism of protein synthesis. In this connection some other investigations of the conditions and characteristic features of the transitions 30 S ⇋ 50 S ⇋ 70 S ⇋ 100 S of the ribosomes are attracting a fresh interest and it would seem that direct evidence will soon have been obtained as to the connection between these cycles and the synthesis of polypeptide chains (56, 87, 116). According to the most recent evidence, the "active state" of the ribosomes depends on their association with "messenger" RNA (see Chapters XV and XX).

### *d) The synthesis of proteins in the ribosomes of plants*

A detailed characterization of the microsomes and other organelles associated with growth and differentiation of plants is set out in a comprehensive review by Setterfield (207).

A number of studies of protein synthesis in microsomes *in vitro* have been made on vegetable materials.  By comparing the ribosomes of pea shoots with ribosomes from the liver in respect of factors necessary for the incorporation of amino acids in protein synthesis, Webster (249) has reached the conclusion that this process is similar, in principle, in animals and plants.  In recent investigations Webster (251) and Raacke (180, 181) found that, in microsomes isolated from pea shoots and incubated *in vitro* in a suitable medium, one can not only observe the "incorporation" of amino acids into proteins but also a perceptible increase of about 10% in the amount of protein present.  This entirely confirms the synthetic nature of the process of incorporation in experiments *in vitro*.  Webster showed that the proteins synthesized by the ribosomes *in vitro* are specific proteins and have enzymic properties (252, 253).

However, it must be noted that Lett & Takahashi (146a) were unable to confirm the large net synthesis of protein by ribosomes *in vitro* claimed by Webster and Raacke.  These authors conducted 30 experiments by the procedure of Webster involving the ribosomes of a number of varieties of pea, but failed to demonstrate net synthesis of protein.  They considered that Webster had observed not synthesis but liberation of ribosomal protein from the bound state.  A further series of experiments was conducted in Webster's laboratory (253a) with various varieties and strains of pea.  It was found the ribosomes of most peas could not bring about net synthesis of protein, but that some lines produced ribosomes very active in this respect.  The authors postulate the existence of some factor stimulating protein synthesis which only passes to the ribosomes in particular varieties of pea.

The work of Ts'o & Sato (244) has demonstrated that the greatest incorporation of labelled amino acids into proteins takes place into just that fraction of the proteins of the ribosomes which is most closely associated with RNA.

The work of Lund (153) demonstrated the synthesis of aldolase by microsomes of shoots of maize *in vitro* in the presence of a mixture of 17 amino acids, ATP, GTP, $Mg^{++}$ and $K^+$ and pH 5 enzymes.

According to Setterfield and colleagues (208) the proteins of the ribosomes of plants have a very complicated composition.  In the acid-soluble fraction of the ribosomal proteins of pea shoots alone the authors established the presence of 17 fractions which behaved differently on electrophoresis.

## 4. Synthesis of proteins in the mitochondria

Cytoplasmic particles of $0.1$-$7\mu$ are usually referred to as mitochondria.  As a rule they are visible under the microscope.  Typically they, like the nucleus, do not form proteins which are secreted.  In most cases the mitochondria only reproduce their own structure during growth.  In this connection we may note the interesting evidence obtained by

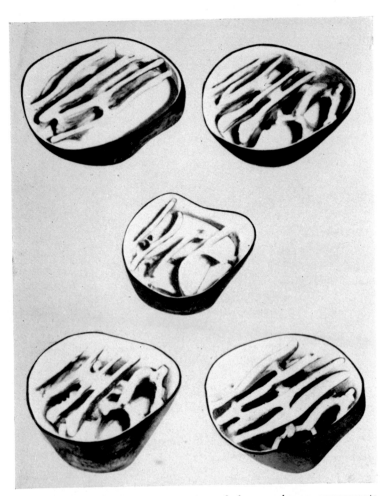

FIG. 26. Three-dimensional reconstruction of the membrane components of a mitochondrion. The reconstruction was made from serial sections. It is divided into three components which are successively piled on top of each other to allow a detailed view of the inner membranes. The gradually assembled model is furthermore demonstrated as viewed from both above and below.

R. B. Khesin (32). This showed that if, under ordinary conditions, the incorporation of labelled amino acids is carried out extremely intensively into the fine cytoplasmic granules of the liver, then, during the regenerative processes, the rate of incorporation into the granules and into the mitochondria becomes more equal. Similar results in respect of plant tissues have been obtained by Szafarz (230).

The comprehensive monograph of Lindberg & Ernster (148) on the chemistry and physiology of the mitochondria and microsomes gives a thorough review of all aspects of the biochemistry of the mitochondria which had been discovered up to 1953. A survey of the materials summarized in this book shows clearly that although secretory mitochondria are sometimes present in particular tissues, yet their essential function consists in bringing about oxidative and glycolytic processes. Detailed surveys concerned with the structure and functions of the mitochondria and other cellular structures are also to be found in other newer publications (5, 39, 35, 51, 117a, 131, 174, 199, 198, 223, 222).

The mitochondria possess a great variety of enzymes and a very complicated microstructure. The electromicroscopic studies of Palade (174), Sjöstrand (221, 222) and many other authors have shown that the microstructure of the mitochondria is characterized by a highly developed system of membranes. These membranes form the external envelope of the mitochondria and a system of internal partitions let into the main body of the mitochondria. Sjöstrand notes that careful observation reveals that the membrane is made up of three layers, the thickness of the membrane and of its layers being the same in all mitochondria within any one cell.

The mitochondrial membrane consists of a double layer of lipid between two layers of protein. The three-dimensional structure of the mitochondria is illustrated in Fig. 26 (43). The amount of nucleic acid present in the mitochondria is insignificant (about 1%) and, as its function is usually connected with the synthesis of protein, this circumstance gives a basis for not believing that the mitochondria participate actively in the synthesis of proteins. Nevertheless the structure of the mitochondria reproduces itself actively during growth and regeneration and there is much evidence that their number is kept up during growth by fission.

*In vitro* the mitochondria of plants assimilate labelled amino acids added to the medium, though less quickly than the granules. This process depends on a system of energy metabolism (246). The ability of mitochondria isolated from animal tissues to synthesize proteins when they are incubated with labelled amino acids has been demonstrated by many authors (33, 36, 82, 119, 120, 160, 186, 192). A. M. Zubovskaya & V. S. Tongur (6) have shown that this ability is retained even after the mitochondria have been destroyed by homogenization in distilled water. In some organs which do not secrete protein, for example muscle, the incorporation of amino acids *in vivo* into the proteins of the mitochondria proceeds almost at the same rate as their incorporation into those of the microsomes (218). However, we still know very little about the biochemical mechanism and structural conditions of protein synthesis in the mitochondria. In almost all of the works referred to above, mention is made of the marked dependence of protein synthesis in the mitochondria on reactions of phosphorylation and on ATP. The mitochondria, being more complicated organelles, are somewhat more autonomous in the synthesis of proteins than the microsomes.

Greengard (119) has succeeded in showing that the mitochondria of the liver depend less on the activity of the cell sap than do the microsomes. While the addition of cell sap to the medium increased the incorporation of [$^{14}$C]leucine into microsomal proteins 15-fold *in vitro*, the rate of incorporation of label into the mitochondria under the same conditions was only increased 2-3-fold.

Bates *et al.* (46) have recently demonstrated the dependence of protein synthesis, in the mitochondria as well as in the microsomes, on the activating system of the supernatant liquid.

Interesting results have also been obtained recently by McLean *et al.* (160). In a study of the synthesis of proteins in the mitochondria of the liver and muscular tissues, they found that this process also requires the activating system of the supernatant liquid. Treatment of the mitochondria with ribonuclease increased the synthetic incorporation of amino acids into their proteins although, in similar experiments with microsomes, the same authors found the ordinary picture of nearly complete inhibition of protein synthesis after treatment with ribonuclease. This paradoxical phenomenon of the stimulating effect of ribonuclease on the incorporation of amino acids into the mitochondria was studied further by Rendi (189, 190) who made a number of extremely interesting observations. He found that there are special ribonucleoprotein particles within the mitochondrion which actively synthesize proteins and are similar to ribosomes. Ribonuclease does not penetrate into the mitochondrion and therefore its stimulating action would appear to be connected with the formation of some degradation product of the RNA of the stroma which stimulates the process of protein synthesis occurring within the mitochondrial particles. The presence of systems for activating amino acids within the mitochondria has recently been demonstrated in the works of Reis *et al.* (187).

Bates, Simpson and their colleagues (46) have studied in detail the process of incorporation of labelled amino acids *in vitro* into the cytochrome *c* synthesized in the mitochondria of the liver and heart. The authors noted that the incorporation of labelled amino acids into a particular, specific protein is the best indicator of the specificity of the process of incorporation. Thus they showed that labelled amino acids really were incorporated *in vitro* into non-terminal loci of the molecules of cytochrome *c*. The process of protein synthesis by the mitochondria did not occur unless the medium contained magnesium ions and ATP or ADP as well as creatine phosphate and creatine kinase. All the other necessary conditions for the synthesis were provided by the mitochondrial structure themselves. However, in a subsequent communication (218a), Simpson and colleagues failed entirely to repeat this supposed biosynthesis of cytochrome *c* in mitochondria.

### 5. The synthesis of proteins in the coarse "intermediate" granules of the cytoplasm

By using differential centrifugation R. B. Khesin (32, 33) divided the cytoplasmic homogenate into three fractions, mitochondria, microsomes and intermediate granules which were called light, large granules. It was found that these granules, isolated from liver tissues, differed from mitochondria in their high RNA content. These results were confirmed by Laird & Barton (144) who also divided the mitochondrial precipitate into two fractions of different specific gravity (light and heavy) and different RNA content. In R. B. Khesin's experiments *in vivo* the incorporation of labelled tyrosine and methionine occurred to far the greatest extent into the intermediate granules. The synthesis in this case was mainly that of albumin by the liver. In these experiments one of the fractions of the albumin of these granules, the so-called "firmly bound albumin" showed a very high specific activity in the earliest period of the experiment. Coarse, light granules isolated from the liver and pancreas displayed great activity *in vitro* and kept their ability to synthesize amylase and albumin when they were incubated with a mixture of amino acids. Active synthesis in these structures *in vitro* has recently been demonstrated with complete reliability by quantitative methods (37).

The evidence obtained in these investigations is of still more importance because what was determined in Khesin's experiments was not overall protein synthesis but the synthesis of individual proteins, albumin and amylase, and this improved his opportunities for localizing their synthesis.

In his monograph (35) R. B. Khesin brings forward the evidence of a number of authors as to the fine structure of the intermediate "coarse" granules. This shows that they consist essentially of elements of ergastoplasm and, it would appear, contain ribosomes. Thus the intermediate granules and microsomes can be assigned, essentially, to the same category of cytoplasmic structures.

### 6. The synthesis of proteins in the chloroplasts (plastids) of plants

The plastids of plants are specific structures which specialize, primarily, in carrying out one of the fundamental biological processes, namely photosynthesis, as well as a whole series of other biochemical processes. We shall pass over, without discussion, the actual process of photosynthesis and its connection with the structural elements of the plastids. We are only interested in the synthesis of proteins, while the photosynthetic cycle mainly covers reactions of a primary nature.

According to electron-microscopic observations (147, 227) the plastids have a very complicated internal structure which is reproduced by the growth and division of "protoplastids" to form plastids after undergoing a series of phases of development (102). It has been found that there is a compact central core surrounded by a multitude of lamellae between which there is a granular substance. Systems of such lamellae form sharply demarcated laminated grana. These grana can form long columns with 20-30 or more grana in each. Lamellar structures may also be found in the stroma of the plastids. Thus in plastids, as in chondriosomes, there is a very highly developed system of surfaces and membranous formations which are capable of growth within certain limits.

When different forms of plant are crossed (e.g. the ordinary and variegated forms of *Oenothera*) their plastids retain their specificity in the hybrids, which is evidence for the continuity of the plastid formations (205). The continuity of the grana has been demonstrated, that is to say, the formation of new grana from the material of the original ones (52). We cannot dwell in detail on a review of such facts concerning the morphology of the plastids. However, even from our cursory acquaintance with the materials in this field, it is clear that the plastids are very complicated formations. It is quite clear that this complexity of structure is, in the first instance, associated with their multifarious physiological functions and does not give us a key to the examination of the mechanisms of protein synthesis itself, though this is a function which the plastids can carry out to a very high degree.

A systematic study of many aspects of the biochemistry of the plastids has been made by N. M. Sisakyan and his colleagues (19-23, 25, 220). In recent years these workers have paid great attention to the problem of protein synthesis in plastids both *in vivo* and *in vitro*. These studies have made it quite clear that the synthesis of proteins in the plastids is connected with systems of energy metabolism and with RNA metabolism.

The influence of the supernatant fraction of the homogenate on the process of incorporation of [$^{14}$C]glycine into the plastids, as revealed in the experiments of N. M. Sisakyan and N. N. Filippovich (26) was, however, rather unexpected. This fraction did not stimulate the synthesis but inhibited it, and this, the authors believe, was due to the presence in the fraction of some sort of inhibitor, the nature of which has not yet been studied. As the supernatant fraction obtained from a homogenate of green leaves includes, not only the substances dissolved in the plasma juice, but also excretory substances from the vacuolar sap, it is obviously quite possible that it contains inhibiting substances. Nevertheless, one cannot exclude the possibility that, as the microsomes were not precipitated out of the "supernatant" fraction in these experiments, the introduction of these microsomes may have led to competition between them and the plastids. At the same time it is quite clear that the activation of amino acids, necessary for protein synthesis, can also take place within the complicated plastid structures.

Stephenson, Thimann & Zamecnik (228) have obtained some very interesting evidence concerning the comparative rates of protein synthesis *in vivo* and *in vitro* in the various cellular fractions isolated from tobacco leaves. These authors carried out two lines of experimental work. In the first place they were studying the introduction of amino acids into the so-called "leaf discs" when they are removed from the leaves and grown on a nutrient medium (like the culture of tissue slices). In the second place they used isolated centrifuged fractions of a homogenate of leaf tissues for their experiments.

In the experiments with the leaf discs it was found that protein synthesis slows down markedly after the discs have been kept in the dark for 60 min. In the light, during the first 60 min. of their exposure, the incorporation of amino acids into proteins was greatest in the plastids while in the dark it was greatest in the microsomes. The microsomes distinguished themselves especially in experiments in which the exposure was short (2-10 min.), which agrees with corresponding results obtained in experiments with material of animal origin. Nevertheless, in experiments carried out *in vitro*, in which labelled amino acids were incubated with isolated fractions of a homogenate of tobacco leaves, the microsomes showed the greatest loss of their ability to synthesize protein while the chloroplasts retained their powers of active protein synthesis for 60 min. One must suppose that the plastids, being more complicated and autonomous formations, could sustain the internal biochemical and energetic regime required for protein synthesis better and longer than the microsomes.

An interesting possibility of studying the local synthesis of protein in the plastids *in vivo* has been found by Deken-Grenson (93). This author has shown that, if detached etiolated leaves of *Cichorium intybus* are exposed to the light, intensive protein synthesis begins in them. It is associated with the transformation of the leucoplasts into chloroplasts. The amount of soluble nitrogen-containing compounds in the leaf falls during the time while it is turning green to almost a tenth, owing to their utilization in the synthesis of protein structures. The synthesis of proteins was also observed in other fractions of the cell (granules, supernatant liquid) but the main bulk of the increase of protein occurred in the plastids, in which grana were seen to be formed in the light. The author considers that light is not itself a necessary factor for the synthesis, but it induces a reorganization of the plastids, that is, it acts like a chemical inductor in the formation of adaptive enzymes. The local synthesis of protein directly in the plastids in these experiments is, however, indubitable, and this must be emphasized as there is a suggestion that pure protein synthesis only occurs in the microsomes, which secrete the proteins, not only outside the cells, but also into the intracellular area so that it may be used to build up specialized structures which are more complicated than the ribosomes.

The selective synthesis of proteins by plastids when they are turning green, discovered by Deken-Grenson, might be a suitable test-case in which to study the effects of various factors on protein synthesis.

The synthesis of the proteins of the chloroplasts may, however, also take place at the expense of the proteins of other fractions under conditions in which exposure to the light is not accompanied by an increase in total protein (61).

The problem of the relative rates of protein synthesis in the various intracellular structures of plant tissues has also been the object of our own studies (13, 14). In these experiments it was shown that, when the plants took in $^{35}SO_4^{--}$ from the nutrient medium, the rates of synthetic incorporation of $^{35}S$ into the proteins of the cytoplasmic granules and into the plastids of the leaves were about equal. In these experiments the greatest activity was found to occur in the proteins of the vacuolar and plasma juices of the leaves although, as was shown later, this effect is determined by the transport of insignificant amounts of highly active soluble proteins from the root system (10, 11).

Racusen & Hobson (183) have found recently that the rates of incorporation of labelled amino acids into the proteins of the plastids and cytoplasm are nearly the same.

On infiltration of labelled sulphates and [$^{35}S$]methionine directly into leaves, the incorporation of labelled compounds into the plastids during relatively short exposures may be higher than in the rest of the cytoplasm (17).

However, the plastids of plants are, as we have seen above, extremely complicated organelles in respect of their morphology and biochemistry and this makes it difficult to use them as examples on which to study the mechanisms of protein synthesis. Tens of different proteins with different functions are produced within the plastids. The plastids have been found to contain DNA as well as RNA (75, 81). This may indicate (in the absence of contamination by nuclear material) the existence within the plastids of an autonomous genetic control over the synthetic processes. The autonomous cytoplasmic genetic control of the multiplication of plastids has actually been demonstrated recently by Brawerman & Chargaff (60, 62, 63). It may be that this is just the reason why, after removal of the nucleus, fragments of the cells of the giant unicellular alga *Acetabularia mediterranea* which have no nucleus but do contain plastids, retain their power of growth, while similar parts of *Amoeba*, which is a unicellular animal, gradually lose their power of growth and die (cf. Chapter XIII).

Stocking & Gifford (229) have recently observed the active incorporation of one component of DNA (thymidine) into the plastid structures of the alga *Spirogyra*.

It is interesting to note that the chloroplasts have a biochemical system for the activation of amino acids (158). It is not impossible that they can reproduce the whole cycle of protein synthesis within themselves without requiring the help of other organelles.

## Conclusion

In finishing this brief review of contemporary evidence concerning the localization of protein synthesis in the cellular structures, it must be concluded that all the organelles of the cytoplasm can synthesize proteins. However, in spite of the generality of the treatment the outcome of the work on the localization of protein synthesis is clear. Studies in this field have given us extremely valuable and direct evidence as to the mechanisms of protein synthesis and, above all, as to the outstanding part in protein synthesis played by the high-molecular nucleic acids situated in these structures. These investigations have given us some of the facts we require in order to draw the conclusion that, while the low-molecular soluble RNA of the intergranular phase is associated in some definite way with the primary reaction of activation of amino acids and possibly small peptides, the high-molecular RNA in the cellular organelles plays the part of a matrix in the final stages of protein synthesis.

The evidence we have surveyed also indicates that the syntheses of RNA and protein are specifically interrelated, but only in respect of whole molecules of the opposite type. The synthesis of cytoplasmic RNA depends on proteins (enzymes) localized in the plasma juice and apparently secreted into it by the microsomes. The final stage of the synthesis of a protein also requires polymeric molecules of RNA and the process of their synthesis is important, primarily because it ensures the reconstitution of the matrices and leads to the formation of fresh RNA which can carry out new syntheses of protein. The newness of these molecules can, in a number of cases, be assessed, not only by the period of their formation, but also by qualitative features as, for example, in the case of the synthesis of adaptive enzymes or the proteins of plastids.

A review of the factual material shows clearly that the system of RNA in some way organizes the synthesis of proteins and the reproduction of their specificity. However, these facts do not yet make it clear what is the exact biochemical mechanism by which the high-molecular RNA of the cellular organelles participates in accomplishing the synthesis of proteins. There is a very large number of factors which provide a sufficiently firm basis for the hypothesis that the RNA of the organelles serves as a matrix for protein synthesis but which do not yet provide a key wherewith we may decipher the mechanism by which these matrices work. It cannot be said that this mechanism is a complete enigma. On the contrary, we now have no shortage of hypotheses, theories and suggestions which try, though only theoretically, to explain the biologically possible means by which the reproduction of the specificity of proteins is brought about.

As we have already mentioned, special interest attaches to the facts and considerations which give us a more precise idea of the connection between the structural peculiarities of the organelles and the mechanism of protein synthesis (the arrangement of the matrix RNA on the surfaces of the particles, the possible dove-tailing of the protein into the groove between the 30 S and 50 S sections of the 70 S ribosomes, etc.).

There is, however, still one aspect of this problem which we have not yet dealt with which, although it is directly connected with the mechanism of protein synthesis, often escapes the attention of the authors of reviews. We are thinking of the intermediate products of protein synthesis. Up to now we have surveyed the biochemical system of synthesis, which is of the nature of a factory making a specific product, a peculiar continuous production line from which there emerge finished components which proceed to reproduce the entire system. However, for a more objective, clear and profitable analysis of the problem it is necessary to study, not only the system of synthesis but also the products of the synthesis at various stages of their formation within the system.

It is quite clear that a finished protein molecule cannot be formed all of a sudden from amino acids even if the process is carried out on matrices. There must be a "peptide phase", even if only for a very short time, between the finished molecules and the individual amino acid even if the peptides are associated with their adaptors. In other words there must exist in the cells, even though only in very small amounts, unfinished proteins, intermediate products of the synthesis, and a study of their localization, kinetics, composition and metabolism might cast some light on the processes which are still concealed from us, which begin with the transfer of activated amino acids to the organelles and end with the secretion of the finished molecules of the protein.

## REFERENCES

1. BELOZERSKIĬ, A. N. (1959). *Nukleoprotidy i nukleinovye kisloty rastenii i ikh biologicheskoe znachenie.* (*Bakhovskaya Lektsiya*). Moscow: Izd. Akad. Nauk S.S.S.R.
2. BLAGOVESHCHENSKIĬ, A. V. (1958). *Biokhimiya obmena azotsoderzhashchikh veshchestv u rastenii.* Moscow: Izd. Akad. Nauk S.S.S.R.
3. GAVRILOVA, L. P., SPIRIN, A. S. & BELOZERSKIĬ, A. N. (1959). *Doklady Akad. Nauk S.S.S.R.* **126**, 1121.
4. GEL'MAN, N. S. (1959). *Uspekhi sovremennoĭ Biol.* **47**, 152.
5. ZBARSKIĬ, I. B. (1957). *Vestnik Akad. Nauk S.S.S.R.* no. **8**, 26.

6. ZUBOVSKAYA, A. M. & TONGUR, V. S. (1959). *Byull. eksp. Biol. Med.* **47**, 56.
7. KARPAS, A. M. (1959). *Tsitologiya*, **1**, 153.
8. KEDROVSKIĬ, B. V. (1958). *Uspekhi sovremennoĭ Biol.* **46**, 3.
9. — (1959). *Tsitologiya belkovykh sintezov v zhivotnoĭ kletke.* Moscow: Izd. Akad. Nauk S.S.S.R.
9a. KISELEV, N. A., GAVRILOVA, L. P. & SPIRIN, A. S. (1961). *Doklady Akad. Nauk S.S.S.R.* **138**, 692.
10. MEDVEDEV, ZH. A. (1957). *Izvest. Timiryazev. sel'skokhoz. Akad.* No. 3, 186.
11. — (1958). In collective work *Fiziologiya rastenii, agrokhimiya, pochvovedenie. Trudy Vsesoyuznoĭ Konferentsii po ispol'zovanii izotopov v nauchnykh issledovanyakh.* (ed. V. M. Klechkovskiĭ *et al.*), p. 43. Moscow: Izd. Akad. Nauk S.S.S.R.
12. — (1959). *Izvest. Timiryazev. sel'skokhoz. Akad.* No. *2*, 57.
13. MEDVEDEV, ZH. A. & FEDOROV, E. A. (1956). *Fiziol. Rastenii*, **3**, 547.
14. MEDVEDEV, ZH. A. & TSZYUN', U. (1956). *Doklady Moskov. sel'skokhoz. Akad. im. K. A. Timiryazeva*, No. *26*, Pt. 1, 273.
15. MIL'MAN, L. S. (1961). *Voprosy med. Khim.* **7**, 212.
16. PEREVOSHCHIKOVA, K. A. & ZBARSKIĬ, I. B. (1957). *Doklady Akad. Nauk S.S.S.R.* **114**, 150.
17. PLESHKOV, B. P. & IVANKO, SH. (1956). *Biokhimiya*, **21**, 496.
18. SABININ, D. A. & POLOZOVA, L. YA. (1957). *Fiziol. Rastenii*, **4**, 38.
19. SISAKYAN, N. M. (1954). *Biokhimiya obmena veshchestv.* Moscow: Izd. Akad. Nauk S.S.S.R.
20. — (1956). *Izvest. Akad. Nauk S.S.S.R., Ser. biol.*, No. *6*, 3.
21. — (1954). In collective work *Voprosy botaniki* (ed. V. N. Sukachev *et al.*), Pt. *1*, p. 195. Moscow: Izd. Akad. Nauk S.S.S.R.
22. — (1955). *Sessiya Akad. Nauk S.S.S.R. po mirnomu Ispol'zovaniyu atomnoĭ Energii, Zasedaniya Otdel. biol. Nauk*, p. 172. (Cf. *Chem. Abs.* **49**, 16082*f* (1955).)
23. — (1959). *Proceedings of the first international symposium on the origin of life on the Earth* (ed. A. I. Oparin *et al.*), p. 400. London: Pergamon.
24. — (1961). *Uspekhi sovremennoĭ Biol.* **51**, 129.
25. SISAKYAN, N. M. & ODINTSOVA, M. S. (1960). *Izvest. Akad. Nauk S.S.S.R., Ser. biol.* **25**, No. 6, 817.
26. SISAKYAN, N. M. & FILIPPOVICH, I. I. (1955). *Doklady Akad. Nauk S.S.S.R.*, **102**, 579.
27. SPIRIN, A. S. (1961). *Proc. V int. Congr. Biochem., Moscow*, **9**, 141.
28. SPIRIN, A. S., GAVRILOVA, L. P. & BELOZERSKIĬ, A. N. (1959). *Doklady Akad. Nauk S.S.S.R.* **125**, 658.
28a. SPIRIN, A. S., KISELEV, N. A., SHAKULOV, R. S. & BOGDANOV, A. A. (1963). *Biokhimiya*, **28**, 920.
29. SPIRIN, A. S. & MIL'MAN, L. S. (1960). *Doklady Akad. Nauk S.S.S.R.* **134**, 717.
30. TONGUR, V. S. (1960). *Uspekhi sovremennoĭ Biol.* **49**, 156.
31. FRAENKEL-CONRAT, H., SINGER, B. A. & WILLIAMS, R. C. (1957). Preprint noted in *Chem. Abs.* **52**, 1331*h* (1958). Cf. Fraenkel-Conrat & Singer, *Proceedings of the first international Symposium on the origin of life on the Earth* (ed. A. I. Oparin *et al.*), p. 303. London: Pergamon (1959).
32. KHESIN, R. B. (1954). *Biokhimiya*, **19**, 407.
33. — (1953). *Biokhimiya*, **18**, 462.
34. — (1959). *Proceedings of the first international Symposium on the origin of life on the Earth* (ed. A. I. Oparin *et al.*), p. 460. London: Pergamon.
35. — (1960). *Biokhimiya tsitoplazmy.* Moscow: Izd. Akad. Nauk S.S.S.R.

36. KHESIN, R. B. & PETRASHKAÏTE, S. K. (1955). *Biokhimiya*, **20**, 597.
37. KHESIN, R. B., PETRASHKAÏTE, S. K., TOLYUSHIS, L. E. & PAUPAUSKAÏTE, K. P. (1957). *Biokhimiya*, **22**, 501.
38. CHARGAFF, E. (1959). *Proceedings of the first international Symposium on the origin of life on the Earth* (ed. A. I. Oparin *et al.*), p. 297. London: Pergamon.
39. SHABADASH, A. L. (1959). *Tsitologiya*, **1**, 15.
40. STRAUB [SHTRAUB], F. B. (1960). *Voprosy med. Khim.* **6**, 115.
41. ENGEL'GARDT, V. A. (1959). *Uspekhi Khim.* **28**, 1011.
42. ABDUL-NOUR, B. & WEBSTER, G. C. (1960). *Exptl. Cell Research*, **20**, 226.
43. ANDERSSON-CEDERGREN, E. (1959). *J. Ultrastructure Res.*, Suppl. **1**, 1.
43a. ARONSON, A. I. (1962). *J. mol. Biol.* **5**, 453.
44. ARONSON, A. I., BOLTON, E. T., BRITTEN, R. J., COWIE, D. B., DUERKSEN, J. B., McCARTHY, B. J., McQUILLEN, K. & ROBERTS, R. B. (1960). *Carnegie Institution, Washington. Department of Terrestrial Magnetism: Annual Report 1959-60*, p. 229.
45. ASKONAS, B. A., SIMKIN, J. L. & WORK, T. S. (1957). *Biochem. Soc. Symposia*, **14**, 32.
46. BATES, H. M., CRADDOCK, V. M. & SIMPSON, M. V. (1958). *J. Amer. chem. Soc.*, **80**, 1000.
47. BEER, M., HIGHTON, P. J. & McCARTHY, B. J. (1960). *J. mol. Biol.* **2**, 447.
48. BEN-ISHAI, R. (1957). *Biochim. biophys. Acta*, **26**, 477.
49. BEN-ISHAI, R. & VOLCANI, B. E. (1956). *Biochim. biophys. Acta*, **21**, 265.
50. BERNHARD, W., GAUTIER, A. & ROUILLER, C. (1954). *Arch. Anat. microscop. et Morphol. exptl.* **43**, 236.
51. BEST, J. B. (1960). *Intern. Rev. Cytol.* **9**, 129.
52. BÖING, J. (1955). *Protoplasma*, **45**, 55.
53. BOREK, E., RYAN, A. & PRICE, T. D. (1957). *Fed. Proc.* **16**, 156.
54. BOCK, R. M. & YIN, F. H. (1961). *Proc. V int. Congr. Biochem., Moscow*, **9**, 122.
55. BOSCH, L., BLOEMENDAL, H. & SLUYSER, M. (1960). *Biochim. biophys. Acta*, **41**, 444, 454.
56. BOWEN, T. J., DAGLEY, S., SYKES, J. & WILD, D. G. (1961). *Nature (Lond.)*, **189**, 638.
57. BOWEN, T. J., DAGLEY, S. & SYKES, J. (1959). *Biochem. J.* **72**, 419.
58. BRACHET, J. (1957). *Biochemical cytology.* New York: Academic Press.
59. — (1960). *The biological role of nucleic acids.* Amsterdam: Elsevier.
60. BRAWERMAN, G. & CHARGAFF, E. (1960). *Biochim. biophys. Acta*, **37**, 214.
61. — (1959). *Biochim. biophys. Acta*, **31**, 164.
62. — (1959). *Biochim. biophys. Acta*, **31**, 172.
63. BRAWERMAN, G., HUFNAGEL, D. A. & CHARGAFF, E. (1962). *Biochim. biophys. Acta*, **61**, 340.
64. BRAWERMAN, G. & CHARGAFF, E. (1959). *Biochim. biophys. Acta*, **31**, 178.
65. BURTON, D., HALL, D. A., KEECH, M. K., REED, R., SAXL, H., TUNBRIDGE, R. E. & WOOD, M. J. (1955). *Nature (Lond.)*, **176**, 966.
65a. BUTLER, J. A. V., COHN, P. & SIMSON, P. (1960). *Biochim. biophys. Acta*, **38**, 386.
66. BUTLER, J. A. V., CRATHORN, A. R. & HUNTER, G. D. (1958). *Biochem. J.* **69**, 544.
67. CAMPBELL, P. N. (1960). *Biol. Rev. Cambridge phil. Soc.* **35**, 413.
68. CAMPBELL, P. N., GREENGARD, O. & KERNOT, B. A. (1958). *Biochem. J.* **68**, 18P.
69. — (1960). *Biochem. J.* **74**, 107.
69a. CAMPBELL, P. N. & KERNOT, B. A. (1962). *Biochem. J.* **82**, 262.

70. CHANTRENNE, H. (1956). *Arch. Biochem. Biophys.* **65**, 414.
71. — (1956). *Nature (Lond.)*, **177**, 579.
72. — (1961). *The biosynthesis of proteins.* Oxford: Pergamon.
73. CHANTRENNE, H. & DEVREUX, S. (1959). *Exptl. Cell Research*, Suppl. **6**, 152.
74. CHENG, P.-Y. (1960). *Biochim. biophys. Acta*, **37**, 238.
75. CHIBA, Y. & SUGAHARA, K. (1957). *Arch. Biochem. Biophys.* **71**, 367.
76. COHEN, G. N. & COWIE, D. B. (1957). *Compt. rend. Acad. Sci., Paris*, **244**, 680.
77. COHN, P. (1959). *Biochim. biophys. Acta*, **33**, 284.
78. COHN, P. & BUTLER, J. A. V. (1957). *Biochim. biophys. Acta*, **25**, 222.
79. COMMONER, B., LIPPINCOTT, J. A., SHEARER, G. B., RICHMAN, E. E. & WU, J.-H. (1956). *Nature (Lond.)*, **178**, 767.
80. CONNELL, G. E., LENGYEL, P. & WARNER, R. C. (1959). *Biochim. biophys. Acta*, **31**, 391.
81. COOPER, W. D. & LORING, H. S. (1957). *J. biol. Chem.* **228**, 813.
82. CRADDOCK, V. M. & SIMPSON. M. V. (1960). *Biochem. J.* **74**, 10P.
83. CRAMPTON, C. F. & PETERMANN, M. L. (1959). *J. biol. Chem.* **234**, 2642.
84. CREASER, E. H. (1955). *Nature (Lond.)*, **176**, 556.
85. — (1955). *J. gen. Microbiol.* **12**, 288.
86. — (1956). *Biochem. J.* **64**, 539.
87. DAGLEY, S. & SYKES, J. (1958). In *Microsomal particles and protein synthesis, Papers Symposium Biophys. Soc., Cambridge, Mass.* (ed. R. B. Roberts), p. 62. London: Pergamon.
88. DALY, M. M., ALLFREY, V. G. & MIRSKY, A. E. (1952). *J. gen. Physiol.* **36**, 173.
89. DANON, D., MARIKOVSKY, Y. & LITTAUER, U. Z. (1961). *J. biophys. biochem. Cytol.* **9**, 253.
90. DAVIDSON, J. N. (1957). *Biochem. Soc. Symposia*, **14**, 27.
91. — (1960). *The biochemistry of the nucleic acids.* (4th edn.) London: Methuen.
92. DECKEN, A. VON DER & HULTIN, T. (1958). *Exptl. Cell Research*, **14**, 88.
93. DEKEN-GRENSON, M. DE (1954). *Biochim. biophys. Acta*, **14**, 203.
94. DOTY, P., BOEDTKER, H., FRESCO, J. R., HASELKORN, R. & LITT, M. (1959). *Proc. nat. Acad. Sci. U.S.* **45**, 482.
95. DOTY, P., BOEDTKER, H., FRESCO, J. R., HALL, B. D. & HASELKORN, R. (1959). *Ann. N.Y. Acad. Sci.* **81**, 693.
96. DUSTIN, J. P., SCHAPIRA, G., DREYFUS, J. C. & HESTERMANS-MEDARD, O. (1954). *Compt. rend. Soc. biol.* **148**, 1207.
97. ELSON, D. (1959). *Biochim. biophys. Acta*, **36**, 372.
98. ELSON, D. & TAL, M. (1961). *Proc. V int. Congr. Biochem., Moscow*, **9**, 148.
99. FRAENKEL-CONRAT, H., SINGER, B. & WILLIAMS, R. C. (1957). *Biochim. biophys. Acta*, **25**, 87.
100. FRAENKEL-CONRAT, H. & WILLIAMS, R. C. (1955). *Proc. nat. Acad. Sci. U.S.* **41**, 690.
101. FRESCO, J. R., ALBERTS, B. M. & DOTY, P. (1960). *Nature (Lond.)*, **188**, 98.
102. FREY-WYSSLING, A., RUCH, F. & BERGER, X. (1955). *Protoplasma*, **45**, 97.
103. GALE, E. F. (1956). *Intern. Symposium on Enzymes, Units of biol. Structure and Function, Detroit, 1955*, p. 49. New York: Academic Press.
104. — (1957). *Biochem. Soc. Symposia*, **14**, 47.
105. — (1956). In *Ciba Foundation Symposium on ionizing radiations and cell metabolism* (ed. G. E. W. Wolstenholme & C. M. O'Connor), p. 174. London: Churchill.
106. — (1955-6). *Harvey Lectures*, Ser. **51**, 25.
107. GALE, E. F. & FOLKES, J. P. (1953). *Biochem. J.* **53**, 483.

108. — (1953).  *Biochem. J.* **53**, 493.
109. — (1953).  *Biochem. J.* **55**, 721.
110. — (1954).  *Nature (Lond.)*, **173**, 1223.
111. — (1955).  *Biochem. J.* **59**, 661.
112. — (1955).  *Biochem. J.* **59**, 675.
113. — (1958).  *Biochem. J.* **69**, 611.
114. GARFINKEL, D.  (1958).  In *Microsomal particles and protein synthesis, Papers Symposium Biophys. Soc., Cambridge, Mass.* (ed. R. B. Roberts), p. 22. London: Pergamon.
115. GIERER, A. & SCHRAMM, G.  (1956).  *Z. Naturforsch.* **11B**, 138.
116. GILLCHRIEST, W. C. & BOCK, R. M.  (1958).  In *Microsomal particles and protein synthesis, Papers Symposium Biophys. Soc., Cambridge, Mass.* (ed. R. B. Roberts), p. 1.  London: Pergamon.
117. GORMAN, J. & HALVORSON, H.  (1959).  *Arch. Biochem. Biophys.* **84**, 462.
117a. GREEN, M. & HALL, B. D.  (1961).  *Biophys. J.* **1**, 517.
118. GREENBERG, D. M., FRIEDBERG, F., SCHULMAN, M. P. & WINNICK, T.  (1948).  *Cold Spring Harbor Symposia quant. Biol.* **13**, 113.
119. GREENGARD, O.  (1959).  *Biochim. biophys. Acta*, **32**, 270.
120. GREENGARD, O. & CAMPBELL, P. N.  (1959).  *Biochem. J.* **72**, 305.
121. GROTH, D. P.  (1956).  *Biochim. biophys. Acta*, **21**, 18.
122. HAHN, F. E., SCHAECHTER, M., CEGLOWSKI, W. S., HOPPS, H. E. & CIAK, J.  (1957).  *Biochim. biophys. Acta*, **26**, 469.
123. HALL, B. D. & DOTY, P.  (1958).  In *Microsomal particles and protein synthesis, Papers Symposium Biophys. Soc., Cambridge, Mass.* (ed. R. B. Roberts), p. 27.  London: Pergamon.
124. — (1959).  *J. mol. Biol.* **1**, 111.
125. HALL, B. D., STORCK, R. & SPIEGELMAN, S.  (1961).  *Fed. Proc.* **20**, 362.
126. HAMILTON, M. G. & PETERMANN, M. L.  (1959).  *J. biol. Chem.* **234**, 1441.
127. HANCOCK, R.  (1957).  *J. gen. Microbiol.* **17**, 480.
128. HENDLER, R. W.  (1959).  *J. biol. Chem.* **234**, 1466.
129. HOAGLAND, M. B.  (1960).  In *The nucleic acids* (ed. E. Chargaff & J. N. Davidson), Vol. 3, p. 349.  New York: Academic Press.
130. HOKIN, L. E. & HOKIN, M. R.  (1959).  *Fed. Proc.* **18**, 248.
131. HOWATSON, A. F. & HAM, A.  (1957).  *Can. J. Biochem. Physiol.* **35**, 549.
132. HULTIN, T.  (1956).  *Exptl. Cell Research*, **11**, 222.
132a. HULTIN, T., LEON, H. A. & CERASI, E.  (1961).  *Exptl. Cell Res.* **25**, 660.
133. HUNTER, G. D., CRATHORN, A. R. & BUTLER, J. A. V.  (1957).  *Nature (Lond.)*, **180**, 383.
134. HUXLEY, H. E. & ZUBAY, G.  (1960).  *J. mol. Biol.* **2**, 10.
135. JESAITIS, M. A.  (1956).  *Nature (Lond.)*, **178**, 637.
136. KELLER, E. B. & ZAMECNIK, P. C.  (1956).  *J. biol. Chem.* **221**, 45.
137. KIRBY, K. S.  (1960).  *Biochim. biophys. Acta*, **41**, 338.
138. KLUG, A., HOLMES, K. C. & FINCH, J. T.  (1961).  *J. mol. Biol.* **3**, 87.
139. KORNER, A.  (1959).  *Biochim. biophys. Acta*, **35**, 554.
140. — (1960).  *Biochem. J.* **76**, 28P.
140a. KORNER, A.  (1962).  *Biochem. J.* **83**, 69.
141. KRAMER, M.  (1957).  *Acta physiol. Acad. Sci. Hung.* **11**, 125.
142. KRAMER, M. & STRAUB, F. B.  (1957).  *Acta physiol. Acad. Sci. Hung.* **11**, 133.
143. KURLAND, C. G.  (1960).  *J. mol. Biol.* **2**, 83.
144. LAIRD, A. K. & BARTON, A. D.  (1957).  *Biochim. biophys. Acta*, **25**, 56.
145. LAMIRANDE, G. DE, ALLARD, C. & CANTERO, A.  (1959).  *J. biophys. biochem. Cytol.* **6**, 291.
146. LASKOV, R., MARGOLIASH, E., LITTAUER, U. Z. & EISENBERG, H.  (1959).  *Biochim. biophys. Acta*, **33**, 247.

146a. LETT, J. T. & TAKAHASHI, W. N. (1962). *Arch. Biochem. Biophys.* **96**, 569.
147. LEYON, H. (1954). *Exptl. Cell Res.* **7**, 265.
148. LINDBERG, O. & ERNSTER, L. (1954). In *Protoplasmatologia* (ed. L. V. Heilbrunn & F. Weber). Vol. III/A/4, p. 1. Vienna: Springer.
149. LIPSHITZ, R. & CHARGAFF, E. (1960). *Biochim. biophys. Acta*, **42**, 544.
150. LITTAUER, U. Z. & EISENBERG, H. (1959). *Biochim. biophys. Acta*, **32**, 320.
151. LITTLEFIELD, J. W. & KELLER, E. B. (1957). *J. biol. Chem.* **224**, 13.
152. LITTLEFIELD, J. W., KELLER, E. B., GROSS, J. & ZAMECNIK, P. C. (1955). *J. biol. Chem.* **217**, 111.
153. LUND, H. A. (1959). *Biochim. biophys. Acta*, **33**, 347.
154. MAEDA, A. (1960). *J. Biochem. (Tokyo)*, **48**, 363.
154a. MAEDA, A. (1961). *J. Biochem. (Tokyo)*, **50**, 377.
155. MANDEL, H. G. (1957). *J. biol. Chem.* **225**, 137.
156. — (1958). *Arch. Biochem. Biophys.* **76**, 230.
157. MANDEL, H. G. & ALTMAN, R. L. (1960). *J. biol. Chem.* **235**, 2029.
158. MARCUS, A. (1959). *J. biol. Chem.* **234**, 1238.
159. MCCARTHY, B. J. (1960). *Biochim. biophys. Acta*, **39**, 563.
160. MCLEAN, J. R., COHN, G. L., BRANDT, I. K. & SIMPSON, M. V. (1958). *J. biol. Chem.* **233**, 657.
161. MENDELSOHN, J. & TISSIÈRES, A. (1959). *Biochim. biophys. Acta*, **35**, 248.
162. ROBERTS, R. B. (ed.) *Microsomal particles and protein synthesis, Papers Symposium Biophys. Soc., Cambridge, Mass.* London: Pergamon.
163. *Symposia Soc. exptl. Biol.* No. *10*, Mitochondria and other cytoplasmic inclusions (1957). Cambridge: University Press.
164. MITSUI, H. & YOSHIDA, E. (1960). *J. Biochem. (Tokyo)*, **48**, 242.
165. MOLDAVE, K., FESSENDEN, J. & WEINER, M. (1961). *Proc. V int. Congr. Biochem., Moscow*, **9**, 101.
166. MÖLLER, W. & BOEDTKER, H. (1961). *Fed. Proc.* **20**, 357.
167. NEIDHARDT, F. C. & GROS, F. (1957). *Biochim. biophys. Acta*, **25**, 513.
168. NOMURA, M. & WATSON, J. D. (1959). *J. mol. Biol.* **1**, 204.
169. NOMURA, M. & YOSHIKAWA, H. (1959). *Biochim. biophys. Acta*, **31**, 125.
170. NOVELLI, G. D. (1960). *Ann. Rev. Microbiol.* **14**, 65.
171. OGATA, K., HIROKAWA, R. & OMORI, S. (1960). *Biochim. biophys. Acta*, **40**, 178.
172. OGATA, K., SHIMIZU, T. & TOGASHI, K. (1958). *Biochim. biophys. Acta*, **29**, 656.
173. OSAWA, S. & HOTTA, Y. (1959). *Biochim. biophys. Acta*, **34**, 284.
174. PALADE, G. E. (1956). *Intern. Symposium on Enzymes, Units of biol. Structure and Function, Detroit, 1955* (ed. O. H. Gaebler), p. 185. New York: Academic Press.
175. — (1958). In *Microsomal particles and protein synthesis, Papers Symposium Biophys. Soc., Cambridge, Mass.* (ed. R. B. Roberts), p. 36. London: Pergamon.
176. PALADE, G. E. & SIEKEVITZ, P. (1955). *Fed. Proc.* **14**, 262.
177. PARDEE, A. B. (1954). *Proc. nat. Acad. Sci. U.S.* **40**, 263.
178. PARDEE, A. B., PAIGEN, K. & PRESTIDGE, L. S. (1957). *Biochim. biophys. Acta*, **23**, 162.
179. PEABODY, R. A. & HURWITZ, C. (1960). *Biochim. biophys. Acta*, **39**, 184.
180. RAACKE, I. D. (1959). *Biochim. biophys. Acta*, **34**, 1.
181. — (1961). *Biochim. biophys. Acta*, **51**, 73.
182. RABINOVITZ, M. & OLSON, M. E. (1956). *Exptl. Cell Research*, **10**, 747.
183. RACUSEN, D. & HOBSON, E. L. (1959). *Arch. Biochem. Biophys.* **82**, 234.
184. RAMACHANDRAN, L. K. & FRAENKEL-CONRAT, H. (1958). *Arch. Biochem. Biophys.* **74**, 224.

185. REINER, J. M. & GOODMAN, F. (1955). *Arch. Biochem. Biophys.* **57**, 475.
186. REIS, P. J., COOTE, J. L. & WORK, T. S. (1959). *Biochem. J.* **72**, 24P.
187. — (1959). *Nature (Lond.)*, **184**, 165.
188. RENDI, R. (1959). *Biochim. biophys. Acta*, **31**, 266.
189. — (1959). *Exptl. Cell Research*, **18**, 187.
190. — (1959). *Exptl. Cell Research*, **17**, 585.
191. RENDI, R. & CAMPBELL P. N. (1958). *Biochem. J.* **69**, 46P.
192. — (1959). *Biochem. J.* **72**, 34.
193. — (1959). *Biochem. J.* **72**, 435.
194. RENDI, R. & HULTIN, T. (1960). *Exptl. Cell Research*, **19**, 253.
195. RENDI, R. & WARNER, R. C. (1960). *Ann. N.Y. Acad. Sci.* **88**, 741.
196. ROBERTS, R. B. (1960). *Ann. N.Y. Acad. Sci.* **88**, 752 (1960).
197. ROODYN, D. B. & MANDEL, H. G. (1960). *J. biol. Chem.* **235**, 2036.
198. ROSSITER, R. J. (1957). *Canad. J. Biochem. Physiol.* **35**, 579.
199. ROUILLER, C. (1960). *Intern. Rev. Cytol.* **9**, 227.
200. SACHS, H. (1958). *J. biol. Chem.* **233**, 650.
201. SACHS, H. & NEIDLE, A. (1956). *Fed. Proc.* **15**, 344.
202. SACHS, H. & WAELSCH, H. (1956). *Biochim. biophys. Acta*, **21**, 188.
203. SCHACHTSCHABEL, D. & ZILLIG, W. (1959). *Z. physiol. Chem.* **314**, 262.
204. SCHLESSINGER, D. (1960). *J. mol. Biol.* **2**, 92.
205. SCHÖTZ, F. (1954). *Planta*, **43**, 182.
206. SCHWEET, R., LAMFROM, H. & ALLEN, E. (1958). *Proc. nat. Acad. Sci. U.S.* **44**, 1029.
207. SETTERFIELD, G. (1961). *Can. J. Bot.* **39**, 469.
208. SETTERFIELD, G., NEELIN, J. M., NEELIN, E. M. & BAYLEY, S. T. (1960). *J. mol. Biol.* **2**, 416.
209. SIEKEVITZ, P. & PALADE, G. E. (1958). *J. biophys. biochem. Cytol.* **4**, 557.
210. — (1959). *Fed. Proc.* **18**, 324.
211. — (1959). *J. biophys. biochem. Cytol.* **5**, 1.
212. — (1960). *J. biophys. biochem. Cytol.* **7**, 619.
213. — (1960). *J. biophys. biochem. Cytol.* **7**, 631.
214. SIMKIN, J. L. (1957). *Biochem. J.* **66**, 6P.
215. — (1958). *Biochem. J.* **70**, 305.
215a. SIMKIN, J. L. (1958). *Biochem. J.* **70**, 305.
216. SIMKIN, J. L. & WORK, T. S. (1957). *Biochem. J.* **65**, 307.
217. — (1957). *Biochem. J.* **67**, 617.
218. SIMPSON, M. V. & McLEAN, J. R. (1955). *Biochim. biophys. Acta*, **18**, 573.
218a. SIMPSON, M. V., SKINNER, D. M. & LUCAS, J. M. (1961). *J. biol. Chem.* **236**, PC 81.
219. SINGAL, S. A., LITTLEJOHN, J. M. & MARSHALL, E. (1959). *Fed. Proc.* **18**, 324.
220. SISSAKIAN, N. M. [SISAKYAN] (1958). *Advanc. Enzymol.* **20**, 201.
221. SJÖSTRAND, F. S. (1953). *Nature (Lond.)*, **171**, 30, 31.
222. — (1956). *Intern. Rev. Cytology*, **5**, 455.
223. — (1960). *Radiation Research Suppl.* **2**, 349.
224. SMITH, K. C. & DOELL, R. G. (1960). *Fed. Proc.* **19**, 318.
225. SPAHR, P. F. & TISSIÈRES, A. (1959). *J. mol. Biol.* **1**, 237.
226. SPIEGELMAN, S., HALVORSON, H. O. & BEN-ISHAI, R. (1955). In *A Symposium on amino acid metabolism* (ed. W. D. McElroy & H. B. Glass), p. 124. Baltimore, Md.: Johns Hopkins Press.
227. STEINMANN, E. & SJÖSTRAND, F. S. (1955). *Exptl. Cell Res.* **8**, 15.
228. STEPHENSON, M. L., THIMANN, K. V. & ZAMECNIK, P. C. (1956). *Arch. Biochem. Biophys.* **65**, 194.
228a. SPAHR, P. F. (1962). *J. mol. Biol.* **4**, 395.

229. STOCKING, C. R. & GIFFORD, E. M. (jun.) (1959). *Biochem. biophys. Res. Commun.* **1**, 159.
230. SZAFARZ, D. (1953). *Arch. int. Physiol.* **61**, 269.
231. TAKANAMI, M. (1960). *Biochim. biophys. Acta,* **37**, 556.
232. — (1960). *Biochim. biophys. Acta,* **39**, 152.
233. — (1960). *Biochim. biophys. Acta,* **39**, 318.
234. TASHIRO, Y. (1958). *J. Biochem. (Tokyo),* **45**, 803, 937.
235. TASHIRO, Y. & INOUYE, A. (1959). *J. Biochem. (Tokyo),* **46**, 1625.
236. TASHIRO, Y., SHIMIDZU, H., HONDE, S. & INOUYE, A. (1960). *J. Biochem. (Tokyo),* **47**, 37.
237. TASHIRO, Y., OGURA, M., SATO, A., SHINAGAWA, Y., IMAI, Y., HIRAKAWA, K. & HIRANO, S. (1958). *Proc. intern. Symposium Enzyme Chem. Tokyo and Kyoto,* **2**, 436.
238. TASHIRO, Y., SHIMIDZU, H., HONDE, S. & INOUYE, A. (1960). *J. Biochem. (Tokyo),* **47**, 185.
239. TISSIÈRES, A., WATSON, J. D., SCHLESSINGER, D. & HOLLINGWORTH, B. R. (1959). *J. mol. Biol.* **1**, 221.
240. TISSIÈRES, A., SCHLESSINGER, D. & GROS, F. (1960). *Proc. nat. Acad. Sci. U.S.* **46**, 1450.
241. TOSCHI, G. (1959). *Exptl. Cell Research,* **16**, 232.
242. TS'O, P. O. P., BONNER, J. & DINTZIS, H. (1958). *Arch. Biochem. Biophys.* **76**, 225.
243. TS'O, P. O. P., BONNER, J. & VINOGRAD, J. (1958). *Biochim. biophys. Acta,* **30**, 570.
244. TS'O, P. O. P. & SATO, C. S. (1959). *J. biophys. biochem. Cytol.* **5**, 59.
244a. TS'O, P. O. P. & VINOGRAD, J. (1961). *Biochim. biophys. Acta,* **49**, 113.
245. WALLER, J. P. & HARRIS, J. I. (1961). *Proc. nat. Acad. Sci. U.S.* **47**, 18.
245a. WALLACE, J. M. & TS'O, P. O. P. (1961). *Biochem. biophys. Res. Commun.* **5**, 125.
246. WEBSTER, G. C. (1954). *Plant Physiol.* **29**, 202.
247. — (1957). In *A Symposium on the chemical basis of heredity* (ed. W. D. McElroy & B. Glass), p. 268. Baltimore, Md.: Johns Hopkins Press.
248. — (1957). *J. biol. Chem.* **229**, 535.
249. — (1957). *Arch. Biochem. Biophys.* **70**, 622.
250. — (1959). *Arch. Biochem. Biophys.* **82**, 125.
251. — (1959). *Fed. Proc.* **18**, 348.
252. — (1959). *Arch. Biochem. Biophys.* **85**, 159.
253. — (1960). *Arch. Biochem. Biophys.* **89**, 53.
253a. WEBSTER, G. C. & WHITMAN, S. L. (1961). *Proc. V int. Congr. Biochem.,* Moscow, **2**, 30.
254. WEBSTER, G. C. & JOHNSON, M. P. (1955). *J. biol. Chem.* **217**, 641.
254a. YIN, F. H. & BOCK, R. M. (1960). *Fed. Proc.* **19**, 137.
255. YOSHIDA, E., MITSUI, H., TAKAHASHI, H. & MARUO, B. (1960). *J. Biochem. (Tokyo),* **48**, 251.
256. ŽÁK, R. & GUTMANN, E. (1960). *Nature (Lond.),* **185**, 766.
257. ZIEGLER, D. M. & MELCHIOR, J. B. (1955). *J. biol. Chem.* **217**, 569.
258. ZUBAY, G. & WILKINS, M. H. F. (1960). *J. mol. Biol.* **2**, 105.

# CHAPTER VI

# INTERMEDIATE STAGES IN THE FORMATION OF THE POLYPEPTIDE CHAIN: THE FORMATION AND POSSIBLE ROLES OF PEPTIDE, PEPTIDE-NUCLEOTIDE AND PEPTIDE-POLYNUCLEOTIDE COMPOUNDS

## *Introduction*

From the evidence presented in earlier chapters it is clear that in the process of protein synthesis amino acids are activated by specific enzymic systems and then become bound to fractions of soluble RNA and conveyed in this form to the organelles (in most cases to the ribosomes); here they are used for the synthesis of polypeptide structures. In this synthesis, high-molecular fractions of RNA, combined with protein to form ribonucleoprotein particles, play the important part of specific matrices. The study of these stages is certainly extremely important but it does not account for the actual synthesis of proteins, which is a process of polycondensation of amino acids giving rise to polypeptide structures. This stage, properly speaking, represents the actual synthesis of proteins, that is to say the transition from amino acids (activated, specifically accepted and brought to the places where synthesis will occur) to polypeptide chains with characteristic sequences of amino acid residues.

There are many variants of possible pathways and kinetics which may bring about this linking up and the choice of the one which corresponds most nearly with reality is very important in providing the answers to many questions concerning the problem of protein synthesis, particularly in showing how the unique structure of each protein is reproduced.

The combination of amino acids with the formation of specific polypeptides may, for example, occur as a single "explosive" process of "fusion" after they have been arranged on the matrix in the order in which they are to be combined in the protein. It may be, however, that when activated amino acids arrive at the sites destined for them on the matrices or along the matrices on the ribosomes, then, if their next-door neighbours are already in place they react with these to form peptides and these peptides grow until they finally unite to form a single chain. It is also possible that growth of the chain may occur only at one end and that there is only one "rudiment" which gradually lengthens as the chain extends and grows up into an "adult" molecule. It may also be that, not only individual, activated amino acids, but also complete blocks of peptides formed by synthesis from amino acids or by the breakdown of proteins, can react with the matrices. Finally, one may suppose that the matrices work on quite a different principle, according to which specific peptide blocks are formed on one set of matrices and "assembled" on another set into polypeptide chains.

At present it is still hard to say which of these variants corresponds most nearly with reality. The synthesis of proteins is undoubtedly polymorphic and it would seem that in nature the polypeptide chains may be assembled in a number of different ways. If we are to have any basis for assessing the possibilities in this field we must make a careful analysis of the extremely motley and diverse evidence concerning possible intermediate products of protein synthesis at the time of formation of the peptide links.

Up till now we have been looking at systems which were, to some extent, external to the protein being synthesized, systems which would enable the process to take place, somewhat like a conveyor belt for the preparation and assembly of the components.

In this chapter we shall forget, for the time being, about the "preparatory" systems and concern ourselves with the nature and properties of those products which may, *perhaps*, be intermediates in the process because their chemical structure is, in fact, intermediate between that of the original monomers and that of the final polymers.

## 1. The origin of the free peptides present in the soluble fractions and organelles of cells

We have already discussed in two special reviews (9, 13) the occurrence of peptides in cells and tissues and their role and possible utilization in protein synthesis. A review of this material has also been made by Synge (152). However, even in the years which have passed since the publication of the last two reviews, many new and interesting observations have been made in this field and many of the facts which had already been discovered have been reinterpreted. Furthermore, there are some aspects of the matter which we could not touch upon in earlier articles. In view of the fundamental importance of the question of peptides for the succeeding theoretical analysis of the mechanisms for the reproduction of the specificity of proteins and the nature of their ontogenetic alterations and for the general completeness of the picture of the biological synthesis of proteins, we therefore think that it is suitable to return to the survey of these materials while directing particular attention to assessing the validity of the methods used in the study of the question.

If we analyse any plant or animal tissues or cultures of microorganisms

PB K

in active life, we can almost always find a certain amount of low-molecular peptide products which are present in them in larger or small quantities alongside free amino acids. Thus, by purely analytical methods, one may find within the cell compounds which are intermediate in complexity between amino acids and proteins. This peptide fraction may be divided into five main groups in respect of origin and function:

1. Peptides with special functions which sometimes contain amino acids or their derivatives not usually occurring in proteins (antibiotic peptides such as penicillin, bacitracin, tyrocidine, etc., peptide co-factors of some processes such as glutathione, ophthalmic acid, carnosine, anserine, etc., hormonal or functional peptides such as oxytocin, vasopressin, glucagon, etc.).

2. Peptides included in complex compounds such as mucopeptides, lipopeptides, glycopeptides, etc.

3. Peptides formed by the incomplete breakdown of cells and tissues (urinary peptides, the peptide fraction of the blood, especially in patho-logical conditions such as peptidaemia, peptides isolated from starved tissues and tissues undergoing autolysis, etc.).

4. Peptides of a synthetic nature, apparently intermediates in protein synthesis.

5. Peptides bound to nucleotides and polynucleotides.

In this chapter we are mainly interested in peptides of the fourth and fifth groups, though it is not impossible that peptides formed by the break-down of protein might, to some extent, be reincorporated during protein synthesis.

The evidence concerning the distribution and functions of the special peptides has received enough discussion in many reviews and mono-graphs (1, 35, 50, 150, 151) and we shall, therefore, not go over the material again although it is of interest in connection with some parts of the problem of protein synthesis because some of these products carry out their biological functions by becoming involved in synthetic processes as inhibitory antimetabolites (antibiotics) or as stimulants (hormones) or as co-factors of processes associated in some way or another with the syn-thesis of proteins.

There is no need either to review here those papers in which the presence of peptide products in cells and tissues is simply established without any analysis of their metabolism, origin or functions as the ques-tion of the presence of peptide products in the tissues may now be held to be settled in a positive sense.

Experimental evidence concerning the occurrence of processes of sustained, active synthesis of peptides in cells and tissues has been begin-ning to accumulate quite quickly during the last few years. In two reviews (9, 13) we have already given quite a detailed account of a character-istic series of experiments in this field using microbiological material (64,

66, 123, 136, 160, 161, 162). In these investigations various methods were used, including kinetic analysis of the incorporation of various labelled compounds into proteins, peptides and amino acids. It was shown clearly that peptide products are very rapidly formed synthetically in cells and the last four papers showed convincingly (on a kinetic plane) the intermediate role of peptides in protein synthesis. The results of these investigations have been further confirmed by some new work by Turba et al. (163) and especially by some recent experiments of McManus (118) who devised a scheme for the study of the kinetics of the incorporation of radioactive precursors into amino acids, peptides and proteins. In her experiments growing cells of Torula utilis were exposed to [14C]acetate for 30 sec., 3, 10 and 150 min. and the specific radioactivity of the amino acids, peptides and proteins was then studied. The results obtained by this author completely confirmed the similar findings of Turba et al. and the work of Connell & Watson, to which we have referred above, which demonstrates the rapid incorporation of radioactivity into amino acids and peptides and their subsequent use for the formation of proteins.

We have also made reference to some investigations concerning the metabolism of the intracellular peptides of plants which definitely indicate their synthesis and their significance as intermediates in the formation of proteins (42, 47, 133, 143) and also a number of studies in this field carried out on material of animal origin (27, 135, 48, 49, 173).

In recent years several new studies along these lines have been published. They further confirm the occurrence of normal synthesis of various peptides of an intermediate character in cells and tissues (103, 57-59).

For example, it is interesting to note that, in the rice plant which is actively assimilating ammoniacal nitrogen, its detoxication and acceptance is brought about, not by the formation of amides, as in most other plants, but by the synthesis of a large intermediate peptide the amount of which in the plant increases rapidly when its nitrogen requirements become greater (175).

In our work (19) we showed that after bean plants have been supplied with small amounts of calcium nitrate there is an increase in the peptide fraction in the tissues of the leaves. It was peculiar, in these experiments, that while in control plants of haricot beans and Galega the bulk of the peptides were situated in the cellular structures which remained after the vacuolar and plasma juices had been expressed from the leaves, the increase of peptides after the addition of nitrogenous matter to the nutrient medium was mainly concentrated in the plasma juice in which we know that the processes of activation of amino acids also take place. In connection with these results, which certainly need amplification and verification, it seems to us that it is not out of the question that low-molecular activated peptides may be formed in the systems which activate amino acids, and they may be transferred to the microsomes for further use in the building of protein molecules. The existence of activated peptides has, as we shall see later, actually been demonstrated by a number of authors.

Interesting facts concerning the changing amounts of peptides in different stages of the embryonic development of animals have been discovered in a number of investigations (45, 61, 62, 86, 106, 139) and in most cases these changes correspond with synthetic processes. The rapid incorporation of labelled leucine into a leucine-containing peptide of the liver was demonstrated by Markovitz & Steinberg (115). The presence in microsomes of diffusible peptides, capable of actively incorporating into themselves [$^{14}$C]amino acids, has recently been demonstrated by Anderson & Albright (37) in a short note.

Synge and his colleagues (57-59, 153, 154) have recently made a very detailed and systematic study of the peptide and oligopeptide fractions of Italian rye-grass using a number of new and original methods for the isolation and separation of these peptides.

These authors studied the incorporation of [$^{14}$C]valine into free amino acids, peptides and proteins. The fractions containing the acid and neutral peptides incorporated amino acids at the same rate as did proteins (on the basis of the specific activity of the valine) but, in all their experiments, the fractions containing the amphoteric peptides had an activity in incorporating valine which was several times greater than that of proteins.

It is interesting to note that intracellular peptides may be activated in the same way as amino acids by a system of activating enzymes and, in a number of researches, activated peptides have been found in the cells of various experimental materials (106a, 107, 140, 141, 155, 156).

The activating enzymes form adenylates with, and thereby activate, not only those peptides which are normally present in the cells, but also peptides supplied to the systems *in vitro* (138, 159). The second paper describes how, if a large series of synthetic dipeptides were added to the system with pH 5 enzyme, many of them formed peptidyl adenylates.

All this interesting evidence indicates clearly that in the cells one may normally observe the active synthesis of various peptides with a high rate of metabolism and that these peptides may be activated and may even be accepted by soluble RNA (41). However, this evidence still does not constitute absolute proof that these peptides really are the intermediate material of protein synthesis isolated by extraction from the site of the formation of specific polypeptide chains. All these results are only an indirect indication of the possibility that there is a peptide stage in protein synthesis; they are not nearly enough to convince us that this is so. Supplementary results which will enable us to study and define in more detail the character of this stage of protein synthesis belong in other categories and we shall try here to survey systematically all possible aspects of these investigations so as to enable us to reach the firmest possible conclusion in this matter.

## 2. Peptides bound to matrices and localized in the actual sites of the synthesis of specific proteins

In recent years there has been a systematic accumulation of evidence as to the existence in cells of a special peptide fraction associated in various ways with molecules of the RNA of the cellular organelles. When these peptides were found, the suggestion naturally arose that they represented incompletely synthesized proteins, a kind of store which remained combined with the matrix until the molecule was fully formed and dissociated from the matrix into the plasma solution outside the organelles.

The first information about peptides in combination with RNA was given by the Czech biochemists Keil & Hrubešová (6). By isolating RNA from ox pancreas they obtained a preparation containing a tightly bound protein residue. By the use of various methods the authors succeeded in separating this protein from the RNA but attempts to remove the last traces of peptide from the RNA always led to its depolymerization. Hydrolysis of purified RNA still bound to this protein showed that it contained mainly glutamic and aspartic acids, serine and lysine, small amounts of arginine, cystine and glycine and traces of alanine. The other amino acids were not found. Dialysis of RNA which had been disintegrated with ribonuclease showed that all these amino acids are present in peptide form and are found in the diffusate only after acid hydrolysis.

Further experimental material bearing on this subject has been obtained by Potter & Dounce (132) who carried out a detailed analysis of RNA isolated from various sources (pancreas, liver, yeast). They found that on alkaline hydrolysis of RNA by the method of Schmidt & Thannhauser (N-KOH at 37°C) the RNA was not broken down completely into nucleotides. A certain proportion of the nucleotides remained firmly bound in some sort of compounds and could be isolated from the products of hydrolysis of RNA by ion-exchange chromatography. On acid hydrolysis of this fraction, nitrogenous bases and some amino acids were obtained. Purines predominated among the bases. In spite of the difference between the sources from which the RNA had been isolated the amino acids in the various preparations consisted mainly of glutamic acid, valine, alanine, glycine, leucine and threonine. The absence of free carboxyl and amino groups in these combined amino acids and also the finding of a fraction in which amino acids were combined with nucleotides in more than equimolar amounts led the authors to the conclusion that some of these amino acids might be present in the form of small peptides.

We have already referred to the facts concerning the infectivity of viral RNA freed from protein. It is, however, a feature of these experiments that the infective RNA, even after it has been completely freed from protein, contains a certain amount of peptide material (0·3-0·5%)

(134, 167). These peptides have been found to contain serine and aspartic and glutamic acids, which account for 80% of the nitrogen of all the amino acids bound to the RNA. This admixture does not play any significant part in the infectivity of RNA. It would not seem to be a chance contaminant, but rather a residue of the products which were synthesized on the surface of the RNA when it was functioning actively in the cytoplasm of the cells of the host.

The presence of a similar amino acid-peptide fraction in RNA isolated from plants has also been demonstrated in our investigations on the RNA of oats and *Galega* (16). We found that the proportion of amino acids in this fraction of RNA of plant origin was considerably higher than that in RNA derived from other sources. 80-90% of the amino acids in this fraction consisted of glycine, serine, aspartic acid and alanine. Other amino acids were present in insignificant amounts while some (arginine, histidine, phenylalanine and several others) were altogether absent. We also made a very interesting observation which suggests that there is a sharp increase in the peptide fraction in RNA when protein synthesis is stimulated by the addition of nitrogenous material to the nutrient medium. This observation can only mean one thing, namely that these products really are associated in some way with the synthesis of protein and it would seem that they are unfinished products of this synthesis directly combined with RNA. There can be no doubt that further study of the composition and variations of this fraction under different experimental conditions will be very important in solving the problem of the functions of RNA in protein synthesis.

The formation of complexes of RNA and peptides is certainly no artefact or accidental adsorption. It has been shown that pure preparations of nucleic acids react very weakly *in vitro* with amino acids (99, 176) and the isolation of such complexes would therefore seem to be an indication of some biochemical processes going on within the cells. We must, however, note that Mark & Stauff (114) found quite considerable combination of several amino acids with RNA and DNA *in vitro*. In these cases, however, the authors used quite high concentrations of amino acids and prolonged exposure and did not try to imitate the conditions present during the analytical treatment of the tissues of experimental material.

Very interesting results have also been published recently by Habermann (87) who isolated and studied peptides firmly bound to RNA, derived from a number of sources (yeast, liver, brain, tumours), after it had been purified with phenol. The quantity of these peptides amounted to 0·2-1·5% of the weight of the RNA and some of them could be identified in the diffusate after degradation of the RNA with ribonuclease. Peptides containing serine and aspartic and glutamic acids seemed to be most firmly bound to the RNA. The author thinks that these peptides may be intermediate products of protein synthesis or else play a structural part in the macromolecule of RNA.

Some peptide groups can be very firmly bound into the structure of RNA by covalent bonds. This has been demonstrated recently by the detailed studies of Prokof'ev, Bogdanov & Antonovich (20-22). These authors found such peptides in RNA isolated from yeast and pancreas. According to them, these peptides formed firm linkages with the phosphate groups of the polynucleotide chains and could be isolated from products of the degradation of RNA in the form of individual nucleotide-peptide complexes. In his recent researches the Japanese biochemist Ishihara (97) has aimed at studying the character of the linkage between the peptides and RNA in natural preparations of RNA. Having isolated from yeast a preparation of RNA completely free from protein, he found amino acids in the products of its hydrolysis. Not counting glycine, which could have been formed by the breakdown of adenine, the amount of amino acids represented about 1% of the total preparation. They were bound to the RNA in the form of amino acids or peptides and mostly by means of ester linkages and partly by ionic bonds.

In considering the meaning of these results the question naturally arises as to whether these peptide products which are comparatively firmly bound to RNA, often by covalent bonds, may be "incompletely synthesized" proteins which would be easily and freely detached from the place where they were synthesized after the completion of the synthetic cycle. It may be that, in this connection, great interest will attach to peptide compounds of a similar kind which are weakly bound to the organelles by means of adsorption or hydrogen bonds and may be removed from the surfaces of the organelles and matrices as a result of the separation of RNA from protein and its preparative isolation and purification.

The existence of weakly bound amino acids in the cells, as well as free amino acids, has been known for a comparatively long time and it is also a comparatively long time since the idea was put forward that these products were localized actually in the sites of protein synthesis (51, 65). However, the study of the weakly bound peptides has only been begun rather recently. A number of investigations towards this end have recently been undertaken in our laboratory using materials of plant origin (17, 18).

In these experiments we studied the rate of incorporation of [35S]methionine and [32P]phosphates into proteins, RNA and a number of other fractions of the leaves of haricot beans after infiltration. It was found that if the precipitate obtained from plasma juice or a homogenate of the structures by precipitation with alcohol is completely freed of all soluble compounds by washing with 65-70% alcohol, then the further treatment of this precipitate with a 10% solution of NaCl at 100°C for 6 hours to break down its hydrogen bonds will lead to the appearance in the solution, alongside the RNA and a small amount of protein, of a certain amount of polypeptide and polynucleotide products which will not precipitate if treated with alcohol again but will remain in solution. After evaporating the alcohol from the supernatant liquid, these products can easily be separated from the NaCl by slow dialysis through cellophane. This non-diffusible fraction of deproteinized plasma juice or homogenate of structures consisted mainly of peptides and polypeptides with some oligonucleotides.

A solution of these peptides did not give a typical biuret reaction. A similar fraction from the structures gave a weak biuret reaction. On studying the rates of incorporation of [35S]methionine into this fraction in two independent experiments carried out at different times, evidence was obtained that these peptides have a very high rate of renewal (Table 9).

TABLE 9

Radioactivity of various fractions isolated from the leaves of haricot beans two hours after infiltration with [35S]methionine (17)

| Fraction | Radioactivity of preparations (counts/min./2 mg.) | | | |
| --- | --- | --- | --- | --- |
| | First experiment | | Second experiment | |
| | Plasma juice | Structures | Plasma juice | Structures |
| 1. Total proteins | 2,982 | 2,886 | 3,364 | 2,647 |
| 2. Precipitate of protein left behind by dialysis of products soluble in 10% NaCl at 100°C but not precipitated by alcohol | 7,116 | 2,908 | 6,420 | 4,782 |
| 3. Polypeptides and peptides isolated from the total alcoholic precipitate after its treatment with 10% NaCl at 100°C, not diffusible or precipitated with alcohol | 13,244 | 4,180 | 7,463 | 778 |
| 4. RNA + protein before phenol treatment | Not determined | 1,300 | 1,938 | 356 |
| 5. Final alcoholic precipitate of RNA | Not determined | 64 | 1,743 | 19 |

At the same time as the activity was determined, the amount of polynucleotides in these fractions was also determined. The greatest amount of polynucleotides was found in the final precipitate of RNA. However, in the non-diffusible fraction No. 3, the precipitate of which had the highest specific activity in two experiments, a considerable amount of oligonucleotides was also found. According to our unpublished results these oligonucleotides had a higher rate of phosphorus metabolism than the soluble and structural forms of RNA. On chromatograms of un-diffusible polypeptide products of the plasma juice and structures, all the amino acids which could be identified in the proteins showed up clearly. It must be mentioned that the non-diffusible peptide-oligonucleotide fraction of deproteinized tissue-extracts is certainly of great interest but in ordinary schemes of analysis these products are, as a rule, discarded or included among the proteins as a whole.

This fraction contains polypeptide and polynucleotide products which are insoluble in 60-70% alcohol before the deproteinization treatment of the plasma juice and structures and become soluble (in 70% alcohol) after treatment of the precipitates with 10% NaCl at 100°C. This enables us to say that these structures were, in the first place, loosely bound in the protoplasm to some structures and were liberated when their bonds were broken. These compounds, isolated from the plasma juice, were the very ones which had the highest specific activity, greatly exceeding that of the proteins as a whole.

The amount of peptide products in this fraction was about 4-6 times (by weight) that of polynucleotides although it only constituted a fraction of 1% of the total protein. Calculating on the basis of the net weight of the leaves, a considerably greater amount of these products may be isolated from the structures than from the plasma juice (3-4 times as much) though the specific activity of these compounds in the structures is con-siderably lower.

The functional role of this fraction is, however, still not sufficiently clear. The presence of amino acid, peptide and protein complexes in RNA which had been extracted with phenol has, as we have seen, also been noted by many workers. However, in purely analytical studies it always remained uncertain whether these products were intermediates in protein synthesis, whether they were only the result of incomplete de-proteinization of the RNA or whether they represented some special and, perhaps, secondary form of chemical interaction with RNA. Only experi-ments on the metabolic activity of these peptides could give a more definite indication of their possible role in protein synthesis.

An extremely clear and detailed study of the peptide "growth" phase of protein synthesis has recently been made by a group of American bio-chemists (71, 158, 166) who investigated the synthesis of haemoglobin by the reticulocytes of the rabbit in detail in this respect. In the first paper in this series (71) they reported that by heating the microsomes of the reticulocytes it is possible to extract from them peptides and polypeptides which, in experiments with labelled amino acids, were renewed more quickly than the total protein of the microsomes. Continuing these experiments (158, 166) the authors studied this fraction at the level of the ribosomes. In reticulocytes the most active part in the synthesis of haemoglobin is played by the complex 78 S ribosomes. *In vitro*, in conditions of magnesium shortage, these ribosomes dissociate to give particles having sedimentation constants of 60 S and 40 S. Experiments with [$^{14}$C]valine were specially interesting. This is quickly incorporated into the proteins of both components of the 78 S ribosomes. However, if these ribosomes become dissociated into 60 S and 40 S particles shortly after incubation with [$^{14}$C]valine and then become associated again, about 20% of the activity can then be isolated from the medium in peptides in which the valine has a very high specific activity, about 4-7 times the average value for the valine in the protein of the ribosomes. The authors believe that these peptides are "partly built" blocks for the synthesis of protein. In the last chapter we have already referred to the ideas of these authors, who put forward the suggestion that the formation of polypeptide chains may take place in the gap between the coarse and fine components of the 70 S ribosomes (157). In the papers which have been mentioned above this hypothesis receives factual confirmation.

In this same laboratory (70) and also in Schweet's laboratory (46) the synthesis of haemoglobin by the reticulocytes has also been studied recently with regard to the type of "growth" of the polypeptide chain. By studying the specific activity of valine at different stages in the formation of the polypeptides of haemoglobin (valine is the $N$-terminal amino acid of the molecule of globin and occurs in other parts of the polypeptide chain) the authors arrived at the conclusion that the synthesis of the molecule of haemoglobin begins at the $N$-terminal part and proceeds by a consecutive lengthening of the peptide at a rate of about 2 amino acid residues/sec. Almost all the incomplete chains of haemoglobin contained $N$-terminal valine. However, the authors state that this type of growth of the polypeptide chain is not the rule for all proteins.

In fact, Yoshida & Tobita (174), in a study of the synthesis of bacterial α-amylase, found a distribution of [$^{14}$C]leucine in the peptides of the partially hydrolysed polypeptide chain which suggested that the synthesis of this peptide chain begins at the carboxyl end. Evidence in favour of the same view has been obtained in experiments with *E. coli* (83a). Nevertheless, in Campbell's recent, interesting paper (54a) on the synthesis of albumin by the ribosomes of the liver, it is stated that, after incubating the ribosomes for 3 min. with [$^{14}$C]amino acids, the labelled residues were distributed equally throughout the polypeptide chains, which may result from the "random seating" of amino acids on the matrices in such a way that several centres of growth of peptides are formed at once and these peptides later coalesce to form a continuous chain. These results were confirmed by Lingrel & Webster (111a) who studied the synthesis of albumin by the microsomes using a similar experimental arrangement. These differences between bacterial and animal forms of protein biosynthesis were confirmed in some recent studies (80a, 123a).

## 3. Peptide-nucleotide and peptide-polynucleotide compounds

As well as amino acids, nucleotides, peptides, oligonucleotides, and the peptide components of RNA and DNA, there were discovered a few years ago in the cells of animals, plants and microorganisms, various different compounds of a new class, complexes of peptides with nucleotides or oligonucleotides, which show peculiar characteristics as to metabolism and composition. Quantitatively, the new class includes only 1-1·5% of the free nucleotide pool but, nevertheless, it has attracted the attention of many laboratories and some very interesting facts concerning it have been found out in a comparatively short time. One must distinguish these compounds from free nucleotide-amino acid compounds (43, 137), which may well be side products of amino acid activation.

Koningsberger and his colleagues (107, 84) first found, in yeast cells, carboxyl-activated peptides with molecular weights of about 4000. These peptides gave the hydroxamate reaction and were bound by macroergic bonds to nucleotides. These compounds occurred both in the juice and in the microsomes of baker's yeast. A further study of these compounds was made in later experiments in the same laboratory (106a, 140, 141). Cytidine-5'-monophosphate peptides, adenosine-5'-monophosphate peptides and uridine-5'-monophosphate peptides were identified. Several amino acids were also identified in these compounds. Activated peptides containing 4-8 amino acid residues have been found in various materials by Dirheimer and his colleagues (72, 169).

Brown (52, 53) has identified a number of adenyl peptides which he found in cultures of streptococci. He has studied the composition of eight such peptides and found that they contained glutamic acid, glycine and aspartic acid, serine, alanine and cystine in varying proportions.

Bergkvist (44) has described the isolation from a fraction of acid-soluble nucleotides, extracted from *Polyporus squamosus*, of considerable amounts of compounds of cytidylic acid with a peptide in which he identified several amino acids. The peptide was combined with the carbohydrate-phosphate group of the nucleotide.

A series of detailed studies of the nucleotide-peptide compounds of brewer's yeast has been carried out in the last few years by Davies & Harris and their fellow-workers (67-69, 88-90). Among these compounds were mononucleotides combined with peptides as well as oligonucleotides combined with a peptide or with some single amino acid. They isolated and identified Arg. Ala. Arg. Ala-5'-uridylate and Ala-[3'-[5'-adenylyl]]-5'-uridylate and a number of other compounds. Uracil was the most common nucleotide component of these compounds. From a quantitative point of view the nucleotides combined with peptides constituted no more than 1-2% of the total free nucleotides.

The study of the dynamics of the nucleotide-peptides in relation to stage of growth of yeast cultures carried out by these authors (69) did not reveal any obvious regularities either quantitatively or qualitatively. The amino acid composition of these compounds was very varied but the amino acids most commonly encountered were arginine and alanine. The same group of investigators has recently (68) found, in yeasts, a number of oligopeptide-oligonucleotide complexes with high rates of turnover of both the nucleotide and the peptide moieties.

Reports of the finding and investigation of nucleotide-peptide compounds have become particularly numerous in the last year or two. Some Czech biochemists (85, 60, 34) have recently studied an extensive group of these compounds in *E. coli* in the phase of logarithmic growth.

By chromatography on Dowex-1 the nucleotides and nucleotide-peptides were divided into 15 fractions and these gave 34 fractions containing amino acids and bases after further chromatography on paper. The individual fractions contained one (adenine, uracil or cytosine) or two bases (adenine and uracil) and 3-10 amino acids, mainly glutamic acid, cysteine, glycine, lysine, alanine, aspartic acid, threonine, valine and leucine. Determination of the rate of incorporation of [$^{14}$C]lysine into the nucleotide-peptide fraction and into the proteins showed that the formation of the nucleotide-peptides preceded that of the proteins and thus, according to the authors, that the nucleotide-peptides might be intermediate products in the biosynthesis of proteins.

We must also take note of two papers of Ito & Strominger (98) who found in the cells of staphylococci an enzymic system which could combine amino acids in sequence with uridylic nucleotides.

Jones & Lewin (100) isolated from three strains of *Chlamydomonas* about seven different groups of nucleotide-peptides containing nucleotides of all the common bases. The amino acids most frequently encountered in the peptides were cystine, glutamic acid, serine and glycine.

In addition to these researches we must refer to a series of other similar investigations which have led to the reporting of the isolation of nucleotide-peptide components from microorganisms (83, 125, 101, 102). The study of nucleotide-peptide compounds has also extended to materials of animal and plant origin.

In a series of interesting articles Hase *et al.* (91-93) report that they have isolated, from the cells of the alga *Chlorella* and from yeast, a number of complex compounds of amino acids and peptides with nucleotides and polynucleotides.

Sisakyan & Veĭnova (26) recently found such a peptide combined with a nucleotide in the coelomic fluid of the pupa of the mulberry silkworm. Weinstein *et al.* (168) have succeeded in isolating from the liver compounds of peptides with all four of the nucleotides found in RNA.

In later works Veĭnova (2, 3) has given an account of the isolation of eight peptide fractions from freeze-dried preparations of the coelomic fluid of the pupa of the mulberry silkworm. In addition to amino acids one peptide fraction contained phosphorus which was shown to be inorganic. Some of the peptides were combined with nucleotides; this was indicated by the fact that the fractions concerned showed maximum absorption at 255-260 m$\mu$ and contained guanine and adenine as well as phosphorus which was not easily hydrolysed.

The discovery of various complexes of nucleotides with peptides in liver extracts (168) was also extremely interesting. Compounds of peptides with oligonucleotides have also been found recently in such extracts (155, 156). In these latter two studies the authors have shown clearly that protein-free liver extracts contain oligonucleotides each of which is combined with 2-4 carboxyl-activated peptides. Twelve amino acids were identified in these peptides. A special form of peptide-nucleotide compound, namely an adenine ribonucleoside monophosphate peptide, has recently been found in liver (128). Several types of such peptides have been isolated from liver extracts by means of perchloric acid. Eight amino acids were isolated from the extracts, five of which were identified. The authors suggest that such proteins activated by an adenyl component may be used in protein synthesis as well as activated amino acids.

Reports of the finding of nucleotide-peptide compounds in the tissues of animals are to be found in several other papers (36, 128, 129) but in these the linkage between the components is not always an acyl-phosphate one. In some cases there is a suggestion that the peptides may be combined with the ribose residue (129).

Their composition shows that by no means all of the peptide-nucleotide compounds can have any connection with protein synthesis. Wilken & Hansen (171) isolated from liver and studied an adenosine-peptide compound containing D-alanine and taurine as well as ordinary amino acids. Quite a number of complex compounds of bacterial origin have been studied which contain, in addition to nucleotides and peptides, a number of other compounds which play a part in the structure of the cell walls including D-amino acids (63, 148, 149).

The existence of peptide and oligopeptide-nucleotide compounds in the tissues of higher plants has also been established recently (15, 57, 58). A small amount of unidentified peptides has also been observed in the soluble RNA fraction (17, 109a).

It is, however, still difficult to put forward any sufficiently firmly based hypotheses as to the place of nucleotide-peptide compounds in the synthesis of proteins and nucleic acids.

There are, however, a number of facts which make such hypotheses quite tenable. The presence of nucleotide-peptides in the 80 S ribosomes (107, 84), the rapid metabolism of these compounds on administration of either amino acids or nucleotides (68, 69, 34, 27, 28, 90), their increase in quantity during the period of logarithmic growth of bacteria (69) and a number of other facts, provide indirect evidence that they are connected with protein synthesis.  More direct evidence has recently been obtained by Harris & Neal (90) who observed the transfer of the activity of [$^{14}$C]arginine of nucleotide-peptides to proteins when yeast was moved from a radioactive medium to one which was not radioactive.  Harris and colleagues (64a) recently supported the idea of an intermediate role for peptide-nucleotides in protein synthesis by studying the kinetics of incorporation of [$^{14}$C]arginine into the free amino acids, peptide nucleotides and proteins of yeast.  But they showed at the same time that the degradation of protein might lead to the formation of some of these compounds. They suggested that the nucleotide-peptides might fill a dual role, some being involved in *de novo* protein synthesis, and others in protein degradation.

Penn (131) has recently found compounds of coenzyme A (containing nucleotides) with peptides in tissues.  Having labelled these peptides he found that they were rapidly incorporated into proteins in the presence of the mitochondrial fraction.  Ideas about the way in which nucleotide-peptides take part in protein synthesis have also been developed by a group of Polish workers who have observed the incorporation of labelled compounds of this sort into proteins while they are being synthesized (156, 156a).

In this section we shall not yet try to combine all these data into some theoretical scheme of protein synthesis as such schemes will not be discussed until the third part of this book.  However, the ubiquity of these compounds in living nature, their variety and their metabolic characteristics do undoubtedly indicate their biochemical significance.  The problem of the nature, structure and possible functions of nucleotide-peptide compounds has recently been reviewed in detail by Medvedev & Khavkin (18a).

It is interesting, furthermore, that nucleotide-peptide nature has been ascribed to the embryonic inductor of formation of vertebral cartilage in the chick (95a).  This is the first claim to have isolated such an inductor in the pure state.

## 4. Experimental studies of the direct utilization of exogenous and endogenous peptides in protein synthesis

The intermediate "peptide" phase of protein synthesis has been investigated not only by the isolation and study of possible intermediate products, but by introducing supposedly intermediate products, in this case peptides, into the system of protein synthesis and observing their fate.

There are, at present, four such lines of work aimed at revealing the nature of the intermediate products of protein synthesis.  The first of these is the study of the nutrient value of various peptide products by observation of the utilization of peptides for protein synthesis in comparison with the utilization of free amino acids for the same purpose. By using labelled peptides in many experiments and checking whether they are hydrolysed it is possible, in such cases, to judge whether the peptide blocks have really been incorporated into protein structures or not. The second line of work is the study of the incorporation of labelled peptides into proteins when they are "ballasted" with unlabelled amino acids and *vice versa*. The third line is the study of the selectivity of the incorporation of peptides into particular specific proteins and, finally, the fourth line is the study of the uneven distribution of labelled amino acids introduced into the organism among the different peptide components of the protein molecule.

### a) Assimilation of peptides introduced into the organism

*Microbiological material.* In the period between 1950 and 1954 many investigations were made in a number of laboratories into the utilization of peptides by various mutant and normal strains of bacteria. After 1954 the number of such investigations diminished.

These studies were prompted by the observations of a number of authors that some strains of bacteria which require ready-made amino acids for their growth can, in certain cases, assimilate dipeptides of these amino acids more quickly and more completely. Strains were found which could assimilate some amino acids only in peptide form.

The ability of many bacterial cultures to sustain growth at the expense of proteins, polypeptides and peptides is well known. The high activity of bacterial proteases and peptidases suggests that these products are utilized in protein synthesis after their preliminary hydrolysis into amino acids. However, some evidence, mainly obtained from mutant strains, suggests that, in many cases, peptides can be incorporated during protein synthesis without undergoing preliminary hydrolysis. Here we shall only review a few typical investigations of this kind. Among such are, for example, the results of the observations of Fruton & Simmonds (33) on the characteristics of the assimilation of peptides of proline by a mutant strain of *E. coli* which could not synthesize proline. A culture of this sort develops better on media containing dipeptides of proline than on media containing free proline. In such experiments peptides of proline with the most varied other components were generally more effective than proline itself. Generally too, while proline-requiring mutants assimilate the peptides prolyltyrosine and prolylleucine better than free proline, mutants which cannot synthesize the corresponding tyrosine or leucine assimilate these peptides less intensively than equimolar amounts of tyrosine or leucine. Fruton thinks that this fact suggests that proline-requiring strains can assimilate dipeptides even without their previous hydrolysis to amino acids.

The results of Fruton & Simmonds were later confirmed by Stone

& Hoberman (147) who showed that free proline is poorly assimilated because it is quickly oxidized under aerobic conditions while, in the peptide form, it is resistant to the oxidative systems.

The fact of better assimilation of peptides has also been observed in the experiments of other authors (117, 119, 120, 121, 95). Specially interesting observations in this field are to be found in works on the comparative effects of amino acid analogues as inhibitors of the incorporation of amino acids into proteins when they are in free or peptide form. Marshall & Woods (116) found that methyltryptophan inhibits the incorporation of tryptophan into protein by *Streptococcus faecalis*. The addition of excess tryptophan abolishes this inhibition. However, dipeptides of tryptophan are more active in this respect than tryptophan itself and tyrosyltryptophan and tryptophylphenylalanine are especially effective. The authors believe that their results suggest that these dipeptides can enter into the same metabolic pathway in which tryptophan itself is used—Kihara and his colleagues (104, 105) have obtained a lot of evidence pointing in this direction. They tried out a number of amino acid inhibitors and, in all cases, they found that peptides were more active as anti-inhibitors than free amino acids.

Apart from works on the assimilation of synthetic peptides we must also note that generally, in a considerable number of studies of the nutrition of microbes, it has been noticed that products of the incomplete hydrolysis of proteins are assimilated more easily than free amino acids and often stimulate bacterial growth (78, 130). All this material, however, only provides indirect evidence and is susceptible of various interpretations. In most researches of this sort it may be noticed that the strains being studied possess enzymes which can break peptides down to amino acids, and therefore the easier assimilation of the peptides may be connected with some property of amino acids at the time of their liberation by hydrolysis (their different localization, activated state, etc.). Nevertheless the evidence we already have concerning the assimilation of peptides by microbes is of particular interest because, while it does not reveal the mechanism of utilization of peptides it shows that, under some conditions (inhibition, oxidation of amino acids) peptides are the actual material which maintains the growth of cells.

Very interesting results relating to the utilization of peptides of phenylalanine by the bacterium *Lactobacillus arabinosus* for synthesis of a particular specific protein (an adaptive enzyme synthesizing malic acid) have been obtained recently by Ifland *et al.* (96). Study of the relationships of free phenylalanine and its peptides to antimetabolites showed that, in this case, they were very different and this led the authors to conclude that phenylalanine combined in peptide form does not follow just the same metabolic pathway as does free phenylalanine. However, they consider that this does not necessarily mean that peptides are utilized directly but they may be associated with the formation of some activated form of amino acid during the preliminary hydrolysis of the peptides. Leach & Snell (110) have also shown recently that the fact that glycine peptides are better utilized than free glycine may be explained by a difference in the amounts of these compounds gathered at the sites of protein synthesis. A small experiment to study the ability of yeast cells to utilize peptides was carried out several years ago by the author of this book (12). In this experiment peptide products obtained by peptic hydrolysis of yeast proteins which had already been totally labelled with $^{14}C$ were introduced into a yeast culture. In a parallel control experiment the yeast culture was cultivated in a medium containing a complete acid hydrolysate of the same protein. The ratio between the utilization of the products of complete and incomplete hydrolysis in protein synthesis and in oxidative processes leading to the formation of $CO_2$ were studied after determinate intervals of time. During the assimilation of labelled peptides this ratio shifted markedly towards protein synthesis which indicated that the peptides are used in synthesis without first being broken down into amino acids.

It is interesting to note that in the synthesis of physiologically active peptides such as glutathione (41) and penicillin (40) dipeptides corresponding to fragments of the molecules of these compounds are utilized as structural material without first being broken down into amino acids.

*Plant materials.* Among the experimental studies on the assimilation of synthetic peptides by materials of plant origin we may point to the investigations of Sisakyan & Filippovich (25, 29) who observed the synthesis of the proteins of chloroplasts *in vitro* after introducing labelled glycylglycine and leucylglycine into the incubation mixture. Free glycine was only incorporated into the proteins of the chloroplasts to a very slight extent. The weakness of our knowledge of the question of the direct assimilation of peptides by plant tissues was the reason why we undertook a series of experiments aimed at studying the behaviour of peptides generally labelled with $^{14}C$ in the synthesis of plant proteins and in oxidative reactions (10, 11, 14, 19).

The theoretical conditions for the direct utilization of peptides in the normal synthesis of specific proteins is that their structure shall correspond with some component of the protein molecules, that is, the artificial reproduction of the scheme: protein → peptides → new protein. This condition was realized by synthesizing generally-labelled plant proteins by placing the plants in a chamber containing $^{14}CO_2$. Peptide products from the partial degradation of these proteins were then introduced into the leaves of plants of the same species. Apart from the completeness of hydrolysis, different [$^{14}C$]-labelled hydrolysates were equivalent by all physico-chemical and chemical criteria (including their content of tryptophan, which had been broken down by acid hydrolysis). In most of the experiments a comparison was made of the ways in which each of the following hydrolysates was utilized by the plants:

1. complete hydrolysate
2. peptic hydrolysate consisting of large peptides
3. tryptic and pancreatic hydrolysates consisting of small di- and tripeptides
4. incomplete acid hydrolysates in which the peptides are about the same size as those in the tryptic hydrolysate.

All these hydrolysates were introduced into the leaves by vacuum infiltration or through the petiole or cut-off stalk and later the rates of their involvement in the synthesis of proteins and in oxidative reactions leading to the formation of $CO_2$ (precipitated as $BaCO_3$) were measured. In most cases the peptides of enzymic origin became involved in protein synthesis more quickly than the amino acids of complete hydrolysates while their involvement in oxidative reactions was several times slower. The peptides contained in the acid hydrolysates were assimilated in a similar way to complete hydrolysates, which was evidence of the importance of the qualitative composition of the peptides even in the assimilation of products of the degradation of one and the same protein.

The results of such determinations provide indirect evidence of direct utilization of peptides in protein synthesis. If a complete hydrolysate labelled with [14]C is introduced into the leaves and they are kept in the dark, only part of the activity of the amino acid becomes incorporated into the proteins. Owing to the absence of photosynthesis (exposure to darkness) another considerable amount of the amino acids introduced undergoes oxidative breakdown with the elimination of [14]$CO_2$. Assuming that the utilization of peptides in protein synthesis requires their previous breakdown into amino acids then, if the incorporation of [14]C into protein is identical whether incomplete or complete hydrolysates are used, the elimination of activity in the form of carbonic acid should also be similar in amount. If the proteins acquired a higher radioactivity after the introduction of labelled peptides than after the introduction of labelled free amino acids while the oxidative elimination was considerably lower, that would be a definite indication that peptide fragments are used directly in protein synthesis.

In the experiments under discussion we did not only study the direct utilization of peptide products reaching the plant cell, but also the possible selectivity of their incorporation into the protein of particular organelles depending on the peptides being derived from the protein of organelles of the same type. We also studied the alteration with age of the ability of the leaves of the haricot to utilize and work up peptides.

The results of all the experiments undertaken for these purposes with several plants and with yeast cultures have enabled us to draw the following conclusions:

1. The peptide products of enzymic breakdown of proteins are in-

volved more quickly than free amino acids in the synthetic reactions of the system synthesizing these same proteins.

2. The character of this utilization depends on the means of hydrolysis and the peptide products of incomplete acid hydrolysis of proteins seem not to be readily available for direct incorporation during the synthesis of protein and are assimilated less well than those of enzymic hydrolysis, probably only after complete intracellular breakdown. At the same time the extent of enzymic hydrolysis of the proteins (peptic, tryptic or pancreatic) did not substantially alter the way in which the hydrolysates were used.

3. We did not observe any considerable specificity in the incorporation of peptides corresponding with the fractions (plastids, plasma juice, mitochondria + microsomes) from which the original labelled proteins were isolated. This result may, however, be connected with some factors which we have not, as yet, studied: in the first place, with the presence of a considerable assortment of proteins of similar structure in the various fractions; in the second, with the presence of peptide groupings which are repeated in various combinations in the different proteins; and, in the third, with the rather long exposures used in the experiments. It would certainly be a good idea to make a further study of the phenomenon of the specificity of synthesis both in respect of fractions of cells and in respect of organs and species of plants.

4. The ratio between the oxidation of amino acids and peptides and the synthesis of proteins was considerably higher in older leaves although the actual rate of the oxidative processes was less.

*Animal materials.* In the reviews mentioned above concerning the role of peptides in protein synthesis (9, 13) we put forward some evidence for the active assimilation of particular peptides and mixtures of peptide products of enzymic degradation of proteins in tissue cultures the growth of which was often stimulated by the peptides (73, 75, 82). In Morgan's review (122) of the question of the nutrition of tissue cultures many other examples of different observations of this sort are given. Similar results concerning the stimulating effect of peptides on protein synthesis have been obtained in experiments in which tissue slices were incubated, especially slices of the pancreas (24, 126, 127).

Sisakyan & Kuvaeva (27, 28) studied the synthesis of protein in the coelomic fluid of the mulberry silkworm and found that the dipeptide [$^{14}$C]glycylglycine was incorporated into proteins more quickly than free glycine during histogenesis. However, these authors did not show that this incorporation was not preceded by a breakdown of the dipeptide.

Evidence against the possibility of the incorporation of the dipeptide valylleucine into haemoglobin has recently been obtained by Burnett & Haurowitz (54). Although this peptide corresponds with the *N*-terminal segment of haemoglobin it was only found to be utilized by the reticulo-cytes after hydrolysis.

### b) Use of the method of "ballasting" for the study of the character of the reutilization of proteins and peptides without their complete breakdown to amino acids

The principle of the method of "ballasting" is very simple. Proteins or peptides, which are labelled generally or on a single amino acid, are introduced into the organism as such or in mixtures with a considerable excess of unlabelled "ballast" in the form of free amino acids. It is then assumed that if proteins are utilized for new synthesis after complete degradation to amino acids there should be a decrease in the specific activity of the labelled compounds on account of their dilution by the "ballast". If this dilution does not occur, this points to the reutilization of the protein or peptide at an intermediate stage of its breakdown. In a review published in 1955 we gave a detailed description of the results of a number of experiments which actually demonstrated the absence of dilution of labels during the assimilation by tissues of proteins mixed with a "ballast" of free amino acids (4, 7, 79, 172). Similar results were also obtained later by Friedberg et al. (80) and Gavrilova & Konikova (5). "Ballasting" of peptides has also been used in several other experiments (12, 124).

The method of "ballasting" cannot, however, be considered to be unexceptionable, especially for proteins. In the experiments referred to above, the utilization of the proteins of the medium by tissues was studied by analysis of the total tissue proteins but it has now become known that tissues can simply accumulate the proteins of the environment on intra-cellular structures and the study of their reutilization requires determina-tion of the radioactivity of particular individual proteins of the tissues. It is also known that labelled proteins introduced into an organism, especially proteins like albumins and globulins, have quite a different (more selective) distribution from that of free amino acids, and their entry into the cell occurs by means of different forms of permeability. Amino acids being liberated from proteins which are being broken down endo-genously may, therefore, not come into equilibrium with the amino acids circulating in the blood (112).

### c) On the question of the phenomena of selectivity in the utilization of peptides and polypeptides in the synthesis of proteins

If we are to have the clearest possible demonstration of the direct utilization of peptides in the synthesis of proteins we must have some method by which we can directly reveal the presence of peptides from the medium as constituent links of the polypeptide chains of some parti-cular proteins. A method approximating to this might be the demonstration of the occurrence of selective incorporation of peptides into proteins analogous with those which had been partially hydrolysed to produce those peptides. If we could show that a specific peptide fragment of a protein when incorporated in a new protein "finds" its own earlier place in the molecule being formed, that fact would be very significant for solution of the problem of intermediate products in protein biosynthesis. In our review of 1959 we only mentioned one, extremely interesting work by Ebert (74) who tried to decipher the mechanism of the selective stimulation of growth of one organ or another of the chick embryo by planting them out in the chorion allantois and then replacing them with transplants of homologous organs.

In the investigations of many authors, both earlier and more recently, mention is repeatedly made of the interesting fact of the selective influence of some organ transplanted into an embryo, or even of injections of saline solutions of proteins and nucleoproteins from some organ into it, on the growth of the homologous organ in the embryo (23, 30, 31, 38, 111, 170). Ideas as to the mechanism of this effect are very contradictory and the work of Ebert differs from all these investigations in that he tries to establish the nature of the products by which the degenerating transplants affect the homologous organs.

In a number of preliminary experiments Ebert found that the spleen of the chicken transplanted into the chorion allantois of a 9-day embryo of the same breed selectively stimulates the growth of the spleen in the embryo. The increase in the growth of the embryonic spleen was considerable (3-4 times that of the control) and seemed to be very specific. Neither transplants of chicken liver nor of mouse spleen had any effect on the growth of the embryonic spleen. In this connection the following questions presented themselves to the author: Is this phenomenon a result of the selective utilization of the proteins of the transplant or of products of its degeneration or is it the effect of a specific hormonal influence? To clarify this problem the author carried out a series of experiments in which transplants containing [$^{35}$S]proteins were planted out. The labelled organs were obtained from chickens and mice after they had received injections containing [$^{35}$S]methionine. The results of the experiments on the transplantation of these organs are given in Table 10.

The results given in the Table were obtained from analyses of 902 implantation experiments. They demonstrate quite clearly the selectivity of the transfer of the protein material of the transplant to the homologous organ. It must be noted that, when the spleens of mice were implanted in the eggs, no similar selectivity was found and less radioactive material was incorporated in the spleen than in the liver and kidneys. Another control, consisting in the injection of [$^{35}$S]methionine in various concentrations in the presence of an unlabelled implant, did not show any selective incorporation of labels although there was increased growth of the homologous organ. All this points to a selective exchange between

TABLE 10

Incorporation of radioactivity from the proteins of labelled transplants
(specific activity of the tissues of the host/specific activity of the
proteins of the implanted tissue)

Implantation of labelled spleen

Individual series of experiments

|  | | | | |
|---|---|---|---|---|
| Spleen of host | 0·14 | 0·16 | 0·04 | 0·12 |
| Liver of host | 0·04 | 0·07 | 0·018 | 0·05 |
| Kidneys of host | 0·04 | 0·06 | 0·016 | 0·04 |

Implantation of labelled kidneys

|  | | | | |
|---|---|---|---|---|
| Spleen of host | 0·038 | 0·06 | 0·025 | 0·05 |
| Liver of host | 0·038 | 0·05 | 0·031 | 0·06 |
| Kidneys of host | 0·098 | 0·13 | 0·059 | 0·16 |

the implant and the homologous organ of some products more complicated than amino acids. The author's calculations show that after three
days more than 15% of the spleen of the host was constructed of materials
from the transplant. In what form does this transfer take place? The
observations of Ebert proved that the transfer of whole cells is out of the
question, the exchange can only be one of specific proteins or of peptides
or polypeptides. The author thinks the second possibility more probable
but this is still only a speculation and the transfer of whole protein molecules, though less probable, cannot be excluded.

Proof of the selectivity of the incorporation of peptides obtained by
the breakdown of some particular protein during the synthesis of that
same protein might be of theoretical significance in solving the problem
of the direct utilization of peptides and for elucidating the mechanisms
for the reproduction of the specificity of protein molecules. The best way
of investigating this problem would seem to us not to consist in making
observations on the fates of implants which requires that the methods of
implantation should be controlled, but in replacing transplants by injections of labelled products of various stages in the breakdown of the
proteins of homologous organs.

Ebert's results have recently been confirmed by the interesting work
of Mahler et al. (113). In preliminary experiments they found that, when
they injected labelled albumin and the products of its complete and
partial enzymic degradation into an egg in which a chick embryo was
developing, all these components were used for the synthesis of embryonic
proteins at the same rate. When the supernatant fraction of homogenates
of heart and liver containing labelled proteins of these organs were injected
into the embryo it was observed that there was marked stimulation of the
growth of the homologous organs and maximal "deposition" of labelled
material in them. The greatest incorporation of labels was found in the
microsomes and in the supernatant fraction of the acceptor tissues and
the least in the proteins of the nuclei and mitochondria. The authors

think that this selective incorporation is associated with peptide products of the breakdown of the proteins introduced, but experiments on the introduction of peptide lysates of the organs of the donor into the embryo are required to prove this hypothesis and they, unfortunately, have not been done.

It must be noted that, on introduction of homogenates of labelled kidneys and liver of rabbits into the organism of an adult rabbit, selective "deposition" of labels in homologous organs has not been noticed (4). These experiments were, however, conducted for another purpose and neither their plan nor the timing of the analyses were suitable for studying the selectivity of utilization of labelled components introduced into the organism.

A study of the question of the selectivity of the incorporation of peptides into the very proteins from which they were obtained by peptic hydrolysis has been the aim of our work as well (14). We did not study the tissue specificity but the intracellular specificity of their incorporation in synthesis. In our experiments we made separate preparations of peptide-containing lysates labelled with $^{14}C$ of the proteins of the plastids and the cytoplasm of the leaves of haricot beans and studied their incorporation into the proteins of the corresponding organelles. We did not, however, succeed in noticing any marked selectivity in the incorporation of these peptides and they were found to be incorporated to a greater extent into the proteins of cytoplasm irrespective of whether they were obtained by breakdown of plastids or of the cytoplasmic fractions. It would seem that the cytoplasm and plastids have many proteins with a similar specificity. This work must be followed up by a more detailed investigation of the protein fractions and a study of the tissue-, species- and possibly even age-specificity of the utilization of proteins and peptides.

*d) Study of the distribution of labelled amino acids in the internal structure of the proteins synthesized*

An original method of studying the existence and role of intermediate peptide products of the synthesis of proteins was used in several researches by Anfinsen and his colleagues (39, 76, 77, 145, 164) who studied *in vivo* and *in vitro* the ways in which ovalbumin, insulin and ribonuclease are synthesized. The fundamental idea behind this method is the study of the distribution of labelled amino acids introduced into the organisms in the various fragments of the protein molecule. When protein is synthesized from free amino acids only, the specific activity of a labelled amino acid in the various parts of the polypeptide chain should be uniform as the degree of "dilution" of the labelled amino acids by unlabelled amino acids of the same type must be the same for all links of the molecule being synthesized. On the other hand, if peptide and polypeptide products, which had been formed in the tissues before the introduction of the labelled compounds, are used as well as free amino acids, then different parts of the molecule would have a different specific radioactivity. It would be higher in those fragments which were built up from free amino acids and lower in those in which ready-made peptide and polypeptide groups were incorporated. In these experiments the distribution of labelled amino acid in the various parts of the molecule was studied by partial enzymic hydrolysis of the protein, chromatographic separation of the peptides and polypeptides obtained in this way and determination of the specific radioactivity of the amino acid in each component. The results obtained by this investigation showed that the distribution of labelled amino acids relative to unlabelled amino acids of the same sort within the molecules of the protein after they had been synthesized was distinctly uneven, which suggests that there is a peptide stage in the biosynthesis of protein.

The authors have given a detailed theoretical discussion of their results on the uneven distribution of labels in proteins in a review (146). However, they came to the conclusion that the results which they had obtained did not disprove the theory of matrices, as they had thought earlier, because it may be supposed that the intermediate products of synthesis are themselves formed on matrices and that, at any particular moment, the cell may contain a wide range of "unfinished" proteins on matrices. Alternatively, the unevenness of the distribution could occur as a result of the self-renewal of proteins without breakdown of their molecules; however, that possibility seems improbable for the proteins studied by these authors. Furthermore, the uneven distribution of labelled amino acids in the various parts of the protein molecules at times shortly after the injections was observed in proteins in which such partial self-renewal without complete breakdown hardly occurs. We have already (9) gone into these results in detail and should therefore turn to the fresh evidence which has been obtained by these methods.

Gehrmann et al. (81) found that [14C]glycine was unevenly distributed along the polypeptide chain when it was incorporated during the synthesis of the metaplasmatic protein collagen. Similar results were obtained in respect of the incorporation of [14C]glycine into haemoglobin both in vitro and in vivo (108, 109). In all these experiments, as in those of Anfinsen, the irregularity of the incorporation disappeared with an increase in duration of the exposure after introduction of the labelled amino acids.

Some Japanese workers have recently carried out interesting and detailed work in this connection by studying the nature of the distribution of the specific activity of [14C]leucine in α-amylase synthesized by the leucine-requiring strain of Bacillus subtilis (174).

After the bacteria had been cultured on a medium containing [14C]leucine for 24 hours the α-amylase content of their cells rose from 0·08 g. to 0·28 g. in the culture, that is, nearly 3·5 times. The specific activity of the total protein of the cells was, however, considerably higher than that of the α-amylase (more than twice as high) because, according to the authors' results, α-amylase is not formed directly from amino acids in these cells, but from some sort of protein precursor. The authors hydrolysed the α-amylase with trypsin and isolated a large number of peptides from the hydrolysate. They determined the specific activity of the leucine in these compounds and found a very uneven distribution of [$^{14}$C]leucine among the pre-existing molecules of ordinary leucine. The specific activity of the total leucine was 4600 counts/min./μmole while the specific activity of leucine situated near the C-terminal end was 7100 and that of the leucine near the N-terminal end was 2700 counts/min. /μmole. In other parts of the chain (eight peptides were isolated) the activity of the leucine varied between 3400 and 3700 counts/min./μmole.

The authors believe that these results provide definite confirmation of the synthesis of proteins by stages.

## 5. Transpeptidation reactions

In an attempt to explain the synthesis of peptide chains on the basis of the principle of transpeptidation it has been suggested that the main requirement for this synthesis is the presence of peptides (formed by reactions in which ATP takes part). Lengthening of the chains may then occur without increasing the number of bonds and without expenditure of energy by enzymic transfer according to the following scheme:

$$gva + aml + tgp \rightarrow gvamlgp + a + t$$

A detailed survey of work along these lines is to be found in the work of Lestrovaya (8) and it is therefore hardly necessary to repeat here facts which are already generally known. It must, however, be mentioned that reactions of transpeptidation and transamidation are not the essential ones in the synthesis of proteins in cells and fulfil some auxiliary, collateral function. The part played by these reactions is confined to the peptide level, that is to say, they consist of exchanges within the peptide fraction. These reactions may, perhaps, cause some recombination within the free peptide fraction and, in some cases, a certain increase in the size of the peptides. The emergence of long and specific polypeptide chains by this means is not probable. However, when proteins first came into being on the surface of the Earth some such non-enzymic reactions may have formed a good basis for the action of specific biochemical selection of polypeptide structures.

## 6. Critical evaluation of some experimental evidence concerning the direct synthesis of protein from amino acids

Although the idea that peptides are significant intermediate products in protein synthesis is becoming more and more widely accepted, there are, in the literature, accounts of some experimental results which are sometimes interpreted otherwise, suggesting, that is to say, that proteins are synthesized from amino acids at one stroke (32, 55, 56, 94, 65, 112, 142, 144, 165). However, careful analysis of these results shows clearly that they do not give grounds for such a conclusion. In the review to which we have referred above (13) we have already discussed this question and therefore we do not need to go into it again as no works containing any material which is new in principle have been published recently. We remarked that all these investigations do, in fact, indicate that amino acids are the fundamental materials for the synthesis of particular tissue-proteins, adaptive enzymes and milk-proteins and that the peptides which are formed by the incomplete breakdown of other proteins are hardly used for collateral syntheses. However, the methods and arrangements of these experiments did not permit the answering of the question as to whether there are or are not some intermediate products and stages in the synthetic path which begins with amino acids and ends with the formation of macromolecular proteins. They cannot, therefore, serve as grounds for denying the formation of intermediate peptide compounds on the synthetic assembly line.

## Conclusion

In this chapter we have surveyed a very important section of the problem of protein synthesis, namely, that of the presence and nature of intermediate peptide products of this synthesis. A survey of all the material set out here gives rise to a conviction that the first part of this question can be answered affirmatively. As a rule peptides are formed as intermediate products of this synthesis without being set free from the "biochemical conveyor belt" composed of nucleic acids and enzymic systems which, in the last stages of its activity, produces specific proteins. The liberation of peptides may, however, occur, if the final stages in the assembly of the macromolecule are checked (for example, at the end of the lag phase of the growth of bacteria). Peptides corresponding in specificity with those formed on the "conveyor belt" may, in many cases, be used alongside these in the final reactions of synthesis. As we shall see later, the affirmative answer to this problem of the significance of peptides as intermediate products, plays a great part in showing us the rules which underlie the reproducton of the specificity and the mechanisms of variability of proteins. Furthermore, study of these peptide products, these incomplete protein molecules, their metabolism, localization, composition and dynamics in different states, enables us to find methods and approaches to the study of those reactions of protein formation which lie hidden within the cellular organelles, particularly the ribosomes, and thus to complete the molecular picture of the pathway which begins with the activation of amino acids and ends with the production of finished molecules, which are set free from the places where they were synthesized so that they may later take part in carrying out some particular biological functions.

## REFERENCES

1. BLAGOVESHCHENSKIĬ, A. V. (1958). *Biokhimiya obmena azotsoderzhashchikh veshchestv u rastenïi*. Moscow: Izd. Akad. Nauk S.S.S.R.
2. VEĬNOVA, M. K. (1959). *Tezisy Dokladov 8-go Mendeleevskogo S″ezda. Sektsiya Khimii prirodnykh Soedinenïi i Biokhimii*, p. 125.
3. — (1961). *Proc. V int. Congr. Biochem., Moscow*, 9, 86.
4. GAVRILOVA, K. I. (1955). *Byull. eksptl. Biol. i Med.* 39, 36.
5. GAVRILOVA, K. I. & KONIKOVA, A. S. (1954). *Biokhimiya.* 19, 414.
6. KEIL, B. & HRUBEŠOVÁ, M. (1955). *Coll. Czech. chem. Commun.* 20, 713.
7. KUZOVLEVA, O. B. (1954). *Biokhimiya*, 19, 453.
8. LESTROVAYA, N. N. (1958). *Uspekhi biol. Khim.* 3, 97.
9. MEDVEDEV, ZH. A. (1955). *Usp. sovrem. Biol.* 40, 159.
10. — (1956). *Doklady Moskov. sel'skokhoz. Akad. im. K. A. Timiryazeva*, Nauch. Konf., No. 22, p. 345.
11. — (1956). *Biokhimiya*, 21, 288.
12. — (1956). *Uchenye Zapiski Khar'kov. Univ.* 68, 65.
13. — (1959). *Usp. sovrem. Biol.* 47, 3.
14. — (1959). *Biokhimiya*, 24, 94.
15. MEDVEDEV, ZH. A. & FEDINA, A. B. (1961). *Proc. V int. Congr. Biochem., Moscow*, 9, 98.
16. MEDVEDEV, ZH. A. & ZABOLOTSKIĬ, N. N. (1958). *Doklady Moskov. sel'skokhoz. Akad. im. K. A. Timiryazeva*, No. 39, p. 142.
17. MEDVEDEV, ZH. A., ZABOLOTSKIĬ, N. N., SHEN', TS.-S., MO, S.-M., DAVIDOVA, E. G. & DAVIDOV, E. R. (1960). *Biokhimiya*, 25, 1001.
18. MEDVEDEV, ZH. A., ZABOLOTSKIĬ, N. N., SHEN', TS.-S. & MO, S.-M. (1959). *Tezisy Dokladov pervoĭ Konferentsii po nukleinovym Kislotam i nukleoproteidam*, p. 31. Moscow: Izd. Akad. med. Nauk S.S.S.R.
18a. MEDVEDEV, ZH. A. & KHAVKIN, E. E. (1962). *Izvest. Moskov. sel'skokhoz. Akad. im. K. A. Timiryazeva*, No. 2, p. 188.
19. MEDVEDEV, ZH. A. & SHEN', TS.-S. (1959). *Biokhimiya*, 24, 709.
20. PROKOF'EV, M. A., ANTONOVICH, E. G. & BOGDANOV, A. A. (1960). *Biokhimiya*, 25, 931.
21. PROKOF'EV, M. A., BOGDANOV, A. A. & ANTONOVICH, E. G. (1959). *Tezisy Dokladov pervoĭ Konferentsii po nukleinovym Kislotam i Nukleoproteidam*, p. 35. Moscow: Izd. Akad. med. Nauk S.S.S.R.
22. — (1961). *Proc. V int. Congr. Biochem., Moscow*, 9, 71.
23. ROMANOVA, L. K. (1957). *Byull. eksptl. Biol. i Med.* 43, 99.
24. RYCHLIK [RYCHLÍK], I., SHVEĬTSAR, YU. & SHORM [ŠORM], F. (1955). *Doklady Akad. Nauk S.S.S.R.* 104, 283.
25. SISAKYAN, N. M. (1955). In collective work *Sessiya Akademii Nauk S.S.S.R. po mirnomu Ispol'zovanii atomnoĭ Energii, Zasedaniya Otdela biologicheskikh Nauk* (ed. A. I. Oparin *et al.*), p. 172. Moscow: Izd. AN S.S.S.R.
26. SISAKYAN, N. M. & VEĬNOVA, M. K. (1958). *Biokhimiya*, 23, 52.

27. SISAKYAN, N. M. & KUVAEVA, E. B. (1957). *Doklady Akad. Nauk S.S.S.R.* **113**, 873.

28. — (1957). *Biokhimiya*, **22**, 686.

29. SISAKYAN, N. M. & FILIPPOVICH, I. I. (1955). *Doklady Akad. Nauk S.S.S.R.* **102**, 579.

30. TUMANISHVILI, G. D., DZHANDIERI, K. M. & SVANIDZE, I. K. (1956). *Doklady Akad. Nauk S.S.S.R.* **106**, 1107.

31. TUMANISHVILI, G. D., DZHANDIERI, K. M. & SVANIDZE, I. K. (1956). *Doklady Akad. Nauk S.S.S.R.* **107**, 182.

32. FILIPPOVICH, YU. B. (1953). *Uchenye Zapiski Moskov. gosudarst. pedagog. Inst.* **77**, 125.

33. FRUTON, J. S. & SIMMONDS, S. (1949). *Cold Spring Harbor Symposia quant. Biol.* **14**, 55 (1949).

34. ČERNÁ, J., GRÜNBERGER, D. & ŠORM, F. (1961). *Proc. V int. Congr. Biochem., Moscow*, **9**, 116.

35. ABRAHAM, E. P. (1957). *Biochemistry of some peptide and steroid antibiotics.* New York: Wiley.

36. ÅGREN, G. (1961). *Proc. V int. Congr. Biochem., Moscow*, **9**, 79.

37. ANDERSON, N. G. & ALBRIGHT, J. F. (1958). *Fed. Proc.* **17**, 4.

38. ANDRES, G. (1955). *J. exp. Zool.* **130**, 221.

39. ANFINSEN, C. B. & STEINBERG, D. (1951). *J. biol. Chem.* **189**, 739.

40. ARNSTEIN, H. R. V. & MORRIS, D. (1960). *Biochem. J.* **76**, 323.

41. BATES, H. M. (1961). See IV 10, 10a, 78a.

42. BATHURST, N. O. (1953). *J. Sci. Food Agric.* **4**, 221.

43. BERG, P. (1958). *J. biol. Chem.* **233**, 608.

44. BERGKVIST, R. (1958). *Acta chem. Scand.* **12**, 364.

45. BISERTE, G., DRIESENS, J. & DUPONT, A. (1957). *Compt. rend. Soc. biol.* **151**, 1884.

46. BISHOP, J., LEAHY, J. & SCHWEET, R. (1960). *Proc. nat. Acad. Sci. U.S.* **46**, 1030.

47. BORRISS, H. & SCHNEIDER, G. (1955). *Naturwiss.* **42**, 103.

48. BORSOOK, H., DEASY, C. L., HAAGEN-SMIT, A. J., KEIGHLEY, G. & LOWY, P. H. (1948). *J. biol. Chem.* **174**, 1041.

49. — (1948). *J. biol. Chem.* **179**, 705.

50. BRICAS, E. & FROMAGEOT, C. (1953). *Advanc. Protein Chem.* **8**, 1.

51. BRITTEN, R. J., ROBERTS, R. B. & FRENCH, E. F. (1955). *Proc. nat. Acad. Sci. U.S.* **41**, 863.

52. BROWN, A. D. (1958). *Biochim. biophys. Acta*, **30**, 447.

53. — (1959). *Biochem. J.* **71**, 5P.

54. BURNETT, P. & HAUROWITZ, F. (1961). *Fed. Proc.* **20**, 389.

54a. CAMPBELL, P. N. (1961). *Proc. V int. Congr. Biochem., Moscow*, **2**, 195.

55. CAMPBELL, P. N., JONES, H. E. H. & STONE, N. E. (1956). *Nature (Lond.)*, **177**, 138.

56. CAMPBELL, P. N. & STONE, N. E. (1957). *Biochem. J.* **66**, 19.

57. CARNEGIE, P. R. (1961). *Biochem. J.* **78**, 687.

58. — (1961). *Biochem. J.* **78**, 697.

59. CARNEGIE, P. R. & SYNGE, R. L. M. (1961). *Biochem. J.* **78**, 692.

60. ČERNÁ, J., GRÜNBERGER, D. & ŠORM, F. (1961). *Coll. Czech. chem. Commun.* **26**, 1212.

61. CHEN, P. S. & BALTZER, F. (1958). *Nature (Lond.)*, **181**, 98.

62. CHEN, P. S. & KÜHN, A. (1956). *Z. Naturforsch.* **11B**, 305.

63. COMB, D. G. (1961). *Fed. Proc.* **20**, 351.

64. CONNELL, G. E. & WATSON, R. W. (1957). *Biochim. biophys. Acta*, **24**, 226.

64a. COOPER, A. H., HARRIS, G., NEAL, G. E. & WISEMAN, A. (1963). *Biochem. biophys. Acta*, **68**, 68.
65. COWIE, D. B. & WALTON, B. P. (1956). *Biochim. biophys. Acta*, **21**, 211.
66. DAGLEY, S. & JOHNSON, A. R. (1956). *Biochim. biophys. Acta*, **21**, 270.
67. DAVIES, J. W. & HARRIS, G. (1960). *Biochim. biophys. Acta*, **45**, 28.
68. DAVIES, J. W., HARRIS, G. & NEAL, G. E. (1961). *Biochim. biophys. Acta*, **51**, 95.
69. DAVIES, J. W. & HARRIS, G. (1960). *Biochim. biophys. Acta*, **45**, 39.
70. DINTZIS, H. M. (1961). *Proc. nat. Acad. Sci. U.S.* **47**, 247.
71. DINTZIS, H. M., BORSOOK, H. & VINOGRAD, J. (1958). In *Symposium on microsomal particles and protein synthesis* (ed. R. B. Roberts), p. 95. London: Pergamon Press.
72. DIRHEIMER, G., WEIL, J. H. & EBEL, J. P. (1958). *Compt. rend. Acad. Sci., Paris*, **246**, 3384.
73. EAGLE, H. (1955). *Proc. Soc. exp. Biol. Med., N.Y.* **89**, 96.
74. EBERT, J. D. (1954). *Proc. nat. Acad. Sci. U.S.* **40**, 337.
75. FISHER, A. (1950). *Biochem. J.* **50**, 491.
76. FLAVIN, M. (1954). *J. biol. Chem.* **210**, 771.
77. FLAVIN, M. & ANFINSEN, C. B. (1954). *J. biol. Chem.* **211**, 375.
78. FOX, E. N. (1961). *J. biol. Chem.* **236**, 166.
79. FRANCIS, M. D. & WINNICK, T. (1953). *J. biol. Chem.* **202**, 273.
80. FRIEDBERG, W., WALTER, H. & HAUROWITZ, F. (1955). *J. Immunol.* **75**, 315.
80a. FRIEDRICH-FREKSA, H. (1961). In *Protein biosynthesis* (ed. R. J. C. Harris), p. 345. New York: Academic Press.
81. GEHRMANN, G., LAUENSTEIN, K. & ALTMAN, K. I. (1956). *Arch. Biochem. Biophys.* **62**, 509.
82. GERARDE, H. W. & JONES, M. (1953). *J. biol. Chem.* **201**, 553.
83. GILBERT, D. A. & YEMM, E. W. (1958). *Nature (Lond.)*, **182**, 1745.
83a. GOLDSTEIN, A. & BROWN, B. J. (1961). *Biochim. biophys. Acta*, **53**, 438.
84. GRINTEN, C. O. van der, SCHUURS, A. H. W. M. & KONINGSBERGER, V. V. (1958). *Proc. IV int. Congr. Biochem., Vienna*, **15**, 30.
85. GRÜNBERGER, D., ČERNÁ, J. & ŠORM, F. (1960). *Coll. Czechoslov. chem. Commun.* **25**, 2800.
86. GUSTAFSON, T., HJELTE, M. B. & HASSELBERG, I. (1952). *Exptl. Cell Research*, **3**, 275.
87. HABERMANN, V. (1959). *Biochim. biophys. Acta*, **32**, 297.
88. HARRIS, G. & DAVIES, J. W. (1959). *Nature (Lond.)*, 184, 788.
89. HARRIS, G., DAVIES, J. W. & PARSONS, R. (1958). *Nature (Lond.)*, **182**, 1565.
90. HARRIS, G. & NEAL, G. E. (1961). *Biochim. biophys. Acta*, **47**, 122.
91. HASE, E., MIHARA, S. & OTSUKA, H. (1959). *J. gen. appl. Microbiol. (Tokyo)*, **5**, 43.
92. HASE, E., MIHARA, S., OTSUKA, H. & TAMIYA, H. (1959). *Biochim. biophys. Acta*, **32**, 298.
93. — (1959). *Arch. Biochem. Biophys.* **83**, 170.
94. HEIMBERG, M. & VELICK, S. F. (1954). *J. biol. Chem.* **208**, 725.
95. HIRSCH, M.-L. & COHEN, G. N. (1953). *Biochem. J.* **53**, 25.
95a. HOMMES, F. A., LEEUWEN, G. van, & ZILLIKEN, F. (1962). *Biochim. biophys. Acta*, **56**, 320.
96. IFLAND, P. W., BALL, E., DUNN, F. W. & SHIVE, W. (1958). *J. biol. Chem.* **230**, 897.
97. ISHIHARA, H. (1960). *J. Biochem. (Tokyo)*, **47**, 196.
98. ITO, E. & STROMINGER, J. L. (1960). *J. biol. Chem.* **235**, PC 5, PC 7.
99. JARDETZKY, C. D. (1958). *J. Amer. chem. Soc.* **80**, 1125.
100. JONES, R. F. & LEWIN, R. A. (1961). *Exptl. Cell Research*, **22**, 86.

101. JONSEN, J., LALAND, S., SMITH-KIELLAND, I. & SÖMME, R.  (1959).   *Acta chem. Scand.* **13**, 838.

102. JONSEN, J., LALAND, S. & SMITH-KIELLAND, I.  (1959).   *Acta chem. Scand.* **13**, 836.

103. KAUFFMANN, T. & KOSEL, C.  (1959).  *Biochem. Z.* **331**, 377.

104. KIHARA, H., KLATT, O. A. & SNELL, E. E.  (1952).  *J. biol. Chem.* **197**, 801.

105. KIHARA, H. & SNELL, E. E.  (1955).  *J. biol. Chem.* **212**, 83.

106. KONDO, Y. & WATANABE, T.  (1957).  *Nippon Sanshigaku Zasshi,* **26**, 298.

106a. KONINGSBERGER, V. V.  (1961).  In *A Symposium on protein biosynthesis* (ed. R. J. C. Harris), p. 207.  New York: Academic Press.

107. KONINGSBERGER, V. V., GRINTEN, C. O. VANDER & OVERBEEK, J. T. G.  (1957). *Biochim. biophys. Acta,* **26**, 483.

108. KRUH, J., DREYFUS, J.-C. & SCHAPIRA, G.  (1960).  *J. biol. Chem.* **235**, 1075.

109. KRUH, J., DREYFUS, J.-C., SCHAPIRA, G. & PADIEU, P.  (1957).  *J. biol. Chem.* **228**, 113.

109a. LANE, B. G. & LIPMANN, F.  (1961).  *J. biol. Chem.* **236**, PC 80.

110. LEACH, F. R. & SNELL, E. E.  (1959).  *Biochim. biophys. Acta,* **34**, 293.

111. LEVANDER, G.  (1945).  *Nature (Lond.),* **155**, 148.

111a. LINGREL, J. B. & WEBSTER, G.  (1961).  *Biochem. biophys. Research Commun.* **5**, 57.

112. LOFTFIELD, R. B. & HARRIS, A.  (1956).  *J. biol. Chem.* **219**, 151.

113. MAHLER, H. R., WALTER, H., BULBENKO, A. & ALLMANN, D. W.  (1958). *Symposium Inform. Theory Biol., Gatlinburg, Tenn.,* 1956 (ed. H. P. Yockey, R. L. Platzman & H. Quastler), p. 124.  New York: Pergamon Press.

114. MARK, W. & STAUFF, J.  (1958).  *Naturwiss.* **45**, 544.

115. MARKOVITZ, A. & STEINBERG, D.  (1957).  *J. biol. Chem.* **228**, 285.

116. MARSHALL, J. H. & WOODS, D. D.  (1952).  *Biochem. J.* **51**, ii.

117. McFADDEN, M. L. & SMITH, E. L.  (1955).  *J. biol. Chem.* **214**, 185.

118. McMANUS, I. R.  (1958).  *J. biol. Chem.* **231**, 777.

119. MIKEŠ, O., SCHUH, V. & ŠORM, F.  (1958).  *Chem. Listy,* **52**, 1801.

120. MILLER, A., NEIDLE, A. & WAELSCH, H.  (1955).  *Arch. Biochem. Biophys.* **56**, 11.

121. MILLER, H. K. & WAELSCH, H.  (1952).  *Arch. Biochem. Biophys.* **35**, 184.

122. MORGAN, J. F.  (1958).  *Bact. Revs.* **22**, 20.

123. MORTON, A. G. & BROADBENT, D.  (1955).  *J. gen. Microbiol.* **12**, 248.

123a. NAUGHTON, M. A. & DINTZIS, H. M.  (1962).  *Proc. nat. Acad. Sci. U.S.* **48**, 1822.

124. NIZET, A. & LAMBERT, S.  (1954).  *Bull. Soc. Chim. biol.* **36**, 307.

125. O'BRIEN, P. J. & ZILLIKEN, F.  (1959).  *Biochim. biophys. Acta,* **31**, 543.

126. OGATA, K., NOHARA, H. & MORITA, T.  (1957).  *Biochim. biophys. Acta,* **26**, 656.

127. OGATA, K., OGATA, M., MOCHIZUKI, Y. & NISHIYAMA, T.  (1956).  *J. Biochem. (Tokyo),* **43**, 653.

128. ONDARZA, R. N. & AUBANEL, M.  (1960).  *Biochim. biophys. Acta,* **44**, 381.

129. ONDARZA, R. N.  (1961).  *Proc. V int. Congr. Biochem., Moscow,* **3**, 135.

130. MIKEŠ, O. & ŠORM, F.  (1958).  *Chem. Listy,* **52**, 1975.

131. PENN, N. W.  (1961).  *Fed. Proc.* **20**, 388.

132. POTTER, J. L. & DOUNCE, A. L.  (1956).  *J. Amer. chem. Soc.* **78**, 3078.

133. RAACKE, I. D.  (1957).  *Biochem. J.* **66**, 101, 110, 113.

134. RAMACHANDRAN, L. K. & FRAENKEL-CONRAT, H.  (1958).  *Arch. Biochem. Biophys.* **74**, 224.

135. RAMACHANDRAN, L. K. & WINNICK, T.  (1957).  *Biochim. biophys. Acta,* **23**, 533.

136. REINDEL, F. & HOPPE, W.  (1952).  *Chem. Ber.* **85**, 716.

137. REITH, W. S.  (1956).  *Nature (Lond.),* **178**, 1393.

138. RENDI, R., MILIA, A. DI & FRONTICELLI, C. (1958). *Biochem. J.* **70**, 62.
139. RUSSO-CAIA, S. (1957). *Ricerca sci.* **27**, 2757.
140. SCHUURS, A. H. W. M., KLOET, S. R. DE & KONINGSBERGER, V. V. (1960). *Biochem. biophys. Research Commun.* **3**, 300.
141. SCHUURS, A. H. W. M. & KONINGSBERGER, V. V. (1960). *Biochim. biophys. Acta,* **44**, 167.
142. SIMPSON, M. V. & VELICK, S. F. (1954). *J. biol. Chem.* **208**, 61.
143. SNELLMAN, O. & DANIELSSON, C. E. (1953). *Exptl. Cell Research,* **5**, 436.
144. SPIEGELMAN, S., HALVORSON, H. O. & BEN-ISHAI, R. (1955). In *A Symposium on Amino Acid Metabolism* (ed. W. D. McElroy & H. B. Glass), p. 124. Baltimore, Md.: Johns Hopkins Press.
145. STEINBERG, D. & ANFINSEN, C. B. (1952). *J. biol. Chem.* **199**, 25.
146. STEINBERG, D., VAUGHAN, M. & ANFINSEN, C. B. (1956). *Science,* **124**, 389.
147. STONE, D. & HOBERMAN, H. D. (1953). *J. biol. Chem.* **202**, 203.
148. STROMINGER, J. L. (1960). *Physiol. Revs.* **40**, 55.
149. STROMINGER, J. L., SCOTT, S. S. & THRENN, R. H. (1959). *Fed. Proc.* **18**, 334.
150. SYNGE, R. L. M. (1949). *Quart. Revs. chem. Soc.* **3**, 245.
151. — (1953). In *The chemical structure of proteins* (ed. G. E. W. Wolstenholme & M. P. Cameron), p. 43. London: Churchill.
152. — (1959). *Naturwiss. Rundschau,* **1**, 1.
153. SYNGE, R. L. M. & WOOD, J. C. (1958). *Biochem. J.* **70**, 321.
154. SYNGE, R. L. M. & YOUNGSON, M. A. (1961). *Biochem. J.* **78**, 708.
155. SZAFRAŃSKI, P., SULKOWSKI, E. & GOŁASZEWSKI, T. (1959). *Nature (Lond.),* **184**, 1940.
156. SZAFRAŃSKI, P., SUŁKOWSKI, E., GOŁASZEWSKI, T. & HELLER, J. (1960). *Acta biochim. Polon.* **7**, 151.
156a. SZAFRAŃSKI, P. & BAGDASARIAN, M. (1961). *Postępy Biochem.* **7**, 49.
157. TISSIÈRES, A., SCHLESSINGER, D. & GROS, F. (1960). *Proc. nat. Acad. Sci. U.S.* **46**, 1450.
158. TS'O, P. O. P. & VINOGRAD, J. (1961). *Biochim. biophys. Acta,* **49**, 113.
159. TUBOI, S. & HUZINO, A. (1960). *Arch. Biochem. Biophys.* **86**, 309.
160. TURBA, F. & ESSER, H. (1953). *Angew. Chem.* **65**, 256.
161. — (1955). *Biochem. Z.* **327**, 93.
162. TURBA, F., HÜSKENS, G., BÜSCHER-DAUBENBÜCHEL, L. & PELZER, H. (1956). *Biochem. Z.* **327**, 410.
163. TURBA, F., LEISMANN, A. & KLEINHENZ, G. (1957). *Biochem. Z.* **329**, 97.
164. VAUGHAN, M. & ANFINSEN, C. B. (1954). *J. biol. Chem.* **211**, 367.
165. VELICK, S. F. (1956). *Biochim. biophys. Acta,* **20**, 228.
166. WALLACE, J. M., SQUIRES, R. F. & TS'O, P. O. P. (1961). *Biochim. biophys. Acta,* **49**, 130.
167. WECKER, E. (1959). *Z. Naturforsch.* **14B**, 370.
168. WEINSTEIN, C., HAMMOND, D., BERKMAN, J. I., GALLOP, P. M. & SEIFTER, S. (1958). *Fed. Proc.* **17**, 332.
169. WEIL, J. H., DIRHEIMER, G. & EBEL, J. P. (1958). *Proc. IV int. Congr. Biochem., Vienna,* **15**, 21.
170. WEISS, P. (1955). In *Biological specificity and growth* (ed. E. Butler), p. 195. Princeton, N.Y.: Princeton University Press.
171. WILKEN, D. R. & HANSEN, R. G. (1961). *J. biol. Chem.* **236**, 1051.
172. WINNICK, R. E. & WINNICK, T. (1953). *Anat. Record,* **115**, 456.
173. WINNICK, T., WINNICK, R. E., ACHER, R. & FROMAGEOT, C. (1955). *Biochim. biophys. Acta,* **18**, 488.
174. YOSHIDA, A. & TOBITA, T. (1960). *Biochim. biophys. Acta,* **37**, 513.
175. ZSOLDOS, F. (1957). *Naturwiss.* **44**, 566.
176. ZUBAY, G. & DOTY, P. (1958). *Biochim. biophys. Acta,* **29**, 47.

# GENERAL CONCLUSION TO PART I

In the first part of this book we have, so far as possible, reviewed systematically the essential path along which proteins are synthesized from amino acids which undergo selective carboxyl activation, selective acceptance by molecules of S-RNA and transfer to the cytoplasmic organelles, on to matrices which take the form of RNA. The polymerization of amino acids also takes place in the organelles, especially the ribosomes, with the intermediate formation of peptides.

Taking part in this process are varied and specific activating enzymes, a large assortment of molecules of S-RNA, complexes of transfer enzymes and S-RNA exchange enzymes, a number of co-factors such as ATP, GTP, etc. carrying out energy-transfer functions and, finally, ribosomes containing matrices and actively synthesizing proteins under conditions in which aggregates are formed between a fine and a coarse ribosome in the groove between which it would appear that the new polypeptide chain is formed. The formation of this chain is controlled by molecules of RNA in such a way that a particular sequence of nucleotides evokes a particular sequence of amino acids.

However, notwithstanding the rather large amount of factual material which has been set out briefly and which represents a colossal amount of work carried out in this field in hundreds of laboratories, and many different theoretical concepts, notwithstanding the striking progress in understanding a number of the stages of protein synthesis yet, in many sections of the problem, there are still no clear hypotheses and many very important questions have still not been discussed in our monograph. These questions flow logically from analysis of the factors involved in protein synthesis. If messenger RNA attached to ribosomes really determines the sequence of amino acids, then the mechanism by which it does so must be discovered. If the molecules of S-RNA are complementary to particular parts of the matrices, then how is his complementarity created? What is the molecular nature of the selectivity of the activating enzymes, the molecules of RNA and the transfer enzymes? Where are the ribosomes formed? How does the formed protein become separated from the ribosome? How and where are the matrices synthesized and what is the relationship between DNA and RNA? All these problems require solution and all of these solutions are necessary for the formation of logically based objective and valid concepts as to the mechanism of the biological synthesis of proteins.

However, before we go on to discuss all these questions, it seems to us appropriate to take a wider look at the real variety of forms of biosynthesis of proteins in nature, to analyse certain specific peculiarities, at least of a few actual syntheses, and thus gather together more extensive material which would enable us to be deeper and more varied in our approach to the general rules of synthesis and would serve as a wider and more promising basis for some general concepts of protein synthesis.

# PART II

# THE SPECIAL FEATURES OF DIFFERENT FORMS OF BIOLOGICAL SYNTHESIS OF PROTEINS

# INTRODUCTION

In the first part of this book we reviewed the general features of protein synthesis and a number of successive stages in the process, using material and concepts derived from different fields of biochemistry. Such an approach to the problem of protein synthesis may be considered justified, as the metabolism of proteins and nucleic acids is the fundamental process of the vital activity of all forms of living material whatever their complexity. We may find very many general similarities between protein synthesis amongst viruses, bacteria, plants and animals and in the formation of adaptive enzymes and antibodies, albumin, collagen, etc.

At the same time there are specific differences between one synthesis and another because the conditions under which they take place are by no means uniform, and analysis of these peculiarities may play a substantial part in leading to an understanding of the general rules of protein synthesis.

The differences between the various forms of synthesis are very multiform. Although the synthesis of proteins with most viruses is associated only with RNA yet in phages, on the contrary, it has been found to depend on DNA.

Although the action of protein antigens leads to the formation of specific antibodies having structures which bring about the precipitation of the antigens, yet, in the formation of adaptive enzymes, a non-protein inducer evokes the formation of an enzyme for which the inducer itself serves as the substrate.

In some cases proteins are formed and function in the form of fairly small molecules, in others complicated aggregates are formed, etc. Analysis of the biochemical conditions necessary for the manifestation of these specific features of different syntheses is extremely important for the study of the fundamental features of protein synthesis and the accretions acquired during evolution which confer on protein metabolism inexhaustible possibilities of accomplishing and developing the multitudinous complicated and varied physiological and biochemical functions of living matter.

In the second part of this book we shall only discuss the fundamental features of a few forms of protein biosynthesis, as the discussion of even a majority of the forms of synthesis now being studied would be a difficult task.

It must be stated that it would have been possible to write a whole monograph on each of the problems with which we shall deal in the following chapters, as each of the biological phenomena reviewed here with reference to protein synthesis is very many-sided and has been studied intensively from different angles. Therefore we shall limit ourselves to a review of the general rules of these phenomena, concentrating our attention mainly on those peculiarities of these syntheses which will give a better understanding of the general rules and mechanisms of the synthesis of proteins in the living cell.

CHAPTER VII

# SOME SPECIAL ASPECTS OF THE SYNTHESIS OF PROTEINS IN THE REPRODUCTION OF VIRUSES

## Introduction

The extensive literature on the biological and biochemical problems of the reproduction of viruses published up to 1956 has been quite fully set out in the monograph of K. S. Sukhov (18) and in a number of symposia on viruses which have been held in recent years. Many reviews on this subject have been published in various books and periodicals (12, 13, 110, 21, 24, 30, 39, 80, 47, 48, 62, 118, 128, 131, 101 and others). A particularly comprehensive review of all aspects of the biochemistry of viruses has recently been brought out as a monograph in three volumes edited by Burnet & Stanley (152).

The synthesis of viral protein is primarily interesting from the point of view of the part played by RNA and DNA in the process, because the nucleoprotein nature of viruses shows up the relationships between protein and nucleic acid in their "purest" form.

Another aspect of the study of the synthesis of viral protein, which is of considerable theoretical interest for the understanding of the factors determining the specificity of the arrangement of amino acids, consists in the fact that, during the reproduction of viruses, many of the preliminary reactions of the synthesis are carried out by means of the enzymic systems of the cells of the host. These fundamental peculiarities of viral synthesis provide a very great deal of valuable information towards an understanding of the various aspects of protein synthesis.

The comprehensive reviews of Kriviskiĭ (13), Tovarnitskiĭ & Tikhonenko (20), Spirin (16) and Sinsheimer (139) cover the material in this field published up to the beginning of 1960. Nevertheless we shall spend a little time on these questions as they are important for the formulation of contemporary concepts of the synthesis of proteins. In doing so, logical sequence will make it impossible for us to avoid setting out, however shortly, many facts which are already well known but which have been important in determining the direction of investigation in this field and have attracted great attention to the problem of the synthesis of viruses from biologists of the most widely differing specialities. However, we shall try to pay our main attention to a number of the most recent

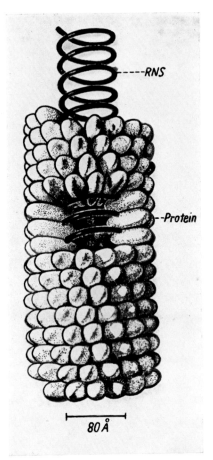

RNS

Protein

80 Å

FIG. 27. Model of a particle of tobacco mosaic virus, constructed by R. Franklin.

achievements in this field and to assess their theoretical significance within the scheme of this monograph, without attempting, in this chapter, any exhaustive synopsis of the synthesis of the proteins and nucleic acids of viruses.

## 1. Some facts concerning the structure and specificity of viral proteins and nucleic acids

A considerable volume of evidence has now been collected in regard to the specificity of the structure of the protein component of various forms and strains of viruses and of the nucleic acids of which they are composed (96, 92, 49, 119, 121, 124, 131, etc.). The minute structure of many forms of viral particles, especially those of the tobacco mosaic virus (TMV), has also been studied rather fully. It has been shown that virus particles of TMV consist of long ribbons of RNA surrounded by a protein envelope which consists of 2100 separate polypeptide chains each of which is formed by 157 amino acid residues arranged in the same way in each of the molecules (64, 65, 84, 95, 128, 131). Another fact of theoretical interest is that the high-polymeric RNA of TMV consists of a long ribbon of a single polynucleotide having the form of a helix and composed of about 6000 individual nucleotides (17, 73, 74, 76, 79, 85, 131).

We have already dealt, in Chapter II, with the results obtained by Reddi (122, 123) who began the investigation of the sequence of nucleotides in the RNA of TMV and established a number of differences in this respect between different strains of virus.

The study of the sequence of amino acids in the protein particles of TMV and in the proteins of other viruses has been begun in many laboratories in the last few years (121, 158, 160, 83). This task was first completed for the protein of TMV in the laboratories of Schramm and Fraenkel-Conrat. In 1960 papers from both laboratories gave an account of the sequence of all the 157 residues in the particles of the viral protein (27, 150a). Thus, if the nucleotide structures of the RNA of TMV were also to be worked out in the next few years, this would enable us to realize a very fascinating idea, namely the comparison of the sequence of amino acids in a protein with the sequence of nucleotides in what is undoubtedly its matrix. It is interesting to note that, according to preliminary evidence, some differences have also been found in the sequences of amino acids in the peptide chains of different strains of TMV (25, 29, 97, 120, 158a, 160).

In the intact particle of TMV the molecule of RNA forms a helix of about 130 turns around an axis, each turn consisting of about 50 nucleotides (125, 157). Fig. 27 shows a model of TMV reconstructed by Franklin (70) (quoted in 157). In the picture one may see the spiral of RNA (black) laid bare in certain places by removal of some of the protein molecules (white) which are closely packed around it. The molecular weight of each such sub-unit of protein is 18,800.

If we compare the length of the molecule of RNA with the length of the polypeptide of the sub-unit of the viral protein it will be seen that one amino acid will correspond with each 40 nucleotides.

A very interesting property of the protein sub-units of TMV and many other viruses is their ability to aggregate specifically without depending on the presence of the ribbon of RNA. In doing so they form structures resembling washers with the holes corresponding with the diameter of the ribbon of RNA. These washers can combine together into rods which resemble the intact virus externally but do not contain the internal filling of RNA (64, 130).

Spirin, Gavrilova & Belozerskiĭ (5, 17) have obtained very important results concerning the physico-chemical structure of viral RNA.

We do not consider it necessary to go into any further detail concerning the structure of viruses, especially TMV, as a full review of the very extensive literature on the structure of viral particles is not part of our task. All that is important for us in this matter is to direct attention to the indubitable fact that both the RNA and the protein of the viral particles are extremely specific compounds, both as regards their composition and structure and, therefore, although crystalline virus particles are not exactly typical living structures, nevertheless, when they are reproduced a characteristic biological synthesis of proteins must take place in the cellular protoplasm involving all the complicated "tasks" which are imposed on biological synthetic mechanisms in regard to the reproduction of specific microstructures and the unique sequences of amino acid and nucleotide linkages in the composition of the proteins and nucleic acids of the viral particles being reproduced.

## 2. First stages in the synthesis of viral particles

A virus multiplies within the protoplasm of the cells of the host using the enzymic system of the host for this purpose. Therefore, before we can ask the question as to the mechanism by which the specificity of viral protein and RNA are reproduced, we must naturally pay attention to the first stages of this protein synthesis and, above all, to the mechanism of activation of amino acids in the synthesis of viral proteins and possible intermediate products of this synthesis.

According to the well-known English virologist Bawden (29) the process of multiplication of viruses must be regarded as a deviant form of the metabolism of nucleoproteins in which the whole cell participates as a synthesizing system while the virus is one of many factors determining the course of the metabolism and its final result.

There is much evidence to show that the formation of new viral-protein particles occurs mainly at the expense of the pool of free non-protein nitrogen in plant and animal cells (cf. the reviews by Sukhov (18, 19)). The original materials from which the peptide structures are built are amino acids activated by a specific enzymic system of the cells of the host. From the review of the processes concerned in this activation, which we made in the first part of this book, we saw that the activating system consists of a complicated collection of enzymes and fractions of soluble RNA localized in the cell sap, the fraction of RNA being low-molecular, non-homogeneous and specific and not being interchangeable with RNA isolated from other organelles of the cell.

Viral protein has no enzymic activity and does not take part in the process of its own synthesis. However, as regards viral RNA, most authors have no doubt that it serves as a matrix for the synthesis of viral protein. Whether this RNA takes part in the formation of the soluble, low-molecular RNA which accepts the amino acids we cannot yet say for certain. There is some evidence (88) that a part of the viral RNA undergoes fragmentation after infection (experiments with labelled RNA of influenza virus) but it is not yet known whether these fragments participate in the transfer of amino acids.

The solution of the problem of the part played by viral RNA in activation is of theoretical importance for the analysis of the factors which guarantee the specificity of protein synthesis. If the transfer of activated amino acids during the synthesis of viral protein is brought about by fractions of the S-RNA of the host, this will indicate quite definitely that this RNA does not play an essential part in determining the sequence of amino acids in the proteins, which must then depend on the RNA of high molecular weight alone. Such a result would also indicate that the nucleotide code was the same for the viral RNA as for that of the host, because the code of the matrices is, in fact "moulded" by the molecules of S-RNA.

The phenomenon of susceptibility of the host to the virus is, to a considerable extent, associated with the properties of the protein of the envelope of the viral particle. Gordon & Smith (79a) have shown recently that the narrow specificity of tobacco mosaic viruses in respect of their hosts is determined by their protein components, though the viral RNA is capable of reproducing and controlling the synthesis of virus in a considerably wider circle of hosts.

There is a lot of evidence to suggest that the protein envelope of viruses has a particular protective function and that after the virus enters the cell there is a process of rupture of the envelope with "liberation" of the RNA (93). The mechanism of the successive reactions is still far from clear. What stands out clearly is only its result, the intensive synthesis of new molecules of RNA and new molecules of protein which then aggregate to form new supplies of viral particles. The nature of these "hidden" reactions in the reproduction of viruses is being studied energetically. When gentle methods are used for the isolation of viruses from plant cells the virus preparations contain microsome-like particles (117) which have recently been given the name of virosomes (32) and this gives the impression that during reproduction within the cell viral RNA must be incorporated into the microsomal fraction. Recently (92a) the presence of infective RNA of TMV has been demonstrated in ribosomes. It has been shown (61, 127, 163) that after the virus has been introduced into the cells of the host what first happens in them is an increase in the synthesis of viral RNA alone. The synthesis of protein only begins some hours after the beginning of the synthesis of RNA and is accomplished very quickly.

We may suppose that when the RNA of the virus enters the cell it becomes involved in all the functions and sites in which the RNA of the host is involved (organelles, nucleus, cell sap) and that its role is not confined to participation in the terminal stages of synthesis.

Another interesting question is that of the localization of the synthesis of viruses in the cell, which varies according to the species of virus. Some viruses are synthesized and accumulated in the cytoplasm of cells, others in the nuclei while, finally, there is a group of viruses which are mainly formed in the cell membranes. A detailed account of the evidence concerning the localization of the synthesis of viruses is to be found in the review of Rose & Morgan (126).

## 3. The role of RNA in the reproduction of the specificity of the proteins of viral particles

The studies of the parts played by RNA and DNA in the reproduction of the specificity of proteins which have been carried out during the past few years on plant, animal and bacterial viruses are classical and we have spent some time on examples from this field and shall spend more in the course of our review of the various theoretical problems with which this work is concerned. Detailed reviews of investigations in this field have recently been published by Kriviskiĭ (13), Tovarnitskiĭ & Tikhonenko (20), Schuster (131) and Herriott (85a). That of Tovarnitskiĭ & Tikhonenko was published in 1960 and is devoted specially to the infectivity of viral RNA. Nevertheless there are some essential stages in these investigations which must be mentioned briefly, because this will make it easier for us to understand why it is that certain aspects of this problem are being studied most intensively at present. The study of the roles of each of the two components of the virus particle was begun long ago. The first thing to be shown was that replacement or blocking of a number of the active groups of the protein component did not impair its infectivity (1-3, 15, 21). It has been shown comparatively recently that if the terminal threonine of all the protein sub-units of TMV is removed with carboxypeptidase it does not lose its power of multiplication while, in the descendants of this virus, the threonine is once more to be found in its proper place (84).  The role of viral RNA in the formation of the active virus particle was further demonstrated clearly by the experiments of Fraenkel-Conrat & Williams (69) on the reconstruction of viral particles from the RNA and protein components which had previously been separated from one another by gentle preparative methods.  The native protein of TMV on its own had no infectivity but this was restored when it was reconstructed with native RNA.  A similar fact was established by Gershenzon (6, 7) independently of Fraenkel-Conrat & Williams.  He reconstituted active polyhedral virus of insects from protein and DNA.*  These broad facts were confirmed by work in many other laboratories (30, 31, 50, 108).

The possibility of reconstituting the viral particle from RNA and protein naturally led to attempts at artificial hybridization of these components by isolating them from different strains of TMV.  A large series of experiments of this sort, carried out in the laboratory of Fraenkel-Conrat (23, 63, 66), showed clearly that, in all cases in which a virus was reconstituted from protein from one strain and RNA from another, while there was a considerable increase in infectivity of the reconstructed particle as compared with the pure RNA, nevertheless the characteristics of the descendants were always those of the strain from which the RNA of the "hybrid" was derived.

However, the determining part played by viral RNA in the reproduction of the specificity of the viral protein which was synthesized by the cell of the host was revealed even more clearly by the work of two laboratories, those of Schramm and Fraenkel-Conrat, which showed that even protein-free viral RNA can reproduce complete virus particles in the cells of plants (63, 68, 78).  In the first experiments which demonstrated the infectivity of the RNA of TMV, the RNA preparations contained a small admixture of protein but in later experiments Ramachandran & Fraenkel-Conrat (120a) succeeded in freeing the RNA almost entirely from protein contamination without loss of its infectivity.  The preparations of RNA which they obtained contained only insignificant amounts of low-molecular peptides ($0 \cdot 1$-$0 \cdot 5\%$) in which only glycine and glutamic and aspartic acids could be identified.

This admixture of amino acids in the viral RNA seems, however, not to be accidental.  In a recent study of the inactivation of the infectivity of the RNA of a number of viruses by hydroxylamine Franklin & Wecker (71) put forward the suggestion that the inactivating effect of the hydroxylamine may be explained by its removal from the RNA of the above mentioned amino acids, which play an important part in its biological activity.  It must be mentioned that ordinary hydrolysis of viral RNA leads to the destruction of most of the amino acids of this admixture, as an excess of

* Publication of this interesting work was delayed for $1\frac{1}{2}$ years because the journal's referee did not believe that the results obtained by the author could possibly be true.

carbohydrate products in the mixture being hydrolysed usually completely distorts the results of the analysis of amino acids owing to numerous secondary reactions.

The infectivity of deproteinized RNA from TMV and other plant viruses was soon confirmed by work in many other laboratories and by later work by the research groups referred to above (22, 64, 65, 67, 79, 111, 41, 61, 33, 4, 129, 155 and others).

The improvement of preparative methods for isolation of viral RNA allowed this RNA to be isolated in a more nearly pure form and with greater retention of its infectivity than in earlier experiments.

The infectivity of the ribonucleic acid particles seems to be characteristic of all viruses and in the past few years similar facts about the infectivity of protein-free RNA have been discovered repeatedly, even in work with many animal viruses (40, 44, 45, 35, 36, 26, 156, 140, 142, 150, 113, 127 and others). In experiments on the infectivity of RNA isolated from animal tissues the RNA was often isolated, not from viral particles but from the infected tissues as a whole, so that the infective preparation contained RNA from the host as well as viral RNA. This feature of the method gave grounds for the suggestion that the high infectivity of the RNA may be associated, not merely with the viral RNA, but also with some fractions of the total RNA of the tissues (89).

The study of the part played by RNA in reproducing the complete viral particle consisting of RNA and a specific protein component is now no longer confined to the establishment of the fact, which is certainly true in essence, of the infectivity of deproteinized RNA. In a recently published paper Commoner et al. (53) have begun the study of the even deeper question as to which parts of the polymeric ribbon of the RNA of TMV are most important for the reproduction of the virus. Using a mild treatment with detergents the authors obtained viral particles from which not all of the protein envelope surrounding the polynucleotide had been removed but only its terminal fragments. As a result of this the terminal parts of the RNA of these particles were laid bare and, as it were, stuck out from their protein wrappings and could be "cut off" by treatment with ribonuclease. The results of this work demonstrated, in the first place, that the different ends of the rod-like viral particles were not of equal importance for its infectivity. For example, if about 0·5% of the mass of the RNA was removed in the form of the terminal nucleotides, the infectivity of the particle was reduced to 12% while removal of 3% of the RNA lowered this infectivity to a very small value (2% of the original). At the same time the further "cutting down" of the RNA (removal of 4-5%) unexpectedly increased its infectivity to 20% and, finally, when more than 5% was removed the virus completely lost its infectivity. The authors believe that their results indicate that there are, in the polynucleotide chain, some fragments which have the greatest effect in determining the specificity of the protein which is reproduced during synthesis and the loss of these parts, even when the greater part of the RNA remains unaffected, deprives the virus of its power of multiplication.

The work on the role of viral RNA in recreating the whole viral particle is very important for very many fields of theoretical and practical biology; nevertheless the question of the mechanism of its action remains, as before, obscure in many respects. If the viral RNA really determines the specificity of the viral protein and organizes its synthesis then the multiplicity of the stages of this process is interesting. In the finished viral particle the amount of protein usually considerably exceeds that of RNA, which only accounts for a few parts percent of the total mass of the virus. Furthermore, quite a large amount of "pure" viral protein, not associated with RNA, accumulates within the cell of the host, having been synthesized in excess and not being of any "use". This sort of process argues that even if the viral RNA regulates the synthesis of viral protein, the protein molecules formed do not remain attached to their matrices but are split off and accumulate in other fractions of the cell and their association with the RNA occurs after this first stage has been completed. There is already quite a large amount of literature showing clearly that the protein which accumulates in the infected cells (the so-called X-protein) without being associated with RNA is the same as the viral protein and that its formation is the result of "excessive" synthesis and not of the breakdown of viral particles (31, 51, 52, 54, 55, 56, 60, 90, 114, 146, 145 and others). The reconstitution of a virus from protein and RNA can obviously occur spontaneously, as it can be reproduced under artificial conditions simply by mixing together the components.

It is interesting to consider what information the viral RNA contains. In searching for regularities in the composition of the proteins and RNA of several viruses Yčas has recently shown (162) that there is a definite correlation between the composition of the RNA and that of the protein, which suggests indirectly that this RNA contains information determining the composition of the protein. However, it is by no means proved whether the RNA of the viral particle is a matrix on the surface of which only the intracellular synthesis of the polypeptide chains of the specific viral protein takes place or whether it carries some additional information. To go on building up different hypotheses, it must, of course, be admitted that of the complementary structures which are formed in the course of the intracellular synthesis of viral RNA, one type of polynucleotide "works" as a matrix for the synthesis of protein and remains within the cell while the other serves as a working matrix for producing more of it. In any case it is clear that if, in general, viral RNA exists in two complementary forms, even if only for a time, only one of them can possess the synthetic information for the synthesis of a particular "purposively" constructed protein. In this case the other would only be a model for the synthesis of the matrix or material for the formation of the molecules of acceptor RNA.

In connection with the discovery of a special information-carrying RNA (messenger RNA) which is formed in the nucleus, and which acts as an active matrix in protein synthesis, attempts have naturally been made recently (in 1962) to study viral RNA in this respect as well. Ofengand & Haselkorn (115a) introduced the RNA of tobacco yellow mosaic virus into the synthetic system of *E. coli* (ribosomes and soluble fraction) and observed stimulation of the incorporation of labelled amino acids into the protein (see also 85b). Tsugita *et al.* (150d) carried out a similar experiment with the RNA of TMV and demonstrated the occurrence of synthesis *in vitro* of immunologically specific viral protein. These experiments demonstrate a definite similarity between viral RNA and the messenger RNA which acts as a matrix in protein synthesis.

The nucleoprotein nature of viruses has enabled us to obtain, in recent years, direct proof that it is the actual sequence of nucleotides in the RNA of the virus which determines the sequence of amino acids. Experiments on the infectivity of viral RNA show the part played by RNA in the synthesis of the proteins of the virus but do not give absolute proof of the recording of "synthetic" information in the form of nucleotide sequences. We are thinking of experiments on the creation of mutations in viruses as a result of direct influences on the sequence of nucleotides in the viral RNA. The occurrence of mutation in TMV under such influences has already been studied for several years (75, 143, 132, 151) but the connection between these mutations and the alteration in the sequence of amino acids in the protein sub-units of the virus has only been established very recently (135, 159). In these experiments the RNA of TMV was treated with graded doses of nitrous acid. Such treatment leads to the deamination of the three bases which contain an amino group, adenine is converted into hypoxanthine, guanine to xanthine and cytosine to uracil. Gierer & Mundry (77) showed earlier that the deamination of even one nucleotide in the molecule of the RNA of TMV can evoke a mutation. According to the evidence of Wittmann (158a, 159), in almost half of the mutations of this sort changes are found in the proteins associated with changes in the sequence of amino acid residues.

The study of the nature of these mutations led the author to the idea that by no means all of the chain of the RNA of TMV acts as a matrix for the synthesis of viral RNA and carries the information which predetermines its specificity. If there is an alteration of the nucleotides outside this zone, changes occur which only lead to alteration of the external features of the infection, without affecting the protein. Experimental results supporting this idea have also been obtained by Siegel (135).

Changes in the sequence of amino acids in mutants of TMV obtained by the action of nitrous acid on the RNA of the virus were found by Tsugita & Fraenkel-Conrat (150b). In the mutants the proline residue of the *C*-terminal group of the protein was replaced by a leucine residue.

Recently Tsugita & Fraenkel-Conrat (150c) have studied the changes in amino acid sequence in twenty-nine strains of TMV in which there had been changes in the RNA. Seventeen of these showed changes in the protein component. This enabled the authors to discover some of the rules of amino acid "coding".

An extraordinary conception of the nature of the synthesis of the RNA and protein of TMV has recently been developed by the well-known virologist Commoner (47, 48, 46). According to this, the synthesis of viral RNA proceeds by linear elongation; only the terminal 48-nucleotide fragments serves as matrix for the synthesis of the 157-residue protein; the RNA is only infectious if it is linked with, at least, a small amount of protein. The experimental material adduced by Commoner in support of this idea of his is so far of a preliminary nature.

While we are dealing with the nature of the synthesis of viral components we must not fail to discuss the results of the interesting experiments of Gershenzon and colleagues (8, 9) on the reproduction of the DNA-containing virus of polyhedral disease of insects by RNA which had been extracted from the host by treatment with phenol.

In these experiments the RNA was isolated by a modified phenol method from healthy and virus-infected larvae of the 5th stage of growth or from pupae of the mulberry silkworm. The RNA preparations were practically free from protein and contained an insignificant admixture of DNA. Solutions of RNA in a pyrophosphate buffer were purified by rapid centrifugation and introduced into healthy larvae in the 5th stage or into pupae of the same species. The RNA from the individuals which were suffering from polyhedral disease was highly infective. Treatment of the RNA with ribonuclease considerably lowered or completely destroyed the infectivity. Control experiments with RNA from healthy larvae or pupae to which large amounts of infective polyhedra or virus had been added before isolation of the RNA would seem to eliminate the possibility that active virus might have remained in the RNA after the procedures used for its isolation. The possibility that latent virus was activated in the healthy individuals was also excluded by experiments on pupae of the oak silkworm. In these experiments the authors used RNA from pupae infected with a mutant strain of the virus characterized by an aberrant form of the polyhedra.

This work is, however, only of a preliminary nature, as the amount of DNA in the preparations of RNA was sometimes 3-6% and the action of desoxyribonuclease on the infectivity has not yet been studied. Further experiments by these authors will certainly clear up the obscure questions in this extremely interesting research.

The experiments of Gershenzon & colleagues are of great significance with regard to our understanding of the nature of the connection between the synthesis of RNA and that of DNA and we shall return to a consideration of them in the chapter on the reproduction of nucleic acids.

True understanding of the mechanism of viral synthesis cannot be separated from questions of the source of viral activity.

On thinking over this question we agree to a considerable extent with the view that, as regards phylogenesis, the viruses arise endogenously within the cell itself and later acquire the power of independent evolution. There exist in nature a whole series of transitional forms between imperceptible "viral cycles", which are almost entirely non-pathological and which are present in the cell through all phases of its development and are transmitted to its descendants, through to highly pathogenic viruses which destroy the cells in a very short space of time.

Such a picture might be the result of the following peculiarities of biochemical evolution. It is certain that, in biochemical evolution, there must occur, not only the emergence of entire new biochemical cycles for the synthesis of some new enzymic protein (DNA → RNA → protein), but also a loss of functional importance by certain other cycles which become peculiar biochemical atavisms. At this stage the course of evolution divides, as it were, into two. On the one hand there arises a tendency towards the elimination of such "superfluous" cycles while, on the other, the variability of these same cycles leads to some of them acquiring a certain autonomy in their peculiar struggle for existence. On this basis there may also arise partially autonomous viral cycles which can multiply only in those cells in which the nuclei contain or, more correctly, still retain the information controlling the synthetic process.

A hypothesis similar to this one was recently put forward by Yamafuji (161) at the Fifth International Congress of Biochemistry. Having found, in the polyhedra of the DNA-containing virus which infects the silkworm, a protease and a desoxyribonuclease with properties similar to those of the same enzymes in the silkworm, the author proposed a scheme according to which the virus of polyhedral disease can arise from fragments of chromosomes which contain the genes controlling the action of both enzymes. (Fig. 28).

Cells containing normal chromosomes
| mutation
↓
Cells containing pro-viral genomes
| induction
↓
Cells containing viral particles

FIG. 28. Scheme of production of the virus of polyhedral disease (after Yamafuji (161)).

## 4. Some features of the synthesis of proteins and DNA among bacterial viruses (bacteriophages)

Two comprehensive reviews (139, 149a) illuminate all the fundamental aspects of the role of nucleic acids in the reproduction of phages and enable us to deal quite shortly with certain biochemical peculiarities of this process.  Further details may be found in the recent book by Gol'dfarb (9a).

Morphologically the bacterial viruses or phages possess a considerably more complicated structure and a more complicated cycle of development within the cells of the host than, for example, plant viruses.  The structure of a typical phage particle is shown in Fig. 29.  Nobody has been able to bring about the spontaneous reconstitution of these structures outside the organisms from the DNA and protein of which the phages are composed.  In the case of many phages it has not yet been possible to bring about the transmission of infection by means of DNA completely freed from protein.  Only recently has it been possible to establish, in the case of two phages, the infectivity of pure DNA freed from protein by gentle preparative methods.  In the first of these cases the DNA was that of the small and primitive phage $\phi$X174 (81, 133).  The DNA of this phage, however, possesses a number of peculiar features (single-chain structure) and we shall deal with their significance later.  The second case in which pure DNA was infective was described recently by Meyer *et al.* (112) who isolated the DNA of a bacteriophage.  The DNA of this phage has the

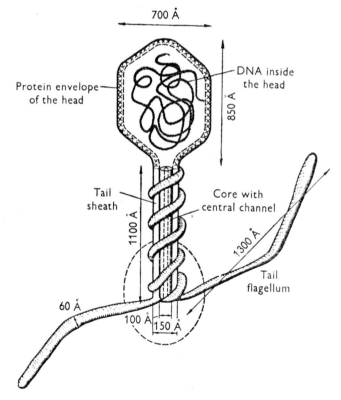

FIG. 29.  Structural scheme of a phage particle.

usual double helix.   Preparations of DNA from T2 phages in their experiments were not infective.   However, it has been shown comparatively recently that even in the case of more complicated phages of the T-group, if the cells are infected with phage in the normal way it is mainly DNA which is introduced into the cytoplasm, while by far the greater part of the protein of the phage particle (up to 98%) remains on the surface of the cell and does not participate in the reproduction and development of new phage particles within the bacterial cell.   This discovery was made by Hershey & Chase (87) using phages which had their proteins and DNA separately labelled with $^{35}$S and $^{32}$P and it supplied the first substantial experimental confirmation of the importance of DNA for the reproduction of the specificity of the proteins of the phage although, in this case, the DNA was not perfectly "pure" and a small amount of protein seems to have entered the infected cell in association with the DNA.   Essentially similar results have been obtained repeatedly by other workers and have been reported in detail in numerous reviews of the multiplication of phages (11-13, 14, 47, 34, 38, 118).

In these researches great interest attaches to the results concerning the later fate of the DNA of the phage after it has been introduced into the cell.   Many experiments on these lines have been carried out with DNA labelled with $^{32}$P or $^{14}$C.   They showed that a considerable part of this DNA breaks down into low-molecular products and only 25-50% of the labelled DNA of the original particles is included in the composition of later generations of phages.   It is interesting that the DNA of the original particle is distributed unequally in the descendants of the phage and the nature of this distribution is essentially similar to the nature of the distribution of the DNA of the chromosomes in the process of mitosis. The DNA of the phage only enters into the composition of one quarter of the descendants while three quarters of the particles which were formed later do not contain labelled DNA from the original particles (72, 154). A characteristic feature of this phenomenon is that the transfer of part of the material of the DNA only goes as far as the first generation in the form of high-polymeric molecules (105-107).   Considering these facts from the point of view of the complementary model of DNA, Levinthal takes the view that the 40% of the activity of the original DNA which remains in the polymeric fraction is in the form of double helices.   The division of these molecules into single polynucleotide chains which do not divide again would also seem to be the explanation of the fact that, after "transfer" of 20% of the activity of the original phage to the first generation, no further dilution of the activity occurs in the succeeding cycles of multiplication.

The same result has been achieved recently by Tessman (147), studying the distribution of [$^{32}$P]-DNA in T2 and T4 phages during their multiplication in the bacterial cell, by recording the tracks of disintegrating particles on sensitive "nuclear emulsions".

The autoradiographic evidence obtained for the total population of bacteria and the results of preliminary experiments with the phage produced from individual cells showed that, in the particle of phage, $\sim 40\%$ of all the DNA is to be found in one structural sub-unit or in several large molecules while the remaining 60% is distributed between smaller fragments which are too small to be studied by this method. When the phage multiplies, the large structural block of the original DNA undergoes replication, most probably in accordance with the scheme of Watson & Crick, so that, at the end, there are two blocks each containing about 20% of the original activity. The method of autoradiography thus enables us to study the replication of DNA at the molecular level.

Stent, Sato & Jerne (144) have recently studied the distribution of [$^{32}$P]-DNA in the progeny of T4 phages using a method based on different principles (the kinetics of the inactivation of phage owing to radioactive decay) and they also showed that the DNA of T4 bacteriophage consists of two fractions, large sub-units which are transferred to progeny without fragmentation, and small blocks of DNA which were distributed among many daughter particles. The authors suggest that only the large blocks carry out genetic functions. According to them the small particles of DNA play some non-genetic, physiological part in bringing about the earliest stages of the intracellular growth of phages.

Very recently, however, some interesting exceptions have been found during a study of the way in which the DNA of phage is distributed in the particles of the progeny. These observations were made after Sinsheimer had found that in the phage with the smallest particles (phage $\phi$X174) there is an unusual single-chained DNA with a composition and structure which do not conform either with Chargaff's rules or with the scheme of the double complementary structure of DNA (136-138).

Bacteriophage $\phi$X174 belongs to the smallest kind of bacteriophages. The "molecular" weight of one particle of this phage is 6,200,000, that is to say nearly 40 times less than that of phage T2. The amount of DNA in the phage represents about 25% of its weight. The molecular weight of this DNA is 1,600,000, which nearly corresponds with the molecular weights of the RNA of plant and animal viruses.

Study of the physical and chemical properties of this DNA has shown that it is a polymer which exists in the particles as a single polynucleotide in which the molar ratio of the bases does not conform to Chargaff's rules, which apply to DNA from other sources.

It is obvious that a study of the way in which this unusual DNA is reproduced during the multiplication of the phage would be of great theoretical interest.

Such work has been done by Kozinski & Szybalski (102-104) who have worked out an extremely sensitive method for the purpose, which enables them to identify one particle of the DNA of the original phage among 1000 particles of progeny.

The results of these authors' work have revealed an interesting and unusual picture of the distribution of the material of the original DNA among successive generations, which is quite unlike that which would be predicted from the rules which have been established for other phages. It seems that the continuity of the single-chained polynucleotide of the DNA of the phage particle of $\phi$X174 is lost during transfer from the original phage to its descendants. Furthermore, the labelled material of this DNA becomes dispersed in the general metabolic pool of DNA from which the descendants of the phage are synthesized. Some dispersal of phage DNA in succeeding generations has also been found in the large phages (e.g. T2) but it may have been connected with the presence of a large number of molecules of DNA within a single phage particle. Each particle of $\phi$X174 phage contains only one molecule of DNA.

According to Tessman (148) the disintegration of even one $^{32}$P atom and the resulting break in the chain will give rise to inactivation. In T2 phages only one in ten disintegrations of $^{32}$P in the DNA is lethal.

The loss of integrity of molecules of DNA during the reproduction of this phage raises a number of questions as to the means of transfer of hereditary information from one generation to the next for, according to the widely-accepted scheme, this transfer is associated with the complementary synthesis of a new polynucleotide on the surface of the parental one.

Kozinski & Szybalski have suggested a number of variations in the means of transmission of hereditary information in phage $\phi$X174, of which the following seems to provide the most plausible explanation of the dispersal of the parental DNA:

(*a*) The synthesis of new complementary polynucleotides on the surface of the parental one is followed by the breakdown of the parental polynucleotide into blocks and their transfer to the general stock. If this is so, however, we must assume that only a part of the parental polynucleotide is broken down at each replication, otherwise true multiplication of the phage would be impossible.

(*b*) The parental DNA breaks down into large blocks, some of which later specify the composition of sub-units while others determine the synthesis of the phage-specific proteins on which the aggregation of the fragments of DNA will later depend.

(*c*) There is an exchange of polynucleotides between the parts by a "crossing over" process.

In later work from Kozinski's laboratory it has actually been shown that the dispersal of phage DNA is brought about by its being broken down into fragments with a molecular weight of about 300,000 (cf. Chapter XIV & Fig. 46).

In a recent publication Sinsheimer (139a) has brought forward new evidence about the replication of the DNA of phage $\phi$X174. According to this, the single-chained DNA on entering the cell first forms an additional complementary chain, thus passing into the so-called "replicative form". After this has been formed, the process of reproduction of the virus begins.

A comparative study of mutagenesis in the phages $\phi$X174 and T4 carried out by Tessman (149) showed that the process of emergence of mutations under the influence of nitrous acid also occurred in different ways in these two phages. The nature of the process of mutation in $\phi$X174 is such as to indicate that this phage has only one "copy" of the genetic information while in T4 the hereditary characters are encoded in duplicate.

The DNA of a phage, when introduced into a cell with a small amount of phage proteins, may, perhaps, have a rather more complicated function than the RNA of the ribonucleoprotein viruses. Jeener (91) was the first to find evidence suggesting that the effect of the DNA of the phage on the synthesis of proteins may not be direct but through the medium of RNA which is newly formed within the cell. In his experiments the synthesis of phage proteins in lysogenic cells of *Bacillus megaterium* was impeded by inhibitors of RNA synthesis (such as thiouracil) which did not have any considerable effect on the synthesis of the proteins of the host cell itself. The author believes that the formation of a phage-specific protein is determined by and occurs with the participation of RNA which is quickly synthesized in induced lysogenic bacteria under the influence of the DNA of the phage.

The dependence of the multiplication of a phage on the preceding rapid synthesis of some new forms of RNA within the infected cell has recently been demonstrated very clearly by Pardee & Prestidge (116) and by Volkin and colleagues (28, 153).

Interesting observations have been made recently by Nomura, Hall & Spiegelman (115). They found that, under the influence of DNA from the phage T2, two new "phage-specific" types of RNA were synthesized in the cells of *E. coli*. One of these was of the high-polymeric, ribosomal type while the other resembled soluble, low-molecular RNA. This "phage-specific" S-RNA differs in nucleotide composition from the S-RNA of the infected bacteria.

According to the results obtained by Hall & Spiegelman in another research (82) the sequences of the nucleotides in the DNA of T2 phage and in the RNA produced under its influence are complementary. A later study of the synthesis within the cells of the host of a phage-specific RNA complementary to the DNA was carried out in 1961 with the discovery of an essentially new type of RNA, differing from other forms in its molecular weight, rate of renewal and functions. The functions of the RNA consist in the "transfer" of the genetic information of DNA and the "direction" of the synthesis of phage-specific proteins "on the foundation" of the ribosomes of the host-cells. Further development of this line of research, resulting from the discovery of "messenger" RNA, was clarified by work of Brenner, Jacob & Meselson (37). They grew *E. coli* in a medium in which $^{15}N$ and $^{13}C$ were the only N and C sources, and then infected these cells in a medium containing ordinary $^{14}N$ and $^{12}C$, as well as radioactive $^{32}P$. Sedimentation of the $^{32}P$-labelled RNA confirmed that the post-infection RNA was associated with ribosomal particles, but density analysis showed clearly that the phage-induced post-infection RNA enters old ribosomes that are already present in the cell before infection. This result led to a new stage in the study of protein synthesis, in which similar studies of plant and animal materials were made. We do not wish to assess here the importance of these experiments which are still incomplete, but in a special section of Chapter XV we shall compare them with analogous material concerning the functions of this "messenger RNA" obtained from other sources.

In studying the mechanism of the formation of phages biochemists have come close to the solution of the problem of the mechanism of the effect of DNA in organizing intracellular metabolism. In this connection the brilliant work of Kornberg *et al.* (100) on the effect of the DNA of T2 phage on the direction of syntheses in the bacterial cell is particularly significant. The DNA of this phage contains a substituted nucleotide (hydroxymethylcytosine) as well as ordinary cytosine. The synthesis of this phage within the cell would therefore be impossible if it did not itself induce the formation, within the bacterial cells, of enzymes carrying out the synthesis of hydroxymethylcytosine. Such inductive activity of the phage in the formation of three specific enzymes bringing about the formation and phosphorylation of hydroxymethylcytosine was also demonstrated in the work of the group of authors mentioned above. Similar cases of rapid synthesis of new enzymes within the bacterial cell after infection with phages have also been observed in several other laboratories (98, 99, 94).

In their recently published work, Setlow & Setlow (134) have established by sensitive methods that the DNA of T2 phage which enters the cell of *E. coli* in the form of a double-stranded polynucleotide is broken down within about 3 minutes into single-stranded polynucleotides and by the 11th minute there begin to appear again within the cell double, complementary, helical molecules of DNA. Comparing these facts with evidence concerning the sequence of other processes associated with phage infection, the authors concluded that the synthesis of the new enzymes determined by the phage (desoxycytidylate hydroxymethylase, hydroxymethyldesoxycytidylate kinase) and of the coat proteins of the phage begins in the presence of double helices of DNA. This synthesis is continued even when the DNA is broken down into single-chain polynucleotides but, by now, according to the authors' hypothesis, it is under the influence of RNA. The reproduction of the DNA itself only begins after it has been broken down into single-chain polynucleotides.

The formation of a special phage-specific RNA in the cells of the "host" and the role of this RNA in the synthesis of the proteins of the phage, including its enzymes which assure the synthesis of a number of phage-specific substances, are now being studied very intensively because it is in this cycle of reactions that the interaction between DNA, RNA and proteins is most conspicuously manifest.    In his recent communication to the Symposium on evolutionary biochemistry of the Fifth International Congress on Biochemistry Cohen (42) stated that, in the case of the T-phages, the specific RNA formed within the cells of the host under the influence of the phage DNA is a matrix and mediator in the synthesis of sixteen proteins which are formed after the beginning of the infection and before the beginning of the formation of the proteins of the phage itself, of which there are at least seven.    The phage-specific RNA formed in the cells does not, however, later enter into the composition of the virus as such.    Its components, however, take part in the creation of the new DNA of the phage in the course of its reproduction.    This was shown clearly by Cohen and colleagues (42, 43) who found that, on completing its "mission", the phage-specific RNA breaks down and the ribonucleo-tides of which it was composed are converted into desoxyribonucleotides which then participate in the synthesis of phage DNA.    Cohen (42) has recently put forward the following general scheme of the biochemical cycles which occur during the synthesis of phage (Fig. 30).    It is interesting to notice that Loeb & Zinder (109) have recently found a very small phage $f_2$, the particles of which contain RNA but not DNA.

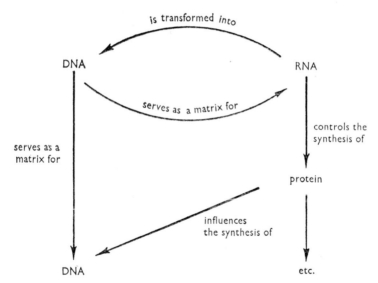

Fig. 30. Relationships of DNA, RNA and protein in *E. coli* injected with bacteriophage $T_2$.

Using RNA-containing phage, Weissmann, Simon & Ochoa (156a) suggested that the DNA-dependent polymerase, which controls "messenger" RNA synthesis in the host, could not organize viral RNA synthesis for the RNA phages. This suggestion led to the prediction that the viral RNA must contain the structural programme for a new RNA-dependent ribonucleotide polymerase. Such a form of polymerase was actually extracted from *E. coli* infected by the RNA phage.

What part is played by the proteins of the phage which enter the bacterial cell together with the DNA in insignificant amounts (according to recent evidence they comprise only about 1% of the mass of the DNA (86))? Most authors are inclined to deny that these proteins have any genetic functions. Nevertheless these proteins do play some part in the synthesis of phages of the T-group. In their reviews Burton (38) and Kriviskiï (13) report a series of researches by different authors in which it was shown that inhibition of protein synthesis by means of specific antibiotics or by deprivation of some amino acids at the moment of infection by the phage completely prevents its reproduction. The inhibition of this synthesis a few minutes after infection does not exert an impeding effect on the synthesis of the DNA of the phage. Kriviskiï suggests that this may be explained by the fact that, in the first minutes after infection, the new enzymes which will later bring about the synthesis of the phage DNA have to be synthesized in the cell. Such dependence is not, however, universal. Crawford (57) has published a short paper in which he reports the normal reproduction of DNA of T1, T3, T5 and other phages when there is a considerable (but not complete) initial inhibition of protein synthesis. According to him the phages T2, T4 and T6, with which most authors have been mainly concerned, are not suitable for these experiments because they contain the unusual pyrimidine base hydroxymethylcytosine (instead of cytosine) and the metabolism of this base therefore requires the preliminary synthesis of a number of adaptive enzymes within the cells. This explanation seems, however, to be erroneous and in later work Crawford himself (58, 59) obtained definite inhibition of the synthesis of DNA when he produced initial inhibition of protein synthesis by means of chloramphenicol and other inhibitors of protein synthesis in phages of the M line (M4) which contained ordinary cytosine in their DNA and in T5 phages.

The need to introduce a certain amount of phage protein as well as DNA into the host cells has been demonstrated specially clearly by Spizizen (141) who did experiments on the multiplication of T2 phage by observing the action of phage particles, which had been disrupted by osmotic shock, on cells of *E. coli* which were complete except that they had no cell wall. The cell walls were removed by means of lysozyme and their absence made unnecessary the special equipment of the phage for injecting DNA into the cells which is provided by the greater part of its superficial protein structures. It was found that the proteins of the phage are necessary components for its multiplication and that disintegration of the phage proteins by trypsin completely arrests the reproduction of phage particles. The author of these experiments thinks that the proteins of the phage may be necessary for its multiplication, not only in the reproduction of particles, but also as factors preventing the massive aggregation of DNA or even for getting through the cell membrane. The fact that proteins are necessary for the reproduction of phage has also been demonstrated strikingly in the work of Il'yashenko & Bass (10).

## Conclusion

Although it is very short, the survey which we have made of the evidence now available concerning the biochemical picture of the synthesis of the proteins of viruses demonstrates clearly the tremendous theoretical significance of these facts for our understanding of the fundamental rules of protein synthesis. Although, for example, the tobacco mosaic virus in its crystalline form does not manifest the characteristic features of life and cannot reproduce itself *in vitro*, nevertheless, when it enters into a plant it shows itself able to involve itself in the intracellular synthetic systems and divert their activity towards the formation of its own specific molecules of protein and RNA. We have already seen that the biochemical system of protein synthesis is very complicated and appears to consist of tens of enzymes, so that the activation and perhaps the transfer of each amino acid has its own enzyme. This synthesis of protein also occurs in many stages and a number of fractions of low-molecular and high-molecular-weight RNA are required for its accomplishment.

Nevertheless, if highly polymerized viral RNA or phage DNA is introduced by itself into the cell, this will cause a complete reorganization of the system in so far as concerns the specificity of its products. Instead of or as well as the synthesis of intracellular proteins it begins to produce the specific proteins and specific RNA or DNA of the viral particles. This phenomenon provides a most convincing demonstration of the ability of the viral nucleic acids themselves to ensure the reproduction of the specificity of the viral proteins and their own simultaneous autosynthesis by attaching themselves to all those enzymic systems which are necessary for the synthesis of proteins and amino acids. Thus it has been shown clearly that DNA may contain the information needed for the synthesis of RNA and, if we accept the hypothesis of Gershenzon, *vice versa*. Nevertheless, the synthesis of proteins in animal and plant viruses, as well as in phages, is ensured by RNA—the immediate matrix of protein synthesis.

In spite of all this there is much about the intracellular mechanism of viral synthesis which remains obscure. If the hypothesis of the endogenous origin of viruses is correct, then it is quite possible that the genetic system of the cells may contain vestiges of the mechanisms which regulated the synthesis of the "ancestors" of the viral RNA and which are brought into action after the beginning of the infection. In this respect phages are more autonomous, as they contain an autonomous genetic system within themselves.

Of course, it is possible that the hypothetical mechanism which we have described of the evolutionary loss or mutation of the RNA-protein cycle is not the only means of biochemical elimination of useless characteristics. Maybe the more complete cycles DNA → RNA → protein can isolate themselves endogenously and evolve into phage particles.

All these hypotheses are still, of course, pure speculation but they show clearly that evolutionary biochemists should direct their attention, not merely to the ways in which new biochemical properties and cycles appear and develop, but also to the means of loss or elimination of some evolutionarily obsolescent syntheses which would appear sometimes to evolve secondarily and give rise to a number of parasitic particles. The assessment of these factors may be important in connection with the problem of genetic immunity.

## REFERENCES

1. AGATOV, P. (1941). *Biokhimiya*, **6**, 269.
2. — (1943). *Doklady Akad. Nauk S.S.S.R.* **38**, 139.
3. — (1947). *Doklady Akad. Nauk S.S.S.R.* **58**, 429.
4. GAVRILOVA, L. P., SPIRIN, A. S. & BELOZERSKIĬ, A. N. (1959). *Doklady Akad. Nauk. S.S.S.R.* **124**, 933.
5. — (1959). *Doklady Akad. Nauk S.S.S.R.* **126**, 1121.
6. GERSHENZON, S. M. (1956). *Byull. Mosk. Obshchestva Ispytateleĭ Prirody, Otd. biol.* **61**, 99.
7. — (1956). *Dopovidi Akad. Nauk Ukr. R.S.R.* p. 489.
8. GERSHENZON, S. M., KOK, I. P., VITAS, K. I., DOBROVOL'S'KA, G. M. & SKURATOVS'KA, I. N. (1960). *Dopovidi Akad. Nauk Ukr. R.S.R.* p. 1638.
9. — (1961). *Proc. V int. Congr. Biochem., Moscow*, **9**, 150.
9a. GOL'DFARB, D. (1961). *Bakteriofagiya.* Moscow: Medgiz.
10. IL'YASHENKO, B. N. & BASS, I. A. (1958). *Tezisy Dokladov Konferentsii. "Izmenchivost' Mikroorganizmov i Bakteriophaga"* (ed. A. A. Imshenetskiĭ et al.), p. 67.
11. KRIVISKIĬ, A. S. (1953). In collective work *Deĭstvie Izlucheniĭ i Primenenie Izotopov v Biologii* (ed. V. M. Klechkovskiĭ et al.), No. 1 (13), p. 136. Moscow.
12. — (1957). In collective work *Bakteriofagiya* (ed. V. S. Antadze), p. 35. Tbilisi: Gruzmedgiz.
13. — (1958). *Uspekhi sovremennoĭ Biologii*, **45**, 286.
14. KRISS, A. E. & TIKHONENKO, A. (1957). *Uspekhi sovremennoĭ Biologii*, **44**, 121.
15. RYZHKOV, V. L. & VOVK, A. M. (1943). *Doklady Akad. Nauk S.S.S.R.* **38**, 247.
16. SPIRIN, A. S. (1960). *Uspekhi sovremennoĭ Biologii*, **50**, 261.
17. SPIRIN, A. S., GAVRILOVA, L. P. & BELOZERSKIĬ, A. N. (1959). *Doklady Akad. Nauk S.S.S.R.* **125**, 658.
18. SUKHOV, K. S. (1956). *Virusy.* Moscow: Izd. Akad. Nauk S.S.S.R.

19. — (1957). *Agrobiologiya*, No. *1*, p. 36.
20. TOVARNITSKIĬ, V. I. & TIKHONENKO, T. I. (1960). *Uspekhi sovremennoĭ Biologii*, 49, 19.
21. TOVARNITSKIĬ, V. I. (1950). *Uspekhi biol. Khimii*, 1, 143.
22. FRAENKEL-CONRAT, H. & SINGER, B. [A.] (1959). In *Proc. 1st internatl. Symposium on the Origin of Life on the Earth* (ed. A. I. Oparin *et al.*), p. 303. London: Pergamon.
23. FRAENKEL-CONRAT, H., SINGER, B. A. & WILLIAMS, R. C. (1957). In *A Symposium on the Chemical Basis of Heredity* (ed. W. D. McElroy & B. Glass), p. 501. Baltimore, Md.: Johns Hopkins Press.
24. HERSHEY, A. D. (1956). In *Currents in biochemical Research* (ed. D. E. Green), p. 1. New York: Interscience.
25. AACH, H. G. (1960). *Nature (Lond.)*, 187, 75.
26. ADA, G. L. & ANDERSON, S. G. (1959). *Nature (Lond.)*, 183, 799.
27. ANDERER, F. A., UHLIG, H., WEBER, E. & SCHRAMM, G. (1960). *Nature (Lond.)*, 186, 922.
28. ASTRACHAN, L. & VOLKIN, E. (1958). *Biochim. biophys. Acta*, 29, 536.
29. BAWDEN, F. C. (1957). In *Ciba Foundation Symposium on the Nature of Viruses* (ed. G. E. W. Wolstenholme & E. C. P. Millar), p. 170. London: Churchill.
30. — (1959). *Proc. roy. Soc.* 151B, 157.
31. BAWDEN, F. C. & PIRIE, N. W. (1956). *J. gen. Microbiol.* 14, 460.
32. BELLETT, A. J. D. & BURNESS, A. T. H. (1961). *Nature (Lond.)*, 190, 235.
33. BOEDTKER, H. (1959). *Biochim. biophys. Acta*, 32, 519.
34. BOYD, J. S. K. (1956). *Biol. Revs. Cambridge phil. Soc.* 31, 71.
35. BROWN, F., SELLERS, R. F. & STEWART, D. L. (1958). *Nature (Lond.)*, 182, 535.
36. BROWN, F. & STEWART, D. L. (1959). *Virology*, 7, 408.
37. BRENNER, S., JACOB, F. & MESELSON, M. (1961). *Nature (Lond.)*, 190, 576.
38. BURTON, K. (1957). *Biochem. Soc. Symposia*, 14, 60.
39. BURTON, D., HALL, D. A., KEECH, M. K., REED, R., SAXL, H., TUNBRIDGE, R. E. & WOOD, M. J. (1955). *Nature (Lond.)*, 176, 966.
40. CHENG, P.-Y. (1958). *Nature (Lond.)*, 181, 1800.
41. COCHRAN, G. W. & CHIDESTER, J. L. (1957). *Virology*, 4, 390.
42. COHEN, S. S. (1961). *Proc. V int. Congr. Biochem., Moscow*, 3, 323.
43. COHEN, S. S., BARNER, H. D. & LICHTENSTEIN, J. (1960). *Science*, 132, 1489.
44. COLTER, J. S., BIRD, H. H. & BROWN, R. A. (1957). *Nature (Lond.)*, 179, 859.
45. COLTER, J. S., BIRD, H. H., MOYER, A. W. & BROWN, R. A. (1957). *Virology*, 4, 522.
46. COMMONER, B. (1959). *Nature (Lond.)*, 184, 1998.
47. — (1959). In *Plant Pathology, Problems and Progress 1908-1958* (ed. C. S. Holton *et al.*), p. 483. Madison: Univ. Wisconsin Press.
48. — (1958). *Proc. IV int. Congr. Biochem., Vienna*, 7, 17.
49. COMMONER, B. & BASLER, E. (jun.) (1956). *Virology*, 2, 477.
50. COMMONER, B., LIPPINCOTT, J. A., SHEARER, G. B., RICHMAN, E. E. & WU, J.-H. (1956). *Nature (Lond.)*, 178, 767.
51. COMMONER, B., LIPPINCOTT, J. A. & SYMINGTON, J. (1959). *Nature (Lond.)*, 184, 1992.
52. COMMONER, B. & RODENBERG, S. D. (1955). *J. gen. Physiol.* 38, 475.
53. COMMONER, B., SHEARER, G. B. & STRODE, C. (1958). *Proc. natl. Acad. Sci. U.S.* 44, 1117.
54. COMMONER, B., SCHIEBER, D. L. & DIETZ, P. M. (1953). *J. gen. Physiol.* 36, 807.

55. COMMONER, B. & YAMADA, M. (1955). *J. gen. Physiol.* **38**, 459.

56. COMMONER, B., YAMADA, M., RODENBERG, S. D., WANG, T.-Y. & BASLER, E. (jun.) (1953). *Science*, **118**, 529.

57. CRAWFORD, L. V. (1957). *Biochem. J.* **65**, 17P.

58. — (1958). *Biochim. biophys. Acta*, **28**, 208.

59. — (1959). *Virology*, **7**, 359.

60. DELWICHE, C. C., NEWMARK, P., TAKAHASHI, W. N. & NG, M. J. (1955). *Biochim. biophys. Acta*, **16**, 127.

61. ENGLER, R. & SCHRAMM, G. (1959). *Nature (Lond.)*, **183**, 1277.

62. EVANS, E. A. (jun.) (1954). *Ann. Rev. Microbiol.* **8**, 237.

63. FRAENKEL-CONRAT, H. (1956). *J. Amer. chem. Soc.* **78**, 882.

64. — (1957). *Virology*, **4**, 1.

65. — (1957). *Fed. Proc.* **16**, 810.

66. FRAENKEL-CONRAT, H. & SINGER, B. (1957). *Biochim. biophys. Acta*, **24**, 540.

67. FRAENKEL-CONRAT, H., SINGER, B. & VELDEE, S. (1958). *Biochim. biophys. Acta*, **29**, 639.

68. FRAENKEL-CONRAT, H., SINGER, B. & WILLIAMS, R. C. (1957). *Biochim. biophys. Acta*, **25**, 87.

69. FRAENKEL-CONRAT, H. & WILLIAMS, R. C. (1955). *Proc. natl. Acad. Sci. U.S.* **41**, 690.

70. FRANKLIN, R. E., CASPAR, D. & KLUG, A. (1958). Quoted by Wildman (157).

71. FRANKLIN, R. M. & WECKER, E. (1959). *Nature (Lond.)*, **184**, 343.

72. FRENCH, R. C., GRAHAM, A. F., LESLEY, S. M. & ROOYEN, C. E. VAN (1952). *J. Bact.* **64**, 597.

73. GIERER, A. (1961). *Naturwiss.* **48**, 283.

74. — (1957). *Nature (Lond.)*, **179**, 1297.

75. — (1958). *Z. Naturforsch.* **13B**, 485.

76. — (1958). *Z. Naturforsch.* **13B**, 788.

77. GIERER, A. & MUNDRY, K. W. (1958). *Nature (Lond.)*, **182**, 1457.

78. GIERER, A. & SCHRAMM, G. (1956). *Z. Naturforsch.* **11B**, 138.

79. GINOZA, W. (1958). *Nature (Lond.)*, **181**, 958.

79a. GORDON, M. P. & SMITH, C. (1960). *J. biol. Chem.* **235**, PC 28.

80. GOTTSCHALK, A. (1957). *Physiol. Revs.* **37**, 66.

81. GUTHRIE, G. D. & SINSHEIMER, R. L. (1960). *J. mol. Biol.* **2**, 297.

82. HALL, B. D. & SPIEGELMAN, S. (1961). *Proc. natl. Acad. Sci. U.S.* **47**, 137.

83. HARRIS, J. I. & HINDLEY, J. (1961). *Proc. V int. Congr. Biochem.*, Moscow, **9**, 154.

84. HARRIS, J. I. & KNIGHT, C. A. (1955). *J. biol. Chem.* **214**, 215.

85. HART, R. G. (1958). *Biochim. biophys. Acta*, **28**, 457.

85a. HERRIOTT, R. M. (1961). *Science*, **134**, 256.

85b. HASELKORN, R., FRIED, V. A. & DAHLBERG, J. E. (1963). *Proc. natl. Acad. Sci. U.S.* **49**, 511.

86. HERSHEY, A. D. (1957). *Virology*, **4**, 237.

87. HERSHEY, A. D. & CHASE, M. (1952). *J. gen. Physiol.* **36**, 39.

88. HOYLE, L. (1957). In *Ciba Foundation Symposium on the Nature of Viruses* (ed. G. E. W. Wolstenholme & E. C. P. Millar), p. 211. London: Churchill.

89. HUPPERT, J. & SANDERS, F. K. (1958). *Nature (Lond.)*, **182**, 515.

90. JEENER, R. (1954). *Arch. intern. Physiol.* **62**, 297.

91. — (1959). *Biochim. biophys. Acta*, **32**, 106.

92. JESAITIS, M. A. (1956). *Nature (Lond.)*, **178**, 637.

92a. KAMMEN, A. VAN. (1961). *Biochim. biophys. Acta*, **53**, 230.

93. KASSANIS, B. (1959). *J. gen. Microbiol.* **20**, 704.
94. KECK, K., MAHLER, H. R. & FRASER, D. (1960). *Arch. Biochem. Biophys.* **86**, 85.
95. KNIGHT, C. A. (1955). *J. biol. Chem.* **214**, 231.
96. — (1957). In *Ciba Foundation Symposium on the Nature of Viruses* (ed. G. E. W. Wolstenholme & E. C. P. Millar), p. 69. London: Churchill.
97. KNIGHT, C. A. & WOODY, B. R. (1959). *Fed. Proc.* **18**, 263.
98. KOERNER, J. F., SMITH, M. S. & BUCHANAN, J. M. (1959). *J. Amer. chem. Soc.* **81**, 2594.
99. — (1960). *J. biol. Chem.* **235**, 2691.
100. KORNBERG, A., ZIMMERMAN, S. B., KORNBERG, S. R. & JOSSE, J. (1959). *Proc. natl. Acad. Sci. U.S.* **45**, 772 (1959).
101. KOVÁCS, E. (1961). *Experientia*, **17**, 153.
102. KOZINSKI, A. W. (1961). *Virology*, **13**, 124.
103. — (1961). *Virology*, **13**, 377.
104. KOZINSKI, A. W. & SZYBALSKI, W. (1959). *Virology*, **9**, 260.
105. KOZINSKI, A. W. & BEER, M. (1962). *Biophys. J.* **2**, 129.
106. LEVINTHAL, C. (1956). *Proc. natl. Acad. Sci. U.S.* **42**, 394.
107. LEVINTHAL, C. & THOMAS, C. A. (jun.) (1957). *Biochim. biophys. Acta*, **23**, 453.
108. LINDNER, R. C., KIRKPATRICK, H. C. & WEEKS, T. E. (1956). *Plant Physiol.* **31**, 1.
109. LOEB, T. & ZINDER, N. D. (1961). *Proc. natl. Acad. Sci. U.S.* **47**, 282.
110. MARKHAM, R. (1953). In *The nature of virus multiplication, 2nd Sympos. Soc. gen. Microbiol.* (ed. P. Fildes & W. E. van Heyningen), p. 85. Cambridge: University Press.
111. McLAREN, A. D. & TAKAHASHI, W. N. (1957). *Radiation Res.* **6**, 532.
112. MEYER, F., MACKAL, R. P., TAO, M. & EVANS, E. A. (jun.) (1961). *J. biol. Chem.* **236**, 1141.
113. MOUNTAIN, I. M. & ALEXANDER, H. E. (1959). *Proc. Soc. exp. Biol. Med. (N.Y.),* **101**, 527.
114. NEWMARK, P. & FRASER, D. (1956). *J. Amer. chem. Soc.* **78**, 1588.
115. NOMURA, M., HALL, B. D. & SPIEGELMAN, S. (1960). *J. mol. Biol.* **2**, 306.
115a. OFENGAND, J. & HASELKORN, R. (1962). *Biochem. biophys. Res. Commun.* **6**, 469.
116. PARDEE, A. B. & PRESTIDGE, L. S. (1960). *Biochim. biophys. Acta*, **37**, 544.
117. PIRIE, N. W. (1957). *Biokhimiya*, **22**, 140.
118. PUTNAM, F. W. (1956). *Ann. Rev. Biochem.* **25**, 147.
119. RAMACHANDRAN, L. K. (1958). *Virology*, **5**, 244.
120. — (1960). *Biochim. biophys. Acta*, **41**, 524.
120a. RAMACHANDRAN, L. K. & FRAENKEL-CONRAT, H. (1958). *Arch. Biochem. Biophys.* **74**, 224.
121. RAMACHANDRAN, L. K. & GISH, D. T. (1959). *J. Amer. chem. Soc.* **81**, 884.
122. REDDI, K. K. (1959). *Proc. natl. Acad. Sci. U.S.* **45**, 293.
123. — (1959). *Biochim. biophys. Acta*, **32**, 386.
124. REDDI, K. K. & KNIGHT, C. A. (1956). *J. biol. Chem.* **221**, 629.
125. RENDI, R. & HULTIN, T. (1960). *Exptl. Cell Res.* **19**, 253.
126. ROSE, H. M. & MORGAN, C. (1960). *Ann. Rev. Microbiol.* **14**, 217.
127. SANDERS, F. K. (1960). *Nature (Lond.),* **185**, 802.
128. SCHRAMM, G. (1958). *Ann. Rev. Biochem.* **27**, 101.
129. SCHRAMM, G. & ENGLER, R. (1958). *Nature (Lond.),* **181**, 916.
130. SCHRAMM, G. & ZILLIG, W. (1955). *Z. Naturforsch.* **10B**, 493.
131. SCHUSTER, H. (1960). In *Nucleic Acids* (ed. E. Chargaff & J. N. Davidson), Vol. *3*, p. 245. New York: Academic Press.

132. SCHUSTER, H., GIERER, A. & MUNDRY, K. W. (1960).  *Abhandl. deut. Akad. Wiss. Berlin, Kl. Med.* p. 76.
133. SEKIGUCHI, M., TAKETO, A. & TAKAGI, I. (1960).  *Biochim. biophys. Acta,* 45, 199.
134. SETLOW, J. K. & SETLOW, R. B. (1960).  *Proc. natl. Acad. Sci. U.S.* 46, 791.
135. SIEGEL, A. (1961).  *Proc. V int. Congr. Biochem., Moscow,* 9, 152.
136. SINSHEIMER, R. L. (1959).  *J. mol. Biol.* 1, 37.
137. — (1959).  *J. mol. Biol.* 1, 43.
138. — (1959).  *Brookhaven Symposia in Biol.* 12, 27.
139. — (1960).  In *Nucleic Acids* (ed. E. Chargaff & J. N. Davidson), Vol. 3, p. 187.  New York: Academic Press.
139a. SINSHEIMER, R. L., STARMAN, B., NAGLER, C. & GUTHRIE, S. (1962).  *J. mol. Biol.* 4, 142.
140. SOKOL, F., LIBIKOVÁ, H. & ZEMLA, J. (1959).  *Nature (Lond.),* 184, 1581.
141. SPIZIZEN, J. (1957).  *Proc. natl. Acad. Sci. U.S.* 43, 694.
142. SPUHLER, V. (1959).  *Experientia,* 15, 155.
143. STAEHELIN, M. (1960).  *Experientia,* 16, 473.
144. STENT, G. S., SATO, G. H. & JERNE, N. K. (1959).  *J. mol. Biol.* 1, 134.
145. TAKAHASHI, W. N. (1959).  *Virology,* 9, 437.
146. TAKAHASHI, W. N. & ISHII, M. (1953).  *Amer. J. Bot.* 40, 85.
147. TESSMAN, I. (1959).  *Lab. Invest.* 8, 245.
148. — (1959).  *Virology,* 7, 263.
149. — (1959).  *Virology,* 9, 375.
149a. THOMAS, G. A. (1963).  In *Molecular genetics* (ed. J. H. Taylor), Pt. 1, Chap. 3, p. 113.  New York: Academic Press.
150. THOMAS, J. A. & LECLERC, J. (1959).  *Compt. rend. Acad. Sci., Paris,* 248, 606.
150a. TSUGITA, A., GISH, D. T., YOUNG, J., FRAENKEL-CONRAT, H., KNIGHT, C. A. & STANLEY, W. M. (1960).  *Proc. natl. Acad. Sci. U.S.* 46, 1463.
150b. TSUGITA, A. & FRAENKEL-CONRAT, H. (1960).  *Proc. natl. Acad. Sci. U.S.* 46, 636.
150c. — (1962).  *J. mol. Biol.* 4, 73.
150d. TSUGITA, A., FRAENKEL-CONRAT, H., NIRENBERG, M. W. & MATTHAEI, J. H. (1962).  *Proc. natl. Acad. Sci. U.S.* 48, 846.
151. VIELMETTER, W. & WIEDER, C. M. (1959).  *Z. Naturforsch.* 14B, 312.
152. BURNET, F. M. & STANLEY, W. M. (Eds.) (1959).  *The Viruses* (3 vols.).  New York: Academic Press.
153. VOLKIN, E. (1960).  *Proc. natl. Acad. Sci. U.S.* 46, 1336.
154. WATSON, J. D. & MAALØE, E. (1953).  *Biochim. biophys. Acta,* 10, 432.
155. WECKER, E. (1959).  *Virology,* 7, 241.
156. WECKER, E. & SCHÄFER, W. (1957).  *Z. Naturforsch.* 12B, 415.
156a. WEISSMANN, C., SIMON, L. & OCHOA, S. (1963).  *Proc. natl. Acad. Sci. U.S.* 49, 407.
157. WILDMAN, S. G. (1959).  *Proc. natl. Acad. Sci. U.S.* 45, 300.
158. WITTMANN, H. G. & BRAUNITZER, G. (1959).  *Virology,* 9, 726.
158a. WITTMANN, H. G. (1960).  *Virology,* 12, 609.
159. — (1961).  *Proc. V int. Congr. Biochem., Moscow,* 1, 240.
160. WOODY, B. R. & KNIGHT, C. A. (1959).  *Virology,* 9, 359.
161. YAMAFUJI, K. (1961).  *Proc. V int. Congr. Biochem., Moscow,* 9, 154.
162. YČAS, M. (1960).  *Nature (Lond.),* 188, 209.
163. ZECH, H. (1960).  *Virology,* 11, 499.

CHAPTER VIII

# CHARACTERISTICS OF THE BIOSYNTHESIS OF INDUCIBLE ENZYMES

## *Introduction*

The problem of the inducible (adaptive) synthesis of enzymes is one of the most important and interesting problems in the general field of protein synthesis. The interests of many departments of biology and biochemistry are involved in the analysis of the facts of the adaptive synthesis of enzymes, especially those of genetics, enzymology, microbiology, the study of evolution, the study of the biosynthesis of proteins and nucleic acids as well as a number of others. In this work, however, we naturally cannot undertake such an extensive and many-sided analysis of the problems of adaptation, although this would be valuable for an understanding of these phenomena as an aspect of biological adaptation.

We shall limit ourselves to the consideration of one purely biochemical question of the mechanism of synthesis of adaptive enzymes, a question which is associated very closely with the main lines of this book and which reveals some new properties of the systems of protein synthesis, their association with the genetic apparatus and with the systems which ensure the suitability and co-ordination of the biochemical processes carried out within the cell.

Questions of the wider aspects of adaptive synthesis have been surveyed in detail in the past few years in many reviews and monographs (1, 18, 2, 4, 5, 6, 9, 32 and others), so we can confine ourselves to just a few introductory remarks which are necessary because they enable us to make a rather better assessment of the value of those hypotheses concerning the mechanisms of adaptive synthesis of enzymic proteins which are now being worked out and to find our way about among the arguments used in attempts to use the phenomenon of adaptive synthesis in support of genetic concepts of one sort or another. In this chapter, however, we shall be very cautious in our use of purely genetic terminology, trying to confine ourselves within the framework of biochemical concepts. Later, however, in the chapters devoted to heredity and morphogenesis, we shall return to a consideration of questions of the genetic significance of adaptive synthesis, especially in connection with our discussion of the hypothesis "one gene—one enzyme".

## 1. Biological prerequisites for the synthesis of adaptive enzymes

Like all the characteristics of organisms, the synthesis of adaptive enzymes must certainly have a definite evolutionary history. It is quite obvious that the ability of some organisms to form adaptive enzymes is a well-marked adaptive reaction enabling them to make better use of the environment under changing conditions. If, under normal conditions, some organism which does not synthesize a particular enzyme (such as galactosidase) begins to form it quickly when the appropriate substrate is added to the medium (in this case galactose) then the process will, according to present-day ideas, primarily consist in the manifestation of a latent ability of the organism in question to form the enzyme as a result of the reorganization of the synthesizing systems.

Even in cases in which not even the smallest amounts of the adaptive enzyme can be found in the cells before the new substrate is present in the medium, its manifestation in the form of a reaction on a substrate is not a process which is entirely determined by the substrate itself. There must surely have existed in the cell a latent ability to synthesize the particular enzyme and the influence of the substrate did not create this latent ability of the cellular organization but merely activated it. It must be emphasized that we consider that it is only permissible to apply this approach to adaptive synthesis to one form of adaptive variability, namely the formation of adaptive enzymes. It should not be extended to other forms of variability of microorganisms. Other forms of variability conform to other rules and they cannot be brought together within some general genetic scheme.

The carrying out of any enzymic reaction requires the backing of many biochemical conditions, first among them being the synthesis of the appropriate enzymes. Even if the substrate for a particular reaction is absent, the enzymic apparatus for carrying it out can still be reproduced although, in this case, it would be "unemployed". The ability of the cell to retain a particular enzymic system, no matter how little work it does, cannot be favourable in all cases. It can only be justified, from an evolutionary point of view, for reactions which are necessary to life, which are accomplished in the cells under all normal conditions (phosphorylation, activation of amino acids, synthesis of proteins, RNA, DNA, etc.). It would not, however, be appropriate for the interchangeable reactions of carbohydrate, fat, amino acid and other categories of metabolism. If, for example, a cell could use galactose and glucose equally effectively but there was only glucose in the medium in which it was growing, then the continual resynthesis within the cells of a galactosidase system for which there was no use would not be a very economical or purposive characteristic. The system is cut down but the ability to synthesize galactosidase under the influence of galactose is retained and therefore the rapid formation of adaptive galactosidase on the introduction of galactose cannot be a complete biochemical surprise for the cells requiring an appropriate mutation. It is a process of reactivation of a synthesis which the cell had not completely lost the ability to perform and for which the mechanism had been retained in a latent form which was yet susceptible to reactivation.

Although these introductory observations on the methodological approach to the analysis of the mechanisms of adaptive synthesis of enzymes are far from original, yet they are very important from the point of view of theory. If we approach the formation of an adaptive enzyme as though it were a case of new formation in all respects (including genetic ones) then, in analysing the mechanism of this synthesis, we shall have to discover the nature of the determinants of the specificity of all stages of the synthesis (the sequence of nucleotides in particular parts of RNA and DNA, the sequence of amino acid residues in the proteins of the enzymic system) on the basis of a simple substrate (such as galactose or, to take the simplest example, oxygen). We are not in a position to set up any reasonable hypothetical schemes as to such determination and it seems to us quite impossible that we should be, just as it is impossible to reconstruct the parameters of a complicated machine from the outline of a simple article which it has made. If we approach the formation of such enzymes in terms of the manifestation and activation of latent powers of the cell, then the mechanism of their activation and specificity could, one would think, be worked out in terms of real biochemical reactions.

Thus, we hold that it is correct to take the view that the power of adaptive synthesis (i.e. the power to synthesize a new and purposive enzymic system in response to some chemical influence, the power of retaining this system through a number of generations and, furthermore, the apparent loss of it when conditions alter) represents an absolutely definite genetic characteristic of the organism determining its hereditary, evolutionary adaptation and not a power of altering heredity. This view of adaptive synthesis is supported convincingly by two fundamental features of adaptive synthesis; in the first place by the fact that the cell can only form adaptive enzymes in response to the action of a certain very small number of substrates and, in the second place, by the existence of a wide range of transitional forms between a simple increase in the production of enzymes which already occurred within the cells and the formation of apparently new enzymes, i.e. enzymes which had not been present in appreciable amounts in the culture in question but which were generally present in other cultures and strains of the microorganism concerned.

It must be pointed out that a genetically controlled mechanism for the power of formation of adaptive enzymes has been demonstrated clearly in many researches on the genetics of bacteria. In this connection we may refer, for example, to the recent extremely interesting work of Pardee, Jacob & Monod (38). These authors worked out a method for obtaining very closely related mutations of E. coli. Some of these mutations included loss of the power of adaptive synthesis of β-galactosidase by the organism. Other mutations, on the contrary, converted the enzyme into a constitutive form, the changes being due to the loss of the power of synthesizing an active "repressor" of the synthesis of galactosidase. In other experiments Pardee & colleagues (39) crossed two mutant strains of *E. coli* one of which could synthesize β-galactosidase only in the presence of the inducer while the other had lost by mutation the power of adaptive synthesis of the enzyme but retained the power of constitutive synthesis of β-galactosidase. In the new population with a mixed genotype the constitutive synthesis of β-galactosidase only proceeds during the first hour after genetic recombination and is then quickly lost whereas the capability for induced synthesis is a dominant character. A repressor, inhibiting the synthesis of β-galactosidase in the absence of the inducer, begins to be synthesized in the hybrid cells under the control of some gene. At first the authors supposed that the repressors took the form of special types of RNA which blocked certain zones of DNA. More recently, however, in 1962, Jacob & Monod stated at a symposium on "coding problems" that the protein nature of the repressors had been established.

The suggestion that the formation of adaptive enzymes is an example of the activation of a latent hereditary characteristic is also supported by the fact that, as a rule, only strictly determinate substances can evoke the formation of an adaptive enzyme in any particular organism. As an example we may refer to the fact that the cells of *Torulopsis utilis* could not become adapted to the utilization of arabinose (3) as this organism does not contain the enzymic systems built up over many generations which would be capable of metabolizing this carbohydrate.

We have spent a little time on this concept of the genetic control of the synthesis of adaptive enzymes (although it is almost universally accepted), in order to oppose it to arbitrary attempts to confuse the adaptive synthesis of enzymes with the phenomenon known as the inheritance of acquired characteristics and to suggest that the functional appearance of an adaptive enzyme in the form of a quick reaction to certain substrates represents a new formation which is genetically equivalent to the appearance of new qualities in evolution.

## 2. The role of nucleic acids in the inducible synthesis of enzymes

Hundreds of enzymes, the synthesis of which can be increased adaptively, have already been discovered. Studies of this sort embrace plants and animals as well as microorganisms. However, only a comparatively small number of very intensively studied systems have so far been used for the investigation of the mechanism of formation of induced enzymes and a number of related theoretical problems.

The connection between adaptive synthesis and the metabolism of nucleic acids has been fairly thoroughly dealt with recently in a number of reviews (7, 14, 25, 32). Also, in our chapter on the role of nucleic acids in protein synthesis we have already brought forward many facts which indicate the importance of the part played by nucleic acids in the formation of adaptive enzymes. There we devoted ourselves mainly to the work of Gale's laboratory (19-23) which showed most clearly the specificity of the activity of nucleic acids in this process and the connection of the new enzymic synthesis, not with the nucleic acids of the whole cell, but with a particular fraction of RNA which was synthesized and renewed at a rate proportional to that of the adaptive synthesis.

In this chapter we shall only deal very shortly with certain theoretical questions involved in this matter.

As concerns the history of the problem, it must be emphasized that the most obvious feature was the relationship between the synthesis and renewal of RNA and the formation of adaptive enzymes. This was first discovered in 1946 by Spiegelman & Kamen (54) whose work laid the basis for many later researches along these lines. Spiegelman has summarized the later work of his laboratory in several reviews (10, 52, 53). He brought to light new facts concerning this interdependence. In the first place he showed that the formation of adaptive enzymes by microorganisms really is a process of protein synthesis which is mainly accomplished at the expense of the pool of free non-protein nitrogen in the cells and external medium and cannot take place by the reconstitution or activation of proteins already present within the cell. The dependence of the synthesis of adaptive enzymes upon the availability to the cell of low-molecular forms of nitrogen compounds has been repeatedly demonstrated by other authors (36, 48, 60). Spiegelman and colleagues also showed clearly that the use of free amino acids for the synthesis of adaptive enzymes depended on RNA and disruption of the RNA (by ultraviolet radiation, for example) completely inhibited the utilization of these amino acids and the production of enzymes, although the incorporation of $^{32}P$ into the disrupted RNA was only diminished by half. Inhibitors and antimetabolites of nucleic acid synthesis also had an analogous and rapid effect, the synthesis of adaptive enzymes being the most sensitive of all protein syntheses occurring within the cell. This feature was most marked for adaptive $\beta$-galactosidase, the synthesis of which competed very weakly against other syntheses for the precursors of RNA. This lack of competitiveness in respect of RNA is not, however, the rule for all adaptive syntheses. In some cases (e.g. the formation of hydrogenlyase in $E.\ coli$) the adaptive synthesis is the most successful competitor for RNA though a very weak competitor in the use of the pool of non-protein nitrogen. Different competitive powers in respect of the requisites for synthesis have been observed, in particular, as between two different adaptive enzymes ($\beta$-galactosidase and nitrate reductase) formed in the same cells (45).

A similar observation has also been made recently in respect of the competition between β-galactosidase and the enzyme responsible for the hydrolysis of D-xylose. Competition between them is only observed when there is nitrogen deficiency (59).

Clear evidence as to the part played by nucleic acids in the synthesis of the adaptive β-galactosidase of *E. coli* has been obtained by Pardee (37, 39a). The repression of the synthesis of DNA in his experiment was obviously not a reflection of the formation of the enzyme, while the synthesis of RNA was necessary for the maintenance and exercise of this synthetic ability by the cell. Suppression of the synthesis of RNA, in this case, primarily affected the formation of the adaptive but not of the constitutive enzymes. Similar results have also been obtained in respect of the synthesis of adaptive glucokinase (47) the formation of which in the presence of gluconate is markedly stimulated by products of the incomplete breakdown of RNA (oligonucleotides).

In his detailed work on the formation of adaptive enzymes by yeast Chantrenne (12, 13) tried to find certain specific associations between the synthesis of proteins and that of nucleic acids. Using a culture of the anaerobic "dwarf mutant" of yeast in which adaptive cytochrome, cytochrome peroxidase and catalase are formed under the influence of oxygen, this author found that the synthesis of these enzymes was accompanied by increased incorporation of [$^{14}$C]purines and pyrimidines into the RNA. The increase in synthesis of RNA was not equal in all fractions. It was greatest in respect of the RNA of the cell supernatant and in that fraction of the total RNA which is precipitated in a half-saturated solution of ammonium sulphate. As a result of this the author believes that the formation of adaptive enzymes is specifically associated with only some fractions of RNA but, in the light of the evidence given above as to the inequality of the rates of metabolism of the RNA in different fractions of the cell even under normal conditions, these new results do not seem to be a direct proof of his hypothesis. A good deal of indirect evidence for the association between the synthesis of adaptive enzymes and RNA metabolism has been obtained by other workers (26, 30).

It is clear now that this effect is connected with the transfer to the unadapted cells of "messenger RNA" (see Chapter XV). It was shown recently that the newly induced synthesis of β-galactosidase in *E. coli* is based on existing ribosomes (36a), which accept messenger RNA programmed for the synthesis of a new enzyme. Hayashi *et al.* (24b) have found that the presence of inducer increases production of messenger RNA from a restricted genetic locus (see also 44a, 18a).

Very striking evidence about the role of RNA in adaptive synthesis has been obtained by Kramer & Straub (29, 16a) who succeeded in transferring the ability to form penicillinase to an unadapted culture of *B. cereus* by means of RNA isolated from an adapted culture. Analogous transfers of the ability to synthesize certain adaptive enzymes by means of RNA isolated from adapted cultures have also been described by other workers (46).

All these extremely interesting facts, which disclose the biochemical conditions required for the synthesis of adaptive enzymes, indicate clearly that their synthesis depends in a definite way on all those systems which also regulate the formation of the constitutive proteins within the cell. In particular, the structural specificity of adaptive enzymes is determined mainly by means of matrices in the form of molecules of RNA the appearance and "work" of which are stimulated in some way by the action of the substrates of the reactions. However, it has not yet been established with any certainty in what way the three indispensible links in the adaptive reaction, the enzyme, the RNA and the substrate, are connected with one another. Any ideas in this field are of the nature of more or less well-founded hypotheses which must be discussed, as they represent as important a stage in our understanding of the peculiarities of this biological phenomenon as the collection of factual material in the field.

## 3. Contemporary theoretical concepts of the mechanism of synthesis of induced enzymes under the influence of specific substrates

Many different hypotheses have been put forward to explain the formation of adaptive enzymes. A critical and methodical analysis of many of the hypotheses in this field has been made in the last few years in a number of reviews (10, 16, 40, 24a, 25, 31, 15, 33, 34, 44, 50).

As we have already pointed out, it is not our task to survey all aspects of the problem of adaptive synthesis and we shall select for critical analysis only those questions which are directly connected with the general features of protein synthesis and enable us to reveal more precisely the causes which determine the specificity of this synthesis. In this connection we shall only look at a few of the contemporary hypotheses as to the mechanism of adaptive synthesis which seem to us most soundly based.

The synthesis of adaptive enzymes is a field of study in which it would seem that there cannot be complete agreement about theoretical concepts, as it has been shown clearly that the formation of each different enzyme has its own specific peculiarities and therefore the initial stage in which the substrate reacts with the systems synthesizing enzymic proteins may, obviously, show considerable variety.

The most studied of adaptive syntheses are those of penicillinase in the cells of B. cereus and of β-galactosidase in the cells of E. coli, which are very different in many respects. Most of the hypotheses about adaptive synthesis are based on factual material obtained by the study of these processes.

In the first place we should like to spend some time on the hypotheses of Pollock (44) who seems to us to approach an understanding of adaptive synthesis from the standpoint of the general biochemical rules regulating the synthesis of all forms of protoplasmic proteins in the cell.

In studying the adaptive synthesis of enzymes it has been found that the formation of a specific enzyme may be evoked, not only by the specific substrate for that enzyme, but also by compounds which have a similar structure to the substrate but which do not have any effect in the cell connected with the enzyme being formed.  On the other hand, many analogues of the substrate, on which the enzyme being formed has almost the same effect as it has on the substrate, could not stimulate the synthesis of the adaptive enzyme (24, 35, 51, 38, 41, 43, 55).  This feature of enzymic induction is very important for an understanding of its mechanism because it is a clear indication of the activating, permissive rather than creative role of the substrate-inducer.

The reports from the literature concerning the genetics of the synthesis of adaptive enzymes which the author adduces show that the ability of the bacterial cell to react to the introduction of some substrate into the medium by the formation of enzymes is a genetic characteristic and thus this synthesis depends on a store of genetic information which already existed within the system.  The genetic character of ability to form adaptive enzymes allows us to look upon enzymic adaptation as a reinforcement or manifestation of a particular power of organization.  Of course, in many cases, when sensitive methods are used, it may be shown that the microorganism contains a small amount of the enzyme before treatment with the inductor.  These insignificant quantities of enzymes which were formed alongside the constitutive ones have been called the "basic enzyme".  A study of its biological properties and serological reactions shows that this fraction of "basic enzyme" does not differ from the fraction formed as a result of adaptation.  The origin of the basic fraction has not been studied in detail and it may be the result of endogenous induction though, in many cases (e.g. the formation of penicillinase), endogenous induction would seem to be out of the question.

On the basis of these theoretical premises and the results of his own experiments on the synthesis of adaptive penicillinase, Pollock believes that the evidence we now have about the $\beta$-galactosidase and penicillinase systems indicates that the inducer, or some product of its metabolism, combines with a specific receptor molecule (R) with the formation of an organizer (O) which then catalyses the synthesis of the enzyme from amino acids.  The catalytic nature of the process is clearly shown in the case of penicillinase.  In this system the inducer is specifically and irreversibly combined when it enters into a short-term interaction with the cells of *B. cereus*.  All the free penicillin may later be removed but the bacteria continue to synthesize the enzyme.  It has been shown that in one hour about forty molecules of enzymes are formed for every molecule of inducer bound by the cells.  The specificity of the penicillin receptor is confirmed by the fact that there is competition at the site of fixation between penicillin and its analogues. This shows that uninduced cells have a limited number of receptor molecules (about 1000/cell) which correspond in their specificity with penicillinase, also reacting selectively with penicillin.

According to Pollock the formation of adaptive enzymes is only one example of the more general dependence of enzyme synthesis on the concentration of intermediate products of some particular enzymic chain. This is only one example which shows the possibility of affecting an enzymic reaction by means of smallish molecules which are metabolically associated with that reaction. In such reactions the inhibiting effect of the product on the formation of the enzyme is no less widely distributed in nature than the stimulating effect, and facts concerning this inhibition are now beginning to be accumulated in the literature. Having discussed some examples of this inhibition, Pollock puts forward the suggestion that the weakness of the formation of particular enzymes in uninduced cells may be explained in terms of their endogenous inhibition by some particular compounds. In such a case the inducer may either combine with the endogenous inhibitor or may occupy the site at which the endogenous inhibitor acts. As a result of this, the surface of the catalyst or matrix associated with the synthesis of the enzyme is freed from inhibition and this markedly increases the formation of the enzyme in question. In such a case the catalytic surface is also the "organizer" which determines the synthesis of the adaptive enzyme.

In another paper Pollock (42) sets out a number of supplementary contributions to the development of this interesting hypothesis. He notes that there are three kinds of penicillinase in the cells, $\alpha$-, $\beta$- and $\gamma$-, and that the $\alpha$- and $\beta$-forms are identical in properties but differ in localization. $\alpha$-Penicillinase, which is formed in the greatest quantity, is secreted into the medium, while $\beta$-penicillinase is bound to the cell wall and is present in insignificant amounts. $\gamma$-Penicillinase differs from the $\alpha$- and $\beta$-forms in its immunological and physico-chemical properties. It is localized within the cell and would seem to be synthesized independently of the $\alpha$- and $\beta$-forms.

A very similar approach to the explanation of the mechanism of adaptive synthesis has been developed in the "unified" hypothesis of Vogel (57, 58). Vogel, like Pollock, believes that the formation of adaptive enzymes is only a partial manifestation of the more general ability of cells to regulate their enzymic reactions by stimulation and inhibition when the need for particular reactions ceases temporarily. This flexibility would certainly be an important evolutionary adaptation which would seem to have guided natural selection.

According to Vogel, the reactions of protein synthesis may provisionally be classified in three groups: the unchanged synthesis of constitutive enzymes, induced synthesis and inhibited synthesis. There may be a functional connection between the induction and inhibition of synthesis and the main question is how the relatively small molecules of inducers and "repressors" (but not inhibitors) can specifically alter the formation of particular proteins of high molecular weight (the molecular weight of $\beta$-galactosidase is about 700,000). Vogel thinks that the inducers or repressors react specifically with the macromolecular matrices for the synthesis of some particular enzyme. In this case the inducer does not act as a steric prototype for the formation of the enzyme but as an augmentor of the system of synthesis of the enzymic protein in question on matrices which are already present in the cell. In discussing the possible nature of the mechanism of this augmentation Vogel suggests that the inducer may affect the rate of liberation of the product formed on the matrix from the matrix itself. In this case inhibition might be a result of the binding of newly formed enzymic protein to the site of its synthesis by means of the corresponding "repressor". Induction might also represent the neutralization of the binding effect which, in the absence of an inducer, has a tendency to retain the newly formed protein beside its matrix. Certainly rapid removal of the synthesized product from the surface of the matrix would accelerate the formation of enzyme while occupation of the matrix by its own product would prevent the participation of the matrix in any further synthetic processes. There is no need to suggest that the "repressor" binds the enzyme to the matrix by the formation of "bridges" between them. It is quite possible that it may alter the configuration of the macromolecules, thus securing their attachment. Similarly, the inducer may either act as a local neutralizer of these bridges or it may affect the macromolecular configuration. Thus, the unified hypothesis of Vogel involves acceptance of the idea that all enzymes, whether they are constitutive or adaptive, induced or inhibited, are formed by the same means common to the synthesis of all proteins, that is to say, on matrices. The differences between them consist merely in their sensitivity to regulators. The main means of regulating synthesis is by affecting the dissociation of the synthesized product from its matrix.

A similar hypothesis has been put forward by Monod (34) who thinks that a precursor of the active enzyme is first formed on the matrix of protein synthesis. This pre-enzyme remains attached to the matrix till the activated inducer begins to act on the matrix-pre-enzyme complex. As a result of this the pre-enzyme separates from the matrix and becomes an active enzyme. When the matrix is freed a new molecule of pre-enzyme is formed. Almost the same scheme of the synthesis of adaptive enzymes has recently been put forward by Rickenberg (49).

The theoretical concepts concerning the mechanism of adaptation developed by Spiegelman (10, 52) are also of considerable interest in helping to draw a general picture of adaptive synthesis. In principle they are similar to the hypotheses of Vogel and Pollock which we have already discussed. Spiegelman considers that rapid enzymic adaptation is an induced change of enzymic activity occurring against some constant genetic background. In Spiegelman's scheme the role of the inducer is also to activate the matrices but this process is not one of unblocking but of increasing the number of active matrices. He suggests that, in the presence of the inducer, cells containing one or a few active matrices and many inactive ones in the form of RNA give rise, in the course of a few hours, to progeny in which a large proportion of the matrices responsible for the synthesis of the enzyme in question are active. In his opinion the transition of the matrix to the active state consists in combination of the matrix with the inducer, the matrix only being active in the form of a complex with the inducer and the enzyme. It is hardly necessary, however, to postulate such a complex structure for the active matrix, especially as the power of adaptive synthesis may be continued for several generations in the absence of the substrate-inducer.

An interesting attempt at a theoretical, mathematical analysis of the most widely supported hypotheses about enzymic adaptation has recently been made by Szilard (56).

We must pay special attention to the profound and well-thought-out scheme of the synthesis of adaptive enzymes which was recently developed by Jacob & Monod (28) in their paper at the Fifth International Congress of Biochemistry. In this paper the authors introduced a new idea as to the existence of two types of genetic determinant. According to them the genetic system comprises, on the one hand, genes (molecules of DNA) which determine the structure of proteins and the sequences of amino acids. On the other hand, they think that the cell possesses special "gene regulators" and "gene operators" which can direct the rate of synthesis of proteins by means of compounds of some sort. In this case the synthesis of adaptive enzymes would take place under a double genetic control. According to this hypothesis the gene regulators are responsible for the synthesis of special substances, the so-called "repressors". In addition, the authors have introduced into their scheme a group of "gene operators" forming "operator substances" which, on combining with the "repressors", suppress the reactions of enzyme formation. When the substrate of the enzyme appears in the system it causes liberation of the "repressor" and "removes" the suppressive effect on the formation of some particular enzyme, because the "gene-operator" then "incorporates" a group of structural genes which synthesize proteins. This group of structural genes, together with the "gene-operator" forms a single functional unit, the operon, i.e. a locus in a single chromosome. This is an interesting scheme for which the author brings forward a good deal of genetic evidence, but so far it has been difficult to extend it to cover any form of induced synthesis. It is, however, the most widely accepted scheme at present.

Any further discussion of this scheme would require the use of genetic concepts on which we have not yet touched, so we shall return to the question in Chapter XVIII when we have dealt with certain aspects of the problem of heredity.

None of these, nor of several other hypotheses concerning adaptation, can, however, be applied to all cases of adaptive synthesis, the evolutionary character of which undoubtedly contributed to the utilization of all possible means of regulating the intracellular enzymic processes. It is, therefore, not impossible that the formation of some adaptive enzymes may occur in accordance with the following scheme:

$$\text{inactive precursor} \longrightarrow \text{enzyme}$$
$$\uparrow$$
$$\text{inducer}$$

Although this scheme has not been confirmed in the case of $\beta$-galactosidase there are some facts which argue in its favour in regard to several other enzymes (invertase, nitroreductase, citrate desmolase (8, 17)). In some cases adaptive synthesis seems to be more complicated and includes increase in the activity of particular enzymes which increase the permeability of cell membranes to particular substrates.

A very interesting case of the synthesis of an adaptive enzyme (invertase) has been described by Yurkevich & Gorodovich (11). In these experiments the adaptive invertase isolated from yeast cells had, in the presence of amino acids, the same stimulating and inducing effect on the synthesis of invertase in a non-adapted culture as had the sucrose which is the substrate of this enzyme.

## Conclusion

In evaluating many of the facts and ideas about enzymic adaptation which have been discussed in this short review, it seems to us that we should emphasize that the formation of adaptive enzymes is a special case of protein synthesis. The specificity of these enzymes is determined by the same hereditary biochemical mechanisms, including nucleic acids, acting in the same way as the systems whereby any other proteins are synthesized. At the same time, by studying adaptation we may be able to discover a new group of phenomena concerned with the influence of substrates and products of reactions on the rate of synthesis of specific proteins and, possibly, also on the manifestation of activity in the systems determining their synthesis (perhaps the activity of matrices). Such regulators might take the form of one of the links in the chain of protein synthesis, primarily regulating the rate at which it occurs. However, the rate of a reaction is a most important indicator of metabolism and the particular relationship between the rates of different reactions determines, to a considerable extent, the harmony of the co-operation between the various processes and functions of the cell. It is therefore certain that further experimental and theoretical study of this problem will provide a key to the analysis of the mechanism of the intracellular regulation of the metabolism of both exogenous and endogenous substances.

In the study of this problem biochemists will undoubtedly come closest to the problem of the manifestation during ontogenesis of latent hereditary characteristics which only appear at a particular moment and under particular biochemical, physiological and morphological circumstances. In this connection we should mention numerous cases of endogenous chains of induction in which the effect of one inducer is to cause the sudden development of a series of adaptive enzymes carrying out a whole cycle of transformations of the inducer. At the same time we must point out that there are also cases of endogenous chains of inhibition, a striking example being the reaction of *E. coli* to the administration of tryptophan (27) which suppresses the activity of the whole group of enzymes taking part in its own synthesis. However, the authors obtained mutants in which tryptophan did not suppress the synthesis of these enzymes. They consider that this is an indication that there is a gene in the cell which in some way regulates the reactions between tryptophan and the genes which govern the synthesis of the tryptophan-synthesizing enzymes.

Further study of the mechanisms of exogenous and endogenous biochemical adaptation will undoubtedly play a very great part in providing a single unified concept of protein synthesis corresponding to the actual biochemical situation within the cell.

## REFERENCES

1. Vizir', P. E. (1957). *Indutsirovannaya Izmenchivost' Bakterii*. Kiev: Izd. Akad. Nauk Ukr. S.S.R.
2. Ierusalimskiĭ, N. D. & Kosikov, K. V. (1957). *Mikrobiologiya*, **26**, 614.
3. Karasevich, Yu. N. (1959). *Mikrobiologiya*, **28**, 364.
4. Kosikov, K. V. (1954). *Genetika Drozhzheĭ i Metody Selektsii drozhzhevoĭ Kultury*. Moscow: Izd. Akad. Nauk S.S.S.R.
5. Kosikov, K. V. (1957). *Napravlennaya nasledstvennaya Izmenchivost' fermentativnykh Svoĭstv Drozhzheĭ pod Vliyaniem spetsificheskogo Substrata*. Moscow: Izd. Akad. Nauk S.S.S.R.
6. Krasil'nikov, N. A. (1958). *Mikroorganizmy Pochvy i vysshie Rasteniya*. Moscow: Izd. Akad. Nauk S.S.S.R.
7. — (1959). *Izvest. Akad. Nauk S.S.S.R., Ser. biol.* No. 6, 814.
8. Oparin, A. I., Gel'man, N. S. & Zhukova, I. G. (1955). *Biokhimiya*, **20**, 571.
9. Timakov, R. D. & Skavronskaya, A. G. (1958). *Vestnik Akad. med. Nauk S.S.S.R.* No. 4, 12.
10. Spiegelman, S. (1956). *Proc. III int. Congr. Biochem., Brussels, 1955* (ed. C. Liébecq), p. 185. New York: Academic Press.
11. Yurkevich, V. V. & Gorodovich, L. T. (1958). *Doklady Akad. Nauk S.S.S.R.* **118**, 146.
12. Chantrenne, H. (1956). *Arch. Biochem. Biophys.* **65**, 414.
13. — (1956). *Nature (Lond.)*, **177**, 579.
14. — (1958). *Rec. Trav. chim.* **77**, 586.

15. COHN, M. (1957). *Bact. Revs.* **21**, 140.
16. COHN, M. & MONOD, J. (1953). In *Adaptation in Micro-organisms (3rd Symp. Soc. gen. Microbiol.)* (ed. R. Davies & E. F. Gale), p. 132. Cambridge: University Press.
16a. CSANYI, V., KRAMER, M. & STRAUB, F. B. (1960). *Acta physiol. Acad. Sci. Hung.* **18**, 171.
17. DAGLEY, S. & SYKES, J. (1956). *Arch. Biochem. Biophys.* **62**, 338.
18. DEAN, A. C. R. & HINSHELWOOD, C. N. (1953). In *Adaptation in Micro-organisms (3rd Sympos. Soc. gen. Microbiol.)* (ed. R. Davies & E. F. Gale), p. 21. Cambridge: University Press.
18a. EISENSTADT, J. M., KAMEYAMA, T. & NOVELLI, G. D. (1962). *Proc. natl. Acad. Sci. U.S.* **48**, 652.
19. GALE, E. F. (1957). *Biochem. Soc. Symposia,* **14**, 47.
20. — (1956). In *Ciba Foundation Symposium on ionizing Radiations and Cell Metabolism* (ed. G. E. W. Wolstenholme & C. M. O'Connor), p. 174. London: Churchill.
21. — (1955-6). *Harvey Lectures,* Ser. **51**, 25.
22. GALE, E. F. & FOLKES, J. P. (1955). *Biochem. J.* **59**, 661.
23. — (1955). *Biochem. J.* **59**, 675.
24. GROSS, S. R. & TATUM, E. L. (1955). *Science,* **122**, 1141.
24a. HALVORSON, H. O. (1960). *Advanc. Enzymol.* **22**, 99.
24b. HAYASHI, M., SPIEGELMAN, S., FRANKLIN, N. C. & LURIA, S. E. (1963). *Proc. natl. Acad. Sci. U.S.* **49**, 729.
25. HOGNESS, D. S. (1959). *Revs. modern Phys.* **31**, 5 (1959).
26. HUNTER, G. D. & BUTLER, J. A. V. (1956). *Biochim. biophys. Acta,* **20**, 405.
27. JACOB, F. & MONOD, J. (1959). *Compt. rend. Acad. Sci., Paris,* **249**, 1282.
28. — (1961). *Proc. V int. Congr. Biochem., Moscow,* **1**, 132.
29. KRAMER, M. & STRAUB, F. B. (1957). *Acta physiol. Acad. Sci. Hung.* **11**, 133.
30. LØVTRUP, S. (1956). *Biochim. biophys. Acta,* **19**, 433.
31. MANDELSTAM, J. (1956). *Intern. Rev. Cytology,* **5**, 51.
32. — (1960). *Bact. Revs.* **24**, 289.
33. MONOD, J. (1956). In *Henry Ford Hospital Intern. Symposium on Enzymes: Units of biol. Structure and Function, Detroit, 1955* (ed. O. H. Gaebler), p. 7. New York: Academic Press.
34. — (1958). *Rec. Trav. chim.* **77**, 569.
35. MONOD, J. & COHN, M. (1952). *Advanc. Enzymol.* **13**, 67.
36. MONOD, J., PAPPENHEIMER, A. M. (jun.) & COHEN-BAZIRE, G. (1952). *Biochim. biophys. Acta,* **9**, 648.
36a. NAKADA, D. (1963). *Biochim. biophys. Acta,* **72**, 432.
37. PARDEE, A. B. (1954). *Proc. natl. Acad. Sci. U.S.* **40**, 263.
38. PARDEE, A. B., JACOB, F. & MONOD, J. (1959). *J. mol. Biol.* **1**, 165.
39. PARDEE, A. B. & PRESTIDGE, L. S. (1959). *Biochim. biophys. Acta,* **36**, 545.
39a. — (1961). *Biochim. biophys. Acta,* **49**, 77.
40. POLLOCK, M. R. (1953). In *Adaptation in Micro-organisms (3rd Sympos. Soc. gen. Microbiol.)* (ed. R. Davies & E. F. Gale), p. 150. Cambridge: University Press.
41. POLLOCK, M. R. (1957). *Biochem. J.* **66**, 419.
42. — (1958). *Proc int. Symposium on Enzyme Chemistry, Tokyo & Kyoto, 1957* (ed. Katashi Ichihara), p. 369. Tokyo: Maruzen.
43. POLLOCK, M. R. & KRAMER, M. (1958). *Biochem. J.* **70**, 665.
44. POLLOCK, M. R. & MANDELSTAM, J. (1958). *Symposia Soc. exp. Biol.* **12**, 195.
44a. RACHMELER, M. & PARDEE, A. B. (1963). *Biochim. biophys. Acta,* **68**, 62.
45. RAMSEY, H. H. & WILSON, T. E. (1957). *Nature (Lond.),* **180**, 761.
46. REINER, J. M. (1960). *J. Bact.* **79**, 166.

47. REINER, J. M. & GOODMAN, F. (1955). *Arch. Biochem. Biophys.* **57**, 475.
48. RICKENBERG, H. V., YANOFSKY, C. & BONNER, D. M. (1953). *J. Bact.* **66**, 683.
49. RICKENBERG, H. V. (1960). *Nature (Lond.)*, **185**, 240.
50. SPIEGELMAN, S. & CAMPBELL, A. M. (1956). In *Currents in Biochemical Research 1956* (ed. D. E. Green), p. 115. New York: Interscience.
51. SPIEGELMAN, S. & HALVORSON, H. O. (1954). *J. Bact.* **68**, 265.
52. — (1953). In *Adaptation in Micro-organisms (3rd Sympos. Soc. gen. Microbiol.)* (ed. R. Davies & E. F. Gale), p. 98. Cambridge: University Press.
53. SPIEGELMAN, S., HALVORSON, H. O. & BEN-ISHAI, R. (1955). In *A Symposium on Amino Acid Metabolism* (ed. W. D. McElroy & H. B. Glass), p. 124. Baltimore, Md.: Johns Hopkins Press.
54. SPIEGELMAN, S. & KAMEN, M. D. (1946). *Science*, **104**, 581.
55. STOEBER, F. (1957). *Compt. rend. Acad. Sci., Paris*, **244**, 950.
56. SZILARD, L. (1960). *Proc. natl. Acad. Sci. U.S.* **46**, 277.
57. VOGEL, H. J. (1957). *Proc. natl. Acad. Sci. U.S.* **43**, 491.
58. — (1957). In *A Symposium on the chemical Basis of Heredity* (ed. W. D. McElroy & B. Glass), p. 276. Baltimore, Md.: Johns Hopkins Press.
59. WEINBAUM, G. & MALLETTE, M. F. (1959). *J. gen. Physiol.* **42**, 1207.
60. WILLIAMS, A. M. & WILSON, P. W. (1954). *J. Bact.* **67**, 353.

CHAPTER IX

# THE PROBLEM OF THE PARTIAL RENEWAL
# OF INDIVIDUAL PROTEINS *IN VITRO*
# AND *IN VIVO*

*Introduction*

In the previous chapters we have surveyed the different types of protein synthesis. These form the basis for the biological self-renewal of proteins and account for the synthesis and resynthesis of whole molecules from non-protein precursors (amino acids or peptides) for purposes of growth. Each molecule of protein synthesized in the organism has a particular life-span which may vary within very wide limits and depends on a variety of factors. Some protein molecules only "live" for seconds but others can carry out their functions for months or even years.

The reasons for the breakdown of protein molecules within the organism are likewise different. In some cases they may be connected with the death and disintegration of the cells (e.g. the haemoglobin of erythrocytes); in others with the rate of assimilation of the proteins by the tissues (e.g. blood albumin); and in yet others with the breakdown of particular complexes of proteins with non-protein products. They may also be connected with functional processes and even with the physico-chemical changes of denaturation of the molecules, which alter their resistance to the action of proteolytic enzymes.

In this book we shall not undertake a detailed discussion of the question of what factors determine the "lifetime" of the molecules of various proteins, as it is very complicated and intimately associated with the general physiological and biochemical functions of the various tissues and organs. In this chapter we want to review the present state of one aspect of the study of the renewal of protein on which people have worked and about which they have thought for many years but which still remains debatable in several respects. We have in mind the data and speculations as to the possibility that proteins may be able to renew themselves without any extensive breakdown of their polymeric molecules and without loss of their "individuality". Such self-renewal would occur by the partial replacement of particular amino acids and peptides in the polypeptide chain, by the dynamic alternation of breakdown and synthesis of peptide bonds. If the alternation of the processes of breakdown and synthesis can be called the "self-renewal" of proteins, then this type of metabolism is essentially the self-renewal of individual molecules by exchange.

200

The author has already made a review of the factual and speculative materials available in this field in 1956 (15) and there is no need to set out again the materials which were brought forward then. In that article we noted that there are three main ways of approaching the study of the self-renewal of protein molecules. Some authors deny the possibility of partial exchange and consider that the only possible means of renewal is by alternation of the processes of complete breakdown and resynthesis of the macromolecules of proteins (24, 38, 14).

The second approach, which is found in a number of investigations, consists in recognition that two forms of renewal of protein may occur, the self-renewal of proteins without complete breakdown taking second place, according to the results and ideas of the proponents of this hypothesis, and accounting for merely an insignificant fraction of the total amount of intracellular synthesis (15, 27-33, 42, 44). If this is true it must be reckoned that the process of partial renewal of the proteins is certainly enzymic and is connected with the functioning of the complicated enzymic systems which determine the breakdown and synthesis of peptide bonds. Finally, a third group of authors adopts the standpoint that self-renewal and autocatalysis are properties inherent in the protein molecules and that, even in the purified state, they are capable of "autometabolism" (the specific incorporation of amino acids) and even of the reproduction of their own specificity independently of the specific activity of RNA, DNA and various enzymic systems. According to this school of thought the molecules of any protein are themselves the enzymes responsible for their own metabolism, self-renewal and even growth (3-6, 9-11, 20). From this point of view the main method of studying this self-renewal consists in experiments on the ability of pure proteins and nucleoproteins to bind a certain amount of labelled amino acid which is not removed by subsequent washing.

In this chapter we propose to trace the development of each of these lines of research since 1956.

## 1. Characteristics of non-enzymic binding of amino acids by proteins *in vitro*

In analysing the evidence on this subject available in 1956 we came to the conclusion that the process of incorporation of amino acids into pure proteins *in vitro* (at 37°C and 100°C) was not a specific, biological phenomenon and was brought about by purely chemical and adsorptive reactions by which the amino acids and peptides became bound by active groups of the polypeptide chain. This does not exclude the possibility that new peptide bonds might be formed at the ends of the polypeptide chains and with free amino and carboxyl groups, but these processes are not directly related to the real phenomenon of biological renewal and specific synthesis of proteins under the conditions existing in protoplasm. These are based on other and more complicated rules.

In recent years Konikova, Kritsman and their colleagues have carried on their researches on the non-enzymic interactions of isolated proteins with amino acids and have studied several interesting features of this process (7, 8, 11-13). They found that glycine and other amino acids can combine with proteins with the formation of peptide bonds. When the concentration of amino acids is increased their "combination" with proteins is also increased. For example, on addition of 25 mg./ml. of glycine to a reaction mixture containing 1 mg./ml. of myosin, one molecule of the protein with a molecular weight of about 1,000,000 had combined with 47 molecules of glycine after 2 hours. If the concentration of the amino acid was decreased to 25 $\mu$g./ml. while the concentration of protein remained the same the amount of glycine entering into combination with the myosin corresponded to one molecule of amino acid for every four molecules of protein.

From a quantitative point of view the "incorporation" of amino acids into proteins only approached that occurring *in vivo* in cases in which the concentration of amino acids in the medium was considerably higher than that of the proteins. When the ratio between the amounts of proteins and amino acids was the same as is usual under physiological conditions (marked preponderance of protein) the combination of amino acids with isolated proteins was quite insignificant.

Similar work on the binding of amino acids by the proteins of hens' eggs has been carried out by Shnol' (21). He found that if the protein of hens' eggs is incubated with [$^{35}$S]methionine and [$^{14}$C]glycine or [$^{14}$C]tyrosine and then precipitated with trichloroacetic acid and washed 7-8 times, an increase in the radioactivity of the protein may be observed. This increase is at first almost linear but later it falls off and finally almost ceases. The author considers that the "inclusion" which may be observed is due to combination of the amino acids with the terminal groups of the protein molecules, affecting both the amino and the carboxyl groups, with the formation of peptide bonds and is not true renewal. The binding is not accompanied by a corresponding elimination of unlabelled amino acids from the protein. Calculations were carried out which showed that when equilibrium was reached there was one molecule of bound amino acid for every 50-100 molecules of protein.

An interesting investigation of the combination of amino acids with isolated proteins has recently been made by Cornwell & Luck (25). This was a continuation of the series of analogous studies made in their laboratory from 1952 onwards (23). Even in their earlier work they found that the "incorporation" of amino acids into isolated proteins (histone or desoxypentose-nucleohistone) depended on the concentration of the amino acids, pH and temperature but not on metabolic sources of energy. In later work the authors investigated this process again, paying special attention to methodological details, using [14C]lysine and [14C]phenyl-alanine and pure preparations of several proteins (histone, insulin, desoxy-pentose-nucleohistone). In these experiments the dependence of the combination of amino acids with the proteins on their relative concentra-tions and on pH and temperature was again confirmed. However, after a detailed study of the products of this "binding" by dialysis and by the use of dinitrophenyl derivatives, the authors found that about 30-80% of the "incorporation" was accounted for by stable adsorption, the rest of the amino acids were chemically combined with the protein molecule but the authors did not observe the formation of peptide bonds of both the carboxyl and the amino groups of the amino acids which were bound; that is to say, they did not observe "insertion" of amino acids into a poly-peptide chain but only their combination with some part or another of the protein molecules.

Adsorptive binding of amino acids by proteins *in vitro* has also been observed by Salganik (18) who found age-changes in the ability of proteins to bind amino acid products.

McMenamy & Oncley (40) have obtained some interesting results concerning the non-enzymic binding of amino acids by proteins. They showed that serum-albumin is very selective and mainly binds tryptophan, while hardly binding other amino acids at all.

In their recent work Zbarskiĭ and colleagues (19) found that the binding of labelled amino acids by isolated proteins is markedly accentu-ated if the protein is combined with RNA or DNA to form a nucleoprotein. Even pure, commercial preparations of DNA and RNA actively bound amino acids (glycine, tyrosine, lysine, etc.). However, although the bond between the amino acid and the RNA and DNA was mainly adsorptive in nature and was easily broken when the preparation was treated with ninhydrin or in alkaline solution, the bond between the amino acid and the protein was chemical in nature, that is to say, it was more stable. The authors consider that such binding is not a process of biological synthesis but that the occurrence of such processes under the conditions obtaining within the cell cannot be ruled out.

The formation of stable bonds between amino acids and insulin by reaction with free amino and carboxyl groups of the side chains has been reported recently by Godin (35). Pasynskiĭ and colleagues (1) have studied the process of binding of [35S]methionine and the products of its radiochemical decomposition by various proteins.

It is important that we should also take note here of certain unusual results recently reported by Konikova and colleagues in the form of a preliminary communication (2). These authors report that when labelled glycine and tyrosine reacted with pure proteins, they were incorporated in the same peptide groups as in experiments *in vivo*. We cannot assess the value of these results until they have been published in more detail.

We ourselves (16) have also studied the adsorptive combination of amino acids with plant proteins. We found stable binding of [$^{35}$S[methionine by the proteins of the cellular protoplasm both before and after denaturation. We found that, if the cellular protoplasm of the leaves of several plants was "pressed out" into filter paper and then incubated with a solution of [$^{35}$S]methionine, then there was a residue of about 0·1-0·8% of the labelled amino acids which could not be washed out of the protein whether the protein had been inactivated (by heating to 100°C and treatment with trichloroacetic acid) before incubation or not.

In another series of experiments on the metabolism of peptides (17) we carried out controls in which [$^{14}$C]amino acids and peptides were added to a homogenate of leaves in the presence of dinitrophenol and here again we found that some of them were closely bound to proteins even when there was complete suppression of protein synthesis. The amount of [$^{14}$C]-products combining with the protein under these conditions amounted, in our experiments, to about 4-6% of the amount used in the synthesis of proteins over the same period *in vivo* when they were introduced directly into the leaves.

In evaluating all these new results concerning the combination of amino acids with isolated proteins we think that we can maintain the opinion which we held at first and which we expressed in our discussion of this question (15). We consider that such "incorporation" has nothing in common with the true renewal and biosynthesis of proteins which takes place in living systems. The study of these phenomena is, however, important from a methodological point of view as it shows that, in all work on the "incorporation" of amino acids *in vivo* and *in vitro*, it is necessary to keep a constant check on the amount of non-specific combination of amino acids with proteins which may be going on both in the course of the experiment itself and during the analytical isolation of the protein from the biological system from which it was obtained. This checking is necessary because the non-specific chemical combination of amino acids with proteins, and even their simple adsorption by proteins, often depends on the same physical and chemical factors as the processes of biological synthesis (temperature, pH, concentration, etc.).

In such experiments it is also important that the amino acids should only be administered in physiological concentrations.

## 2. The possibility of partial renewal of proteins *in vivo*

In our survey in 1956 of the problem of the forms of self-renewal of proteins we mentioned a number of results obtained by Gale & Folkes (28-32) which, in the opinion of the authors, suggested the possibility of the biological incorporation of amino acids into the proteins of cells of *Staphylococcus aureus* which had been disrupted by supersonic waves, not only as a result of new synthesis but also as a result of exchange of amino acids in the proteins without whole new molecules being synthesized. This exchange only occurred in particular fractions which, the authors thought, were associated with nucleic acids. It should be mentioned that, up till now, the observations of Gale & Folkes have been the chief examples used as arguments in favour of the occurrence of partial renewal of protein molecules by the substitution of individual amino acids or peptides within their chains. In the three or four years following their studies, though, several other results have been obtained which suggest that such a form of partial renewal of proteins may occur, but they were indirect rather than direct evidence.

The theoretical idea that two forms of renewal of proteins may occur in *Staph. aureus*, which was advanced by Gale & Folkes and exerted a great influence on the theoretical propositions of other authors, has later been completely revised as a result of new experiments in Gale's laboratory (34), which completely reversed the conclusions which this group of research workers had arrived at over the preceding years.

The first suggestion that two forms of renewal of protein might occur within the cells of *Staph. aureus* was based on the occurrence of substantial differences in the incorporation into proteins of amino acids according to whether they were introduced into the incubation mixture individually or in a mixture with other amino acids. If individual amino acids were added to the medium they were quickly incorporated into the proteins without there being any general increase in the protein nitrogen and this was the process which was interpreted as the partial renewal of the proteins. However, later experiments by the authors showed that in fact under these conditions the individual amino acids were not bound by being incorporated into the cytoplasmic proteins. It was found that the "binding" of the [$^{14}$C]glutamic acid in the absence of other amino acids was, in the first place, due to their labile acceptance by S-RNA in the systems of activation and, in the second place, to their combination with the polypeptide structures of the cell wall. In this process an unidentified factor which augments incorporation is important for the "working" of the activating systems (as a specific co-factor). Thus it is clear that all the earlier experiments of Gale & Folkes, which at first seemed to indicate that, alongside the ordinary synthesis of proteins in living cells, there was also a specific form of partial breakdown and resynthesis of their molecules, are now integrated into the generally accepted idea that there is only one way in which proteins are broken down and synthesized.

It would seem that a similar revision of the conclusions drawn about the occurrence of partial exchange of amino acids in proteins might be made in respect of the work of Nisman *et al.* (41), who also noticed a difference in the nature of the incorporation of labelled amino acids into the proteins of two species of bacteria according to whether they were introduced into the medium individually or in a mixture. When the amino acids were incorporated individually into the proteins the reactions of their combination were reversible; when they were incorporated in growth the reactions were irreversible.

Certain facts suggesting the possibility of partial renewal of the plasma proteins have also been found in two other laboratories (26, 43) but they have been reported in the form of short notes from which it is impossible to judge the reliability of the experimental methods. At the same time, in their recently published work, Orekhovich and colleagues (14) obtained quite different results. These authors found that, in rats in which the liver has been functionally isolated by operation, radioactive methionine is not incorporated into the albumin of the blood. The incorporation of amino acids into the serum albumin only occurs at the time of its formation.

The absence of metabolic incorporation of amino acids into preformed proteins has also been demonstrated by the work of King et al. (37).

It should also be mentioned that, as regards a number of proteins, no conditions have been found in which they manifest any great degree of lability. This is particularly true of the proteins of the nucleus into which, according to Allfrey, Mirsky & Osawa (22), the incorporation of labelled alanine and methionine is irreversible and from which they cannot be displaced even if the nuclei are incubated with a 200-fold excess of these amino acids in the unlabelled form. The truly synthetic nature of the original incorporation of these amino acids into proteins has been demonstrated by the selectivity of the process in respect of the L-form of the amino acids and its inhibition by the introduction of the D-form into the medium.

Suppositions as to the possible occurrence of partial renewal of proteins have, however, been made by a group of biochemists in the University of Texas (45). On introducing serum albumin labelled with both [$^{14}$C]valine and [$^{35}$S]methionine into the blood stream of rats they found that, as the albumin circulated, the [$^{14}$C]valine was lost more quickly than the [$^{35}$S]-methionine, which does not accord with the idea of the complete destruction of proteins by catabolism. They suggest that the shift in the ratio $^{14}$C:$^{35}$S is possible owing to the lability of the valine bonds and the possibility of the valine leaving the protein. In another research using dogs as the experimental animals and a different pair of amino acids ([$^{35}$S]cystine and [$^{14}$C]lysine) as the double labels in the protein, no similar shift in the ratio between the two labelled residues was observed (36). It is, however, possible that the results obtained by the Texan biochemists might be associated, not with a difference in the rate of breakdown, but with a difference in the rate of reutilization of the breakdown products in the synthesis of new portions of albumin, which goes on continuously in the liver of the animal.

In a recent review of the renewal of proteins and nucleic acids Mandelstam (39) also arrives at the conclusion that there is no really sound proof of the occurrence of partial renewal of proteins either in vivo or in vitro.

## Conclusion

The comparatively small number of facts and ideas which have just been adduced virtually exhausts the results of all the investigations of the process of partial renewal of proteins occurring both within biological systems and as a non-enzymic exchange *in vitro*. If we compare this rather small collection with the extensive material which has been obtained over the same period as a result of investigations of the main way in which proteins are synthesized, it becomes obvious that it is total and not partial synthesis which forms the basis for the self-renewal of proteins, while the process of partial replacement, if it occurs at all, would appear only to represent an incidental or perhaps only apparent offshoot of protein metabolism.

As concerns the phenomenon of non-enzymic binding of amino acids by individual proteins, this process certainly does not represent the specific biological incorporation of amino acids into proteins and it is scarcely suitable to base any general concepts of the ways in which proteins are renewed on this sort of binding. Nevertheless, the study of these processes is of methodological importance in showing that many supplementary control analyses are required in experiments on the synthesis of proteins.

REFERENCES

1. VOLKOVA, M. S., KOMAROVA, L. V. & PASYNSKIĬ, A. G. (1960). *Biokhimiya*, **25**, 422.
2. KONIKOVA, A. S., KOROTKINA, R. N. & POGOSOVA, A. V. (1961). *Proc. V int. Congr. Biochem., Moscow*, **9**, 94.
3. KONIKOVA, A. S. & KRITSMAN, M. G. (1953). *Voprosy Filosofii*, No. *1*, 143.
4. — (1954). *Voprosy Filosofii*, No. *1*, 210.
5. — (1957). In preprint referred to (Ref. 1) by M. G. KRITSMAN & A. S. KONIKOVA (1959). In *Proceedings of the first international Symposium on the Origin of Life on the Earth* (ed. A. I. Oparin *et al.*), p. 275. London: Pergamon.
6. KONIKOVA, A. S., KRITSMAN, M. G. & SAMARINA, O. P. (1954). *Biokhimiya*, **19**, 440.
7. — (1956). *Doklady Akad. Nauk S.S.S.R.* **109**, 593.
8. KONIKOVA, A. S., SUKHAREVA, B. S. & KRITSMAN, M. G. (1958). *Doklady Akad. Nauk S.S.S.R.* **119**, 749.
9. KRITSMAN, M. G. & KONIKOVA, A. S. (1955). *Voprosy Filosofii*, No. 6, 172.
10. — (1959). *Uspekhi sovremennoĭ Biol.* **48**, 136.
11. KRITSMAN, M. G., KONIKOVA, A. S. & OSIPENKO, Ts. D. (1952). *Biokhimiya*, **17**, 488.
12. KRITSMAN, M. G., SUKHAREVA, B. S., KONIKOVA, A. S. & KOROTKINA, R. N. (1960). *Biokhimiya*, **25**, 17.
13. KRITSMAN, M. G., SUKHAREVA, B. S., SAMARINA, O. P. & KONIKOVA, A. S. (1957). *Biokhimiya*, **22**, 449.

14. LEVYANT, M. I., LEVCHUK, T. P. & OREKHOVICH, V. N. (1959). *Biokhimiya*, **24**, 177.
15. MEDVEDEV, ZH. A. (1956). *Biokhimiya*, **21**, 627.
16. — (1957). *Doklady Akad. Nauk S.S.S.R.* **117**, 860.
17. MEDVEDEV, ZH. A. & SHEN', TS.-S. (1959). *Biokhimiya*, **24**, 709.
18. SALGANIK, R. I. (1956). *Voprosy med. Khim.* **2**, 424.
19. SAMARINA, O. P., ZBARSKIĬ, I. B. & PEREVOSHCHIKOVA, K. A. (1960). *Biokhimiya*, **25**, 443.
20. SYSOEV, N. (1956). *Voprosy Filosofii*, No. *1*, 152.
21. SHNOL', S. E. (1957). In *Trudy vsesoyuz. Konf. med. Radiol. eksptl. med. Radiol.* (ed. P. D. Gorizontov), p. 244. Moscow: Medgiz.
22. ALLFREY, V., MIRSKY, A. E. & OSAWA, S. (1957). In *A Symposium on the Chemical Basis of Heredity* (ed. W. D. McElroy & B. Glass), p. 200. Baltimore, Md.: Johns Hopkins Press.
23. BRUNISH, R. & LUCK, J. M. (1952). *J. biol. Chem.* **198**, 621.
24. BULMAN, N. & CAMPBELL, D. H. (1953). *Proc. Soc. exp. Biol. Med., N.Y.*, **84**, 155.
25. CORNWELL, D. G. & LUCK, J. M. (1958). *Arch. Biochem. Biophys.* **73**, 391.
26. CRATHORN, A. R. & HUNTER, G. D. (1958). *Biochem. J.* **69**, 47P.
27. GALE, E. F. (1956). In *Ciba Foundation Symposium on ionizing Radiations and Cell Metabolism* (ed. G. E. W. Wolstenholme & C. M. O'Connor), p. 174. London: Churchill.
28. GALE, E. F. & FOLKES, J. P. (1953). *Biochem. J.* **53**, 483.
29. — (1953). *Biochem. J.* **53**, 493.
30. — (1953). *Biochem. J.* **55**, 721.
31. — (1954). *Nature (Lond.)*, **173**, 1223.
32. — (1955). *Biochem. J.* **59**, 675.
33. — (1958). *Biochem. J.* **69**, 611.
34. GALE, E. F., SHEPHERD, C. J. & FOLKES, J. P. (1958). *Nature (Lond.)*, **182**, 592.
35. GODIN, C. (1960). *Can. J. Biochem. Physiol.* **38**, 805.
36. GOLDSWORTHY, P. D. & VOLWILER, W. (1958). *J. biol. Chem.* **230**, 817.
37. KING, D. W., BENSCH, K. G. & HILL, R. B. (jun.) (1960). *Science*, **131**, 106.
38. MACHEBOEUF, M. (1954). *Exptl. Med. Surg.* **12**, 163.
39. MANDELSTAM, J. (1960). *Bact. Revs.* **24**, 289.
40. MCMENAMY, R. H. & ONCLEY, J. L. (1958). *J. biol. Chem.* **233**, 1436.
41. NISMAN, B., HIRSCH, M. L. & MARMUR, J. (1955). *Compt. rend. Acad. Sci., Paris*, **240**, 1939.
42. RABINOVITZ, M., OLSON, M. E. & GREENBERG, D. M. (1954). *J. biol. Chem.* **210**, 837.
43. RUMSFELD, H. W. (jun.), BURR, W. W. (jun.) & WIGGANS, D. S. (1957). *Fed. Proc.* **16**, 240.
44. SIMPSON, M. V. & TARVER, H. (1950). *Arch. Biochem.* **25**, 384.
45. WIGGANS, D. S., BURR, W. W. (jun.) & RUMSFELD, H. W. (jun.) (1957). *Arch. Biochem. Biophys.* **72**, 169.

# CHAPTER X

# FEATURES OF CERTAIN FORMS OF PROTEIN SYNTHESIS IN PLANTS

## *Introduction*

In the preceding chapters we have surveyed the general aspects of protein synthesis and in so doing we have repeatedly cited results obtained from work on plant materials (activation of amino acids, synthesis in plastids, the role of peptides, etc.). There is no need for us to go over this material again in the present chapter or to try to embrace all aspects of protein synthesis in plants, especially since they have recently been surveyed in several reviews (3, 4, 14, 15, 17, 23, 24, 44, 36, 43, 45, 48). Here we shall only dwell on a few facts which fail in some way to fit obviously into the general arrangement of protein synthesis which we have already described and would rather seem to be incompatible with it and apparently peculiar to plants. It should be mentioned that the synthesis of proteins in plastids, which is peculiar to plants, has already been discussed in the chapter on the localization of protein synthesis.

## 1. "Transport synthesis" of proteins in plants

One of the forms of protein synthesis peculiar to plants is the intensive synthesis which occurs in the phloem of the vascular system and provides one of the keys to the mechanism of the transport of organic materials in plants. In the animal organism the transport of water, salts, organic substances, proteins, oxygen and other substances is combined in the single blood-circulatory system. In plants the transport of water and mineral substances in the first place, that of proteins and other organic substances in the second and that of air in the third, are each carried out through their own independent system of channels, the xylem, the phloem and the air cavities respectively. This separation is not absolute; for example, some organic substances are to be found in the xylem liquid, while, on the other side, inorganic compounds can move from place to place in the sieve tubes of the phloem. A certain difference of functions and of mechanisms of transport between these systems is, however, an undisputed fact. Here we shall not deal with the mechanisms of transport of substances within the phloem, if only because their nature is still by no means established.

We must, however, mention that the transport of substances in the phloem, even over long distances, is transport by living protoplasm and is functionally connected with the very intensive synthesis of proteins within this system. In fact, for example, the process of upward movement of substances, beginning in the root system, is accompanied by intense processes of synthesis, not only of proteins, but also of many other compounds such as amino acids, organic acids, etc. The occurrence of synthesis of amino acids in the vascular system of plants has been demonstrated clearly in a very large number of investigations (5, 7, 8, 20, 26). Active synthesis of amino acids may also occur directly in the vascular systems of leaves as we have shown recently (14) by the use of the ninhydrin reaction on imprints of bean leaves on filter paper after various inorganic substances had been introduced into the leaves via the stalks (Fig. 31).

It is quite possible that one of the ways in which nitrogen is transported is in the form of peptides which are found particularly in the root system (38) and in the xylem liquid of a number of plants (29). We found that when *Galega* plants are kept in the dark and the catabolic processes are strengthened, peptide products, as well as amino acids, accumulate in the leaves (9). Movement of RNA by means of the vascular system has been reported by Oota & Osawa (39).

Another very interesting possibility is that soluble proteins may be transported in the plant for the purpose of carrying out the same trophic functions which are fulfilled by the proteins of the blood and lymph in animals. There are a number of cytological observations with which we cannot deal here, which indicate, quite definitely, that the vascular system of plants can transfer polymeric molecules. A striking example of this is provided by the dispersal of viral RNA throughout the plant. Kretovich and colleagues have observed the transport of small quantities of albumin in pumpkin juice (6).

We have also obtained a certain amount of evidence for the possibility that small amounts of soluble proteins may be transported from the root system to the leaves. In our experiments we made an analytical study of the localization of protein synthesis in various fractions of the root system and leaves of a number of plants (11, 16). We found that when small quantities of [$^{35}$S]methionine were introduced into the region of the root system it was quickly incorporated into the proteins being synthesized on the spot almost without reaching the leaves (the specific activity of the proteins of the roots was 100-500 times that of the leaves). Several hours after the beginning of the experiment the most active fraction was that containing the soluble plasma proteins of the roots. At the same time it was found by radioautography that a highly active fraction of soluble proteins occurred in the vascular system of the leaves. It was also found that the phloem of the vascular system had a great power of synthesizing proteins from amino acids.

The possibility of local synthesis and transport of proteins in the vascular systems of the leaves has been studied in more detail since then

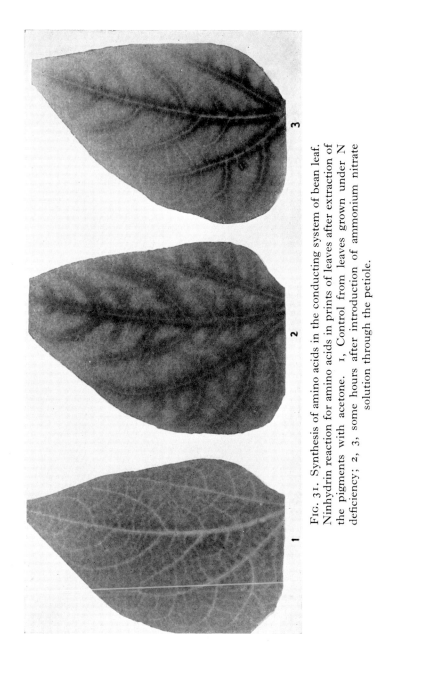

Fig. 31. Synthesis of amino acids in the conducting system of bean leaf. Ninhydrin reaction for amino acids in prints of leaves after extraction of the pigments with acetone. 1, Control from leaves grown under N deficiency; 2, 3, some hours after introduction of ammonium nitrate solution through the petiole.

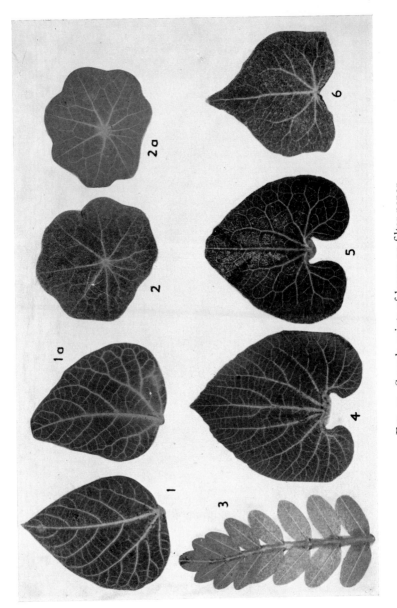

FIG. 32. Sample prints of leaves on filter paper.

by the method of radioautography of imprints of leaves on filter paper which we have worked out (10, 11, 12, 13, 37).

The theoretical background and assumptions of this method are as follows. When bulk objects such as the leaves of plants, are radioautographed, difficulties arise in washing them free from extraneous substances. Ordinary radioautography of plants will, therefore, only show localization and concentration of radioactive isotopes in various organs and will not reveal the nature of the actual compounds (proteins, nucleic acids, lipids, etc.) in which they occur, nor the fractions of the cells (vascular sap, plasma, organelles, etc.) in which the radioactively labelled elements are concentrated to greater or lesser extents. This sets a sharp limit to the applicability of radioautography to the study of the physiology and biochemistry of plants and means that it cannot be used for analysing the transformation of substances in plants or for studying the localization and rate of the synthesis of, for example, proteins and nucleic acids. Furthermore, ordinary radioautography often gives false pictures of the distribution of radioisotopes because there is darkening of the emulsion of the film, not only at the places where the radioisotopes are concentrated, but also at the thickest parts of the plants (stems, petioles, the veins of leaves, particular zones of the parenchyma of leaves, etc.).

With the hope of overcoming these limitations we have worked out a new method of preparative treatment of plants which seems to us to provide us with several additional possibilities for the study of biochemical and physiological processes in plants by means of radioactive isotopes.

In this method we do not use ordinary leaves containing radioactive isotopes for making our radioautographs, instead we use prints of these leaves made under high pressure on filter paper. In making these prints the protoplasm of the leaves is squeezed out on to or sucked up by the filter paper so that the characteristic topographical picture of the leaf is retained in the print (Fig. 32). It should be noted that there is a possibility of making differential studies, from a biochemical and physiological point of view, of the localization, synthesis and transport of different substances. This may be achieved, in the first place, by varying the intensity and duration of the pressure (separating the plasma juice from the vacuolar sap etc.) and, in the second place, by selective washing of the leaf prints with various solutions and solvents. If prints of leaves containing compounds labelled with $^{35}S$ or $^{32}P$, for instance, are washed with a solution of trichloroacetic acid, acid-soluble compounds of sulphur and phosphorus will be washed out of them, while the proteins, nucleic acids and lipids remain. Washing with alcohol, ether or chloroform will remove fats, lipids and pigments from the prints. Washing with a hot solution of trichloroacetic acid will remove nucleic acids from the print but leave proteins etc. Thus one may study the localization of particular substances by radioautography.

In particular, the use of the methods of radioautography and colorimetric evaluation of the prints has enabled us to establish the fact that there is active synthesis and transport of proteins in the vascular system, that there is adsorption of [$^{35}$S]methionine and [$^{32}$P]phosphates by the proteins of the leaves *in vitro* and several other phenomena.

Fig. 33 shows a radioautograph showing a typical picture of local synthesis of protein in the vascular system of leaves.

In experiments using isotopes greater attention has been paid during the last few years to the transport of compounds of comparatively low molecular weight such as sugars and other soluble carbohydrates, amino acids, organic acids, inorganic compounds, etc. It seems to us, however, that it would be extremely valuable to make a careful study also of the transport within plants of the polymeric substances, proteins and nucleic acids, because the discovery of the way in which these polymers are carried about within the plant would enable us to explain many important physiological and biochemical features of plants as whole systems.

Finally, elucidation of the connection between this transport and the local synthesis of proteins and nucleic acids in the cells of the phloem might enable us to analyse the mechanism by which the transport of substances within plants takes place.

## 2. Rhythm of protein synthesis in leaves and its connection with photosynthesis

One of the important forms of functional synthesis of proteins in plants is their formation in developing leaves, which will then provide products of protein metabolism for the use of growing organs. A considerable part of the protein synthesized in the leaves is not used locally but is transported into developing organs, and the means by which the protein emigrates from the site of its synthesis has not yet been studied enough. Most of the evidence suggests that this emigration precedes the breakdown of the proteins into amino acids and it would appear that it is this alternation of synthesis and breakdown of proteins associated with the phenomenon of migration which explains the very high rate of renewal of the proteins of leaves, which has been demonstrated in several researches (27, 28, 41). Another mechanism peculiar to plants is that whereby the stock of non-protein nitrogen is replenished. On the one hand amino acids are synthesized actively as a result of the reactions of photosynthesis (18, 19, 32) while, on the other, a considerable quantity of amino acids is formed in the root system and reaches the leaves through the vascular system of the plant.

It has not yet been fully established whether there is formed in the leaves any special reserve form of protein destined later to undergo periodic breakdown and migration into developing organs and not having other specific functions in cellular metabolism, or whether the migration

FIG. 33. Synthesis of proteins in the vascular system of bean leaves. Radioautographs of prints of leaves on filter paper during 50 hr. after introduction of [$^{35}$S]-methionine into region of root system. Upper row—total activity of prints. Lower row—radioactivity of proteins after washing out all non-protein compounds. In each row the prints are from two consecutive layers of filter paper used for making the print.

mainly takes place at the expense of the partial breakdown of all the proteins of the cell.

Without lingering over the varied and partly contradictory hypotheses on this question which have been put forward during the last few years, we should, nevertheless, mention that the results obtained recently suggest that some one form of protein is predominantly synthesized within the cytoplasm of the cell. It has been found that there is considerable electro-phoretic heterogeneity among the proteins of the plastids (21, 22, 25) whereas, in the rest of the cytoplasm, in several species of plants, a single electrophoretic fraction preponderates to the extent of comprising 50-60% of the total protein of the cytoplasm. The remaining 40-50% is com-prised of a very heterogeneous mixture (47).

Wildman and colleagues (33) made a detailed study of the composition and kinetics of this electrophoretically uniform fraction which they refer to as "protein fraction I". In doing so they found that while the synthesis of this fraction occurs considerably more rapidly than that of the protein as a whole during the formation of the leaves of the tobacco plant, during the ageing of the leaves the breakdown of this fraction proceeds faster and more actively than that of the total protein of the leaves. In leaves which were beginning to develop, the cytoplasmic protein fraction I only repre-sented 17% of the total cytoplasmic proteins but, during the later growth of the leaves, the synthesis of this fraction outstripped that of the proteins of all the other fractions and, after 17 days, the proportion of this protein in the cytoplasm had risen to 55% of the total cytoplasmic proteins. In the later development of the leaves the amount of this protein began to fall off sharply and by the 45th day after the beginning of the experiment it had fallen to one eighth of this value, constituting only 26% of the total cytoplasmic proteins, and this quantity itself had also fallen somewhat. The protein fraction I studied by Wildman and colleagues and by Lyttle-ton would seem to be of great biochemical interest. The fact that it can be synthesized and broken down rapidly is certainly important for the storage of nitrogenous products and their migration from the leaves. It is interesting that there is great physical similarity between the proteins of this fraction in the leaves of different species of plants. It has recently been shown that light has a strong effect upon the synthesis of this protein (34) and that a large quantity of this protein is formed in the chloroplasts (35). This protein only appears in the leaves of wheat and ryegrass after exposure of the plants to light. If shoots of wheat were kept in the dark the amount of this protein in them fell rapidly. While the ratio of the pro-teins of the uniform fraction I to the mixture of other proteins in the cytoplasm was 66:100 in green leaves, after the plants had been kept in the dark for five days it fell to 18:100. After exposure of these etiolated plants to the light again for 20 hours the proportion of these proteins rose again also threefold (52:100). Characteristically, neither albino leaves nor roots contain proteins belonging to this electrophoretically homogeneous fraction.   On the basis of his own results, Lyttleton came to the conclusion that the proteins of this fraction are synthesized in the leaves only when active photosynthesis is going on.   There are indirect indications that there may be a connection between this protein and the binding of carbon dioxide by carboxylation (46) but the fact that it had enzymic properties or contained an admixture of carboxylating enzymes would not prevent it from playing the part of a reserve and transport mechanism, which may be supposed from its kinetics, its dependence on age and photosynthetic conditions and its considerable quantity.   It seems to us that the formation of this particular protein is the reason for the marked stimulation of protein synthesis in leaves in the light observed by Andreeva (1, 2).

In this connection we must also mention the recent results of Brawerman & Chargaff (31) who found indirect evidence for the formation of specific forms of RNA in the cells of *Euglena* under the influence of light.

## 3. Some specific peculiarities of the synthesis of proteins in ripening seeds

Another form of functional synthesis of proteins in plants which has been studied in great detail is their formation in the endosperm and cotyledons of ripening seeds and also in various storage organs (tubers, rhizomes, etc.).   The great practical importance of these proteins provided the motive for the very considerable number of researches on their synthesis, so in most of this work attention is confined to the kinetics of the accumulation of these proteins, the dependence of their synthesis on the stage of development, external conditions and so forth, and not on the mechanism of the synthetic reactions.

As concerns the mechanism of synthesis of proteins during ripening, we must first discuss the interesting investigations of Snellman & Danielsson (42) and of Raacke (40) which showed clearly that there are peptide intermediate stages in these syntheses.   The work of Raacke was specially detailed.   Snellman & Danielsson found that when peas are ripening large polypeptides accumulate in them.   The amount of these polypeptides decreases later with the corresponding formation of reserve proteins. The authors deduce from this that the synthesis of the reserve proteins of the seed is brought about by the condensation of the large oligopeptides. The formation of a large polypeptide has also been observed at a particular stage in the ripening of the seeds of *Agrostemma githago* (30).   At this stage the peptide accounted for a considerable part of the soluble nitrogenous compounds of the seed.

The most detailed and precise studies of the kinetics of amino acids, peptides and proteins during the ripening of the seeds and pods of *Pisum sativum* have, however, been carried out by Raacke (40) who followed up the work of Danielsson mentioned above, in the same laboratory.

In making a critical analysis of the results of a number of authors who failed to find any considerable quantity of peptides in the tissues of plants, Raacke concluded that this failure was due to inadequacies in their methods (adsorption of peptides by protein precipitates, their precipitation along with the proteins by a number of precipitants, their insolubility in alcohol and the difficulty of identifying them on chromatograms). For the process of deproteinization she used the more gentle method of dialysis. Four main fractions were determined quantitatively, i.e. free amino acids, amides, peptides and proteins. The kinetics of these compounds demonstrated clearly the intermediate role of peptides in protein synthesis. During the ripening of the pods amino acids accumulated first; later came the onset of the phase of increase in peptides when the amino acid content began to decline; finally, the curve representing the peptide content, having reached its maximum, began to fall. This coincided with the period of rapid increase in the protein content and with the drying up of the seeds, which stimulates the process of condensation. In this connection the author quite justifiably states that during progressive dehydration it is hard to imagine an increase in the activity of proteases and peptidases which would split the peptides before they were used for protein synthesis. Interesting results have also been obtained by separate analyses of the kinetics of nitrogenous products in the pods and in the cotyledons. As is well known, the protein material of the pods is broken down during ripening and is used for the growth of the cotyledons. Raacke showed that this breakdown did not go right down to amino acids but only as far as peptides. The "pumping over" of nitrogenous products from the pods to the seeds takes place mainly as peptides without their later breakdown. In this connection, their reconstitution into proteins in the cotyledons requires considerably less energy than would be needed for synthesis from free amino acids.

In evaluating these most odd and interesting results from the point of view of the mechanism of protein synthesis, it should be remembered that the physico-chemical, colloid-chemical and biochemical conditions under which proteins are synthesized in the ripening seed are most unusual. In the first place this synthesis is associated to a considerable extent with the reutilization of products of the breakdown and migration of proteins from the vegetative organs and from other parts of the generative system. In the second place it occurs against a background of progressive dehydration of the protoplasm and its impoverishment in respect of active cytoplasmic elements. Having studied these conditions and the results of Danielsson & Raacke concerning the part played by polypeptides and peptides in the synthesis of these proteins, one may suggest that the synthesis of the reserve proteins of the seed may occur by a somewhat different mechanism from that of the active cytoplasmic proteins and we cannot exclude the possibility that it may be based on the simple enzymic polycondensation of polypeptides and peptones which migrate

there from other parts of the plant.  It must be mentioned in this connection that, in several species of plants, peptones constitute a considerable
part of the nitrogen reserves of the seeds (26).

Recently detailed work on the synthesis of protein in ripening maize
grains has been done by Rabson, Mans & Novelli (40a).  They studied
the synthetic systems of the grain *in vitro*, and isolated from homogenates
of endosperm and embryos both cytoplasmic particles (sedimenting at
105,000 $g$) and a supernatant fraction.  They determined the speed of
incorporation of labelled amino-acids into the proteins of these fractions.
Pollination and fertilization markedly stimulated the synthetic system of
the endosperm; however, as ripening proceeded, the capacity for protein
synthesis in the organelles fell, and together with it the capacity for activation of amino-acids of the supernatant fraction.  However, the authors
omitted to study whether, as the activity of the general synthetic system
fell off, there was any shift in the ratio of protein to non-protein nitrogen;
it is therefore difficult to relate their results to those of Raacke.

## Conclusion

From a careful analysis of this evidence, however limited, concerning the
synthesis of proteins in plants the obvious question arises: Is it true that
all syntheses carried out within the organism always follow the extremely
long and complicated route evolved during phylogenesis for reproduction
of specificity from amino acids through activating enzymes, selective
transfer to S-RNA, matrices and the microsomal membranes?  For
the formation of specific enzymic proteins and many others this pathway,
in which there is the most accurate possible reproduction, is inevitable,
but is it necessary, for example, for the synthesis of the large masses of
reserve proteins of the seeds which are only formed in order that they may
be broken down later and which really only amount to a polymerized
form of the stock of amino acids.  The results of Raacke and others on
the rise and fall of peptides during the ripening of seeds suggest that
there may also be another, perhaps less specific but more economical
method of simple polycondensation of peptides in which the specificity of
the proteins depends on the specificity of the general composition of the
peptides.  The question of the multiplicity of possible means of protein
synthesis is very interesting from a theoretical point of view.  For the
present we can only point out the existence of the question.  It will be
considered in more detail in later chapters after we have become familiar
with a number of other relevant facts.

REFERENCES

1. ANDREEVA, T. F. (1955). *Doklady Akad. Nauk S.S.S.R.* **102**, 165.
2. — (1956). *Fiziol. Rastenii*, **3**, 157.
3. BLAGOVESHCHENSKII, A. V. (1950). *Biokhimicheskie Osnovy evolyutsionnogo Protsessa u Rastenii.* Moscow: Izd. Akad. Nauk S.S.S.R.
4. — (1958). *Biokhimiya Obmena azotsoderzhashchikh Veshchestv u Rastenii.* Moscow: Izd. Akad. Nauk S.S.S.R.
5. DADYKIN, V. P. (1956). *Doklady Akad. Nauk S.S.S.R.* **106**, 923.
6. KRETOVICH, V. L., EVSTIGNEEVA, Z. G., ASEEVA, K. B. & SAVKINA, I. G. (1959). *Fiziol. Rastenii*, **6**, 13.
7. KURSANOV, A. L. (1954). *Voprosy Botaniki*, No. *1*, p. 129.
8. — (1957). *Izvest. Akad. Nauk. S.S.S.R.*, *Ser. biol.* No. *6*, p. 689.
9. MEDVEDEV, ZH. A. (1956). *Doklady Moskov. sel'skokhoz. Akad. im. K. A. Timiryazeva*, No. *23*, p. 214.
10. — (1957). *Priroda*, **46**, No. *8*, p. 90.
11. — (1957). *Izvest. Timiryazev. sel'skokhoz. Akad.* No. *3*, p. 186.
12. — (1958). *Bot. Zhur.* **43**, 61.
13. — (1958). In Sbornik *Fiziologiya Rastenii, Agrokhimiya, Pochvovedenie. Trudy vsesoyuzn. nauchno-tekh. Konf. po Primeneniyu Izotopov* (ed. V. M. Klechkovskii), p. 43. Moscow: Izd. Akad. Nauk S.S.S.R.
14. — (1959). *Priroda*, **48**, No. *2*, p. 93.
15. — (1959). *Izvest. Timiryazev. sel'skokhoz. Akad.* No. *2*, 59.
16. MEDVEDEV, ZH. A. & TSZYUN', U. (1956). *Doklady Moskov. sel'skokhoz. Akad. im K. A. Timiryazeva*, No. *26*, Pt. 1, p. 273.
17. MOLCHANOVA, V. V. (1957). *Doklady Akad. Nauk S.S.S.R.* **112**, 1119.
18. NEZGOVOROVA, L. A. (1956). *Fiziol. Rastenii*, **3**, 497.
19. NICHIPOROVICH, A. A. (1958). In Sbornik *Fiziologiya Rastenii, Agrokhimiya, Pochvovedenie. Trudy Vsesoyuzn. nauchno-tekh. Konf. po Primeneniyu Izotopov* (ed. V. M. Klechkovskii), p. 56. Moscow: Izd. Akad. Nauk S.S.S.R.
20. SABININ, D. A. (1949). *O Znachenii kornevoi Sistemy v Zhiznedeyatel'nosti Rastenii.* (*9-e Timiryazevskoe Chtenie*). Moscow: Izd. Akad. Nauk S.S.S.R.
21. SISAKYAN, N. M. (1953). *Uspekhi sovremenno Biol.* **36**, 332.
22. — (1956). *Izvest. Akad. Nauk S.S.S.R.*, *Ser. biol.* No. *6*, p. 3.
23. — (1958). In Sbornik *Fiziologiya Rastenii, Agrokhimiya, Pochvovedenie. Trudy vsesoyuzn. nauchno-tekh. Konf. po Primeneniyu Izotopov* (ed. V. M. Klechkovskii), p. 22. Moscow: Izd. Akad. Nauk S.S.S.R.
24. — (1959). In *Proceedings of the first international Symposium on the Origin of Life on the Earth* (ed. A. I. Oparin *et al.*), p. 400. London: Pergamon.
25. SISAKYAN, N. M. & MELIK-SARKISYAN, S. S. (1956). *Biokhimiya*, **21**, 329.
26. TOKARSKAYA [MERENOVA], V. I. & KUZIN, A. M. (1956). *Biokhimiya*, **21**, 816.
27. TURCHIN, F. V., GUMINSKAYA, M. A. & PLYSHEVSKAYA, E. G. (1953). *Izvest. Akad. Nauk S.S.S.R. Ser. biol.* No. *6*, p. 66.
28. — (1955). *Fiziol. Rastenii*, **2**, 3.
29. BOLLARD, E. G. (1956). *Nature (Lond.)*, **178**, 1189.
30. BORRISS, H. (1955). *Naturwissenschaften*, **42**, 103.
31. BRAWERMAN, G. & CHARGAFF, E. (1959). *Biochim. biophys. Acta*, **31**, 172.
32. CHAMPIGNY, M. L. (1956). *Compt. rend. Acad. Sci.*, *Paris*, **243**, 83.
33. DORNER, R. W., KAHN, A. & WILDMAN, S. G. (1957). *J. biol. Chem.* **229**, 945.
34. LYTTLETON, J. W. (1956). *Nature (Lond.)*, **177**, 283.
35. LYTTLETON, J. W. & TS'O, P. O. P. (1958). *Arch. Biochem. Biophys.* **73**, 120.

36. McKee, H. S. (1958). In *Handbuch der Pflanzenphysiologie* (ed. W. Ruhland), Vol. *8* (ed. K. Mothes), p. 516. Berlin: Springer.

37. Medvedev, Zh. A. (1958). In *Proc. 1st UNESCO internat. Conf. on Radioisotopes in sci. Res.* Vol. *3*, p. 648.

38. Morgan, C. & Reith, W. S. (1954). *J. exp. Bot.* **5**, 119.

39. Oota, Y. & Osawa, S. (1954). *Experientia*, **10**, 254.

40. Raacke, I. D. (1956). *Biochem. J.* **66**, 101, 110, 113.

40a. Rabson, R., Mans, R. J. & Novelli, G. D. (1961). *Arch. Biochem. Biophys.* **93**, 555.

41. Racusen, D. & Foote, M. (1960). *Arch. Biochem. Biophys.* **90**, 90.

42. Snellman, O. & Danielsson, C. E. (1953). *Exptl. Cell Research*, **5**, 436.

43. Steward, F. C. & Pollard, J. K. (1958). *Proc. IV int. Congr. Biochem.*, *Vienna*, **6**, 193.

44. Steward, F. C. & Thompson, J. F. (1954). In *The Proteins* (ed. H. Neurath & K. Bailey) Vol. *IIA*, p. 513. New York: Academic Press.

45. Webster, G. C. (1959). *Symp. Soc. exp. Biol.* **13**, 330.

46. Weissbach, A., Horecker, B. L. & Hurwitz, J. (1956). *J. biol. Chem.* **218**, 795.

47. Wildman, S. G. & Jagendorf, A. T. (1952). *Ann. Rev. Plant Physiol.* **3**, 131.

48. Yemm, E. W. (1958). In *Handbuch der Pflanzenphysiologie* (ed. W. Ruhland), Vol. *8* (ed. K. Mothes), p. 437. Berlin: Springer.

# CHAPTER XI

# SOME FUNCTIONAL SYNTHESES OF PROTEINS IN ANIMALS

## *Introduction*

In studying the problems of the synthesis of proteins one may use various methods of analysis and classification of the experimental and theoretical material. In the preceding chapters we have already surveyed the ways in which this synthesis is accomplished in relation to the sequence of reactions by which the polymeric and specific molecules of protein are formed, in relation to the localization of the synthesis and the roles of the various organelles in accomplishing it, in relation to the metabolism of RNA and evolutionarily adaptive reactions (adaptive synthesis) and in relation to several other factors. However, these aspects do not by any means exhaust the variety of forms of synthesis or the possibilities of comparing and contrasting them. Of course, we cannot cover all these possibilities but it does seem worth while to give some account, though only a very short one, of the great variety of forms of functional syntheses of protein which constitute individual links in the accomplishment of the ordinary physiological and biochemical functions of organs and tissues.

By functional protein synthesis we mean a form of protein synthesis which is not brought about to achieve the growth or the reproduction of the cell or organelle in question but in order that the cell may fulfil some functions which form a part of the life of the organism and of its development as a whole. Typical examples of such syntheses are the synthesis of antibodies, serum albumin, digestive enzymes, the proteins of wool, silk, milk and eggs, haemoglobin, etc.; that is to say, synthesis of proteins which are often extracellular and which differ from one another very markedly as to their role, amino acid composition, structure and physicochemical properties.

A survey of the problem of the synthesis of protein by analysing the peculiarities of individual actual forms of synthesis is very interesting in that it shows clearly that the mechanism of synthesis of proteins and the factors which determine its specificity do not take the form of a process of agglomeration. Like any other complicated biological process, the synthesis of protein has developed by evolution and has undergone changes and become specialized in accordance with the requirements of particular functions. We have no doubt that there exist in nature at present many mechanisms of synthesis which differ from one another to some extent according to the character of the protein being formed, the starting material for the synthesis, the role of the protein in the organism and many other properties which are specific to it.

These remarks will be clearly illustrated by many actual examples.

## 1. Peculiarities of the formation of protein antibodies in reactions of immunity

The synthesis of antibodies in the tissues of the animal organism is one of the most remarkable and specific of biological phenomena. The synthesis of antibodies is only one link in the complicated biological phenomenon of immunity which has arisen in the course of evolution in the form of a biochemical adaptive reaction to the introduction of foreign high-molecular compounds or foreign cells into the organism. Here we shall not survey the whole complex of immune reactions and discuss all aspects of the role of proteins in reactions of immunity. Detailed reviews of the evidence on this subject have been published recently (21, 3, 43, 9, 29, 26, 28, 44, 67, 75, 49, 99a) and this will enable us to limit ourselves to the single, though central, question concerning the participation of proteins in reactions of immunity, namely the mechanism of synthesis of antibodies. In doing so we shall only touch briefly on the fundamental theoretical ideas in this field and shall concentrate our attention on those recent experimental and theoretical works which, to some extent, link up the synthesis of antibodies with the general concepts of protein synthesis.

If this problem is to be analysed systematically the different features of the synthesis of antibodies must be discussed in a consequential order.

### a) The nature of antibodies and the mechanism of their reactions with antigens

The formation of antibodies is, by and large, a manifestation of the ability of the organism to distinguish between "its own" and "foreign" substances and to react to the introduction of compounds of foreign origin into its internal medium in such a way as to prevent them from having a harmful effect on its vital processes. In the organism there is usually a wide selection of means of elimination of any "extraneous" substances, such as by excretion, splitting by enzymes, binding in certain organs with gradual detoxication, adaptation or phagocytosis. The formation of antibodies is one of the multitude of these reactions and is specialized, in the main, for the elimination of foreign proteins and, to a lesser degree, other biopolymers (nucleic acids, polysaccharides) which, in this connection, are called antigens. The essential conditions for the manifestation of antigenic properties are that the compound must be polymeric and foreign. This means that various different types of cells and sorts of compounds (proteins, DNA, etc.) have antigenic properties. Proteins with varied structures such as albumins, proteids, prolamines, etc. are all equally capable of acting as antigens and each of them stimulates the formation of specific antibodies. Proteins of the same kind but having a different origin and belonging to a different species (the albumins of horses, birds, rats, etc.) also induce the formation of different antibodies. Nevertheless, all antibodies belong to the same main type of protein, being γ-globulins and, although they have different powers of combining with different proteins, their amino acid compositions are the same and the sequence of five amino acids which has been established as forming the N-terminal peptide seems also to be the same in all of them (66, 85). The sequence of amino acids in the other parts of the polypeptide chains of different antibodies has not yet been determined and probably, owing to their high molecular weight, it will not be established soon. However, the most that can be hoped for, in view of the similarity of the chemical properties of the various antibodies, is to find alterations in the arrangement of particular amino acids in short sections of their molecules.

It is, however, interesting to note the recent preliminary communication of Gurvich and colleagues (6) in which they established differences between particular antibodies and non-specific γ-globulins using the so-called peptide "map" method after hydrolysis with trypsin and chymotrypsin. These results provide tentative evidence of the existence of differences in amino acid sequence.

However, whether or not the various antibodies differ in the sequence of the amino acids composing their reactive groups, if we are to study their reactions with antigens we must take into account the secondary and tertiary structures of both antibody and antigen as these determine their peculiar form of relationship, that peculiar complementarity of their superficial groups which gives rise to the rapid formation of a specific complex.

In view of the fact that the question of differences in the amino acid sequences of antibodies has not been sufficiently studied, we must first discuss those hypotheses which explain the specificity of immune reactions not only in chemical, but also in stereochemical terms and assume that the essential factor in the specific interaction betweeen antibodies and antigens is the configuration of the surfaces of the molecules of the antibodies which is complementary to that of the molecules of the antigenic substances. Two very similar hypotheses in this connection were put forward long ago (24, 73) and these are still being developed and occupy the most prominent position among explanations of the interactions between antibodies and antigens. The most widely supported of these hypotheses is that of Haurowitz which he treated in detail in his monograph on the biology of proteins and in other works (41-43).

Haurowitz considers that antibodies are compounds the configuration of which is geometrically complementary to that of the determinant groups of the molecules of the antigen. In his view, the polar groups of the antigen affect the process of formation of globulins from amino acids, so that the normal course of this process is changed and molecules of globulin with a different three-dimensional configuration are formed.

According to this hypothesis the complementary form of the surface of the antigen is determined, on the one hand, by the way in which the peptide chain is folded and, on the other, by the presence of ionic groups of opposite sign in those parts of the antibody and antigen which combine directly with one another. The dependence of the specificity of the reaction of the antibody on characteristic determinant groups arranged on its surface rather than on the structure of the molecule as a whole has been demonstrated clearly by the fact that two or more specific antibodies may be formed on injection of a single antigen which has several determinant groups. For example, immunization with globulin substituted with iodine and azophenylarsonic acid leads to the appearance of two types of antibodies, namely antibodies against diiodotyrosine and antibodies against azophenylarsonic acid. These two types of antibody may be separated from one another by precipitating the one with iodoovalbumin and the other with arsonylazoovalbumin (58). If arsonylglobulin of the sheep is used for immunization at least three types of antibody are formed, antiarsonyl, antisheep and antiarsonylsheep (45).

Several other facts have now been collected which confirm that it is only specific groups of the molecule of the antigen which have a determinant effect on the synthesis of antibodies (21, 43).

According to Haurowitz the mechanism of formation of the specific configuration of antibodies depends on the inclusion of the antigen in the process of formation of the globulins at the moment when the polypeptide chain has already been formed on the matrix but when it has not yet been desorbed and turned into a three-dimensional globule. At this moment the polar groups of the antigen disturb the normal process of folding of the peptide chain in such a way that the configuration of the globular particles being formed becomes, as it were, complementary to the configuration of the polar determinant groups of the antigen. It is assumed to be possible that the electrostatic forces of the determinant polar groups of the antigen would give rise to a definite field of force which would be able to alter the character of the globules of antibody at particular points.

An interesting attempt to demonstrate this principle experimentally has been made by Vyazov & Tseĭtlin (2). Assuming that the antibody is a geometrical reflection of the antigen which was introduced, they decided to use a protein antibody as an antigen by introducing it into the blood of animals of another species and expected thus to obtain a protein with a structure similar to that of the antigen used first.

The stereochemical similarity between the final antibody B and the original antibody A was revealed by an anaphylactic reaction. Guinea pigs sensitized to the serum of fowls which had been immunized to "anti-human-albumin" rabbit serum responded to the internal administration of human albumin with anaphylactic shock. In other cases sensitization shock did not occur. The authors regard these results as direct confirmation of the geometrical complementarity of antigen and antibody.

Although this concept is very simple and very striking it is still extremely tentative, especially as concerns the factors regulating the synthesis of antibodies. Analysis of many facts indicates that the way in which this happens is considerably more complicated than this.

The facts set out above and the discussions of many authors seem to us to indicate that the synthesis of antibodies is a complicated, biochemical adaptive reaction associated with a definite reorganization of the synthetic apparatus and, above all, with some alteration in the matrices of protein synthesis which regulate the specificity of the products being synthesized. If we are to get a clearer conception of the mechanism and character of this reorganization we must first get some idea of the nature of the differences between different antibodies, that is to say, the nature of their specificity.

It seems to us very doubtful whether the different antibodies are alike apart from a difference in the folding of each polypeptide chain of the globule. Changes in configuration would seem to involve the whole large molecule of the antibody. In our survey of the facts concerning the structure of proteins we saw that the nature of the configuration of the polypeptide chain is a specific characteristic which is correlated with the amino acid sequence and the arrangement of active groups. This being the case, the formation of the three-dimensional globules, even if it occurred spontaneously, would still be subject to definite rules. It is very hard, not to say impossible, to imagine that there could be an unending multitude of forms of globule, all at the same energy level, which would reflect the configurations of all possible proteins. The deformation of the polypeptide chain into a shape which is markedly unlike that which it takes up spontaneously requires energy and will inevitably stop if steric and energic factors do not act on the molecule. Therefore, if the molecules of antibodies have the same amino acid composition and sequence and only differ from one another in the configuration of their chains, then these differences cannot be very great. This, however, is only true for an uninterrupted polypeptide chain. The high molecular weight of antibodies would seem to indicate that their molecules are made up of a large number of polypeptide chains and this would certainly allow the systems of synthesis to vary their arrangement and configuration within very wide limits at the same energy level. There is no direct proof of the geometrical complementarity of the molecules of antibodies and antigens and experiments such as those of Vyazov & Tseĭtlin represent an indirect approach, the results of which can be interpreted in other ways than on the basis of geometrical complementarity. However, if the configuration of the surface of the molecules of the antibodies is not the key which would enable us to discover the way in which they act, then the only other element in their specificity which could do so is the sequence of amino acids in some parts or other of their molecules which will later react with the antigens. In such a case we should also have to assume that there was some connection between the antigen and the synthesis of the antibody mediated by the matrices.

After many years of study of the mechanism of immune reactions, although many hypotheses have been put forward, we have still only made a very small amount of progress towards understanding what determines the specific relationship between the antigen and the antibody and which groups and parts of the molecules of the interacting pair dominate the reaction.

Very recently, however, Arnon & Sela (17, 81, 82) have worked out a very promising method for studying this question. The authors noted that there was a certain paradox in the immune reaction. On the one hand the antigen must have a definite complexity of structure and a high molecular weight while, on the other, only a comparatively small part of the protein molecule determines its antigenic specificity. This led to the question of what chemical structure proteins must have to ensure that they will have antigenic properties and what is the chemical nature of the groups, the presence of which can either augment their antigenicity or affect their antigenic specificity.

For their study of this question the authors chose the protein gelatin, which has a very weak antigenicity. The immunological inertness of gelatin is such a characteristic property that for many years this protein has been thought to be non-antigenic. This weak antigenicity is associated with the absence of any definite firm structure in this denatured protein, with a large amount of glycine and with a deficiency of aromatic amino acids.

Using gelatin as a polymeric basis Arnon & Sela added to it peptide and polypeptide groups of known composition by reaction with $N$-carboxy-$\alpha$-amino acid anhydride. In this way peptides of different sizes were attached to the free amino groups of gelatin.

Even in their first experiments they found that the amino acids were not all equally effective in producing an antigenically active complex and that the aromatic amino acids were specially important. Polytyrosyl, polyphenylalanyl, polytryptophyl and, to a lesser extent, polycysteinyl derivatives of gelatin behaved as strong antigens. The combination of polyglutamyl, polylysyl and polyarginyl chains with the gelatin did not lead to any significant increase in its antigenic properties. In these experiments the peptide chains which were added to the gelatin represented about 75% of the quantity of the gelatin itself (calculated on the number of amino groups).

In later experiments the authors made quantitative changes in the tyrosine peptides which they combined with the gelatin. By doing so they found that the combination of 2% of tyrosine with the gelatin converted it into a strong antigen without any considerable change in its serological specificity. Increasing the tyrosine chain to 10% of the gelatin was enough to alter the specificity of the complex markedly. On the other hand, the combination of mixed tyrosylglutamyl peptides with the gelatin also altered the specificity of the antigen. In this case the presence of glutamyl groups had a pronounced effect on the specificity, although pure polyglutamyl chains were inert. In later experiments (17) the authors worked out an interesting method for studying the nature of the bond between antigen and antibody. Having isolated the complex of gelatin and its antibody they selectively hydrolysed the gelatin by means of collagenase, which is an enzyme acting only on collagen-like proteins. They were thus enabled to obtain, after dialysis, a pure and native protein with the properties of an antigelatin antibody. In this case, when a complex was formed between this antibody and gelatin containing about 2% of additional tyrosyl peptide groups, the parts of the antigen which reacted with the antibody were not broken down during hydrolysis and remained bound to the antibody which thus lost its power to react with new portions of antigen. We shall see that this enables us to study the active groups of both antibodies and antigens and further investigations along this line are sure to be extremely fruitful.

Nikolaev (12) has recently made an interesting observation. He showed that the antigenic activity of keratins has a species specificity which depends on the configuration of their molecules as determined by disulphide bonds. The antigenic activity of these proteins changed with changes in their cysteine groups.

The discovery of the molecular mechanisms of the reaction between antigen and antibody is also very important for our understanding of the nature of antibody formation. Different antibodies have one common property and that is their ability to form complexes with very varied antigens. It seems to us that this common property must have some common structural basis in the molecules of any analogous antibodies. There must be some sort of combination of active, perhaps disulphide, groups which confer on the antibody the power to form complex compounds under particular conditions. Karush (53-55) found that, if half of the disulphide bonds of an immune globulin were reduced, this decreased its power to react specifically with antigen to one seventh of its former value. In a recent communication to the Fifth International Congress of Bio-chemistry in Moscow Karush (56) brought out some new and interesting facts concerning the possible significance of disulphide bonds in the determination of the specificity of antibodies. He directed attention to the fact that in $\gamma$-globulins there is a comparatively large number of cysteine residues which makes it possible that there is a large number of disulphide bonds. According to his evidence univalent fragments of rabbit antibodies contain about sixteen half residues of cystine. If there is a corresponding number of SH groups, then the number of different possible ways in which four —S—S— bridges could be formed is $1 \cdot 3 \times 10^6$.

Karush considers that the antigen determines the type of —S—S— bond in the antibody and that the structure of the reactive groups of the antibody is complementary to the structure of the antigenic determinant.

We mentioned earlier that the ability of the antibody to react may have a common molecular basis in the form of some "active" group. In this case it is not impossible that the antigen might be like a unique key which opens a unique lock and reveals the latent universal ability of antibodies to combine with different polymers and that when the antigen and antibody are separated the "lock" is again shut. The dual nature of the reaction and the definite difference between the antigenic specificity and the actual reaction between antigen and antibody have been demonstrated clearly by Arnon & Sela in the work discussed above. It also seems to us that only the occurrence of two stages in the reaction of antibody with antigen could explain the formation of complicated complexes in which several molecules of antibody can suddenly become bound to one molecule of antigen. There is a well-known case in which 50-60 molecules of antibody are combined with one of antigen in precipitates (43) and this is obviously only possible because of the power of the antibody, which has been specifically activated by the antigen, to combine with parts of the antigen other than the determinant group. Possibly this is the explanation of the "multivalency" of some antibodies (11).

It would be premature to go into any further detail now about these ideas but it seems to us that a solution of the problem of antibodies will, in the first place, depend on elucidating the fine details of their structure and the mechanism of their reaction with antigens.

### b) Short characterization of the biochemical system of antibody synthesis

We shall not dwell in detail on the question of the localization of antibody synthesis in tissues. Detailed information on this subject is to be found in many studies and reviews (15, 21, 29, 39, 40, 67, 86, 95, 101). The formation of antibodies occurs in many organs, though it does not take place in all the cells of the organs and tissues, but only in the specialized reticulo-endothelial and lymphoid cells. Furthermore, the formation of antibodies cannot be based on the synthesis of any protein but only on that of the $\gamma$-globulin fraction, that is to say that it must be based on the synthesis of one type of protein. This suggests that the formation of antibodies is a result of the activity of a functionally and biochemically specialized system which has become differentiated in the process of evolution to fix and precipitate foreign polymers and extraneous bodies and which has a completely specific form of sensitivity towards these bodies. It is also characteristic that the cells which can synthesize antibodies only develop at particular stages in ontogenesis and, for example, newly born animals and embryos are not usually yet capable of immune reactions. Šterzl (15a) has reviewed the evolutionary, phylogenetic aspects of immune reactions and antibody formation.

An interesting method for determining the formation of antibodies in individual cells has recently been worked out by Nossal (70). This author has clearly demonstrated the existence of particular cells, specialized for the production of antibodies. By means of a micromanipulator Nossal placed individual cells in microdrops of a physiological solution. Motile, ciliated bacteria such as *Salmonella* species served as the antigen. One cell from a suspension of cells from the lymph gland of the leg of an animal which had been immunized against the bacterium was introduced into the droplet, then, after various intervals (1-12 hours) a few living bacteria were introduced. Loss of motility served as an indicator of the presence of antibodies. Repeated introduction of new bacteria into the drop enabled the author to titrate the amount of antibody formed. Of 601 cells studied in this way 93 produced antibodies and of these 91 were plasma cells, which shows the predominant if not exclusive role of plasma cells in the formation of antibodies.

It must also be mentioned that, in many cases, antigens do not re-organize pre-existing syntheses of $\gamma$-globulin but stimulate new syntheses in addition to those going on in the cell. This was demonstrated clearly in, for example, the works of Gurvich *et al.* (7, 8) using quantitative determinations of the synthesis of $\gamma$-globulins and antibodies during immunization of animals with one and two antigens. The authors studied the synthesis of antibodies in rabbits by observing the incorporation of [$^{14}$C]glycine into the proteins of the antibodies and $\gamma$-globulins. It was found that the incorporation of [$^{14}$C]glycine into the proteins of the anti-bodies during the period while the antibody content of the blood was increasing proceeded considerably faster than its incorporation into non-specific $\gamma$-globulins. The faster the antibody content of the blood in-creases the more rapid the incorporation of [$^{14}$C]glycine into the antibodies. The intensiveness of the incorporation of [$^{14}$C]glycine into two antibodies being formed simultaneously depends on the rate of increase of the amount of each antibody in the blood and is not connected with their immunological specificity.

According to these authors the formation of the main mass of the proteins of the antibodies occurs shortly before their appearance in the blood. During the first 2-3 days after the introduction of the antigen (the latent period) there is no formation of any appreciable amount of anti-bodies and also it would seem that their protein and polypeptide pre-cursors are not formed either.

Relating their own results to the hypothesis of Haurowitz, Gurvich & Smirnova (8) noticed that, on the basis of that hypothesis, it was rather to be expected that a considerable increase in the biosynthesis of one antibody would hinder the formation of both other antibodies and non-specific $\gamma$-globulin because the formation of the antibody in question would use up some of the polypeptide chains which previously went towards the formation of those other antibodies and non-specific γ-globulins. On the basis of the information available in the literature as to the "half-life" of γ-globulins (4-5 days) they calculated that in the organism of the rabbit 1-1·5 mg. of polypeptide chains of γ-globulins is formed daily for every ml. of serum and this may be used for the formation of any antibodies or non-specific γ-globulins. In their own experiments, however, the daily increase in antibodies was considerably greater than this, reaching as much as 10 mg. in 24 hours.

A considerable augmentation of the synthesis of the proteins of antibodies during immunization as compared with the synthesis of non-specific globulins and other proteins of the cells has also been found in the work of other authors (48, 71). It has also been demonstrated clearly that the synthesis of antibodies is a complete protein synthesis, starting from amino acids and not from some protein precursors which were already present in the cell (88, 95, 96). The sensitivity of this synthesis to the harmful action of irradiation and to inhibitors of nucleic acids (33, 95) indicates that it depends on normal systems of protein synthesis.

The work of Šterzl & Hrubešova (90) is also very interesting. They managed to transfer the power of forming antibodies to non-immunized recipients by means of nucleoproteins isolated from the spleens of immunized rabbits. However, we must mention that in a later and more detailed research which was carried out jointly with Askonas & Humphrey, Hrubešova (47) was not able to confirm her earlier idea that the appearance of antibodies in the organism of the recipient really represented, in this case, the synthesis of new proteins on the matrices which had been introduced into the organisms in the form of nucleoproteins. In this research it was shown that the antibodies which appear in the blood of the recipient were not formed *de novo* but were preformed in the nucleoprotein in some masked condition. After injection these antibodies were liberated and assumed the active state. These results indicate that, although isolated nucleoproteins may lose their biochemical activity during preparation, yet they are, in fact, the site at which the original formation of antibodies takes place in immunized animals.

Many facts which confirm the connection between the specificity of antibodies and the actual system synthesizing them have been brought out in the recent comprehensive review by Šterzl (15) of the characteristics of the inductive phase of antibody formation.

All this evidence indicates that, in exerting their effect on the formation of specific antibodies, antigens do not interact with the actual protein during its formation but have some sort of effect on the system of protein synthesis which can then function irrespective of the presence or absence of antigen. This would seem to explain the formation of a number of antibodies long after not even a trace of the original antigen can be found in the cells.

The recently published work of Askonas & Humphrey (18) contains some extremely interesting observations which also indicate the specialization of the function of synthesis of different antibodies. They found that when a rabbit was immunized with two antigens simultaneously (egg albumin and pneumococci) different tissues display very different properties in regard to the relative rates at which they synthesize each of the two antibodies. While the cells of the lymph glands (apart from the bronchial ones) mainly formed antibodies to the albumin, the synthesis of anti-pneumococcal antibodies predominated in the bronchial lymph glands, lungs and bone marrow. Non-specific $\gamma$-globulin was formed alongside the antibodies but, in this case, the relative amounts of this and of the antibodies varied with the nature of the tissues. The synthesis of $\gamma$-globulins was not decreased but increased in comparison with what occurred before immunization. The authors explain this phenomenon in terms of a difference in the distribution of the antigens in the organism depending on their nature and the way in which they are introduced. Similar results have been obtained by Stavitsky and colleagues (86, 87).

From this evidence one may suppose that the individual cells do not undertake the synthesis of a very large assortment of antibodies but that there is a certain "division of labour" among cells and groups of cells. The connection between immune reactions and the DNA system which also controls heredity has been shown clearly by the work of Dutton *et al.* (34). These authors observed an inhibition of the formation of antibodies under the influence of an analogue of thymine—5-bromouracil desoxy-riboside—which impedes the formation of DNA. At the same time this inhibitor did not affect the incorporation of amino acids into ordinary proteins. Furthermore, it only exerted its inhibitory action during a particular "critical" early period when the antigen and the inhibitor were introduced at the same time. When the inhibitor was introduced 17 hours after the introduction of the antigen it did not affect antibody synthesis. The authors, therefore, think that the inhibitor of DNA metabolism does not directly inhibit the synthesis of antibodies but affects the early, preparatory "form-determining" stages of their synthesis. These results have received further confirmation from the work of Nikolaev & Akhma-dieva (13) who have shown that when the synthesis of nucleic acids is inhibited by radiation the synthesis of antibodies is also slowed. However, this effect is only observed during the first 48 hours of the immune reaction. If the irradiation is carried out 48 hours after the introduction of the antigen, then the synthesis of the antibody proceeds normally.

The connection between the formation of antibodies and the synthesis of RNA and of DNA has also been shown in other experiments on the inhibition of antibody synthesis by analogues of ribo- and desoxyribo-nucleotides (89, 35, 78). From an analysis of the facts concerning the dependence of the synthesis of amino acids on nucleic acids Bernard (19) suggests that the beginning of the synthesis of antibodies occurs at the particular stage of the mitotic cycle, known as the interphase of cell-division, during which new DNA is being formed.

### c) Contemporary theories of the mechanism of antibody synthesis

The unusual nature of the interaction between the synthesis of proteins and external agents in immune reactions has for long attracted the attention of many scientists to this phenomenon and they have tried to unravel the mechanism of these interactions. In no other field of protein chemistry does there seem to be such a superabundance of hypotheses, speculations, theories and schemes as is to be found in the biochemistry of immune reactions and this variety is, in itself, a very reliable indication that the real picture of this synthesis as it actually occurs is still a long way from having been worked out.

There are reviews of contemporary concepts, of the synthesis of anti-bodies in many papers (15a, 15b, 26-28, 30, 61, 4, 44, 75, 76, 51) but, during the last few years alone, more than ten new hypotheses and modifications of old ideas have been published and this means that we must again analyse the present state of this problem.

Most of the theories of antibody synthesis can be assigned to one or other of two groups known as "instructive" and "selective" theories. According to the "instructive" theories antigens are somehow involved in the synthesis of $\gamma$-globulins and they modify this synthesis so that the power of forming protein molecules which can react specifically with the antigens is brought into being de novo. The "selective" theories, on the other hand, are based on the assumption that the ability to form a wide range of antibodies is already present in the genetic structure and antigens only bring out and selectively stimulate the production of particular anti-bodies, acting like the inducer in the formation of adaptive enzymes.

One of the first "instructive" theories is the classical one of Breinl & Haurowitz (24) and of Pauling (73) which was worked out many years ago but has later been modified by other authors.

We have already discussed briefly the essential ideas behind this theory. One of the new variants of it is that worked out recently by Karush (54). He thinks that the antigen provides the information which determines the occurrence of a particular type of folding of the polypeptide chain. This is fixed by —S—S— bonds and is complementary to particular groups in the antigen. In their earliest forms, "instructive" theories assumed that the synthesis of antibodies in fact requires the continued presence of the antigen actually within the system in which the new protein is synthesized, and its participation as an indispensible component of this system. In this form the theory ran into a number of serious difficulties and was unable to explain several features of immune reactions (the lasting nature of immunity, the inheritance of immunity over many generations of cells, differentiation between antibodies and homologous proteins with a similar tertiary structure, the relationship between the synthesis of new types of antibodies and that of other antibodies and γ-globulins etc.).

In this connection some authors have put forward the suggestion that the "instructive" influence of the antigen may not affect the form of folding of the polypeptide chain but may affect some features of the matrices of protein synthesis, i.e. RNA and DNA.

The suggestion with the best theoretical justification is that, if antibodies really differ from one another in some parts of their molecules, then these differences are induced and maintained by some differences in the systems of synthesis and reproduction of the specificity of proteins. Schweet & Owen (79) believe that the first thing that happens after the introduction of an antigen is that it reacts with the DNA at the loci in the chromosomes which are associated with the formation of globulins. The action of the antigen on the DNA is, according to their ideas, to stimulate the formation of altered DNA and of a cell which only inherits such altered DNA. The DNA of these cells can form matrices consisting of RNA and manufacturing proteins which combine specifically with the antigen. Regarding this hypothesis from an immunological point of view the authors mention that the penetration of a number of antibodies into the nucleus has actually been demonstrated (31).

Furthermore, the authors have put forward evidence that increasing synthesis of antibodies is accompanied by intensive multiplication of the cells which produce these antibodies. At the same time the basophilia of the cytoplasm of these cells increases noticeably, which indicates that they are producing more ribonucleoprotein particles. The most puzzling property of these cells is their ability to distinguish between their own and foreign substances. According to Schweet & Owen the factors responsible for this ability are associated with some functions of the organism as a whole. The ideas of these authors are certainly very interesting but they do no more than indicate a line of investigation and do not really provide an explanation of the mechanism of the specific synthesis of antibodies. Naturally the failure to solve the more general problems of protein synthesis and of the mechanisms whereby DNA and RNA matrices act in this synthesis interferes with the presentation of any simple and obvious theory of this kind. The absence of evidence as to the nature of the differences between antibodies also hinders theoretical analysis of the nature of their synthesis. However, it is obvious that the solution of these general problems will be made very much easier by rapid progress in a more limited field. Nevertheless, this must work both ways.

While studying the specificity and the unusual nature of the actual phenomenon of antibody synthesis we must always be looking for material which will facilitate the discovery of the more general laws of protein synthesis. The assumption that the effect of the antigen could be such as to change the matrix in an appropriate way is a very weak point in schemes

of this sort.  It is assumed that, under the influence of the antigenic
protein, the matrices are altered in such a way as to give rise to a new
protein which will react only with the antigen.  Such "foresight" on the
part of the matrices about the way in which they would have to alter
themselves under the influence of an external factor in order to cause a
reaction back on that factor is logically impossible and factually unrealistic.
If such alterations of the matrices under the influence of the antigen
really are observed they can only be a realization of pre-existing properties
of the matrices, an activation of information already present in them, and
revealed rather than created by the antigen.  This is the very assumption
which lies behind the "selective" theories which are better worked out
from a theoretical standpoint.

One of the best known of the "selective" theories of antibody formation
is that worked out by Jerne (50).  He put forward the idea that the $\gamma$-
globulin fraction of blood is extremely polymorphic and heterogeneous
and that among the tremendous variety of molecules there are always
some which will be complementary to any particular antigen.  Any
"foreign" proteins which may enter the blood will eventually react with
these complementary antibodies and this will lead in some way to an
increase in the production of just those matrices and cells which are responsible for the synthesis of these forms of $\gamma$-globulin molecule.  This
principle was later modified by Burnet who worked out what he called a
theory of "clonal selection" (25-28) which has been widely accepted by
immunologists.  This theory assumes that there is a wide variety of clones
of mesenchymal cells within the organism, each of which has immunologically active centres which correspond with some complementary
features of one or several potential antigenic determinants.  When they
reach a certain stage of development these populations of cells can produce
populations of globulin molecules which are potential antibodies.  Introduced antigens react with the cells of the corresponding clone, mainly
lymphocytes, and stimulate the multiplication and activity of the clone,
the production of which becomes sufficient to eliminate the antigen from
the organism.  In this case the synthesis of antibodies would proceed in
the same way as ordinary protein synthesis and the information required
for its specificity would be supplied by the genome of the cell.  Thus,
the "selective" theories do not modify our ideas on the mechanism of the
synthesis of proteins and nucleic acids, the reproduction of their specificity,
genetic control, etc.  They only try to demonstrate the specific character
of a function which is carried out by the systems at a higher level of
organization.

The same "selective" ideas of the synthesis of antibodies are, in fact,
reproduced in the schemes of a number of other authors.  Thus, for
example, in trying to explain the immunological specificity of a variety of
antibodies towards one antigen, immunological tolerance and other
phenomena, Talmage (97) puts forward the possibility that there may be

an almost unlimited heterogeneity among normal $\gamma$-globulins. This non-uniformity reflects differences in the chemical and physico-chemical properties of the individual types of $\gamma$-globulins, the number of which amounts to more than 5000. Each type is characterized by a definite relationship to other substances of the nature of haptens, with which they enter into combination. All types of $\gamma$-globulins exist already and are formed each in its own particular lymphoid cells. The action of the hapten on particular parts of the cells provides the stimulus for their formation, hence the selection of cells and the specificity of the $\gamma$-globulins formed. Thus, there arise immune antibodies which are necessarily partial variants of normal $\gamma$-globulins.

Another worker, Lederberg (62), is also trying to find a connection between the specificity of immune reactions and the pre-existing, genetically determined properties of the cells. He thinks that the stereospecific segment of each antibody is determined by a particular sequence of amino acids. The cell which produces the antibody will, accordingly, have the corresponding unique sequence of nucleotides in its chromosomal apparatus, i.e. in the gene for the synthesis of $\gamma$-globulin. The genetic variety of the precursors forming the antibodies in the cells is, in his opinion, a consequence of the high rate of spontaneous mutation of particular stages in the proliferation of the cells in question. Lederberg believes that each ripening plasma cell spontaneously produces small amounts of antibodies corresponding with its own genotype. Reaction with antigens stimulates the formation of the homologous antibodies and the development of the clones of cells which produce the particular antibody.

Boyden (23) has recently developed an original "evolutionary" concept of the formation of antibodies. He assumes that the ability of the organism to differentiate biochemically between what is "its own" and what is "foreign" is the essential feature of immunological reactions and that this ability is present even in unicellular organisms at a stage of organizational development at which there is not yet any production of antibodies. In Boyden's opinion the existence of this property among the simplest organisms and in phagocytes is connected with the properties of their surfaces, particularly with the presence there of proteins (agglutinins) which can react selectively with foreign bodies such as the encapsulated bacteria which become "stuck" to the surfaces of amoebae, for example. The same properties of the surfaces of the cells are also to be found in the phagocytes of invertebrates and this is why they can "recognize" foreign bodies. Phagocytes can "feel" a considerably wider range of bodies and their surfaces must therefore be considerably richer in molecules with varying steric configurations. These molecules are synthesized by means of specific matrices.

According to Boyden, therefore, the first step in the synthesis of antibodies would appear to be the sorption of the antigens by particular types of cells. According to the hypothesis of this author these cells are covered with special receptors for a wide assortment of foreign molecules (antigens). These receptors are $\gamma$-globulins and their synthesis is determined by matrices. Cells which have adsorbed antigens begin to augment their production of the corresponding receptor which, properly speaking, is also the antibody. Thus, it is only certain abstract considerations as to the way in which the receptor molecules of the cell wall are formed, which we shall not discuss now, which differentiate this hypothesis from the "selective" hypotheses. In a recent symposium on the cellular basis of immunity (29) there was a special discussion devoted to the "selective" hypotheses (72) and most of the participants were very favourably disposed towards these ideas.

In all these, as in earlier theoretical speculations as to the nature of the synthesis of antibodies, it is important to note that, as in the case of the synthesis of adaptive enzymes, it is assumed that, even before the introduction of the antigen, the organism possessed the power of forming the corresponding antibody and this power was merely strengthened. It seems to us that one might go even further in this respect and assume that the organism has not a weak but often a latent power of forming antibodies. Some of the assumptions of the "selective" theories are, however, not above criticism. Šterzl (89) has recently remarked that, while analogues of nucleotides suppress the formation of antibodies, substances which inhibit the division of cells do not have the same effect on the process. According to Šterzl, this indicates that the multiplication of cells of a particular clone cannot be the main process in the inductive phase of the formation of antibodies but that what is fundamental is that a functionally new system of nucleic acids arises. This objection to the theory of clonal selection can, however, be overcome if we assume that selection and stimulation occur, *not at the cellular level but at the level of, say, the ribosomes*. Selection at the level of the ribosomes, which are the immediate producers of proteins, is in many ways more attractive as the reaction of the organism to the antigen would, in this case, be faster, the polymorphism of possible antigens would be wider and the possibilities of the cells and genomes would be more multivalent.

Another objection to the "selective" theories has recently been raised by Haurowitz (44), who considers that they cannot give a convincing explanation of the presence of preformed antibodies to azophenylarsonate, the azophenyltrimethylammonium group and other non-biological and unnatural products of organic chemistry. According to his reckoning the theory of clonal selection allows for the existence of no more than $10^6$ types of cell able to form antibodies and gives rise to the problem of the impossibility of providing for antibodies against all possible antigens.

However, if we admit that selection may occur at the level of the ribosomes, then the figure of $10^6$ may be raised (see 70a). Even in such a case, of course, the number of possible variants of $\gamma$-globulin can certainly not compete with the number of possible variants of all other biopolymers.

However, under real biological conditions the number of possible antigens is very limited and the postulated mechanism for the formation of antibodies completely fulfils the requirements of phylogenesis. As concerns the antigenicity of certain organic compounds, this is quite comprehensible if we take into account the random nature of the process of mutation, which gives natural and synthetic compounds equal opportunities to be antigens. Also, the fact that by no means all artificial organic polymers possess an antigenic function indicates that we do not face the task of explaining *unlimited* possibilities of immunity.

*d) The biochemical mechanism of the ability to differentiate between "own" and "foreign" in reactions of immunity*

The question of the recognition of "foreign molecules" as such in immune reactions is the crucial one for the whole phenomenon. The random nature of the process of mutation makes it inevitable that antibodies must be formed which react with normal components of the living body but such immune reactions do not in fact occur, apart from certain pathological processes. To use Jerne's (51) picturesque simile, the system of antibody formation is like an electronic machine for translation from a foreign language into English. Such a machine only produces an English text but it only "recognizes" the foreign one. On this analogy the system of antibody formation, in its fully-developed aspect, resembles a dictionary in which all the "words" corresponding to the configurations of the organism itself have been removed and only the foreign "words" are left. How, then, is this elimination of the words of the "native language" of the organism brought about? The most ingenious answer to this question is that given by the selective theory of Burnet (26-28) on the basis of consideration of the phenomenon of tolerance. The discovery of the phenomenon of immunological tolerance is a great biological achievement and Medawar who had the honour of making it was awarded the Nobel Prize in 1960, sharing it with Burnet for his working out of the theory of immunity.

According to Medawar's definition (68), immunological tolerance is a state of non-reactivity or "indifference" to substances which would normally evoke an immune reaction. Tolerance may develop during embryonic life and although, for example, a foreign protein $D$ introduced into the organism of a rat will stimulate the formation of the antibody anti-$D$, the introduction of the protein $D$ during embryonic life evokes the phenomenon of tolerance so that the adult organism will not react with the protein $D$, acting towards it as though it were its own.

We are not at present concerned with the tremendous practical importance of this discovery. From the theoretical point of view it enables us to understand the selectivity of immunity. Burnet (26) suggests that during embryonic life the genetic loci responsible for the immunological

PB R

reactions pass through a special "sensitive" stage during which they can be disturbed or "arrested" by any antigen whatever. This stage coincides with or occurs later than the stage of "mutagenesis" of the system and, as a result of this, all genomes which can evoke autoimmune reactions go out of action, leaving only those clones active which produce antibodies which can fix "foreign" substances.

Burnet (28a) has recently portrayed diagrammatically the theory of clonal selection, showing how it explains both the phenomenon of tolerance and the origin of the capability of the immunological system to distinguish "own" and "foreign".

Such an approach to the explanation of the phenomenon seems to us quite justifiable. If the cells of the organism really do possess a definite biochemical system controlling heredity and containing the biochemical information which determines the sequences of amino acids and nucleotides and the specificity of the polysaccharides of "its own" polymers which, themselves, make their appearance in various stages of ontogenesis, then this system is the only one which could theoretically serve as the basis for differentiating between what substances are the organism's "own" and what "foreign". In this connection it is interesting that the appearance of immune properties occurs after active morphogenesis has finished when the main potentials of the "biochemical programme of development" have already mainly been realized.

## 2. The synthesis of haemoglobin

In earlier chapters we have already referred more than once to the facts and phenomena of the system whereby haemoglobin is synthesized, as the reticulocytes are a classical experimental object for the study of protein synthesis. In the present short section we shall not repeat all this material but shall only deal with certain special features of this synthesis in which it differs from the synthesis of other proteins.

For example, the finding of sequential growth of the polypeptide chain from the amino end to the carboxyl end (20, 32) is a characteristic feature which, so far, has only been found in the synthesis of haemoglobin.

Schweiger et al. (80) arrived at the conclusion that when the reticulocyte becomes an erythrocyte there is an intensive conversion of the proteins of the stroma of the cells into haemoglobin, not directly but through an intermediate stage in which it occurs as non-protein nitrogenous material.

In the study of the synthesis of haemoglobin there arises, alongside the problems which are common to all protein syntheses, a new question which is also common to many protein syntheses. This is the question as to the nature of the mechanisms which determine the specific combination of the protein with some non-protein prosthetic group. The haem of haemoglobin is synthesized independently from the protein part or globin, although amino acids are used in synthesizing it too (22, 57, 59).

The independent synthesis of the haem and globin parts of the enzyme was established by studying the synthesis of the cytochromes, which are formed in all aerobic tissues (65). It is not yet known how these components react with one another but the mechanism of this process would seem not to be very complicated. Under normal conditions the synthesis of haem and that of globin proceed at the same rate (65), but under certain forms of inhibition, by cobalt for example, the synthesis of the haem group is mainly affected. This suggests that, although the syntheses of the two components of these conjugated proteins are independent, yet their occurrence is co-ordinated by some biochemical mechanism, probably the rate of formation of the terminal complex. The possibility of stimulating the formation of haemin by the action of globin (77) is especially suggestive in this respect. The site of the globin synthesis in the erythrocytes is the microsomes, which quickly secrete the formed globin into the cell sap (16, 64, 74). As to the synthesis of haem, according to Minakami et al. (69) the mitochondria exhibit by far the greatest activity in this respect.

The synthesis of haemoglobin in mammalian reticulocytes proceeds, as is well known, in the absence of nucleus and chromosomes, i.e. in non-nucleated cells. At the same time in the reticulocytes messenger-RNA is present both on the ribosomes (99b) and in the soluble fraction (60, 64). This has been formed before the loss of the nuclear substance of the cell. This shows that the messenger-RNA of mammalian cells can serve as a relatively stable matrix, functioning over rather long periods of time.

Several unusual features of the synthesis of haemoglobin in the nucleated erythrocytes of ducks have been reported recently by Wiggans et al. (100). The authors studied the alteration of the specific activity of the haemoglobin after the cells had been transferred from a solution containing [$^{14}$C]leucine into a medium which contained no labelled component. In spite of the absence of any external sources of labelled leucine the specific activity of the haemoglobin continued to increase for several hours. The authors consider that this is evidence for the existence of a quickly formed "precursor" of haemoglobin which is later gradually transformed into haemoglobin.

### 3. The synthesis of fibrin during clotting of blood

The biochemical system for clotting of blood consists of many proteins, but we shall only touch very briefly here on the essential final product of the process, namely fibrin. The monograph of Belik & Khodorova (1) on the biochemistry of the clotting of blood makes it superfluous for us to give a detailed review of the evidence on this question. We shall simply mention the fact that the formation of fibrin from fibrinogen occurs by polymerization in which, in essence, the synthetic process of polymerization takes place without the participation of nucleic acids or other specific factors. The spontaneity of this process does not prevent it from being specific in that the combination of the fibrin monomers formed enzymically from fibrinogen occurs by means of particular bonds which have been set free as a result of the detachment of certain peptide groups from the fibrinogen. The example of this synthesis demonstrates that the formation of large protein polymers out of polypeptide fragments with a definite specificity can occur spontaneously, or at any rate, not on matrices, which are necessary to determine the specificity of the arrangement of amino acids in individual, discrete, monomeric protein molecules. Some details of this polymerization have been surveyed in the work of Jorpes, Blombäck & Yamashima (52). A recent short review by Laki, Gladner & Folk (63) gives quite a clear idea of the unique mechanism of the synthesis of fibrin. The picture of this synthesis which is being built up from a large number of detailed investigations shows that, under the influence of thrombin, which is a narrowly specific protease, two peptides are split off from each molecule of fibrinogen. These are attached to the main part of the fibrinogen molecule through arginine and are situated at the end and on the side of the fibrinogen molecule (Fig. 34). These strongly charged peptides mask the combining groups and their removal quickly brings about the association of the monomers by the interaction of the end of one with the lateral part of another, thus ensuring the formation of a molecular tangle of fibrin.

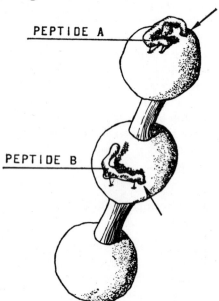

FIG. 34. Diagrammatic representation of a fibrinogen molecule. The arrows show the position of the Arg. Gly bonds attacked by thrombin. The 'legs', on which the peptides stand, represent hydrogen bonds (63).

## 4. The special form of functional synthesis of amylase—the addition of amino acids to the formed protein molecule

We have already referred in previous chapters to the large quantity of data concerning the synthesis of the proteins of the pancreas and here we shall only deal with one aspect.

As a result of many researches by Straub and colleagues (5, 38, 91, 98) it has been found that if a soluble system from acetone-dried pancreas is incubated in a medium containing a Krebs solution of ATP and amino acids, synthesis of amylase will occur and may be demonstrated by an increase in the activity of this enzyme.  It appeared that there was no need for all the amino acids to be present for the stimulation of this synthesis to occur; arginine and threonine alone were enough.  The addition of other amino acids did not increase the synthesis.  Inhibitors of protein synthesis hindered this process.

In view of these facts the authors put forward the suggestion that the synthesis of amylase takes place in two stages.  In the first stage a precursor of amylase is formed and in the second it is simply finished off by a partial synthesis in which only arginine and threonine participate.

If these facts are really confirmed they might form the basis for interesting generalizations about the possibility of two-stage building of protein molecules, each stage depending, according to the authors, on RNA.

In a more recent work Straub *et al.* (92) have found a number of new and unusual features of this process.  They have succeeded in showing that the transition from the protein precursor to amylase really depended on arginine and threonine, ATP and RNA, but the arginine, threonine and ATP were required for some sort of transformation of RNA (polymeric) into specific RNA.  The second stage of the process, namely the formation of the enzyme, depended only on the specific RNA.  Only the RNA of the mitochondria and coarse granules were active in this respect.

## 5. Synthesis of the fibroin of silk

We have already discussed several investigations in which the general rules of protein synthesis were studied using systems functionally associated with the synthesis of the proteins of silk.  It is, however, interesting also to pay attention to the specific features of this unusual synthesis of a fibrillar protein which is the simplest, with regard to the sequence of amino acids, of all proteins produced by living organisms.

In the first place, it is a peculiar fact that the ribonucleoproteins of the silk-secreting glands have an amino acid composition which is markedly different from that of the proteins of silk (84) and this indicates that the main protein component of the intracellular centres of synthesis of the proteins of silk is probably a component of the synthesizing system.

The amino acid composition of the pool of free amino acids reaching the silk-secreting gland from the body fluid of the silkworm does not correspond with the amino acid composition of silk either, although, quantitatively speaking, the proteins of silk constitute the main product of the gland (36).

It is interesting that the activating enzymes of the gland can also activate amino acids in proportions which are far from being the same as those in which they are found in fibroin molecules, and glycine, which is the main component of this protein, is very weakly activated (46). These findings have not, however, been confirmed by the recent work of Faulkner & Bheemeswar (37) who observed intensive activation of glycine in the glands of the silkworm and accomplished a partial purification of the glycine-activating enzyme.

Interesting observations on the incorporation of [$^{14}$C]glycine into the proteins of silk have been made recently by Shimura and colleagues (83). They found that the glycine which was incorporated during the synthesis of fibroin was arranged irregularly in the molecule which, they thought, suggested that the protein was synthesized by stages, the final stage being the combination of peptide and polypeptide products. The synthesis of fibroin was localized in the microsomes (93).

The idea that fibrillar proteins consist of repeating fragments of a lower molecular weight is very widely held (99). This is especially characteristic of the proteins of silk, in the chains of which the same long sequence repeats itself (10). It is very likely that, in this case, the matrices only produce the sub-units of the molecule, which later polymerize in some other way.

The silk-secreting gland is very rich in RNA. According to the evidence of Takeyama and colleagues (94), this RNA is important for the synthesis of fibroin in the polymeric form. In this work the authors studied the incorporation of [$^{14}$C]glycine into the proteins of the gland and of [$^{14}$C]orotic acid into the uracil of the RNA. The incorporation of the glycine into the protein of the macerated gland was extremely active, nearly 100 times as active as incorporation into liver tissue under the same conditions. This incorporation was not slowed by inhibition of the synthesis of RNA by means of antimetabolites. It was, however, completely halted by splitting the RNA which was already in the gland with ribonuclease. It is well-known that the silk-secreting gland of the mulberry silkworm synthesizes two proteins of different amino acid composition, fibroin in the posterior part and sericin in the reservoir. No difference has, however, been established between the nucleotide composition of the RNA of these two parts (14).

## Conclusion

In this chapter we have made a very short survey of only a very few functional syntheses for, if we had tried to give a more complete review of this field, our present work might have had to be increased to several times its size. The literature of the present day contains accounts of a large number of researches on such functional syntheses as those of the casein, albumins and globulins of milk, collagen and procollagen, elastin, myosin, myogen and actin, various enzymes, hormones and toxins and so on. However, the short review which we have just given shows that their synthesis exhibits many specific features associated with differences in the structure, evolutionary history and role of the proteins. The unusual feature of the synthesis of amylase, namely the small addition which completes it, is not, however, really synthesis but rather activation. The organism had to react very quickly to the arrival of food by producing and excreting enzymes and this led to the development of processes of activation of stored enzymes as their formation *de novo* would probably not be an "operative" enough process. The biochemical nature of activation is very different in different enzymes. In most proteases activation is associated with the removal of some masking groups but in amylase a synthetic mechanism of activation has been evolved.

The synthesis of high-molecular fibrin has to be even more rapid and a quick and spontaneous process of polymerization has therefore been evolved. The synthesis of antibodies is quite anomalous. The system by which it is carried out has an important feature which promises to be of great interest; this is the existence of labile matrices in the cells which either change their properties under the influence of antigens or can undergo a wide range of mutations over a short time. All this shows clearly that the systems and mechanisms of protein synthesis are not unchanging. In nature there is a great variety of all possible forms of synthesis because, in all these cases, it is necessary not only to produce proteins which are different in structure, form and size, but also to produce them at different rates and with different demands on their specificity. Furthermore, the actual process of breakdown and resynthesis of proteins may have a direct functional significance and must therefore conform to the rules of variability, heredity and selection.

## REFERENCES

1. BELIK, YA. V. & KHODOROVA, E. L. (1957). *Biokhimiya svertyvaniya krovi.* Kiev: Izd. Akad. Nauk Ukr. S.S.R.
2. VYAZOV, O. E. & TSEĬTLIN, P. I. (1956). *Doklady Akad. Nauk S.S.S.R.* **110,** 119.
3. GOSTEV, V. S. (1959). *Khimiya spetsificheskogo immuniteta.* Moscow: Medgiz.

4. GASHEK [HAŠEK], M. (1961). *Uspekhi sovremennoĭ Biol.* **51**, 153.
5. GARZÖ, T. (1959). In collective work *Aktual'nye problemy sovremennoĭ biokhimii*, Vol. **1** *Biokhimiya belkov* (ed. V. N. Orekhovich), p. 178. Moscow: Izd. Akad. med. Nauk S.S.S.R.
6. GURVICH, A. E., GUBERNIEVA, L. M. & MYASOEDOVA, K. N. (1961). *Proc. V int. Congr. Biochem., Moscow*, **9**, 473.
7. GURVICH, A. E. & KARSAEVSKAYA, N. G. (1956). *Biokhimiya*, **21**, 746.
8. GURVICH, A. E. & SMIRNOVA, N. P. (1957). *Biokhimiya*, **22**, 626.
9. ZIL'BER, L. A. (1958). *Osnovy immunologii.* Moscow: Medgiz.
10. IOFFE, K. G. (1954). *Biokhimiya*, **19**, 495.
11. NIKOLAEV, A. I. (1959). *Byul. eksperim. Biol. i Med.* **48**, 79.
12. — (1961). *Vopr. med. Khim.* **7**, 74.
13. NIKOLAEV, A. I. & AKHMADIEVA, A. (1961). *Proc. V int. Congr. Biochem., Moscow*, **9**, 478.
14. RAMENSKAYA, G. P., ZBARZKIĬ, I. B. & MIL'MAN, L. S. (1960). *Doklady Akad. Nauk S.S.S.R.* **132**, 1206.
15. SHTERTSL', YA. [ŠTERZL, J.] (1959). *Uspekhi sovremennoĭ Biol.* **48**, 356.
15a. — (1961). *Uspekhi sovremennoĭ Biol.* **51**, 337.
15b. EFROIMSON, V. P. (1961). In collective work *Problemy Kibernetiki* (ed. A. A. Lyapunov), No. *6*, p. 161. Moscow: Izd. Fiz. Mat. Lit.
16. ALLEN, E. H. & SCHWEET, R. S. (1960). *Biochim. biophys. Acta*, **39**, 185.
17. ARNON, R. & SELA, M. (1960). *Bull. Res. Council Israel Sect. A (Chemistry)*, **9A**, 111, 112.
18. ASKONAS, B. A. & HUMPHREY, J. H. (1958). *Biochem. J.* **68**, 252.
19. JAROSLOW, B. N. (1960). *J. infect. Diseases*, **107**, 56.
20. BISHOP, J., LEAHY, J. & SCHWEET, R. [S.] (1960). *Proc. natl. Acad. Sci. U.S.* **46**, 1030.
21. BOYD, W. C. (1954). In *The Proteins* (ed. H. Neurath & K. Bailey) Vol. 2, Pt. B, p. 755. New York: Academic Press.
22. BORSOOK, H., FISCHER, E. H. & KEIGHLEY, G. (1957). *J. biol. Chem.* **229**, 1059.
23. BOYDEN, S. V. (1960). *Nature (Lond.)*, **185**, 724.
24. BREINL, F. & HAUROWITZ, F. (1930). *Z. Physiol. Chem.* **192**, 45.
25. BURNET, F. M. (1957). *Australian J. Sci.* **20**, 67.
26. — (1960). *Perspectives Biol. Med.* **3**, 447.
27. — (1961). *New Engl. J. Med.* **264**, 24.
28. — (1959). *The clonal selection theory of acquired immunity.* Nashville, Tenn.: Vanderbilt University Press.
28a. — (1962). *Scient. American*, **207**, no. 5, p. 50.
29. *Ciba Foundation Symposium on cellular aspects of immunity* (ed. G. E. W. Wolstenholme & M. O'Connor). London: Churchill (1960).
30. CAMPBELL, D. H. & GARVEY, J. S. (1960). *J. infect. Diseases*, **107**, 15.
31. COONS, A. H. (1956). *Intern. Rev. Cytology*, **5**, 1.
32. DINTZIS, H. M. (1961). *Proc. natl. Acad. Sci. U.S.* **47**, 247.
33. DUTTON, R. W., DUTTON, A. H. & GEORGE, M. (1958). *Nature (Lond.)*, **182**, 1377.
34. DUTTON, R. W., DUTTON, A. H. & VAUGHAN, J. H. (1960). *Biochem. J.* **75**, 230.
35. — (1959). *Fed. Proc.* **18**, 219.
36. FUKUDA, T., KIRIMURA, J., MATSUDA, M. & SUZUKI, T. (1955). *J. Biochem. (Tokyo)*, **42**, 341.
37. FAULKNER, P. & BHEEMESWAR, B. (1960). *Biochem. J.* **76**, 71.
38. GARZÓ, T., PERL, K., SZABÓ, M. T., ULLMANN, Á. & STRAUB, F. B. (1957). *Acta physiol. Acad. Sci. Hung.* **11**, 23.

39. ATTARDI, G., COHN, M., HORIBATA, K. & LENNOX, E. S. (1959). *Bact. Revs.* **23**, 213.
40. HARRIS, T. N. & HARRIS, S. (1960). *Ann. N.Y. Acad. Sci.* **86**, 948.
41. HAUROWITZ, F. (1951). *Symposia Sect. Microbiol. N.Y. Acad. Med.* No. *5, Nature and significance of the antibody response* (ed. A. M. Pappenheimer), p. 3 (Pub. 1953). New York: Columbia Univ. Press.
42. — (1955). *Scientia (Asso. Italy)*, **90**, 335.
43. — (1950). *Chemistry and biology of proteins.* New York: Academic Press.
44. — (1960). *Ann. Rev. Biochem.* **29**, 609.
45. HAUROWITZ, F. & SCHWERIN, P. (1942). *Brit. J. exptl. Path.* **23**, 146.
46. HELLER, J., SZAFRAŃSKI, P. & SUŁKOWSKI, E. (1959). *Nature (Lond.)*, **183**, 397.
47. HRUBEŠOVA, M., ASKONAS, B. A. & HUMPHREY, J. H. (1959). *Nature (Lond.)*, **183**, 97.
48. HUMPHREY, J. H. & McFARLANE, A. S. (1955). *Biochem. J.* **60**, xi.
49. ISLIKER, H. C. (1957). *Advanc. Protein Chem.* **12**, 387.
50. JERNE, N. K. (1955). *Proc. natl. Acad. Sci. U.S.* **41**, 849.
51. — (1960). *Ann. Rev. Microbiol.* **14**, 341.
52. JORPES, J. E., BLOMBÄCK, G. E. B. & YAMASHINA, J. (1958). *Proc. internatl. Sympos. on Enzyme Chemistry, Tokyo & Kyoto,* 1957 (*I.U.B. Symposium Series No. 2*), p. 400. London: Pergamon.
53. KARUSH, F. (1957). *J. Amer. chem. Soc.* **79**, 5323.
54. — (1958). *Trans. N.Y. Acad. Sci.* (Ser. 2), **20**, 581.
55. — (1960). *Science,* **132**, 1494.
56. — (1961). *Proc. V int. Congr. Biochem., Moscow,* **9**, 474.
57. KASSENAAR, A., MORELL, H. & LONDON, I. M. (1957). *J. biol. Chem.* **229**, 423.
58. KOOYMAN, E. C. & CAMPBELL, D. H. (1948). *J. Amer. chem. Soc.* **70**, 1293.
59. KRUH, J. & BORSOOK, H. (1956). *J. biol. Chem.* **220**, 905.
60. KRUH, J., ROSA, J., DREYFUS, J.-C. & SCHAPIRA, G. (1961). *Proc. V int. Congr. Biochem., Moscow,* **9**, 95.
61. KUNKEL, H. G., FUDENBERG, H. & OVARY, Z. (1960). *Ann. N.Y. Acad. Sci.* **86**, 966.
62. LEDERBERG, J. (1959). *Science,* **129**, 1649.
63. LAKI, K., GLADNER, J. A. & FOLK, J. E. (1960). *Nature (Lond.),* **187**, 758.
64. LAMFROM, H. & GLOWACKI, E. (1961). *Proc. V int. Congr. Biochem., Moscow,* **9**, 96.
65. MARSH, J. B. & DRABKIN, D. L. (1958). *J. biol. Chem.* **230**, 1073.
66. McFADDEN, M. L. & SMITH, E. L. (1955). *J. biol. Chem.* **214**, 185.
67. McMASTER, P. D. (1961). In *The cell: biochemistry, physiology, morphology* (ed. J. Brachet & A. E. Mirsky), Vol. 5, p. 323. New York: Academic Press.
68. MEDAWAR, P. B. (1961). *Science,* **133**, 303.
69. MINAKAMI, S., YONEYAMA, Y. & YOSHIKAWA, H. (1958). *Biochim. biophys. Acta,* **28**, 447.
70. NOSSAL, G. J. V. (1959). *Brit. J. exptl. Path.* **40**, 301.
70a. NOSSAL, G. J. V. & MAKELA, O. (1962). *J. Immunol.* **88**, 604.
71. OGATA, K., OGATA, M., MOCHIZUKI, Y. & NISHIYAMA, T. (1956). *J. Biochem. (Tokyo),* **43**, 653.
72. Discussion in (29), pp. 157-171.
73. PAULING, L. (1940). *J. Amer. chem. Soc.* **62**, 2643.
74. RABINOVITZ, M. & OLSON, M. E. (1956). *Exptl. Cell Research,* **10**, 747.
75. RITTENBERG, M. B. & NELSON, E. L. (1960). *Amer. Naturalist,* **94**, 321.
76. SCHULTZE, H. E. (1959). *Clin. chim. Acta,* **4**, 610.

77. SCHWARTZ, H. C., HILL, R. L., CARTWRIGHT, G. E. & WINTROBE, M. M. (1959). *Biochim. biophys. Acta*, **36**, 567.
78. SCHWARTZ, R., EISNER, A. & DAMESHEK, W. (1959). *J. clin. Invest.* **38**, 1394.
79. SCHWEET, R. S. & OWEN, R. D. (1957). *J. cell. comp. Physiol.* **50**, Suppl. 1, 199.
80. SCHWEIGER, H. G., RAPOPORT, S. & SCHÖLZEL, E. (1956). *Nature (Lond.)*, **178**, 141.
81. SELA, M. & ARNON, R. (1960). *Biochem. J.* **75**, 91.
82. — (1960). *Biochem. J.* **77**, 394.
83. SHIMURA, K. (1956). *Nippon Seikagaku Kaishi*, **28**, 197.
84. SHIMURA, K., SATO, J., SUTO, S. & KIKUCHI, A. (1956). *J. Biochem. (Tokyo)*, **43**, 217.
85. SMITH, E. L., McFADDEN, M. L., STOCKELL, A. & BUETTNER-JANUSCH, V. (1955). *J. biol. Chem.* **214**, 197.
86. STAVITSKY, A. B. (1957). *Fed. Proc.* **16**, 652.
87. STAVITSKY, A. B., AXELROD, A. E. & PRUZANSKY, J. (1957). *J. Immunol.* **79**, 200.
88. STAVITSKY, A. B. & WOLF, B. (1958). *Biochim. biophys. Acta*, **27**, 4.
89. ŠTERZL, J. (1961). *Proc. V int. Congr. Biochem., Moscow*, **9**, 485.
90. ŠTERZL, J. & HRUBEŠOVA, M. (1956). *Fol. biol. (Prague)*, **2**, 21.
91. STRAUB, F. B. & ULLMANN, Á. (1957). *Biochim. biophys. Acta*, **23**, 665.
92. STRAUB, F. B., ULLMANN, Á. & VENETIANER, P. (1960). *Biochim. biophys. Acta*, **43**, 152.
93. SUTO, S. & SHIMURA, K. (1961). *J. Biochem. (Tokyo)*, **49**, 69.
94. TAKEYAMA, S., ITO, H. & MIURA, Y. (1958). *Biochim. biophys. Acta*, **30**, 233.
95. TALIAFERRO, W. H. (1957). *J. cell comp. Physiol.* **50**, Suppl. 1, 1.
96. TALIAFERRO, W. H. & TALMAGE, D. W. (1955). *J. infectious Diseases*, **97**, 88.
97. TALMAGE, D. W. (1959). *Science*, **129**, 1643.
98. ULLMANN, Á. & STRAUB, F. B. (1957). *Acta physiol. Acad. Sci. Hung.* **11**, 31.
99. WAUGH, D. F. (1957). *J. cell comp. Physiol.* **49**, Suppl. 1, 145.
99a. WEIGLE, W. O. (1961). *Advanc. Immunol.* **1**, 283.
99b. WEISBERGER, A. S. (1962). *Proc. natl. Acad. Sci. U.S.* **48**, 68.
100. WIGGANS, D. S., BURR, W. W. (jun.) & RUMSFELD, H. W. (jun.) (1960). *J. biol. Chem.* **235**, 3198.
101. WORMALL, A. (1955). *Brit. J. Radiol.* **28**, 33.

# CHAPTER XII

# ON THE POSSIBLE EXISTENCE OF SUPPLEMENTARY SYSTEMS OF PROTEIN SYNTHESIS WITHOUT PARTICIPATION OF THE "ACTIVATING ENZYME-S-RNA COMPLEX"

*Introduction*

Alongside the large number of investigations of the part played by S-RNA and the activating enzymes in the synthesis of specific proteins, a small number of investigations have been carried out in recent years on a small number of objects in which it has been found that there was a certain difference in the system for incorporation of amino acids during protein synthesis. In our survey of the peculiarities of the synthesis of various proteins (those of viruses, adaptive enzymes, antibodies, etc.) we reviewed the factual peculiarities and variations in the working of the ordinary fundamental system of protein synthesis described in the first chapters. Only two cases (the synthesis of the reserve proteins of peas and that of fibrin) did not fit into the ordinary scheme, but they were also not cases of true synthesis but rather of the assembly of large blocks formed in the ordinary way. Here we wish to deal shortly with researches which indicate the possibility that there may be ways of incorporation of amino acids into the protein chain, based on different principles.

## 1. The enzymic system of "direct" incorporation of amino acids into proteins

What we have primarily in mind here is a series of papers by Beljanski & Ochoa (1-5). These authors reported that they had found in the microorganism *Alcaligenes faecalis* a special "enzyme for incorporating amino acids into proteins" which was quite different from the ordinary activating enzymes. They isolated from these cells a granular fraction which brought about the incorporation of amino acids into proteins without the ordinary exchange of ATP and pyrophosphate and formation of aminoacyladenylates. This enzyme was purified and, in two later works by Beljanski, it was found that, in the presence of magnesium, it could catalyse the transfer of phosphate from the triphosphates of nucleosides to form the corresponding diphosphates as follows: ATP $\rightarrow$ ADP or UTP $\rightarrow$ UDP.

This enzyme was found to contain four components, one for each pair of nucleosides. The preparation was free from diphosphonucleoside kinase and monophosphonucleoside kinase. Each of the enzymes was named after the nucleoside it affected e.g. adenosinetriphosphate—adenosidediphosphate kinase, etc. It would seem that this transfer of phosphate was associated in some way with the incorporation of amino acids in protein synthesis. It is interesting to note that the 18 amino acids which were studied individually formed four main groups, each being specific for one nucleoside triphosphate. In another paper Beljanski (3) reported a study of the artificial formation of peptides from amino acids (glycine, glutamic acid, leucine or phenylalanine) in the presence of an enzymic system isolated from *Alcaligenes faecalis* and of nucleoside triphosphates (ATP, GTP, UTP and CTP). The course of the reaction was judged by the formation of free orthophosphate.

It was established that in experiments with glycine, glutamic acid, leucine and phenylalanine there was a considerable liberation of orthophosphate (by glutamic acid in the presence of ATP, GTP and UTP, by glycine in the presence of ATP and GTP, by leucine in the presence of UTP and CTP and by phenylalanine in the presence of CTP). The addition of chloramphenicol (as a specific inhibitor of the biosynthesis of proteins) completely suppressed the formation of orthophosphate. Experiments with valine, leucine, phenylalanine, glutamic acid and glycine labelled with $^{14}C$ (chromatographic analysis followed by autoradiography) have established that under these conditions radioactive compounds are formed which break down on hydrolysis into amino acids. The author draws the conclusion that the reactions of formation of peptides catalysed by the enzymic system probably proceed according to the scheme

$$X—R—P—P—P + \text{amino acids} \rightarrow X—R—P—P + P + \text{peptide}$$

where X is a purine or pyrimidine base, P a phosphate residue and R a ribose residue. In a recent communication Beljanski and colleagues (3a) made a further study of the reactions catalysed by this enzyme, which they called "polypeptide synthetase".

Nisman & Fukuhara (10) isolated an enzyme which they called "amino acid polymerase" from the membrane of another bacterium, *Escherichia coli*. This enzyme also incorporated amino acids into proteins without the ordinary system of activation.

Interesting evidence as to the possible existence of two ways of incorporating amino acids into the proteins of the microsomes of the liver has been obtained by Rendi & Campbell (12). They made a deep study of the role of S-RNA in protein synthesis but, as well as the ordinary system of activation of amino acids and their transfer to molecules of S-RNA, they also studied the special protein fraction known as S-protein, previously discovered by Sachs (14), which augmented the incorporation of amino acids into the proteins of the microsomes in the presence of glutathione. This protein brings about the incorporation of labelled amino acids into proteins without the formation of intermediate products combined with S-RNA.

In laterstudies Rendi & Hultin (13) studied the part played by this protein in more detail using, instead of microsomes, ribosomes obtained by treating microsomal material with detergents. The ability of the S-protein (S-enzyme) to alter the cell sap was confirmed but the non-participation of small polynucleotides in the reaction has not yet been fully established.

Among other experiments along these lines we must mention work (9, 15) in which it was found that the "incorporation factor" of Gale (an unidentified substance, cf. Chapter X) can "replace" the activating enzymes in the systems of *Staphylococcus aureus*.

Weissbach (16) has also recently described the occurrence in *E. coli* of a somewhat peculiar system of incorporation of amino acids in protein synthesis which is not inhibited by chloramphenicol. However, S-RNA still participates in the reactions of "incorporation" in this case. Finally, we must include in this series of investigations the recent work of Zalta and colleagues (17, 18) who reported the isolation, from the microsomes of the livers of rats, of particles containing a system which catalyses the incorporation of amino acids into the proteins of the microsomes but which does not require "pH 5 activating enzymes" and is resistant to ribonuclease. Enzymes were found in these particles which catalysed the exchange of phosphate between di- and triphosphoribonucleosides. This last feature makes the system very like that described by Beljanski & Ochoa. These results were confirmed by the recent work of Prosser *et al.* (11), who also isolated from the liver particles containing a ribonuclease-resistant system incorporating amino acids into protein.

## 2. On the possibility of lipid carriers of activated amino acids

In his recent interesting work Hendler (6) has made it clear that in the cells of the oviducts of hens, where intensive protein synthesis takes place, one may find the very rapid incorporation of labelled amino acids into a special lipid material which very quickly accepts and parts with labelled amino acids. On the other hand, intermediate complexes of amino acids with S-RNA were only renewed very slightly in this case, quite out of proportion to the rate of protein synthesis. The authors came to the conclusion that this lipid material might be identified with the lipid membranes of the microsomes and that it fulfils the function of a carrier of amino acids in protein synthesis.

In other papers Hendler (7, 7a) described the occurrence of similar complexes in the microsomes themselves.

The work of another group of authors (8) provides evidence for the acceptance of amino acids by phospholipids in *Bacillus megaterium*. These amino acids are then incorporated into the proteins of the bacillus without first passing through the pool of free amino acids. In pointing out the multiplicity of the evidence for the occurrence of various forms of amino acid-lipid compounds, the authors indicate their possible role in the accumulation and transport of activated amino acids.

## Conclusion

The results described in this short chapter are very interesting but they are still too meagre to provide a clear picture of the numerous possible supplementary ways in which amino acids may be incorporated in protein synthesis. It is far from being certain that in all these cases we are dealing with a true synthesis of proteins because the "incorporation" of amino acids and the synthesis of protein are not always the same. There are very many cases in the history of the working out of the problem of synthesis in which some particular experimentally studied "protein synthesis" was not really what it was thought to be at first (it was an artefact, the synthesis of peptides of the cell wall, non-specific combination, the synthesis of lipopeptides or glycopeptides or something of that sort). In most of the work referred to above there has not yet been any study of the formation of a definite protein, such as an enzyme, with the incorporation of the amino acids into a specific sequence. So far we are, as a rule, only dealing with the process of "incorporation" in a more general sense and we must therefore wait until these new systems of protein synthesis have been studied in more detail before we base any general concepts on the data which have been obtained.

REFERENCES

1. BELJANSKI, M. (1960). *Biochim. biophys. Acta*, **41**, 104.
2. — (1960). *Biochim. biophys. Acta*, **41**, 111.
3. — (1960). *Compt. rend. Acad. Sci., Paris*, **250**, 624.
3a. BELJANSKI, M., BELJANSKI, MONIQUE & LOVIGNY, T. (1962). *Biochim. biophys. Acta*, **56**, 559.
4. BELJANSKI, M. & OCHOA, S. (1958). *Proc. natl. Acad. Sci. U.S.* **44**, 494.
5. — (1958). *Proc. natl. Acad. Sci. U.S.* **44**, 1157.
6. HENDLER, R. W. (1959). *J. biol. Chem.* **234**, 1466.
7. — (1960). *Fed. Proc.* **19**, 346.
7a. — (1961). *Biochim. biophys. Acta*, **49**, 297.
8. HUNTER, G. D. & GOODSALL, R. A. (1961). *Biochem. J.* **78**, 564.
9. MEHTA, R., WAGLE, S. R. & JOHNSON, B. C. (1960). *Biochim. biophys. Acta*, **39**, 504.
10. NISMAN, B. & FUKUHARA, H. (1959). *Compt. rend. Acad. Sci., Paris*, **248**, 1438.

11. PROSSER, E. J. T., HIRD, H. J. & MUNRO, H. N. (1961). *Biochem. biophys. Research Commun.* **4**, 243.
12. RENDI, R. & CAMPBELL, P. N. (1959). *Biochem. J.* **72**, 435.
13. RENDI, R. & HULTIN, T. (1960). *Exptl. Cell Research*, **19**, 253.
14. SACHS, H. (1957). *J. biol. Chem.* **228**, 23.
15. WAGLE, S. R., MEHTA, R. & JOHNSON, B. C. (1960). *Biochim. biophys. Acta*, **39**, 500.
16. WEISSBACH, A. (1960). *Biochim. biophys. Acta*, **41**, 498.
17. ZALTA, J. P. (1960). *Compt. rend. Acad. Sci., Paris*, **250**, 4058.
18. ZALTA, J. P., LACHURIE, F. & OSONO, S. (1960). *Compt. rend. Acad. Sci., Paris*, **251**, 814.

# GENERAL CONCLUSION TO PART II

In concluding our examination of the peculiarities of some protein syntheses it is appropriate to put a question which is of general theoretical or even of general philosophical significance. Is it, in fact, possible that there is a single method for the synthesis of different proteins in different organisms at different stages of phylogenetic and ontogenetic development? If we approach this question in the abstract we may give either a positive or a negative answer to it. There are many analogies which would justify the giving of a positive answer because there are, in nature, very many biochemical systems and mechanisms which are found satisfactory in principle by almost all lower and higher forms of life and which have been inherent in all living material through almost all the stages of its development with only small alterations, mainly as a result of side branches of evolution at low stages of development. Among these are, for example, the mechanisms of phosphorylation (AMP-ATP) the mechanisms of reduplication of the polynucleotides of DNA, mechanisms of contraction and of glycolysis, the mechanism of photosynthesis and so forth.

There is no less foundation for the negative answer to this question and therefore only a study of the real situation can provide a sufficient basis for forming any conclusion.

Of course, the synthesis of proteins in the complicated form in which we observe it today naturally did not develop all at once. In the earliest stages of the development of life, which are now concealed from us, nature and natural selection tried out many different variants of this synthesis. There is no need for us to deal here with all the theoretical and experimental aspects of the problem of the origin of protein synthesis, especially as this has already been done in several monographs and in the proceedings of the international symposium devoted to the problem of the origin of life which took place in Moscow in 1957. We need only point out that the requirements imposed by evolution on the accuracy and speed of protein synthesis in the lower forms of life, such as microorganisms, are no less but rather even more than in higher organisms, as the rate of growth and multiplication is at its greatest in lower organisms. It is obvious that the main form of selection in the earliest stages of the establishment of life was selection for rate of growth and ability to retain the "accomplishments" which had been attained. In this respect a system of protein synthesis which ensured extremely rapid growth of the microorganisms could fully satisfy, even to excess, all the requirements of the later stages of the development of life.

However, the development of an extraordinary variety of forms and functions in the process of evolution altered the nature of certain stages of protein synthesis in many cases, adapting it to the different physical and chemical conditions in which the synthesis had to take place and to the occurrence of differences between specialized proteins. It cannot be imagined that precisely the same means of synthesis with the participation of the systems of S-RNA and of matrices would occur under conditions which differ as widely from those ordinarily prevailing in cytoplasm as do, for example, those of acute dehydration during the ripening of seeds, or, on the other hand, those obtaining in the blood plasma (synthesis of fibrin), where the very system of the two forms of RNA is absent. In all these cases, however, the reproduction of the specificity of the sequence of amino acids in the original peptide and polypeptide components is undoubtedly carried out on matrices. *There would seem to be a single mechanism for determination of the sequence of amino acid residues on the peptide and polypeptide chains with obligatory participation of nucleic acids. There may, however, be many mechanisms by means of which nature creates complicated protein molecules consisting of polypeptide blocks or even of aggregates of molecules having different structures.*

# PART III

# PROBLEMS OF REPRODUCTION OF THE SPECIFICITY OF PROTEINS AND NUCLEIC ACIDS

# INTRODUCTION

The specific sequence of amino acid residues in the polypeptide structures of proteins is undoubtedly their essential property; it is the "mirror" which reflects both the inter-species differences between homologous proteins and the differences between the catalytic properties of the various enzymes and the structural and architectural features of their molecules and, finally, their morphogenetic growth and pathological alteration, directing ontogenesis on the one hand while, on the other, leading to its orderly cessation. Protein bodies are chemical compounds and in their transformations they are subject to the laws of physics and chemistry. However, the peculiar specificity of the sequence of the fragments of which they are composed is mainly associated with the manifestation of their biological properties and is a typical biological characteristic, the purposiveness of which, in individual protein bodies, is only revealed in the processes of biological metabolism and in the complicated higher forms of manifestation of biological activity by organisms.

It is quite natural that the reproduction of these complicated and biologically purposive forms of the chemical specificity of the structure of proteins should require the setting up, during phylogenesis, of special very complicated biochemical systems which are literally capable of collecting, storing, reproducing and transferring the fantastic amounts of biological information which determine all stages of the synthesis of specific proteins at all stages of ontogenesis. The evolutionary development of this sytem provided the impetus for further rapid development of more and more complicated forms of living nature and formed the material basis for the most important feature of living things—the property of heredity.

The nucleic acids, RNA and DNA, are the main components of this system and the study of their properties, structure, specificity, autosynthesis and role in the reproduction of the specificity of proteins and in the transfer of hereditary characteristics is one of the most brilliant, interesting and significant achievements of biological chemistry, although many of the questions involved in this problem are still far from being solved.

The study of these problems has now become a focal point of biology. Although a large number of major discoveries have been made in this field during the last ten years and there has been a "great leap forward" in comparison with the previous period, yet this has only shown us the colossal extent of the field yet to be explored and revealed many new and still more complicated problems.

254

# THE BIOCHEMICAL FUNCTIONS OF THE CELL NUCLEUS IN CONTROLLING THE REPRODUCTION OF THE SPECIFICITY OF PROTEINS AND NUCLEIC ACIDS

## Introduction

The cell nucleus is an organ of the cell which, for many decades, has engrossed the attention, not only of cytologists and geneticists, but also of biochemists. The synthesis of proteins in cells is very highly differentiated and, although the main share of the direct synthetic work falls on the ribosomes, many other conditions and requisites for this synthesis (the provision of sources of energy, the activation and selective acceptance of amino acids) occur in other parts of the cell, in the mitochondria and cell sap. According to current theories the nucleus has quite a special part to play in this process, providing the matrices and possibly also adaptors of protein synthesis and there is a suggestion that the ribosomes may be formed in the nucleus (17a, 33a, 47, 93). The nucleus is, as it were, the storehouse of the main pool of hereditary information which is recoded in the form of the grouping of nucleotides in desoxyribonucleic acids and which controls the biological syntheses in the cytoplasm according to the scheme: gene → matrix → protein. The nucleus and chromosomes are thus functionally specialized for processes of reproduction of the specificity of biological structures and therefore a survey of the problem of the reproduction of the specificity of proteins and nucleic acids should form the starting point of an analysis of the biochemical activity of the nucleus.

In this chapter we shall deal primarily with the factual material and hypotheses regarding the special biochemical functions of the nucleus as a whole in the synthesis of proteins and nucleic acids, as well as the interaction between the nucleus and the cytoplasm. In later chapters we shall undertake the analysis of the facts, theoretical schemes, hypotheses and ideas relative to the relationships at a molecular level between DNA, RNA and proteins.

## 1. Synopsis of the structure of the nucleus, nucleolus and chromosomes

It is difficult to consider the biochemistry of the nucleus apart from the facts concerning the fine structure of this most important and complicated organelle of the cell.  We shall therefore also consider here certain evidence as to the structure of the nucleus, although of course this will, of necessity, be done very briefly and generally.

The nucleus of the resting cell has the form of a spherical, homogeneous body surrounded by a membrane and containing a denser nucleolus.  After fixation and staining fine chromatin threads become visible in the form of a fine network extending through the nuclear sap.

The literature contains hundreds or even, more probably, thousands of works devoted to the study of the morphology and fine structure of the resting nucleus and chromosomes, which elucidate various details of their structure depending on the type of object, type of tissue or organ, phase of ontogenesis, phase of mitosis, crossing, mutation, polyploidy and many other factors and processes.  The monograph by E. B. Wilson, *The cell in development and heredity*, gives an exhaustive review of all the older work in this field.  The later literature on the structure of the nucleus has been surveyed in the monographs of Brachet (50) and Kedrovskiĭ (22) and also in a number of recent reviews specially devoted to the structure of the nucleus, nucleolus and chromosomes (24, 154, 158, 159, 168a, b, 100, 101).  However, notwithstanding the extensiveness of the literature, there are still many questions as to the structure of the nucleus which are unsolved and which are undergoing intensive study.

Very interesting electron-microscopic observations in this field have been published recently by Georgiev & Chentsov (16) who studied the structure of isolated nuclei from the livers of rats.  On the basis of an analysis of the electron-microscopic results which they had obtained themselves and of results reported in the literature, they put forward the following hypothetical scheme for the resting nucleus (Fig. 35).  Georgiev & Chentsov explain this scheme in terms of the following concepts:

1. The main structural elements of the nucleus are (*a*) the nuclear membrane, (*b*) threadlike molecules of desoxyribonucleoprotein and (*c*) nucleonemes, which are threadlike formations with ribonucleoprotein particles (nuclear ribosomes) on them.

2. The nucleonemes seem to have an axial structure of chromatin and are attached to the nuclear membrane.  Clumps of nucleonemes form the nucleolus.

3. The threads of desoxyribonucleoprotein are attached to the nucleonemes to form a complex of chromatin material.  The threads of nucleoprotein which are attached to the nucleonemes of the nucleolus and which lie on its surface constitute the perinucleolar chromatin.

4. The space between the filamentous structures of the nucleus (the threads of nucleoprotein and the nucleonemes) is filled up with particles (ribosomes) and the soluble proteins of the nuclear sap.  According to the recent evidence of Georgiev (11), the nucleolus does not contain DNA,

* (3rd edn.).  New York: Macmillan, 1928.

only RNA and proteins.  Other cytochemists, however, have observed the presence of DNA also in the nucleolus (118a, 148a).

The membrane of the nucleus is a structure which has been studied in detail by many authors.  In the most varied organisms it usually consists of two parallel membranes at a distance of 200-300 Å from one another

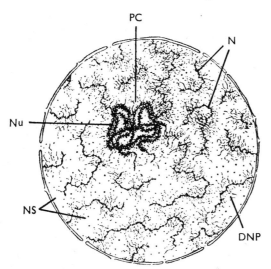

FIG. 35.  Diagram of cell nucleus in interphase.
N = Nucleonemes; DNP = Reticulum of desoxyribonucleoprotein; PC = Perinucleolar chromatin; NS = Nuclear sap; Nu = Nucleolus.

and having many pores with diameters of 400-1000 Å.  These pores seem to be of considerable importance for exchanges between the nucleus and the cytoplasm (168b).

In recent years great interest has been taken in the study of the fine structure of the chromosomes, as the experimental and theoretical material of contemporary genetics indicates that there must be an orderly structure of the molecular loci of hereditary information.

According to the calculations of Freese (75) one chromosome of *Drosophila* contains about 5000 molecules of DNA, each of which has a molecular weight of the order of $10^7$.  There are many reasons to believe that the combination of all these 5000 molecules with one another is extremely specific and that the sequence of this combination must be reproduced at each division of the chromosome during mitosis.  It is therefore necessary, in elucidating the mechanism whereby the molecules of DNA are bound to one another, to reckon also with the need for the reproduction of this structure during mitosis.  The study of this question is still only based on collateral evidence and therefore schemes of the molecular structure of the chromatin threads are of a hypothetical nature.  A careful analysis of the successes and deficiencies of various schemes of the molecular structure of the chromosomes and chromatin threads has been made recently in a number of theoretical and experimental papers (23, 75, 66, 100, 154). We shall not here go into the evaluation of the evidence for or against any given scheme, as the absence of direct proof for any of them makes final selection premature. Until recently there were two main schemes, one of which postulated the attachment of the molecules of DNA to an axial protein structure from which they radiate like the needles on the twig of a spruce tree (Fig. 36.1), while the other postulated the linear attachment of the molecules of DNA to one another through a protein layer with subsequent intensive folding of this long thread (Fig. 36.2). The results of recent studies of both the electron-

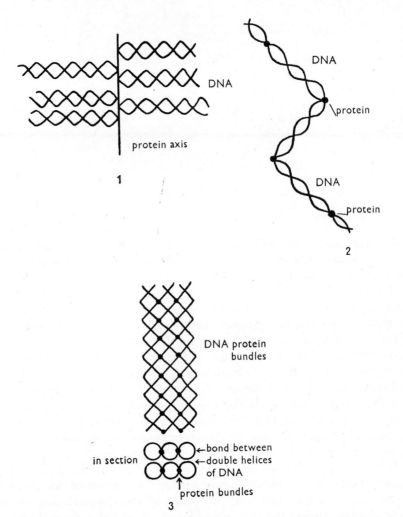

FIG. 36. Three hypothetical arrangements of molecules of DNA in chromosomes.

microscopic structure of the nucleus (100) and the mechanism of replication of DNA under the conditions obtaining in the cell (60) have largely confirmed a third possible variant of the structure of chromosomes in the form of a composite structure, composed of sub-units in which the molecules of DNA are bound together in parallel to form complex ribbon-like structures (Fig. 36.3).

## 2. Characteristics of intranuclear protein syntheses

Although the assortment of enzymes in the nucleus is not so varied as that in the cytoplasm, the nucleus still contains many proteins which are of structural importance and all these proteins have to be reproduced, especially during rapid growth.

In the study of the biochemistry of the nucleus most attention is generally paid to DNA although, quantitatively, proteins constitute the largest fraction of the nuclear material, accounting for 70-80% of the dry matter of the nucleus (65).   A considerable part of these proteins is specific in composition and is not found in the cytoplasm, which suggests that they are probably synthesized within the nucleus.   The nucleus contains, in addition to the typically nuclear basic proteins (the protamines and histones), many other acidic proteins which have been studied by many authors (2, 4, 5, 17, 18, 19, 12, 132).   A detailed review of those studies of intranuclear protein synthesis which had been published up to 1958 is to be found in the monographs of Kedrovskiĭ (22) and Brachet (50) and in a number of reviews (34, 35, 38, 25).   Synthetic processes in the nucleus have recently been the subject of a special symposium (*J. Histochem. Cytochem.* **10**, No. 2, p. 137 (1962)).

*a) The activation of amino acids in the nuclei of cells*

The connection of protein synthesis with ATP and energy metabolism has been demonstrated in many researches (cf. reviews 33, 40).

Hopkins (94-95), Allfrey (33) and Gvozdev & Khesin (9, 10) have all recently reported the finding, within the nucleus, of activating enzymes which bring about the activation of amino acids in the same way in which it has been described as occurring in cytoplasm (cf. Chapter III).   Hopkins, Mirsky & Allfrey (33) isolated these enzymes from the nuclei of the cells of various tissues and partly purified them.   According to Khesin (31a), no differences could be found between the enzymes of the nucleus and of the cytoplasm responsible for the activation of amino acids (formation of aminoacyl adenylates).

*b) Acceptance and transfer of amino acids in nuclei by
nuclear low-molecular S-RNA*

The first evidence for the existence of low-molecular S-RNA in nuclei was obtained in the interesting research of Hopkins (94) who actually found, in isolated nuclei, a system for the activation of amino acids as well as soluble adaptor RNA carrying out the functions of transfer of amino acids.

In later experiments carried out by Hopkins, Allfrey & Mirsky (95, 33a) it was shown that the combination of the aminoacyladenylate with the nuclear S-RNA is brought about at the terminal adenosine group by the nuclear "pH 5 enzyme", which means that this RNA resembles cytoplasmic RNA in the way it reacts. There is, however, a functional difference between nuclear RNA and cytoplasmic RNA. This was established by the interesting work of Webster (166) who studied the interaction between the S-RNA and the activating systems isolated separately from the nucleus and cytoplasm and found definite differences between the nuclear and cytoplasmic systems for the activation of amino acids in a number of objects. Webster's results are given in Table 11.

TABLE 11

Specificity of nuclear S-RNA

| S-RNA isolated from | Moles of [$^{14}$C]alanine combined with S-RNA under the influence of | |
|---|---|---|
| | Nuclear activating enzymes | Cytoplasmic activating enzymes |
| Cytoplasm of pig liver, 2 mg. | 1·0 | 7·4 |
| Nuclei of pig liver, 2 mg. | 10·2 | 1·8 |
| Cytoplasm + nuclei, (2 mg. + 2 mg.) | 11·0 | 8·5 |
| Cytoplasm of muscles, 2 mg. | 1·8 | 8·4 |
| Cytoplasm of calf liver, 2 mg. | 1·0 | 0·8 |
| Yeast, 2 mg. | 0·2 | 0·6 |
| Pea seeds, 2 mg. | 0·4 | 0·8 |

The work of Webster is only a preliminary study but, if it is confirmed, it will enable us to draw certain conclusions as to the organelle specificity of both S-RNA and the system of activating enzymes. Definite differences in the ways in which proteins are synthesized in the nucleus and in the cytoplasm were also found by Breitman & Webster (57) in another research. These authors noticed that, in many organisms, cytoplasmic protein synthesis was augmented by potassium ions while sodium ions had no appreciable effect on it or even, sometimes, inhibited it. The incorporation of amino acids into proteins in isolated nuclei, on the contrary, requires the presence of sodium ions while potassium ions are inactive in this case. On studying this phenomenon in more detail the authors found that sodium was not only necessary for the synthesis of proteins but also for the biochemical utilization of amino acids in other ways, for instance, in the metabolism of purines and pyrimidines. They suggested that the effect of sodium ions in this case was connected with their role in the transport of amino acids through the nuclear membrane. The significance of sodium in the transport of amino acids from the cytoplasm to the nucleus was demonstrated by the work of Allfrey & Mirsky (35-40). From these results it seems that we may consider that the nuclei also have their autonomous system for the accumulation and regulation of the intake of the starting products of protein synthesis. However, in recent work of Khesin, Gvozdev & Astaurova (31b), no differences were found between the properties of the low-molecular tyrosine-accepting RNA of the nucleus and of the cytoplasm.* The presence in the nuclei of cells of both high-molecular and low-molecular RNA has been established by the work of Zbarskiĭ, Mant'eva & Georgiev (13, 15, 79), Ficq (71) and others. The low-molecular RNA shows intensive exchange of phosphate residues and seems in every way similar to S-RNA.

c) *The incorporation of amino acids into the proteins of nuclear ribosomes, chromosomes and the nucleolus*

The study of the overall intranuclear synthesis of proteins has been pursued by many investigators (1, 8, 20, 21, 28, 137, 177-179) but at the present moment these studies are very divergent and pay attention to the peculiarities of different intranuclear structures. It has recently been established that the nuclear sap contains ribonucleoprotein particles of the same sort as microsomes or ribosomes (28, 29, 17a, 77, 13, 76, 167, 168). Samarina & Georgiev (29) found microsomes of two types in the nuclei, sedimenting at 56,000 g and 100,000 g. These ribonucleoprotein particles are the active centres of protein synthesis. The incorporation of amino acids into the proteins of these particles required the usual cofactors, in particular GTP (76). However, Wang (167) found that the requirement for GTP for this synthesis is not so absolute as in the cytoplasmic particles, while cytidine triphosphate (CTP) is necessary for protein syntheses by the nucleus; these syntheses are less strongly inhibited by chloramphenicol than are the cytoplasmic ones.

The chromosomes can also synthesize proteins locally. Some very exciting facts concerning the localization of protein synthesis in the chromosomes of the salivary glands of *Drosophila* have recently been obtained by Sirlin & Knight (150) by means of microradioautography. It was demonstrated quite clearly that the incorporation of [35S]methionine and [3H]leucine into chromosomes does not occur along the whole length

---

* Later work from that laboratory (10a) supports this conclusion. It was shown that incubation of isolated nuclei with [14C]aminoacyl-S-RNA results in transfer of amino acids to nuclear proteins. This is revealed as radioactivity not only in the proteins of the nuclear sap but also in chromosome proteins. Histones, however, did not incorporate the labelled amino acids, and the authors suggest that they are synthesized in a special way.

of the chromosome but locally in particular parts. The possible biological significance of this non-uniformity of the synthetic activity of the chromosomes will be discussed later in connection with the question of the genetic control of the processes of morphogenesis. Workers in this laboratory (151) have also used microautoradiography to establish incorporation of labelled amino acids into the proteins of the nucleolus.*

*d) The role of DNA in the synthesis of the nuclear proteins*

Some years ago Allfrey & Mirsky (35, 38) showed in a series of papers that the synthesis of proteins in isolated nuclei *in vitro* was halted by desoxyribonuclease and restored by the addition of DNA. This led to the development of the idea that DNA plays a direct part in protein synthesis. However, it was shown later (27, 37, 6) that the effect of DNA in increasing the incorporation of amino acids into proteins was indirect and non-specific. It appeared that the isolated nuclei which had been treated with desoxyribonuclease regain their lost ability to incorporate amino acids into proteins, not only on the addition of DNA from other sources, but also on the addition of RNA, synthetic polynucleotides and other polyanionic molecules (polyethylene sulphate, heparin, etc.). However, polycationic molecules (such as protamine and polylysine) had no effect on such nuclei. These results led to the conclusion that the role of DNA in this process is non-specific and is only due to the action of the negative charges on its molecules. This is quite reasonable. The synthesis of the nuclear proteins, like that of all the other proteins in the cell, is connected with RNA which is localized in the nucleus, and the discovery of the specific functions of the nuclear DNA in cellular metabolism demands other methods and other approaches. It is also not impossible that the stimulating effect of DNA in this case is not only associated with the charge but also with the peculiar magnetic properties of DNA. Blyumenfel'd (6), who suggested this possibility, found by the method of paramagnetic resonance that not only DNA, but also its "substitute" heparin, has a powerful cloud of unpaired electrons.

The structure of DNA is hardly such as would enable it to act as a matrix for protein synthesis and there is therefore no basis for some of the suggestions as to the direct participation of DNA in protein synthesis.

Naturally, the proof that DNA does not play a direct and immediate part in the synthesis of protein does not in any way detract from the significance of the nucleus in the phenomena of heredity, nor can it provide grounds for the sceptical opinions in this matter of certain geneticists and biochemists who try to minimize, as much as they can, the central role of

* The chromosomes are now generally considered to be the source of "messenger" RNA (see Chapters XIV and XV). At the same time the nucleolus is very probably a source for ribosomes and for the synthesis of ribosomal RNA (46b, 59a).

DNA as the storehouse of genetical and biochemical information.  The specialized system of DNA is such a complicated, delicate and precise mechanism that attempts to study extracts of total DNA as co-factors in particular systems of protein synthesis can only be expected to have such an effect as might be obtained by loading up a smelting furnace with the components of an electronic programming machine and then studying its effect on the process of smelting iron.

## 3.  The synthesis of RNA in nuclei and cytoplasm and the migration of RNA from the nucleus to the cytoplasm

The nucleus, and especially the nucleolus, is a place where RNA is synthesized with special intensity.  Part of this RNA then migrates out of the nucleus into the cytoplasm.  A fairly full review of the material supporting this statement is given in the monographs of Kedrovskiĭ (22), Brachet (50), Khesin (31) and Chantrenne (61) to which we have often referred before.  We shall, therefore, confine ourselves once more to a brief review of the main facts of the history of these investigations and concentrate the attention of the reader on the work which is of the greatest theoretical importance and on those most recent investigations which have contributed significantly to our knowledge in this field.

According to current theoretical ideas it is assumed with a high measure of probability that the synthesis of RNA in the nuclei and its transfer to the cytoplasm, where it performs its functions of matrix and adaptor, is the physical basis for the transfer of information from DNA to proteins, the RNA acting as the "instrument" for the translation of this information into a particular sequence of amino acids.  It is therefore obvious that all the facts relevant to this cycle require the most careful evaluation and theoretical analysis.

In our discussion of the synthesis of RNA in the nuclei we cannot omit to deal at the same time with the facts concerning the synthesis of RNA in the cytoplasm, as many of the experiments in this series are, in fact, devoted to comparisons between the nucleus and the cytoplasm in this respect.

### a) Experiments on the relative rates of synthesis of RNA in the chromosomes, nucleolus and cytoplasm of the cell

The first intimation of the extremely high specific activity of nuclear RNA when [32]P becomes available to the cell was given by Marshak (107) and this effect has since been studied by many authors.  The fact that RNA is synthesized most quickly in the nucleus and nucleolus has been further confirmed by recent work (15, 14, 7, 39, 43a, 74, 91, 96, 99, 92, 136, 144, 148a, 149, 159, 152, 123, 123a, 133).

Particularly striking evidence for the intranuclear localization of RNA synthesis has been obtained by microradioautography of cells of various kinds at different intervals after the administration of labelled precursors of RNA to the cells. In presenting these data we shall not follow the chronological sequence but shall concern ourselves mainly with the recent series of researches by Zalokar (177, 179) who succeeded in working out a very effective method for studying intracellular syntheses.

The main experimental material used by Zalokar was the multi-nucleate mycelium of the mould *Neurospora crassa*. The hyphae of this mould usually grow radially in straight lines and at the tip of each there is a growing point. Zalokar used this feature of the growth of the hyphae of *Neurospora* and the stability of the cell wall in order to carry out an

centrifugal force
⟶

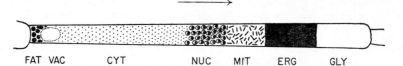

FAT  VAC          CYT               NUC    MIT     ERG      GLY

FIG. 37. Centrifugation of a hypha of *Neurospora* (diagrammatic).
FAT = Fat globules; VAC = Vacuoles; CYT = Cytoplasm; NUC = Nuclei; MIT = Mitochondria; ERG = Ergastoplasm; GLY = Glycogen.

original form of *in vivo* centrifugation of the mould which grew outwards from the centre of a special disc which was then spun at the speed of an ultracentrifuge. The hyphae thus served as microcentrifuge tubes within which the separation of the cellular organelles took place. Fig. 37 gives a diagram of the *in vivo* separation of the organelles into granules, mito-chondria, nucleus and hyaloplasm, which is somewhat different from the sequence usually obtained by sedimenting them in sucrose solutions. After separation in this way the intracellular fractions remained viable and active and rapidly incorporated any precursors of proteins or nucleic acids which were introduced into the incubation medium. If [³H]leucine was introduced into the medium and the preparations were then radio-autographed, it was found that even during the first seconds of the experi-ment the labels were incorporated mainly into the microsome fraction. At the same time the substrate for the synthesis of RNA, labelled uridine, was only incorporated into the fraction containing the nuclei during the first 16 min. and it was only after this, during longer exposures, that it could be detected in the other segments of the hyphae.

In another research the same author (178) demonstrated radioauto-graphically that RNA was primarily formed in the nuclei of the ovarian follicles of *Drosophila*.

As to the localization of RNA synthesis within the nucleus, a number of authors ascribe it to the nucleolus, in which the incorporation of labels into RNA may be observed to occur more quickly than elsewhere. How-ever, facts found in a number of works suggest that RNA is synthesized

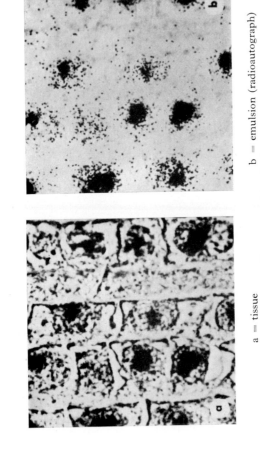

a = tissue          b = emulsion (radioautograph)

Plate exposed for 60 days

FIG. 38. Phase-contrast photomicrograph of a section of root treated with 30 $\mu$C/ml. [³H]cytidine for 30 min. RNA and DNA were not separated.

quickly in the chromatin structures of the nucleus. The work of Goldstein & Micou (82) is specially interesting in this connection. Using short intervals after the contact of the cells with the labelled precursors of RNA, the authors found that the incorporation of labels into RNA occurred more quickly in the chromosomes than elsewhere. They therefore suggest that the primary synthesis of the RNA may occur in the chromosomes, after which it is transported to the nucleolus.

Very interesting and detailed experiments which give a striking demonstration of the important role of the nucleus in the synthesis of RNA have been carried out by Woods (171) using the meristem of the roots of the bean *Vicia faba*. The meristematic cells of the rootlets of this plant are large, which facilitates the radioautographic demonstration of the localization of the incorporation of labelled compounds. The mean diameter of these cells is about 20$\mu$ and that of the nucleus about 10$\mu$ and of the nucleolus about 5$\mu$. Usually about 10-15% of these cells are in various stages of mitosis and this makes it possible to observe the process of incorporation of isotopes in different stages of the mitotic cycle of the cell. [³H]Cytidine was used in most of the experiments. One hour after the exposure of the cells to solutions containing [³H]cytidine the label was found to be incorporated mainly in the nucleolus and it was only later that labelled RNA appeared in the cytoplasm. Fig. 38 shows a typical radioautographic picture of such incorporation while Fig. 39 shows a diagram illustrating the relative rates of incorporation of [³H]cytidine into the RNA of the nucleolus and cytoplasm. In this case the [³H]cytidine was only in contact with the cells for one hour, after which the rootlets were transferred to a solution containing an excess of unlabelled cytidine. When this was done the concentration of the isotopes in the nucleoli rapidly reached a maximum and then fell, which obviously suggests the transport of RNA to the cytoplasm. However, an important feature of this work is that the author did not confine himself to work on the concentration of labels in the nucleus and in the cytoplasm but calculated the total amounts of the labelled compounds in these fractions on the basis of the volumes of the nucleolus and cytoplasm, based on a calculation of the blackening of the grains of the emulsion in each direction. The results are set out in Fig. 40.

The results of these calculations showed that a considerably larger amount of labelled RNA had accumulated in the cytoplasm (which has a volume of 32 times that of the nucleolus) than the amount contained in the nucleolus when the rootlets were transferred to the non-radioactive medium. After the cells had been treated with [³H]cytidine for 1 hour, 20 units of tritium were fixed in the nucleolus (calculated from the blackening of the grains of the emulsion). At this moment about 20 units were also fixed in the cytoplasm, but these were more diffusely arranged. At the end of the experiment, when the rootlets had been taken out of the radioactive solution and kept for 7 hours in a solution containing unlabelled cytidine, it was calculated that the cytoplasm contained about 200 units of tritium while the nucleolus had lost practically all its labelled material. If we assume that the incorporation of [³H]cytidine into the nucleolus continued for some time after the transfer of the cell to the fresh solution, even so only 40 units of the tritium in the cytoplasm could have come

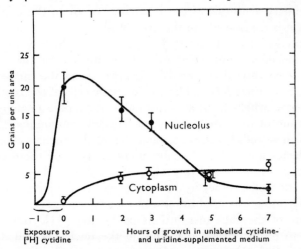

Fig. 39. Relative concentration of tritium in the RNA of the nucleolus and cytoplasm.

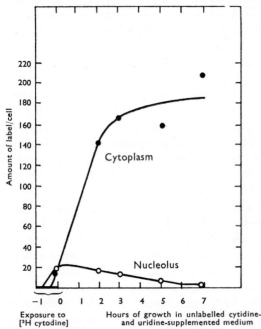

Fig. 40. Total amount of tritium (arbitrary units) in the RNA of the nucleolus and cytoplasm.

from the nucleus. 80% of the incorporation of labels into the RNA of the cytoplasm must, in this case, have occurred locally. Woods suggests that this incorporation may be accounted for by the metabolism of low-molecular S-RNA which, as we have already pointed out, goes on very actively in the cytoplasm.

In other experiments Woods measured the incorporation of [³H]cytidine into the RNA of the meristematic cells over shorter periods of time (1-14 min.). He found that the label appeared very quickly in the chromosomal material at a time when the nucleolus still showed only a very weak activity and the cytoplasm was almost inactive. The author summarizes his observations as follows. The most rapid incorporation of [³H]cytidine is found in the chromatin of the nucleus. The concentration of label in the chromatin reaches a state of equilibrium 14 minutes after the beginning of the experiment. The concentration of label in the nucleolus continues to increase for more than 90 minutes and becomes considerably greater than that in the chromatin. The cytoplasm only becomes labelled very slowly during the first hour of the experiment but, if it is then measured after seven hours, the radioactivity is found to have increased markedly.

In discussing these results the author puts forward the view that rapid synthesis of RNA takes place in the chromosomal parts of the nucleus and that it is then transferred to the nucleolus where it is stored for about an hour before "transfer" to the cytoplasm. He also thinks that part of the RNA may be synthesized de novo in the nucleolus in addition to the RNA which reaches it from the chromatin. The results of Vincent (165), who found two types of RNA in the nucleoli of oocytes, agree with this hypothesis.

Woods also considers that the pool of RNA in the cytoplasm includes some which is synthesized there de novo as well as the transport RNA from the nucleus. This author shows convincingly that the rate of incorporation of [³H]cytidine and other labelled precursors into RNA depends, not only on the rate of synthesis of the polymer, but also on the dilution of the labelled component with unlabelled material, i.e. on the size of the "pool" of unlabelled precursors in each fraction.

Some work by Woods, which we shall discuss later, formed the subject of a discussion at a symposium on the structure and function of the genetic elements, and in discussing this work Swift (159a) brought out a number of facts which show that the rate of incorporation of labelled precursors into nuclear RNA depends on the object being studied. In some cases it is the nucleolus which becomes radioactive first, in others it is the chromosomal structure.

We must also bear in mind the results of Woodard (169) who showed that though the pollen of Tradescantia contains no appreciable amount of nuclear RNA one may nevertheless observe a rapid increase in its cytoplasmic RNA.

Very clearly defined quantitative results indicating that RNA is synthesized *de novo* in the cytoplasm as well as the nucleus have been obtained recently by Harris (91), who used microradioautography to study this aspect of the cells of the connective tissues of the heart of the rat and the macrophages of the rabbit grown *in vitro*. In these cells the rate of renewal of the RNA was roughly the same in the nucleus and in the cytoplasm. Only a small part of the nuclear RNA passed into the cytoplasm in a stable form. A considerable proportion of the nucleotide products which migrated from the nucleus into cytoplasm were acid-soluble. The RNA in the nucleoli of the cells of connective tissue was no more active metabolically than that in the rest of the mass of the nucleus. Preliminary results were also obtained concerning whether the nucleotide composition of the nucleolar RNA of these cells is different in some respects from the RNA of the rest of the nucleus. In a recent communication to the Fifth International Congress of Biochemistry Harris (91a) brought forward much new evidence indicating that the labile nuclear RNA is synthesized in the nucleus while the stable, cytoplasmic RNA is synthesized in the cytoplasm.

Some interesting facts concerning the differences between nuclear and cytoplasmic RNA have been discovered recently by the work of Sacks & Kamarth (141) who have used an interesting method for the study of this subject. When [$^{32}$P]phosphate was injected into rabbits the radioactivity of the nuclei of the liver cells was considerably higher after two hours than the specific radioactivity of the RNA in the mitochondria, microsomes and the soluble fraction of the cytoplasm. The distribution of the $^{32}$P in the various parts and fragments of the RNA was, however, very irregular. When the RNA was partially hydrolysed to oligonucleotides, the individual components of the RNA of each fraction showed a wide range in their specific radioactivity. The products of hydrolysis of the RNA of the microsomes and of the soluble RNA of the cytoplasm included fragments with a considerably greater specific radioactivity than any fragments of RNA from the nuclei. The authors consider that this indicates the independent formation of RNA by the cytoplasmic fractions, though it does not rule out the possibility of partial utilization of RNA formed in the nucleus for this purpose.

On the other hand, in the recent work of Amano & Leblond (41), comparing the specific radioactivity of the RNA of the chromatin, the nucleolus and the cytoplasm of liver cells 1, 3 and 8 hours after the introduction of labelled precursors, the nature of the curves obtained for the incorporation of labels pointed clearly to the primary synthesis of RNA in the chromatin.

Objective analysis of the evidence as to the relationship between the RNA of the different fractions of the cell, based on the rates of incorporation of isotopes, is, however, made more difficult by the fact that in both the nucleus and the cytoplasm the total RNA is heterogeneous and the various fractions differ in their rates of metabolism. The metabolic heterogeneity of the RNA of both nucleus and cytoplasm has been demonstrated in a number of researches (13, 15, 36, 42, 79, 70, 89, 102, 106, 123a, 148, 173, 174) and this fact makes it necessary to use a wider range of methods and experimental schemes for assessing the values of the evidence concerning the localization of RNA synthesis. The setting up of experiments which only take account of the "rate of incorporation" of precursors into different fractions of RNA can only provide the basis for hypotheses, not for absolute certainties. It will show up the characteristic features of structures but will not give an unequivocal demonstration of their relationships. The recent work of McMaster-Kaye (113) is typical in this respect. He made a very careful study of the natures of the curves for the synthesis and breakdown of RNA in chromatin, in the nucleolus and in the cytoplasm of the cells of the salivary glands of *Drosophila*. According to his calculations the rapid incorporation of labelled precursors in the RNA of the nucleolus is due to the rapid metabolism of the nucleolar RNA (synthesis and breakdown) more than to the transport of newly synthesized molecules out of the nucleolus. The author is more inclined to the idea of independent RNA metabolism occurring at different rates in the chromosomes, the nucleolus and the cytoplasm than to the widely held scheme of transfer of RNA synthesized in the chromosomes from them to the nucleolus and thence to the cytoplasm.

In this connection, great interest attaches to recent results of Rho & Bonner (138a), who studied the localization of RNA synthesis in nuclei isolated from pea sprouts. The curves of incorporation of [$^3$H]cytidine into the chromatin and nucleolar fractions led them to conclude that RNA is first synthesized in the chromatin. This fraction of the RNA is originally closely associated with the chromatin. The incorporation of [$^3$H]cytidine into the RNA of the nucleolus exhibited a lag phase, which the authors interpreted as due to transfer thither of RNA formed in contact with DNA in the chromatin part of the nucleus. In another work from the same laboratory (46a), it was shown that nuclear protein synthesis (incorporation of [$^3$H]leucine) proceeds especially actively in the nucleolus.

*b)  Studies of the processes of transfer of RNA and proteins
from the nucleus to the cytoplasm*

The evidence which we have already discussed indicates that the synthesis of RNA occurs, in many cases, in both the nucleus and cytoplasm, but the intensity of the synthesis is usually greater in the nucleus. The rapid local synthesis of any particular polymer is usually compensated for by its equally rapid breakdown or by migration to other organelles or by secretion by the cell. In the case of nuclear synthesis of RNA there is much evidence that the RNA which has been synthesized leaves the nucleus for the cytoplasm.

This was shown specially clearly in the work of Goldstein & Plaut (85). By transplanting a nucleus labelled with RNA from one amoeba into the cytoplasm of another with a micromanipulator they demonstrated by radioautography the admixture of labelled RNA from the nucleus with that of the cytoplasm. In later researches from the same laboratory (128, 129), however, it has been shown that even non-nucleated fragments of amoebae retain the power to incorporate labelled adenine into their RNA, though only very weakly, and therefore they retain the power to synthesize RNA independently of the nucleus. From these observations the authors reckon that only part of the cytoplasmic RNA is of nuclear origin. The transfer of RNA from the nucleus and nucleolus to the cytoplasm has been found recently in several laboratories (74, 134, 145, 172, 176). The transfer of nuclear RNA to the cytoplasm is found to be specially pronounced during mitosis and the disintegration of the nucleus.

A very interesting method of ejecting nuclear RNA into the cytoplasm has been observed recently by Raïkov (26). In studying the cytology of the nuclear apparatus of infusoria he found that special structures were formed periodically in the nucleoli of the macronucleus. These the author called RNA-spherules, as they were made of ribonucleic acid. Raïkov frequently observed the process of ejection of these spherules from macronuclei into the cytoplasm after they had attained a certain size. These RNA spherules go into the cytoplasm through a special opening which is formed in the enveloping membrane of the nucleus. They gradually disintegrate in the cytoplasm. The ejection of RNA-spherules into the cytoplasm occurs shortly before cell division.

The question of the state of the RNA which is transferred from the nucleus to the cytoplasm was the subject of detailed investigation by Goldstein & Micou (84). They studied the process of migration of RNA from the nucleus to the cytoplasm using the method of radioautography on the cells of human foetal membranes grown in tissue culture and exposed to [³H]cytidine. In cells of this type the autonomous cytoplasmic synthesis of RNA is very weak and, according to the evidence of these authors, most of the RNA of the cytoplasm is of nuclear origin. Using ribonuclease treatment of the cells the authors found that a considerable part of the transfer of RNA from the nucleus into the cytoplasm occurred in the form of polynucleotides of high molecular weight.

According to Goldstein (80, 81) proteins can also migrate from the nucleus to the cytoplasm. This phenomenon was demonstrated by the use of the same technique of transplantation of labelled nuclei. Thus Goldstein discovered an interesting phenomenon, namely that, when a new labelled nucleus is transplanted into a complete cell which already has a nucleus, [³⁵S]proteins are given out into the cytoplasm and are then concentrated in the unlabelled nucleus of the "host "cell. The author suggests that these proteins are connected with the transfer of genetic material and that after transferring RNA, for example, to the cytoplasm they are then "returned" again to the nucleus.

c) *The biochemical consequences of the removal of the nucleus
for the processes carried out in the cytoplasm*

The question of the part played by the nucleus in the regulation of
the cytoplasmic processes is answered very clearly by experiments on the
removal of the nucleus from cells and subsequent observation of the biochemical consequences of this operation. The occurrence of active
synthesis of RNA in the nucleus and nucleolus and the migration of part
of this RNA into the cytoplasm admits of no doubt. It is, however, very
important to solve the theoretical problem as to what part of the cytoplasmic RNA is of nuclear origin and what part of the synthesis of RNA
can occur directly in the cytoplasm. Also, if the cytoplasmic RNA is a
mixture of "nuclear" and "local" RNA, then it would be interesting to
find out whether there are any functional and biochemical differences
between them.

The most original and direct way of solving these problems is by
means of experiments on non-nucleated fragments of cells. The use of
experiments designed to study the role of the nucleus by its microsurgical
removal began long ago and extensive experimental data in this field have
now accumulated. They indicate, as a rule, that non-nucleated fragments
of unicellular organisms retain, though in a weakened state, the power of
synthesizing proteins and RNA (87, 88, 110, 120, 155, 163, 164).

The studies of Brachet and his colleagues (48-50, 52, 53-55), extending
over many years, are specially interesting examples of research done on
nucleated and non-nucleated fragments of cells. For the most part they
were carried out on two objects, the cells of the giant green, unicellular
alga *Acetabularia mediterranea* and on *Amoeba proteus*. A comprehensive
review of these investigations is given in Brachet's excellent monograph
on biochemical cytology (50). In experiments with *Acetabularia* it was
found that non-nucleated fragments of this alga retained the power of
synthesis and self-renewal of proteins for a long time (10-12 days) at the
same level as cells with nuclei. A considerable increase of protein was
observed in these non-nucleated cells. The rate of this growth in non-
nucleated cells, as compared with nucleated controls, did not begin to fall
off until two weeks after the beginning of the experiment. However,
although the synthesis of proteins for purposes of growth decreased, the
synthesis for renewal (measured by the incorporation of isotopes into
proteins) remained at the same level. These results, taken together with
other evidence, show clearly that the presence of the nucleus is by no means
always obligatory if the amount of protein in the cytoplasm is to increase.
Nevertheless, the fact that non-nucleated cells of the alga did still fall
behind the controls in protein synthesis two weeks after the beginning of
the experiment gave Brachet grounds for suggesting that the nucleus
exercises some sort of control over protein synthesis in the cytoplasm in
the unicellular algae as well as in other organisms. Not only protein

synthesis but also RNA synthesis takes place in the non-nucleated cells of this alga. This fact was demonstrated clearly in many experiments by Brachet and colleagues (50, 55). The rate of this synthesis, like that of protein synthesis, did not begin to fall for several weeks after the removal of the nucleus. In this connection the suggestion was put forward that if the nucleus does exercise any control over the synthesis of RNA in the cytoplasm of *Acetabularia*, then this control is not of a direct nature.

As well as these experiments on the autotrophic marine alga *Acetabularia*, Brachet and colleagues carried out a series of studies on the removal of the nucleus from another unicellular organism, the giant amoeba *Amoeba proteus*, which is a heterotroph. These experiments had many special features, as the non-nucleated fragments of amoebae lost the power of putting out pseudopodia and engulfing their food. The comparisons between the non-nucleated and nucleated halves of the cells had, therefore, to take place during starvation. Such a situation did not arise in the experiments with the alga because its cytoplasm could photosynthesize. The experiments with amoebae showed that their cytoplasm, like that of the algae, contains all the biochemical conditions for the incorporation of amino acids into proteins, though the rate of this incorporation in non-nucleated fragments of amoebae began to fall off in comparison with that in the controls considerably earlier and more sharply than in the experiments with the alga. It was observed that the breakdown of proteins and especially of RNA occurred more quickly in non-nucleated fragments of amoeba than in the nucleated controls under the same conditions of starvation. In this case, however, results were obtained which showed that the RNA of the non-nucleated parts could incorporate adenine and [$^{32}$P]phosphate autonomously. Brachet has again confirmed this result in recent work (51). The ability of the cytoplasm of non-nucleated parts of cells to synthesize RNA autonomously has also been demonstrated by several other authors (43, 64, 68).

Clear-cut evidence for the possibility of autonomous synthesis of RNA in the cytoplasm has also been obtained in the researches of Plaut & Rustad (125, 130). In their experiments nucleated and non-nucleated halves of amoebae were incubated with the nucleotide precursors [$^{14}$C]orotic acid and [$^{14}$C]uracil. Autoradiography of the cells after various periods of incubation showed that both of these precursors were incorporated into the cytoplasmic RNA in the non-nucleated as well as the nucleated halves of the cells. The results of one such experiment are given in Fig. 41.

In a review specially devoted to an evaluation of experiments on the synthesis of RNA in nucleated and non-nucleated halves of amoebae Plaut (125) notes that, although there is no doubt that cytoplasm can synthesize RNA, the question is really more complicated. It is still not clear what functional relationship exists between the RNA's of nuclear and cytoplasmic origin.

The author notes that the passage of nuclear RNA into the cytoplasm

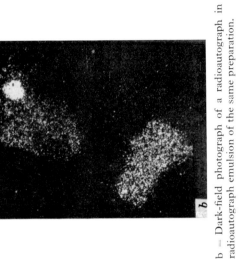

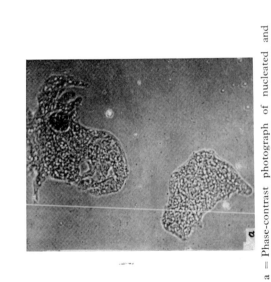

a = Phase-contrast photograph of nucleated and non-nucleated amoebae fixed after 117 hours incubation in the labelled solution.

b = Dark-field photograph of a radioautograph in radioautograph emulsion of the same preparation.

Fig. 41. Incorporation of $^{14}$[C]orotic acid into the cytoplasm of *Amoeba*. The bright granules in both cells indicate the incorporation of labels into the RNA. The nucleus of the nucleated cell shows considerably greater activity than the rest but the cytoplasm of the non-nucleated cell shows greater activity than that of the nucleated cell.

occurs mainly during the period of mitosis. Before cellular division the nucleus of the amoeba is literally packed with RNA while the daughter nuclei contain very small quantities of it. However, notwithstanding the apparent unequivocalness of the experiments with non-nucleated amoebae in demonstrating the theoretical possibility of autonomous synthesis of RNA in the cytoplasm, it would seem that they are not absolutely indisputable. This was made apparent in the recent paper by Prescott (133) who drew attention to the fact that the vacuoles of amoebae usually contain the cells of bacteria and yeasts which have been engulfed by them and that these remain alive for a certain time and are digested over the course of several days. The incorporation of labelled components into the RNA of non-nucleated amoebae may therefore represent incorporation by these bacteria. If the nucleus was removed from the amoebae after starvation for 72 hours the authors found that, in their experiments, they could not observe the incorporation of labelled adenine, uracil or orotic acid into the RNA of the non-nucleated fragments. Prescott recognizes, however, that even his own method with its 72 hours of preliminary starvation is not above suspicion, as the starvation may lead to the disappearance from the cells of some materials for RNA synthesis such as carbohydrates, nucleotides, etc. as well as substrates for the energy-producing processes. In his opinion it is necessary to carry out such experiments under sterile conditions but in the presence of the normal constituents of the diet of the cells.

Such an experiment has been done by Prescott (135) in his later work with the cells of *Acanthamoeba* grown under sterile conditions. Synthesis of RNA did not occur in non-nucleated fragments of these cells, which, the authors consider, argues in favour of the view that the nucleus is the sole source of the cytoplasmic RNA in cells of this kind. While not denying that this may be true, it should yet be noted that the categorical conclusion drawn by the author cannot be regarded as proven because the possibility cannot be excluded that the absence of RNA synthesis in the non-nucleated fragments may, with equal likelihood, indicate a disturbance of the ability of the cytoplasm to synthesize RNA autonomously in non-nucleated fragments. The extirpation of some organ may well affect the rest of the processes but it is hardly right to ascribe the functions which have been lost in this way directly to the organ which has been removed. The same conclusion is also drawn from work (83) in which a sharp inhibition of RNA synthesis was observed in a culture of human amniotic cells.

A very peculiar result was obtained while checking the original experiments on the synthesis of RNA in non-nucleated cells of *Acetabularia* (156). It appeared that, after fragments of this alga had been treated with ribonuclease, any further synthesis of RNA only occurred in cells with nuclei, while non-nucleated fragments had lost the power to synthesize RNA. As a result of this the idea grew up that, although the synthesis of RNA

may take place in the cytoplasm, yet this process must involve the participation of the fraction of RNA which had earlier passed into the cytoplasm
from the nucleus and served as the matrix material for cytoplasmic syntheses (31). We have seen that the migration of RNA from the nucleus
into the cytoplasm is a normal process but it is still not known whether
this RNA is the only source of matrix material for the synthesis of cytoplasmic RNA because no direct proof of this has yet been obtained.

The retention of the ability to synthesize RNA in nucleated cells,
even when they are treated with ribonuclease, suggests that DNA may
act as a matrix for the synthesis of RNA. However, it has not yet been
established whether the intranuclear RNA of such cells is broken down,
i.e. whether the potential ribonucleic material there disappears.

An original method for studying the biochemical role of the nucleus
has been used recently by Keck and colleagues. In their first series of
experiments they found that, after removal of the nucleus, the non-
nucleated fragments of *Acetabularia* continue to synthesize specific
enzymes (103, 104, 63). The authors hold that this shows that the
cytoplasm of these cells can maintain the genetic information which it
originally received from the nucleus. The case of acid phosphatase was
the only partial exception found in these experiments. The synthesis of
this enzyme was diminished after "amputation" of the rhizoid containing
the nucleus. In connection with the peculiarity of this enzyme Keck has
tried in a recent research (103) to carry out interspecies transplantation
of nuclei between *Acetabularia mediterranea* and *Acicularia scheneckii* and
to follow its effect on the synthesis of acid phosphatase. In preliminary
experiments it was found that the acid phosphatases of these two species
differed markedly in their electrophoretic mobility. Transplantation of
the nucleus was carried out by cross-grafting of the rhizoids containing
the nuclei. In the first of these experiments the cells of *Acicularia scheneckii*
had their nuclei removed and immediately replaced by nuclei from
*Acetabularia mediterranea*. The consequences of this operation seemed
to be as follows. During the first day after the grafting operation only
the type of acid phosphatase characteristic of *Acicularia* was formed. On
the second day a small amount of the enzyme characteristic of *Acetabularia*
appeared in the cells and its production increased over the following days.
At the same time the production of the phosphatase of cytoplasmic origin
ceased completely on the fifth day.

The authors found it interesting that the rapid cessation of synthesis
of the cytoplasmic type of phosphatase occurred considerably more
rapidly in the cells in these experiments than in cells which simply lacked
a nucleus. However, in the reverse type of grafting (cytoplasm of *Acetabularia* and nucleus of *Acicularia*) the effect of the nucleus was not manifested and the "hybrids" continued to synthesize only the phosphatase
characteristic of *Acetabularia* which indicated that there are factors which
control this synthesis in the cytoplasm. This feature (autonomy of the

cytoplasm) seemed to predominate in "hybrids" between the two species. These interesting experiments thus indicate that, as well as nuclear control, there is also an autonomous cytoplasmic system for reproducing proteins and their immediate matrices.   The different behaviour of different enzymes in *Acetabularia* after removal of the nucleus has also been mentioned in the work of Brachet (51).

*Acetabularia*, as an object for the study of the localization of the synthesis of RNA, has attracted the attention of many workers.   Schweiger & Bremer (146) have shown that, after non-nucleated fragments of *Acetabularia* have been kept in the dark for 10 days, the synthesis of RNA in their cytoplasm ceases while in nucleated cells it is diminished but does not cease.   Webster and colleagues (157) recently did some interesting work on the effect of removal of the nucleus on the synthesis of RNA in the ribosomes of *Acetabularia*.   In these experiments the incorporation of [$^{32}$P]phosphate and [$^{14}$C]adenine into the ribosomes of non-nucleated fragments was higher than in ordinary cells.   The authors suggest that in this alga the nuclei are not directly involved in the formation of the ribosomal RNA, but it would seem that the only thing which passes from the nucleus into the cytoplasm is a certain amount of RNA which is a carrier of information.   Schweiger & Bremer (146a), who also studied the synthesis of RNA in nucleated and non-nucleated cells of *Acetabularia*, established that the effect of removal of the nucleus depends on the regime of exposure to light.   According to their evidence, some substance of nuclear origin which stimulates the synthesis of RNA accumulates in the cytoplasm in the dark.   This substance disappears in the light.*

## 4.  Some other lines of work on the connection between RNA synthesis in the nuclei and cytoplasm and the dependence of RNA synthesis on DNA

There are a number of statements in the literature that the nuclear and cytoplasmic RNA differ in their nucleotide composition (68, 121). This does not, however, indicate that they are synthesized independently, as the RNA in both the nucleus and the cytoplasm is very heterogeneous and therefore not all of the fractions of nuclear RNA can take the same part in transport from the nucleus to the cytoplasm and this might be the cause of the differences between the nucleotide composition of the RNA of the nucleus and that of the cytoplasm.   Edström's paper (67) is very interesting in this connection.   Using ultramicro methods he showed that there was no difference in nucleotide composition between the RNA of the cytoplasm and that of the nucleolus.   The RNA of the nuclear sap,

---

* In considering these matters it is important to bear in mind that Baltus & Brachet (42a) have found DNA to be present in the chloroplasts of *Acetabularia mediterranea*.

however, had a number of peculiarities and differed in composition from that of the cytoplasm.

A new statistical approach to the assessment of the relationship between DNA and RNA in bacterial cells has been used in the work of Belozerskiĭ and colleagues (3, 3a, 30). These authors studied the differences in the nucleotide composition of DNA and RNA in various species of bacteria and submitted the figures thus obtained to a special statistical comparison to find out if there was any correlation between the nucleotide compositions of the DNA and RNA.

Comparison of the composition of DNA with that of RNA showed that there was no appreciable correlation between them. However, the existence of some correlation must be considered to be very probable. This correlation is expressed in a certain tendency for the ratio $\frac{G+C}{A+U}$ in the RNA to be larger in species in which the ratio $\frac{G+C}{A+T}$ is larger in the DNA.

On the basis of the results obtained Belozerskiĭ (3, 3a) put forward the idea that there may be some part of the cellular RNA in which the correlation with the corresponding DNA is complete. This is quite likely only a very small part, in view of the very low regression of $\frac{G+C}{A+U}$ in the RNA on $\frac{G+C}{A+T}$ in the DNA. The main mass of the cellular RNA does not show any correlation with DNA and its composition is similar in the most diverse species.

The method of studying the statistical correlation between RNA and DNA has also been used in the recent work of Yčas & Vincent (175) on a fraction of RNA which they isolated from yeast and compared with the appropriate DNA in respect of nucleotide composition.

The local action of ultraviolet or ionizing radiation on the nucleus or nucleolus provides a very promising method for studying the role of these organelles in the biochemical functions of the cell. Gaulden & Perry (78, 123) have worked out a method using a microbeam of ultraviolet light having a diameter of $2 \cdot 2\mu$, which enabled them to inhibit synthesis in the nucleoli of cells selectively. It was found that treatment of the nucleoli of HeLa cells with a lethal dose of ultraviolet irradiation led to an almost complete inhibition of the incorporation of [³H]cytidine into the nuclear RNA. In the cytoplasm, however, the synthesis of RNA continued independently of the nucleus, although it was very much retarded. In new experiments Perry and colleagues (69, 124) showed that about two thirds of the cytoplasmic synthesis of RNA depended on the nucleus, while one third of the extranuclear incorporation of labelled cytidine depended on the nucleolus. The synthesis of proteins in the cytoplasm was only altered to an insignificant extent by microirradiation of the nucleus.

Seed (147) used X-rays instead of ultraviolet rays for this purpose (the diameter of the beam was about $2 \cdot 5\mu$ and the dose about 500 röntgens to the nucleolus). Treatment of the nucleolus with this dose (the energy required being far less than that of the dose of ultraviolet light used by Perry and colleagues) led to a considerable inhibition of the synthesis of RNA in the nucleus while similar treatment of parts of the nuclear sap had no appreciable effect on synthesis in the nucleus.

These results show clearly that the nucleolus is not merely a reservoir for RNA which has been synthesized in the chromatin material but plays an active and independent part in the synthesis of RNA.

It must not, however, be thought that the synthesis of RNA in the nucleolus, unlike that in the chromosomes, is completely independent of DNA. Working on the synthesis of DNA, RNA and proteins in the cells of explants of the heart of the 5-day old rat, Harris (90) showed that synthesis of DNA took place in the nucleolus as well as synthesis of protein and RNA, the incorporation of [³H]thymidine into the DNA of the nucleolus beginning earlier than its incorporation into the rest of the nucleus.

It is interesting to note that blocking of protein synthesis in these cells by antimetabolites led to cessation of the synthesis of DNA. Blocking RNA synthesis also affected some processes in the formation of DNA. However, Georgiev (13) did not find DNA in the nucleoli of liver cells.

We shall discuss the relations at a molecular level between the syntheses of DNA, RNA and proteins in later chapters. It must be mentioned that, according to our present evidence, the syntheses of RNA and protein do not require the simultaneous synthesis of DNA. Norcross and colleagues (118) have shown recently that the synthesis of DNA is not necessary for the formation of ribosomes. Nevertheless, the synthesis of DNA, which can only be carried out *in vitro* in the presence of primer, nucleoside triphosphates and polymerase (cf. Chapter XIV), depends on the synthesis of RNA and proteins. This dependence has been studied in detail by Okazaki & Okazaki (119) as well as in the paper by Harris referred to above. The Okazaki's studied the synthesis of DNA by the bacterium *Lactobacillus acidophilus*. In this work it was found that, if the culture of bacteria is deficient in desoxyribosides, the synthesis of DNA in it will come to a complete standstill while the synthesis of RNA and proteins continues. Cell-division is also halted. Deficiency of uracil (which is a substrate for RNA) in the cells led to inhibition of RNA synthesis and protein synthesis, but the synthesis of DNA was not decreased but even increased. Inhibition of protein synthesis by means of chloramphenicol also scarcely affected the synthesis of DNA but a deficiency of amino acids in the cells markedly slowed the synthesis of DNA. It thus seemed that, in this case, the synthesis of DNA did not require the synthesis of proteins but did require the presence of amino acids. In the next chapter we shall see that the synthesis of RNA also requires the presence of certain amino acids, although it too is, in many cases, independent of protein synthesis. In discussing this evidence and the results of their own work the Okazaki's suggested that desoxyribonucleotides may take part in the synthesis of DNA in combination with amino acids. They also think that it is not impossible that the synthesis of DNA may depend on the metabolism of some small fraction of protein in respect of which inhibition by chloramphenicol is ineffective. The dependence of the synthesis of DNA in bacterial cells on some fraction of the proteins which must be synthesized in advance has also been observed by Nakada (117).

As for the problem of the relationship between RNA and DNA we cannot avoid dealing with certain other facts which are, as yet, too isolated for any conclusions to be drawn from them but which, nevertheless, cannot be passed over without attention. In the first place we must mention several reports that the cytoplasm of the eggs of amphibians and the cells of amoebae contain a certain amount of DNA which can incorporate labelled thymidine (a nucleotide present in DNA) and [$^{32}$P]phosphate (126, 131, 72). It has also been reported that, in plants, DNA may occur in such typical cytoplasmic structures as plastids (98). The incorporation of labelled thymidine into the polynucleotides of the cytoplasm does not, however, prove that functional DNA is present in it.

The experiments which we have already mentioned on the incorporation of labelled thymidine into cytoplasmic structures were carried out on amoebae. While checking these results Roth & Daniels (140) recently showed clearly that the cytoplasm of the amoeba contains large numbers of infective microorganisms which contain both DNA and RNA and which multiply in the cytoplasm of the amoeba. These microorganisms retained their activity even in amoebae which had been starved for two weeks. The results of this research cast doubt upon all the evidence concerning the synthesis of DNA in the cytoplasm and it shows how careful one must sometimes be not to come to a conclusion without taking account of all possible factors.

The presence of many enzymes associated with RNA metabolism in the cytoplasm (62) suggests that RNA may be synthesized autonomously there.

Of course, the ability of the cytoplasmic structures to synthesize RNA can be studied *in vitro* on cell-free systems. In this connection we must mention the detailed studies of Work and colleagues (46) who showed that RNA synthesis by microsomes takes place *in vitro*. In such studies of S-RNA synthesis it is hard to distinguish between metabolism of the terminal group and synthesis of the main chain. It is, therefore, difficult to say which of these processes is associated with the intensive renewal of S-RNA in the cytoplasm both *in vitro* and *in vivo*.

In 1961 a new approach to the study of the transfer of RNA from the nucleus to the cytoplasm was opened up by the discovery of a special form of RNA, peculiar to the nucleus, which carries genetic information and performs an important task in protein synthesis. It is also very important to note that in 1960-1 a special enzymic system was found in the nucleus which can bring about the synthesis (*in vitro* and *in vivo*) of specific forms of RNA from nucleoside triphosphates, only in the presence of DNA. The relevant material will be discussed in later chapters concerned with the biochemical aspects of nucleic acid synthesis.

## 5. The connection between the synthesis of DNA and RNA in cells and the different phases in the cellular cycle

The synthesis of new polynucleotide chains of DNA takes place in the nucleus. Studies of the characteristics of this synthesis *in vivo* have been made in many laboratories and quite extensive and very interesting information has been accumulated which it would be very hard to review comprehensively. In our discussion of the matter we shall therefore confine ourselves to a few fundamental questions.

A very interesting observation, first made about ten years ago (97), is that there is a definite rhythm in the rate of synthesis of DNA in the cell, correlated with particular phases in the cellular cycle. Various methods have been used to demonstrate this rhythm (incorporation of [$^{32}$P]phosphate and labelled nucleotides, ultraviolet microscopy) and also various objects (the cells of plants, animals and microorganisms) and the results of the observations made are, therefore, very varied.

Most of the evidence indicates, however, that the period of intensive synthesis of DNA usually precedes cell-division and that no appreciable increase in DNA occurs in the nuclei of cells which are not about to divide (45, 58, 86, 90, 97, 109a, 116, 138, 139, 142, 153, 161, 170). It is quite understandable that the curve of DNA synthesis should be of this type for it is just in this premitotic period that the doubling of the chromosomes occurs.

Experiments on the incorporation of precursors, labelled with isotopes, into the DNA being synthesized also show that although there is no absolute increase in the amount of DNA in the resting nuclei in the period between mitoses, yet a certain renewal of the existing pool of DNA does take place (105, 112). This is seen particularly well in experiments on the incorporation of labelled precursors of nucleic acids into isolated nuclei in which none of the changes of the mitotic cycle occur. A very indicative research along these lines has been carried out by Breitman & Webster (56). Some of their results are given in Table 12.

These results indicate that the DNA of isolated nuclei incorporate labelled precursors, though 20-50 times more weakly than RNA. There seems to be no need for us to introduce here any further evidence to show that DNA is slowly renewed in cells which are not dividing although they are in an active functional state, though it is not renewed so actively as

TABLE 12

Incorporation of various labelled components into the nucleic acids
of isolated nuclei

| Compound (total activity of each 300,000 counts/min.) | Counts/min./mg. after 2 hrs. incubation | |
|---|---|---|
| | RNA | DNA |
| [8-$^{14}$C]Adenine | 9,618 | 153 |
| [4-$^{14}$C]Guanine | 11,097 | 195 |
| [8-$^{14}$C]Adenosine-5′-monophosphate | 5,586 | 312 |
| [2-$^{14}$C]Uracil | 120 | 0 |
| [$^{14}$C]Formic acid | 3,096 | 321 |
| [2-$^{14}$C]Glycine | 405 | 21 |
| [1-$^{14}$C]Alanine | 765 | 51 |

RNA or proteins. This fact has been confirmed tens of times during the past decade and may be considered as firmly established. It is only important to emphasize that the stability of DNA is not absolute, and that the molecules of DNA are renewed to some extent, however slowly, even in resting nuclei. Comparatively active metabolism of DNA *in vivo* without an increase in absolute amount has recently been observed in the experiments of Pelc (122) on several types of cell. There is, at present, no reason to suppose that the mechanism of this renewal is different from that of the synthesis of new polynucleotides during the period of mitosis or that it does not conform with the scheme of synthesis of DNA derived from the model of Watson & Crick. However, in a double helix, partial renewal of the molecule by replacement of individual links in individual polynucleotides without the disruption and complete disintegration of the double helix is quite feasible from a theoretical point of view and we cannot, therefore, exclude the possibility that the metabolism of the resting nucleus, to which we have referred above, may be carried out in this very way.

It must, however, be mentioned that, although we observe a definite rhythm of synthesis of RNA and DNA associated with the cellular cycle in cells which have highly developed nuclei, yet in bacteria, which do not possess a differentiated nucleus, the synthesis of DNA usually goes on without interruption, without any obvious connection with the cycle of cell division. The evidence for these statements was obtained in several laboratories on so-called "synchronized cultures" (32, 143, 111, 122, 59, 115). Nevertheless the synthesis of RNA does occur rhythmically in a number of bacterial cultures (114). It is interesting to note that, in spite of the presence of a very large number of individual molecules of DNA and their complicated and ordered organization, a method exists for distributing the parent and daughter molecules of DNA between the parent and daughter cells and parent and daughter chromosomes (73, 127, 131, 162).

Disagreement arose between the different laboratories in which these experiments were being carried out as to whether the separation of the labelled and unlabelled parts of the chromosome occurred during the first or second division. This difference of opinion appears, however, to be connected with differences between the organisms being studied, and this is, in part, recognized by the disputants. The fact that such a separation occurs is, however, indubitable. Of course one may certainly not draw any close analogy between the division of chromosomes and the division of molecules of DNA, as chromosomes are several degrees more complicated than double helices of DNA and if the chromosomes consisted merely of a single helical complementary molecule of DNA, then, according to the calculations of Taylor *et al.* (162) the length of the molecule would be 1-1·5 metres. Several structural models have been devised to explain the observed ability of the chromosomes to divide and these have been reviewed by Taylor (160). A detailed discussion of this question is outside the scope of this book, however, as it requires the analysis of much evidence as to the structure of the chromosomes, the theory of karyokinesis and other such cytological problems. However, it seems highly probable to us that the amazing coincidence of the theoretically possible power of division of the polynucleotide chain and the actual duplication of the many times more complicated complex of the chromosomes is not a mere coincidence and it is quite possible that the rules governing the division of polynucleotides also determine the behaviour of chromosomes.

Finally we must deal with the so-called "disappearance" of DNA from the nucleus at certain stages in oogenesis. These problems were first taken up in the work of Marshak & Marshak (108, 109) and have since been studied by many authors. As the results of these observations will mainly be used in discussing the role of DNA in the phenomena of heredity, we shall deal with the matter in more detail in the appropriate chapter. Here it is enough to mention that there is no true "disappearance" of DNA in the oocytes and that what is observed is merely the dispersal of DNA and certain other phenomena connected with changes in its localization (44, 50).

## Conclusion

The results discussed in this chapter point to close association of the nucleus with the processes going on in the cytoplasm, but, of all the various processes in which the nucleus and cytoplasm take part jointly, the dominant one is the transfer of the RNA synthesized in the nucleus to the cytoplasm. Although the cytoplasm can synthesize RNA autonomously this synthesis does not usually fulfil all the functional requirements of the cytoplasm. Autonomous synthesis of RNA in the cytoplasm does not guarantee the autonomy of the cytoplasm and if the protein synthesis occurring there is to be carried out properly it requires the access of RNA of nuclear origin. The dominant nature of these processes in nuclear-cytoplasmic metabolism, along with many genetical observations concerning the importance of the nucleus in the phenomena of inheritance and morphogenesis, give us reason to believe that the transfer of nuclear RNA to the cytoplasm is the material expression of the transfer of information from the DNA to the site of protein syntheses in the cytoplasm. The recent discovery of special forms of RNA which transfer information, which we shall consider in later chapters, provides further support for this idea.

During the process of evolution very complicated systems have arisen in organisms for carrying out various functions and we are only now starting to elucidate them. There is a direct correspondence between the functional and the biochemical complexity of any biological system. It may be that there would be no need for the development of such a many-staged system of matrices for the formation of any protein individually. However, the evolutionary development of the organism, which is accompanied by intracellular and intercellular differentiation, leads to the development of a variety of types of proteins and thus the problem of their synthesis becomes far more complicated. If we assume that hundreds of different proteins must be synthesized within the cell in different amounts and that the nature and relative amounts present must vary very greatly at different periods of the ontogenesis of the cell and of the whole organism, then special, automatic, biochemical "administrative bureaux" will be required to direct and co-ordinate these syntheses. There can be no doubt that the central "administrative bureau" of any cell is its nucleus. Not all biologists nowadays agree in accepting this very widely held idea and to some, especially those in fields far from biochemistry, it seems idealistic and "mechanistic". According to their ideas there is still some slight hope that the undoubted, qualitative peculiarity of living things means that living material has found some special biological way of creating extraordinarily complicated and specific systems without any difficulty by means of some mysterious property inherent in all the molecules of the living body. However, the development of biochemistry, cytology and genetics confounds these hopes of finding a simple means of creating these extraordinarily complicated and purposive systems. If, on observing the rapid growth of some tissue or bacterial culture, we marvel at the apparent ease with which these cells accomplish the extremely complicated tasks which require the work of many tens of generations of scientists for their elucidation, then we must be forgetting that, behind this so-called ease, there stand hundreds of millions of years of continuous improvement of living systems through variation and selection.

## REFERENCES

1. BELIK, YA. V. & KRACHKO, L. S. (1959). *Ukraïn. biokhim. Zhur.* **31**, 322.

2. BELOZERSKIĬ, A. N. (1936). *Biokhimiya*, 1, 255.
3. — (1959). *Nukleoproteidy i nukleinovye kisloty rastenii i ikh biologischeskoe znachenie.* (*Bakhovskaya lektsiya*). Moscow: Izd. Akad. Nauk S.S.S.R.
3a. BELOZERSKY [BELOZERSKIĬ], A. N. & SPIRIN, A. S. (1958). *Nature (Lond.)*, 182, 111.
4. BELOZERSKIĬ, A. N. & URYSON, S. O. (1958). *Biokhimiya*, 23, 568.
5. BELOZERSKIĬ, A. N. & CHERNOMORDIKOVA, L. A. (1940). *Biokhimiya*, 5, 133.
6. BLYUMENFEL'D, L. A. (1960). *Doklady Akad. Nauk S.S.S.R.* 130, 887.
7. BRODSKIĬ, V. YA. (1960). *Doklady Akad. Nauk S.S.S.R.* 130, 189.
8. — (1960). *Tezisy Dokladov Pervoi Konferentsii po Voprosam Tsito- i Gistokhimii*, p. 7. Moscow: Inst. Biophys. and Inst. Morphol., Acad. Sci. U.S.S.R.
9. GVOZDEV, V. A. (1960). *Biokhimiya*, 25, 920.
10. GVOZDEV, V. A. & KHESIN, R. B. (1960). *Doklady Akad. Nauk S.S.S.R.* 134, 1226.
10a. GVOZDEV, V. A. & PONOMAREVA-STEPNAYA, M. A. (1963). *Biokhimiya*, 28, 152.
11. GEORGIEV, G. P. (1960). *Tsitologiya*, 2, 186.
12. GEORGIEV, G. P., ERMOLAEVA, L. P. & ZBARSKIĬ, I. B. (1960). *Biokhimiya*, 25, 318.
13. GEORGIEV, G. P. (1960). *Tezisy Dokladov Pervoi Konferentsii po Voprosam Tsito- i Gistokhimii*, p. 5. Moscow: Inst. Biophys. and Inst. Morphol., Acad. Sci. U.S.S.R.
14. — (1961). *Proc. V int. Congr. Biochem., Moscow*, 9, 124.
15. GEORGIEV, G. P. & MANT'EVA, V. L. (1960). *Biokhimiya*, 25, 143.
16. GEORGIEV, G. P. & CHENTSOV, YU. S. (1960). *Doklady Akad. Nauk S.S.S.R.* 132, 199.
17. ZBARSKIĬ, I. B. (1950). *Usp. biol. Khim.* 1, 91.
17a. — (1961). *Proc. V int. Congr. Biochem., Moscow*, 2, 116.
18. ZBARSKIĬ, I. B. & GEORGIEV, G. P. (1959). *Biokhimiya*, 24, 192.
19. ZBARSKIĬ, I. B. & ERMOLAEVA, L. P. (1960). *Biokhimya*, 25, 112.
20. ZBARSKIĬ, I. B. & PEREVOSHCHIKOVA, K. A. (1956). *Doklady Akad. Nauk S.S.S.R.* 107, 285.
21. — (1960). *Vopr. med. Khim.* 6, 23.
22. KEDROVSKIĬ, B. V. (1959). *Tsitologiya belkovykh sintezov v zhivotnoi kletke.* Moscow: Izd. Akad. Nauk S.S.S.R.
23. KIKNADZE, I. I. (1961). *Tsitologiya*, 3, 1.
24. MAKAROV, P. V. (1960). *Usp. sovrem. Biol.* 50, 44.
25. PLATOVA, T. P. (1959). *Usp. sovrem. Biol.* 47, 168.
26. RAĬKOV, I. B. (1959). *Tsitologiya*, 1, 566.
27. SALGANIK, R. I. (1958). *Biokhimiya*, 23, 377.
28. SAMARINA, O. P. (1960). *Tezisy Dokladov Pervoi Konferentsii po Voprosam Tsito- i Gistokhimii*, p. 8. Moscow: Inst. Biophys. and Inst. Morphol., Acad. Sci. U.S.S.R.
29. SAMARINA, O. P. & GEORGIEV, G. P. (1960). *Doklady Akad. Nauk S.S.S.R.* 133, 694.
30. SPIRIN, A. S., BELOZERSKIĬ, A. N., SHUGAEVA, N. V. & VANYUSHIN, B. F. (1957). *Biokhimiya*, 22, 744.
31. KHESIN, R. B. (1960). *Biokhimiya tsitoplazmy.* Moscow: Izd. Akad. Nauk S.S.S.R.
31a. — (1961). *Proc. V int. Congr. Biochem., Moscow*, 2, 257.
31b. KHESIN, R. B., GVOZDEV, V. A. & ASTAUROVA, O. B. (1961). *Biokhimiya*, 26, 807.
32. ABBO, F. E. & PARDEE, A. B. (1960). *Biochim. biophys. Acta*, 39, 478.

33. ALLFREY, V. G. (1960). *Cell nucleus—Sympos. Faraday Soc. Cambridge, Eng., 1959* (ed. J. S. Mitchell), pp. 170.

33a. — (1961). *Proc. V int. Congr. Biochem., Moscow,* **2**, 127.

34. ALLFREY, V. [G.], DALY, M. M. & MIRSKY, A. E. (1953). *J. gen. Physiol.* **37**, 157.

35. ALLFREY, V. G., HOPKINS, J. W., FRENSTER, J. H. & MIRSKY, A. E. (1960). *Ann. N.Y. Acad. Sci.* **88**, 722.

36. ALLFREY, V. G. & MIRSKY, A. E. (1957). *Proc. natl. Acad. Sci. U.S.* **43**, 821.

37. — (1958). *Proc. natl. Acad. Sci. U.S.* **44**, 981.

38. ALLFREY, V. [G.], MIRSKY, A. E. & OSAWA, S. (1957). In *A Symposium on the chemical basis of heredity* (ed. W. D. McElroy & B. Glass), p. 200. Baltimore, Md.: Johns Hopkins Press.

39. ALLFREY, V. G. & MIRSKY, A. E. (1959). *Proc. natl. Acad. Sci. U.S.* **45**, 1325.

40. — (1961). In *Protein biosynthesis* (ed. R. J. C. Harris), p. 49. London: Academic Press.

41. AMANO, M. & LEBLOND, C. P. (1960). *Exptl. Cell Research,* **20**, 250.

42. ANTONI, F., VARGA, L. & HIDVÉGI, E. J. (1959). *Acta physiol. Acad. Sci. Hung.* **16**, 1.

42a. BALTUS, E. & BRACHET, J. (1963). *Biochim. biophys. Acta,* **76**, 490.

43. BARNUM, C. P., HUSEBY, R. A. & VERMUND, H. (1953). *Cancer Res.* **13**, 880.

43a. BATHER, R. & PURDIE-PEPPER, E. (1961). *Can. J. Biochem. Physiol.* **39**, 1625.

44. BELL, P. R. (1961). *Proc. roy. Soc.* **153B**, 421.

45. BERGERARD, J. (1955). *Compt. rend. Acad. Sci., Paris,* **240**, 564.

46. BHARGAVA, P. M., SIMKIN, J. L. & WORK, T. S. (1958). *Biochem. J.* **68**, 265.

46a. BIRNSTIEL, M. L., CHIPCHASE, M. & BONNER, J. (1961). *Biochem. biophys. Research Commun.* **6**, 161.

46b. BIRNSTIEL, M. L., CHIPCHASE, M. I. H. & HYDE, B. B. (1963). *Biochim. biophys. Acta,* **76**, 454.

47. BONNER, J. (1959). *Amer. J. Bot.* **46**, 58.

48. BRACHET, J. (1955). *Nature (Lond.),* **175**, 851.

49. — (1955). *Biochim. biophys. Acta,* **18**, 247.

50. — (1957). *Biochemical Cytology.* New York: Academic Press.

51. — (1958). *Exptl. Cell Research,* Suppl. **6**, 78.

52. — (1959). *Ann. N.Y. Acad. Sci.* **78**, 688.

53. BRACHET, J. & CHANTRENNE, H. (1951). *Nature (Lond.),* **168**, 950.

54. — (1952). *Arch. intern. Physiol.* **60**, 547.

55. — (1956). *Cold Spring Harbor Symposia quant. Biol.* **21**, 329.

56. BREITMAN, T. R. & WEBSTER, G. C. (1959). *Exptl. Cell Research,* **18**, 413.

57. — (1959). *Nature (Lond.),* **184**, 637.

58. BUCHER, N. L. R. & MAZIA, D. (1960). *J. biophys. biochem. Cytol.* **7**, 651.

59. BURNS, V. W. (1961). *Exptl. Cell Research,* **23**, 582.

59a. CASPERSSON, T. O., FARBER, S., FOLEY, G. E., LOMAKKA, G., KILLANDER, D. & CARLSON, L. (1962). *Exptl. Cell Research,* **28**, 621.

60. CAVALIERI, L. F. & ROSENBERG, B. H. (1961). *Biophys. J.* **1**, 317, 323, 337.

61. CHANTRENNE, H. (1961). *The biosynthesis of proteins.* Oxford: Pergamon Press.

62. CHUNG, C. W., MAHLER, H. R. & ENRIONE, M. (1960). *J. biol. Chem.* **235**, 1448.

63. CLAUSS, H. (1959). *Planta,* **52**, 534.

64. CROSBIE, G. W., SMELLIE, R. M. S. & DAVIDSON, J. N. (1953). *Biochem. J.* **54**, 287.

65. DOUNCE, A. L. (1955). In *Nucleic Acids* (ed. E. Chargaff & J. N. Davidson), Vol. 2, p. 93. New York: Academic Press.
66. — (1959). *Ann. N.Y. Acad. Sci.* **81**, 794.
67. EDSTRÖM, J.-E. (1960). *J. biophys. biochem. Cytol.* **8**, 39, 47.
68. ELSON, D., TRENT, L. W. & CHARGAFF, E. (1955). *Biochim. biophys. Acta*, **17**, 362.
69. ERRERA, M., HELL, A. & PERRY, R. P. (1961). *Biochim. biophys. Acta*, **49**, 58.
70. EZEKIEL, D. H. (1960). *J. Bact.* **80**, 119.
71. FICQ, A. (1961). *Exptl. Cell Research*, **23**, 427.
72. FINAMORE, F. J. & VOLKIN, E. (1958). *Exptl. Cell Research*, **15**, 405.
73. FIRKET, H. & VERLY, W. G. (1958). *Nature (Lond.)*, **181**, 274.
74. FITZGERALD, P. J. & VINIJCHAIKUL, K. (1959). *Lab. Invest.* **8**, 319.
75. FREESE, E. (1958). *Cold Spring Harbor Symposia quant. Biol.* **23**, 13.
76. FRENSTER, J. H., ALLFREY, V. G. & MIRSKY, A. E. (1960). *Proc. natl. Acad. Sci. U.S.* **46**, 432.
77. — (1961). *Biochim. biophys. Acta*, **47**, 130.
78. GAULDEN, M. E. & PERRY, R. P. (1958). *Proc. natl. Acad. Sci. U.S.* **44**, 553.
79. GEORGIEV, G. P., MANTIEVA [MANT'EVA], V. L. & ZBARSKY [ZBARSKII], I. B. (1960). *Biochim. biophys. Acta*, **37**, 373.
80. GOLDSTEIN, L. (1958). *Exptl. Cell Research*, **15**, 635.
81. — (1960). *Science*, **132**, 1492.
82. GOLDSTEIN, L. & MICOU, J. (1959). *J. biophys. biochem. Cytol.* **6**, 301.
83. GOLDSTEIN, L., MICOU, J. & CROCKER, T. T. (1960). *Biochim. biophys. Acta*, **45**, 82.
84. GOLDSTEIN, L. & MICOU, J. (1959). *J. biophys. biochem. Cytol.* **6**, 1.
85. GOLDSTEIN, L. & PLAUT, W. (1955). *Proc. natl. Acad. Sci. U.S.* **41**, 874.
86. GRUN, P. (1956). *Exptl. Cell Research*, **10**, 29.
87. HÄMMERLING, J. (1934). *Roux Arch. EntwMech. Organ.* **131**, 1.
88. — (1953). *Internatl. Rev. Cytol.* **2**, 475.
89. HARBERS, E. & HEIDELBERGER, C. (1959). *Biochim. biophys. Acta*, **35**, 381.
90. HARRIS, H. (1959). *Biochem. J.* **72**, 54.
91. — (1959). *Biochem. J.* **73**, 362.
91a. — (1961). *Proc. V int. Congr. Biochem., Moscow*, **2**, 107.
92. HERBERT, E. (1958). *J. biol. Chem.* **231**, 975.
93. HOAGLAND, M. B. (1960). In *Nucleic acids* (ed. E. Chargaff & J. N. Davidson), Vol. 3, p. 349. New York: Academic Press.
94. HOPKINS, J. W. (1959). *Proc. natl. Acad. Sci. U.S.* **45**, 1461.
95. HOPKINS, J. W., ALLFREY, V. G. & MIRSKY, A. E. (1961). *Biochim. biophys. Acta*, **47**, 194.
96. HOTTA, Y. & OSAWA, S. (1958). *Biochim. biophys. Acta*, **28**, 642.
97. HOWARD, A. & PELC, S. R. (1951). *Exptl. Cell Research*, **2**, 178.
98. IWAMURA, T. (1960). *Biochim. biophys. Acta*, **42**, 161.
99. JARDETZKY, C. D. & BARNUM, C. P. (1957). *Arch. Biochem. Biophys.* **67**, 350.
100. KAUFMANN, B. P. (1960). In *The Cell Nucleus* (Faraday Society informal meeting, ed. J. S. Mitchell), p. 251. London.
101. KAUFMANN, B. P., GAY, H. & McDONALD, M. R. (1960). *Intern. Rev. Cytol.* **9**, 77.
102. KAY, E. R. M., SMELLIE, R. M. S., HUMPHREY, G. F. & DAVIDSON, J. N. (1956). *Biochem. J.* **62**, 160.
103. KECK, K. (1960). *Biochem. biophys. Research Commun.* **3**, 56.
104. KECK, K. & CLAUSS, H. (1958). *Bot. Gazette*, **120**, 43.
105. KOENIG, H. (1958). *J. biochem. biophys. Cytol.* **4**, 664.
106. LOGAN, R. & DAVIDSON, J. N. (1957). *Biochim. biophys. Acta*, **24**, 196.
107. MARSHAK, A. (1948). *J. cell. comp. Physiol.* **32**, 381.

108. MARSHAK, A. & MARSHAK, C. (1954). *Nature (Lond.)*, **174**, 919.
109. — (1955). *Exptl. Cell Research*, **8**, 126.
109a. MAZIA, D. (1961). *Ann. Rev. Biochem.* **30**, 669.
110. MAZIA, D. & PRESCOTT, D. M. (1955). *Biochim. biophys. Acta*, **17**, 23.
111. MCFALL, E., PARDEE, A. B. & STENT, G. S. (1958). *Biochim. biophys. Acta*, **27**, 282.
112. MCFALL, E. & STENT, G. S. (1959). *Biochim. biophys. Acta*, **34**, 580.
113. MCMASTER-KAYE, R. (1960). *J. biophys. biochem. Cytol.* **8**, 365.
114. SCOTT, D. B., MCNAIR & CHU, E. (1959). *Exptl. Cell Research*, **18**, 392.
115. MITCHISON, J. M. & WALKER, P. M. B. (1959). *Exptl. Cell Research*, **16**, 49.
116. MOSES, M. J. & TAYLOR, J. H. (1955). *Exptl. Cell Research*, **9**, 474.
117. NAKADA, D. (1960). *Biochim. biophys. Acta*, **44**, 241.
118. NORCROSS, F. C., COMLY, L. T. & ROBERTS, R. B. (1959). *Biochem. biophys. Research Commun.* **1**, 244.
118a. O'DONNELL, E. H. J. (1961). *Nature (Lond.)*, **191**, 1325.
119. OKAZAKI, T. & OKAZAKI, R. (1959). *Biochim. biophys. Acta*, **35**, 434.
120. OSAWA, S. & OTAKA, E. (1959). *Biochim. biophys. Acta*, **36**, 549.
121. OSAWA, S., TAKATA, K. & HOTTA, Y. (1958). *Biochim. biophys. Acta*, **28**, 271.
122. PELC, S. R. (1959). *Lab. Invest.* **8**, 225.
123. PERRY, R. P. (1960). *Exptl. Cell Research*, **20**, 216.
123a. PERRY, R. P., ERRERA, M., HELL, A. & DÜRWALD, H. (1961). *J. biochem. biophys. Cytol.* **11**, 1.
124. PERRY, R. P., HELL, A. & ERRERA, M. (1961). *Biochim. biophys. Acta*, **49**, 47.
125. PLAUT, W. (1958). *Exptl. Cell Research*, Suppl. **6**, 69.
126. — (1960). *Biochem. Pharmacol.* **4**, 79.
127. PLAUT, W. & MAZIA, D. (1957). *Texas Rep. Biol. Med.* **15**, 181.
128. PLAUT, W. & RUSTAD, R. C. (1956). *Nature (Lond.)*, **177**, 89.
129. — (1957). *J. biophys. biochem. Cytol.* **3**, 625.
130. — (1959). *Biochim. biophys. Acta*, **33**, 59.
131. PLAUT, W. & SAGAN, L. A. (1958). *J. biophys. biochem. Cytol.* **4**, 483.
132. POORT, C. (1961). *Biochim. biophys. Acta*, **46**, 373.
133. PRESCOTT, D. M. (1959). *J. biophys. biochem. Cytol.* **6**, 203.
134. — (1957). *Exptl. Cell Research*, **12**, 196.
135. — (1960). *Exptl. Cell Research*, **19**, 29.
136. RABINOVITZ, M. & PLAUT, W. (1956). *Exptl. Cell Research*, **10**, 120.
137. RENDI, R. (1960). *Exptl. Cell Research*, **19**, 489.
138. RÉVÉSZ, L., FORSSBERG, A. & KLEIN, G. (1956). *J. natl. Cancer Inst.* **17**, 37.
138a. RHO, J. H. & BONNER, J. (1961). *Proc. natl. Acad. Sci. U.S.* **47**, 1611.
139. ROELS, H. (1954). *Nature (Lond.)*, **173**, 1039.
140. ROTH, L. E. & DANIELS, E. W. (1961). *J. biophys. biochem. Cytol.* **9**, 317.
141. SACKS, J. & KAMARTH, K. D. (1956). *J. biol. Chem.* **223**, 423.
142. SCHAECHTER, M., BENTZON, M. W. & MAALØE, O. (1959). *Nature (Lond.)*, **183**, 1207.
143. SCHERBAUM, O., LOUDERBACK, A. L. & JAHN, T. L. (1959). *Exptl. Cell Research*, **18**, 150.
144. SCHNEIDER, J. H. (1961). *Biochim. biophys. Acta*, **47**, 107.
145. SCHOLTISSEK, C., SCHNEIDER, J. H. & POTTER, V. R. (1958). *Fed. Proc.* **17**, 306.
146. SCHWEIGER, H. G. & BREMER, H. J. (1960). *Exptl. Cell Research*, **20**, 617.
146a. — (1961). *Biochim. biophys. Acta*, **51**, 50.
147. SEED, J. (1960). *Proc. roy. Soc.* **152B**, 387.
148. SIBATANI, A., YAMANA, K., KIMURA, K. & TAKAHASHI, T. (1960). *Nature (Lond.)*, **186**, 215.

148a. SIRLIN, J. L. (1961). *Endeavour*, **20**, 146.
149. SIRLIN, J. L., KATO, K. & JONES, K. W. (1961). *Biochim. biophys. Acta*, **48**, 421.
150. SIRLIN, J. L. & KNIGHT, G. R. (1960). *Exptl. Cell Research*, **19**, 210.
151. SIRLIN, J. L. & WADDINGTON, C. H. (1956). *Exptl. Cell Research*, **11**, 197.
152. SMELLIE, R. M. S., HUMPHREY, G. F., KAY, E. R. M. & DAVIDSON, J. N. (1955). *Biochem. J.* **60**, 177.
153. SMITH, C. L., NEWTON, A. A. & WILDY, P. (1959). *Nature (Lond.)*, **184**, 107.
154. STEFFENSEN, D. (1959). *Brookhaven Symp. Biol.* **12**, 103.
155. STICH, H. F. & KITIYAKARA, A. (1957). *Science*, **126**, 1019.
156. STICH, H. & PLAUT, W. (1958). *J. biophys. biochem. Cytol.* **4**, 119.
157. SUTTER, R. P., WHITMAN, S. L. & WEBSTER, G. [C.] (1961). *Biochim. biophys. Acta*, **49**, 233.
158. SWIFT, H. (1959). *Brookhaven Symp. Biol.* **12**, 134.
159. — (1959). *Symposium mol. Biol. Univ. Chicago* (ed. R. E. Zirkle), p. 266. Chicago: Univ. Chicago Press.
159a. — (1959). In Discussion following ref. 171.
160. TAYLOR, J. H. (1957). *Amer. Naturalist*, **91**, 209.
161. — (1958). *Amer. J. Bot.* **45**, 123.
162. TAYLOR, J. H., WOODS, P. S. & HUGHES, W. L. (1957). *Proc. natl. Acad. Sci. U.S.* **43**, 122.
163. VANDERHAEGHE, F. (1954). *Biochim. biophys. Acta*, **15**, 281.
164. VANDERHAEGHE, F. & SZAFARZ, D. (1955). *Arch. intern. Physiol.* **63**, 267.
165. VINCENT, W. S. (1957). *Science*, **126**, 306.
166. WEBSTER, G. C. (1960). *Biochem. biophys. Research Commun.* **2**, 56.
167. WANG, T.-Y. (1961). *Biochim. biophys. Acta*, **49**, 108.
168. WEHR, H., SZAFRAŃSKI, P. & GOŁASZEWSKI, T. (1961). *Proc. V int. Congr. Biochem., Moscow*, **9**, 86.
168a. WHITE, M. J. D. (1961). *The Chromosomes*. London: Methuen.
168b. WISCHNITZER, S. (1961). *Internatl. Rev. Cytol.* **10**, 137.
169. WOODARD, J. W. (1958). *J. biophys. biochem. Cytol.* **4**, 383.
170. WOODARD, J. W., RASCH, E. & SWIFT, H. (1961). *J. biophys. biochem. Cytol.* **9**, 445.
171. WOODS, P. S. (1959). *Brookhaven Symp. Biol.* **12**, 153.
172. WOODS, P. S. & TAYLOR, J. H. (1959). *Lab. Invest.* **8**, 309.
173. YAMANA, K. & SIBATANI, A. (1960). *Biochim. biophys. Acta*, **41**, 304.
174. YAMANA, K. & SIBATANI, A. (1960). *Biochim. biophys. Acta*, **41**, 295.
175. YČAS, M. & VINCENT, W. S. (1960). *Proc. natl. Acad. Sci. U.S.* **46**, 804.
176. ZALOKAR, M. (1959). *Nature (Lond.)*, **183**, 1330.
177. — (1960). *Exptl. Cell Research*, **19**, 114.
178. — (1960). *Exptl. Cell Research*, **19**, 184.
179. — (1960). *Exptl. Cell Research*, **19**, 559.

# CHAPTER XIV

# BIOSYNTHESIS AND REPRODUCTION OF THE SPECIFICITY OF NUCLEIC ACIDS

*Introduction*

A survey of all the material which has gone before shows that, although many aspects of the problem are not solved, the hypothesis that proteins are synthesized on RNA matrices seems most likely to be true. The main foundation of this hypothesis is the assumption that the sequence of nucleotides in the polynucleotide chain in some way determines the sequence of amino acid residues in the polypeptides being synthesized. If we accept this assumption as a basis for discussion, then the question arises as to the nature of the factors regulating the replication of a particular sequence of nucleotides in the matrix on which the protein is to be synthesized. And although the solution of the problem of the mechanism of replication of specificity in RNA may be pushed one stage further back because there is experimental and theoretical evidence that DNA may act as a matrix for the synthesis of RNA, yet, as concerns DNA, we are driven to accept the view that it is capable of self-reproduction while retaining its specificity (replication), for DNA represents the ultimate link in this hypothetical chain of transfer of information. Some people postulate the circular passage of specificity (information) in the system

$$RNA$$
$$\nearrow \qquad \searrow$$
$$DNA \leftarrow Protein$$

but without any convincing arguments (13, 16, 17). Of course proteins do take part in the synthesis of DNA and RNA, but only as catalysts of processes of polymerization. They do not by their participation influence the sequence in which the nucleotides are arranged in the polynucleotide chains.

A hypothetical mechanism, giving a fairly logical explanation of the self-replication of the polynucleotides of DNA and particularly applicable to the case of the possible autonomous replication of RNA in the synthesis of RNA under the control of DNA is inherent, in its general form, in the model for the structure of the molecules of DNA as two complementary helical polynucleotides joined by hydrogen bonds between obligatorily complementary pairs of bases (adenine-thymine, guanine-cytosine) and we have already discussed this question in Chapter II.

According to this scheme, when the double helix separates into two single polynucleotides, each of these can again only form on its surface a strictly complementary structure, as no pairs of nucleotides can be formed other than those mentioned above. Thus two identical paired molecules are created. Until recently the scheme of the possible replication of DNA was only a more or less probable idea and it is only during the last few years that it has received fairly convincing experimental support, primarily from experiments on the artificial synthesis of DNA *in vitro*. In this chapter we shall have to give a very short review of the biochemical features of the synthesis of RNA and DNA in the cells from a biochemical and from a cytological point of view because, notwithstanding the many peculiar features of nucleic acid metabolism, the questions with which we shall be dealing are those directly concerned with the problems of protein synthesis, inheritance and ontogenesis. A large number of reviews and monographs has been published, systematically dealing with the synthesis of nucleic acids and their components (1, 2, 9, 8, 12, 10, 11, 16a, 18, 19, 19a, 3, 37, 34, 64a, 67, 69, 74, 86, 95, 115 and others). This enables us to do as we have done in earlier chapters and set out our material on the synthesis of nucleic acids in the most generalized form possible, concentrating on the most important investigations of recent years.

As the synthesis of RNA may be under the control of DNA (the original polymer of the system DNA → RNA → protein) insofar as its specificity is concerned, we must naturally begin our survey of the synthesis of nucleic acids by studying the problem of the synthesis and replication of DNA. We shall not survey the work on the synthesis of the nucleotides, nucleosides and their nitrogenous bases which form the starting material for nucleic acid synthesis just as we did not survey the work of the synthesis and intermediary metabolism of amino acids.

## 1. DNA-synthesis *in vitro*

In the earlier chapters we have seen that the synthesis of proteins *in vivo* and *in vitro* is carried out by an extremely complicated biochemical system formed of a wide range of activating enzymes, a collection of soluble ribonucleic acids, transport enzymes, matrices in the ribosomes, carriers of information from DNA and a group of low-molecular co-factors ATP, GTP and so on.

The synthesis of DNA *in vitro*, on the contrary, seems to be very simple and, apart from the appropriate substrates and the priming DNA, which plays the part of matrix, all that is required for it to be carried out is the one enzyme polymerase, discovered by A. Kornberg, who was awarded a Nobel Prize for this discovery (68, 70). *In vivo* this synthesis is carried out in a rather more complicated way for in this case it requires, in addition, a system for phosphorylating the nucleosides and several other factors.

The enzyme which brings about the synthesis of DNA in the presence of nucleoside triphosphates and primer DNA was first isolated from the bacterium *Escherichia coli*. The author succeeded in purifying it till it was 2000 times as active as in the original extract.

The reaction of polymerization which occurs was studied in particular detail in later researches by the same group of workers (20, 23, 24, 69, 75). The enzyme isolated in Kornberg's laboratory would only bring about polymerization in the presence of triphosphates of all the four nucleosides found in DNA. Magnesium ions were required to make the reaction take place, as well as the primer and the substrates.

In their experiments Kornberg and his colleagues succeeded in obtaining quite a large yield of newly synthesized DNA, amounting to 20 times the original mass of the "primer". The DNA synthesized had a molecular weight of about 5 million and did not differ in its physicochemical properties from ordinary DNA isolated from a living organism.

A further paper from Kornberg's laboratory (76) was specially interesting. The workers there found that the DNA synthesized outside the organism did not differ in its nucleotide composition from that used as a primer. If DNA from the thymus of a calf was used as the "primer" at the beginning of the synthesis, then the newly formed DNA was just like it. If the DNA used as a "primer" was that isolated from bacteriophage or from some bacterium, while all the other conditions remained the same, then the newly formed molecules conformed precisely to the original pattern in spite of differences in the relative proportions of nucleoside triphosphate substrates in the medium; these proportions could be varied from two to a thousandfold.

These results provided yet another definite confirmation of the principle of complementarity in the structure and synthesis of DNA. The experiments of the same authors on the incorporation of analogues of nucleotides into DNA synthesized *in vitro* were also significant in this respect. By replacing the usual precursors by corresponding derivatives of uracil, 5-bromouracil, hypoxanthine, 5-bromocytosine and 5-methylcytosine the authors showed that these also are incorporated into DNA. In these experiments each analogue specifically replaced a base which resembled it very closely in its ability to form hydrogen bonds in the structure of the DNA corresponding with the model of Watson & Crick. This result directly confirms the principles of reproduction of the specificity of DNA which are derived from the hypothesis on which this model is based. The fact that the newly synthesized polynucleotides are complementary to the polynucleotides of the "primer" has been demonstrated recently by the brilliant experiments of Josse, Kaiser and Kornberg (65, 66). These authors studied several parameters relating to the sequence of nucleotides in the newly formed DNA. They used a very ingenious and elegant method for this purpose. In these experiments the synthesis of DNA took place in a system containing purified polymerase from *Escherichia coli*, a special DNA "primer" and all four desoxynucleoside-

5′-triphosphates, one of which was labelled with $^{32}$P on the phosphate group adjoining the sugar residue.  When the triphosphate labelled in this way reacted, during the synthesis, with the last nucleotide on the growing end of the nucleotide chain, the $^{32}$P formed part of the phosphate bridge which was formed between the phosphate nucleoside with which it was combined and the terminal nucleotide mentioned above.  The DNA synthesized under these conditions was isolated and then broken down enzymically to 3′-desoxynucleotides in such a way that the $^{32}$P atom was in the 3′-position of the next-door neighbour nucleotide.  A micrococcal desoxyribonuclease or a diesterase was used to effect this breakdown. Thus, the label which had been introduced in one of the component nucleotide triphosphates, e.g. desoxy-ATP, was, after hydrolysis of the DNA which had been synthesized, "transferred" to a neighbour which might be any one of the four nucleotides.  The $^{32}$P content of each of the four 3′-desoxynucleotides isolated from such a hydrolysate by electrophoresis on paper enabled the authors to calculate to what extent the original nucleotides, which were labelled, in the first place, in the 5′-position and were used in the form of desoxynucleoside-5′-triphosphates, combined with each of the other nucleotides in the polynucleotide chain which was synthesized.  All these operations could be carried out four times using a particular type of "primer" DNA, a different desoxynucleoside-5′-triphosphate being used each time.  In this way it was possible to determine the relative frequency of formation of each of the 16 possible combinations of two adjacent nucleotides in the DNA which was synthesized.

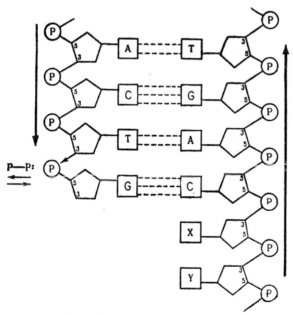

Fig. 42. Mechanism of enzymic replication of DNA.

This method is of wider significance and makes it possible to study the species differences of DNA. In the experiments which we have described the distribution of the frequencies of formation of the different combinations of nucleotides was not random and was appreciably changed in accord with the species characteristics of the primer DNA (fifteen samples of DNA, from animals, plants and microbes, were studied). In all cases the parameter of the distribution of the neighbouring nucleotides of the newly synthesized DNA was identical with that of the DNA which was used as the "primer". Further analysis of these results led the authors to the conclusion that the reaction which occurred really proceeded by the formation of hydrogen bonds between pairs of bases in the two chains of DNA and the new, "growing" polynucleotide increased in length in the opposite direction to that in which its complementary nucleotide or matrix would grow. Josse's (65) scheme of this synthesis is given in Fig. 42.

In the last few years enzymes of the same sort as Kornberg's polymerase have also been found in other objects, in particular in the tissues of mammals (27-29, 80, 81, 112a, 113, 114), which suggests that the synthetic reaction occurring *in vitro* which we described above is analogous to the process which takes place under biological conditions.

## 2. The synthesis of DNA *in vivo*

Polymerases catalysing the synthesis of DNA have been isolated from biological objects and it is therefore quite natural to suppose that they perform the same function in living cells. It is also natural to suppose that the DNA of the cells acts as the "primer" in this case while a system of special enzymes (kinases) which carries out the synthesis of nucleoside triphosphates *in vitro*, is responsible for the synthesis of the corresponding substrates *in vivo*. It has also been shown that, at the level of nucleosides and nucleotides, ribonucleotides can be converted into desoxyribonucleotides and thus the products of the metabolism of RNA may serve as building materials for DNA (15, 35, 42, 97a).

In fact the study of the processes of DNA synthesis which is being carried out in many laboratories shows that this process is completely analogous with the process which has been studied *in vitro* (38, 79, 113, 114, 45, 112, 27-29). In any case the synthesis of DNA consists essentially in replication and requires the same substrates and co-factors as are required *in vitro*.

It has not always been possible to use native DNA as a "primer" for the reactions taking place *in vitro*.

Owing to the existence of the double-helix structure in the "primer" DNA its efficacy is decreased or increased by complete or partial separation of the molecules into single polynucleotides by heating, partial denaturation or even partial hydrolysis by desoxyribonuclease (30, 74, 112a).

However, it is interesting to note that, although native DNA from the liver could not act as a "primer", the DNA from regenerating liver could act in this way for the synthesis of new DNA without any preliminary treatment (81). It would seem that when active synthesis is going on a certain percentage of the molecules of DNA are in the form of single-chain polynucleotides.

In Chapter XIII we saw that the synthesis of DNA in the cells is associated with particular phases of the cellular cycle and it would seem that, when conditions require synthesis, some sort of preparatory molecular change takes place in the molecules, enabling them to act as "primers".

The question of the nature of the molecular phenomena accompanying the synthesis of DNA in the cells continues to attract the attention of biochemists and geneticists, as being the process whereby the genetic information stored in the DNA is replicated. It is quite obvious that, if DNA is the carrier of the information, then the synthesis of DNA *in vivo* must be responsible for the maintenance and "multiplication" of this information.

According to the scheme of the structure of DNA worked out by Watson & Crick (cf. Chapter II) the synthesis of DNA is brought about by the formation of a new polynucleotide on the surface of an already existing one, after the hydrogen bonds in the pairs of bases of the original

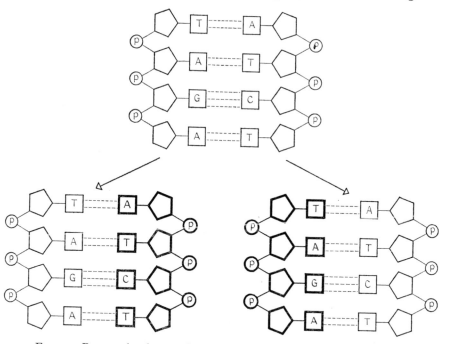

Fig. 43. Proposed scheme of replication of a Watson & Crick DNA model. Bold-lined polynucleotide chains of the two daughter molecules represent newly synthesised strands.

matrix molecule have been broken by some process, giving rise to a surface consisting of nitrogenous bases which have been set free to react with the low-molecular substrates in the medium. Free nucleotides combining with this surface form a new polynucleotide chain on the surface of the old one. A scheme of such a synthesis is given in Fig. 43. The essential feature of this scheme is the division of the molecule of DNA into two polynucleotides and the formation of "progeny" in which each molecule is composed of one "parental" and one newly formed polynucleotide chain. This has been somewhat altered by the recent work of Plishkin, Luchnik & Taluts (14) (Fig. 44). According to these authors the combination of the nucleotides with the original molecule of DNA precedes the disjunction of the strands of these molecules, though the formation of the hydrogen bonds takes place after this disjunction. The scheme put forward by these authors also makes several provisional

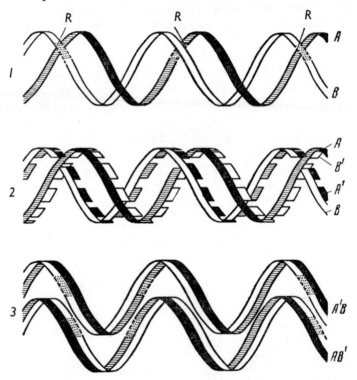

Fig. 44. Diagram of the replication and division of the molecule of DNA. *A* & *B* are the complementary chains of the original molecule. *R* represents a place where a break can occur.

1. Original molecule.
2. New nucleotides have become combined with the original chains by means of hydrogen bonds but the lengthwise bonds have not yet been formed so the molecule can divide.
3. Two molecules of DNA after division and subsequent formation of lengthwise bonds.

assumptions (the preliminary rotation of the bases around the glycoside bonds, the presence of non-coincident discontinuities in the polynucleotide chains, etc.), and this means that the mechanism of "preparation" for replication which they postulate is by no means more simple than the untwisting of the DNA from one end. A number of original ideas as to the possible mechanism of "fission" of the double helix of DNA have been expounded in the excellent review of Delbrück & Stent (40).

As well as the very popular model of the duplication of DNA devised by Watson & Crick, two others have also been put forward. One involves duplication while still retaining the original double helix, the two-stranded molecule building up another two-stranded molecule alongside itself. The other involves a duplication which is accompanied, not merely by the preliminary "fission" of the double helix, but also by fragmentation of each of the polynucleotides.

The scheme of duplication in which the original double helix is retained was worked out by Bloch (26). She bases her ideas on the stabilizing role of histone, which can keep both of the chains alongside it even when the hydrogen bonds are temporarily broken. According to Bloch's idea, the single polynucleotide chains remain combined with the helical histone after the breaking of the complementary hydrogen bonds, but they rotate through 180°. Each of them then builds a complementary chain. After this the new complementary bonds are broken and all four chains again rotate through 180° so that the old structure is completely restored and an identical new nucleotide complex arises alongside it.

The scheme which represents the synthesis of DNA as a process involving the loss of the molecular integrity of both of the polynucleotides and the disorderly distribution of their fragments in the newly synthesized molecules was put forward by Delbrück (39). In this case both of the chains of the daughter pair of polynucleotides of DNA consist of alternating sections of the original and newly formed material.

Bloch's scheme has been called conservative, Watson & Crick's semi-conservative and that of Delbrück dispersive. Fig. 45 demonstrates the difference between these schemes in the form of tentative models. In the last few years several other schemes of the duplication of DNA have been suggested (63, 64) but they are only modifications of previous models.

It would appear that analysis of these three main schemes of the duplication of DNA should lead to the selection of the one which best fitted the facts. However, the facts which are now gradually being collected indicate quite clearly that, in living nature, there are really three types of synthesis and new formation of DNA, a conservative, a semi-conservative and a dispersive which are encountered in different living things and would seem to reflect the evolution of the system of reproduction of DNA. The conservative type of replication of DNA has been studied recently by Cavalieri & Rosenberg (32) in the bacterium *Escherichia coli*. In their study of the molecular make-up of DNA in these bacteria in the various phases of their life cycle, they found that although the DNA in resting cells of *E. coli* occurs as two-stranded molecules corresponding with the model of Watson & Crick, during the period of active growth four-stranded molecules appear in the cells. It was shown in synchronized cultures that during the period of preparation for division these four-stranded molecules break down again into two-stranded molecules which then form a new double helix.

Another and semiconservative method of duplication of chromosomes has been observed in animals and algae. Several authors have used an ingenious method worked out by Meselson, Stahl & Vinograd (85) for studying this question. This method is based on the massive labelling of the polynucleotides with 5-bromodesoxyuridine and the subsequent separation of "heavy" and "light" polynucleotides by centrifugation in solutions of a specially chosen density. Using this method Simon (107) and Sueoka (119), working on animal tissues and *Chlamydomonas* respectively, have obtained results which agreed with the semiconservative method of duplication of DNA corresponding with the model of Watson & Crick. The semiconservative method of duplication of DNA has also been found to operate in the multiplication of various bacteriophages.

An exception, however, is the reproduction of DNA discovered by Sinsheimer in a special phage $\phi$X174, the solitary polynucleotide chain of which seems to be unpaired (cf. Chapter VII). When this phage multiplies, its polynucleotide undergoes fragmentation into smaller sub-

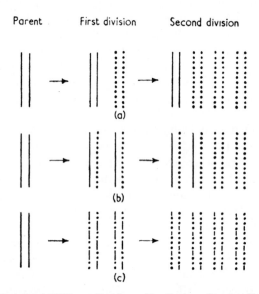

Fig. 45. Models of DNA replication. Predictions for the distribution of parental material between the daughter duplexes for the first and second division according to the three models of replication: (a) conservative, (b) semiconservative, and (c) dispersive.

units which are arranged at random in the DNA of the progeny in accordance with the dispersive scheme of duplication of DNA.

However, in recent work Sinsheimer and colleagues have shown (111a) that the process of replication of this phage is more complicated. After penetration of the cell there occurs the synthesis of a complementary polynucleotide chain; the reproductive form of DNA is formed as a double helix, and only then does the actual reproduction of the true phage DNA take place. Thus the dispersal of the original polynucleotide, which happens after formation of the complementary reproductive chain, does not signify the dispersal of genetic information.

Quite recently, however, Kozinski (71) has published a detailed study showing that dispersive distribution of polynucleotide material may occur in phages of the T group which have molecules of DNA in the form of a double helix with the enormous molecular weight of about $2 \times 10^7$-$4 \times 10^7$.

Kozinski used a very ingenious method for his experiments. The DNA of the host cells was labelled with 5-bromouracil and became "heavy". The DNA of the T4 phages was labelled with $^{32}$P. It is known that the synthesis of phage DNA takes place at the expense of the breakdown of the DNA of the host and therefore, if the duplication is conservative or semiconservative, after the second or third cycle of reproduction the light $^{32}$P polynucleotides will be retained in the progeny without further dispersal of labels in the progeny of the DNA. However, a different pattern of distribution was observed. The molecules of the DNA of all the progeny were "speckled" with fragments of the parental DNA labelled with $^{32}$P in accordance with the dispersive scheme. However, the author notes that there can be two types of dispersive method of duplication. On the one hand the distribution of the parental material may occur at random, as when alcohol is dissolved in water. On the other

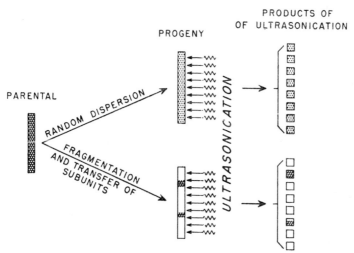

FIG. 46. Two possible modes of parental $^{32}$P dispersion in progeny DNA.

hand the dispersal may take the form of fragmentation, the DNA of the progeny being like a patchwork quilt with fragments of the parental DNA scattered fairly widely through it. In this case the structure of the gigantic molecule of DNA must also be fragmentary, consisting of sub-unit. which constitute the essential, discrete carriers of genetic informations A scheme of two possible types of dispersive replication of DNA is given in Fig. 46.

By partly breaking down the progeny of the DNA ultrasonically and later separating the fragments into a light fraction labelled with $^{32}$P and a heavy one, Kozinski showed that it is in fact "speckled" with large fragments of parental DNA. The molecular weight of these fragments varied *between* $5 \times 10^5$ *and* $5 \times 10^6$, *that is to say, it corresponds more or less with the molecular weight of the RNA which receives from the DNA the information which it contains.*

Kozinski believes that each of these discrete carriers of information is reproduced in accordance with the principles given in the model of Watson & Crick and that the assemblage of such fragments is essentially the template for the synthesis of all the other fragments of which the molecules of the DNA of the progeny are composed.

In recent years there have been a number of further publications, some confirming one, some another, and still others yet a third scheme of the reproduction of DNA (36, 43, 117, 89, 84). Any one of these may dominate the scene. However, what is important to us at the present moment is to note, in the first place, that *all three of the means of reproduction of DNA which have been observed are theoretically capable of ensuring the reproduction and conservation of genetic information.*

Even dispersive reproduction which seemed, until recently, an exception which it was hard to explain, does not involve random, but fragmentary dispersion, which perfectly satisfies the principle of the conservation of information. In this case the information is merely "regrouped".


## 3. The synthesis of RNA *in vitro*

Studies of the synthesis of RNA *in vitro* have been going on intensively for the last few years since 1955 when Ochoa discovered the peculiar enzyme, polynucleotide phosphorylase, which catalyses the formation of phosphoester bonds between the ribose components of nucleotides. Nucleoside diphosphates served as substrates for this reaction. For this discovery Ochoa was awarded a Nobel Prize which he shared with Kornberg who studied the synthesis of DNA.

The discovery of Ochoa served to start a long series of researches on the synthesis of polyribonucleotides *in vitro* (6, 7, 51, 52, 56, 91, 92). As these investigations have been reviewed in detail (3, 59, 110) we can merely deal shortly with the most important aspects of the reaction.

Even outside the organism, polynucleotide phosphorylase will bring about the synthesis of polymeric polynucleotides from nucleoside diphosphates according to the following equation:

$$n(\text{XRP-P}) \leftrightharpoons (\text{XRP})_n + n\text{P}$$

where R = ribose, P-P = pyrophosphate, P = inorganic phosphate and X one of the four nitrogenous bases (A, G, C or U). $\text{Mg}^{++}$ ions are required to carry out the reaction.

The enzyme forms polymers whether it acts on individual nucleoside diphosphates or on mixtures of them; the nucleotide composition of the polymers synthesized from a mixture of diphosphates corresponds almost exactly with the original composition of the incubation mixture. In other words, there was no competition between the nucleoside diphosphates during the processes of polymerization, as the energy of the internucleotide bond seems not to depend on the nature of the nitrogenous bases.

An interesting feature of the enzymic syntheses of polynucleotides described above is their very close similarity to RNA in very many respects (the type of the internucleotide bonds, molecular weight (30,000-1,000,000), physico-chemical properties, relationship to ribonuclease and phosphoesterase, etc.). Hart & Smith (55) found that the artificial biosynthetic polymers resembling RNA can react with the protein of tobacco mosaic virus to give rods which are like the ordinary viral rods in their structure. It has been shown that if solutions of adenylic and uridylic polynucleotides are mixed they form a double helix which gives a well-oriented X-ray photograph similar to those given by DNA.

Enzymes of the same type as polynucleotide phosphorylase have since been found in bacteria, yeasts, plants and animals. However, unlike the synthesis of DNA which we have described above, even when a primer was used in the form of a naturally occurring RNA its structure was not reproduced during the synthesis of RNA and the sequence of nucleotides in the newly synthesized RNA was random (97).

In a series of interesting experiments on the part played by oligonucleotides as primers in the action of polynucleotide phosphorylase Singer and colleagues (57, 108-111) found that the addition of oligonucleotides to a mixture of nucleoside diphosphates and polynucleotide phosphorylase markedly accelerates the process of polymerization. The oligonucleotide itself enters into the macromolecule being synthesized and acts as its "core". The effect of hastening polymerization is observed even when dinucleotides are added to the medium; an increase in the molecular weight of the primer was associated with a sharp decrease in the concentration required. It was also found that the primers were specific, but with exceptions. Thus polymers of adenylic acid only activated the polymerization of adenosine diphosphates, polymers of uridylic acid activated that of uracil diphosphates and so on. The incorporation of primers such as A-A into polynucleotide chains consisting otherwise of uracil was, however, observed.

PB X

It was also found that, although a primer was required for the poly-merization of guanosine diphosphate molecules, such a primer could also serve for the combination of adenylic acid with uracil.

The problem of the significance of polynucleotide primers in *in vitro* synthesis of polynucleotides has taken a new turn as a result of the work of Ochoa & Mii (95a), who showed that even highly purified preparations of *Azotobacter* polynucleotide phosphorylase contain about 3% of firmly bound polyribonucleotide. In these experiments the enzyme was purified by a series of operations to 500 times its original activity; it retained a fairly small polynucleotide chain, corresponding in base composition to that of the total RNA of the microorganism. The polynucleotide could only be removed by procedures which also resulted in inactivation of the enzyme. The authors left the conclusion open whether this polynucleo-tide is a primer of synthesis, a prosthetic group or a firmly bound im-purity. However, it is clear that in all the previous experiments a small amount of polynucleotide material was introduced into the synthetic system along with the enzyme.

These and several other features of the synthesis of polyribonucleo-tides under the influence of polynucleotide phosphorylase indicate clearly that the synthesis of RNA *in vitro* is, as known at present, far from being an imitation of the synthesis of biologically active RNA *in vivo* and that polynucleotide phosphorylase is probably only one component of a complicated biochemical system for synthesizing RNA within the cells.

## 4. Synthesis of RNA *in vivo*

### a) Does polymerization of nucleoside diphosphates to form RNA occur in vivo?

Researches into the incorporation of nucleotides and their pre-cursors during the synthesis of RNA have been carried out by many workers (41, 48, 44, 45, 58, 101-105, 120, 47). However, the single fact of the "incorporation" of nucleotides into RNA does not give any idea of the mechanism of the synthesis of this polymer or of the enzymes which are responsible for the process. Certainly the incorporation of nucleo-tides does suggest the occurrence of RNA synthesis but there are no serious grounds for believing that the synthesis of RNA in the cell is accomplished mainly by polynucleotide phosphorylase in the presence of nucleoside diphosphates and a primer. The role of the polynucleotide phosphorylase in the cells, however, has not yet been established satis-factorily, but it seems to be only a part of some more complicated system which is not only responsible for the synthesis of the polynucleotide chain, but also for the reproduction of the specificity of the RNA in respect of the transfer of information from DNA.

*b) The building up of the polymeric forms of RNA from
smaller sub-units*

In Chapters V and VI we have already dealt with the interesting series
of researches carried out by Roberts, Aronson and their colleagues at the
Carnegie Institute in Washington in which they studied the peculiarities
of the RNA of ribosomes of different sizes.    In recent researches made
in the same laboratory Aronson & McCarthy (21, 82) obtained some very
remarkable evidence as to the kinetics of the synthesis of this RNA.    As we
mentioned earlier, the large ribosomes of *E. coli* which have a sedimentation constant of 70 S are composite and may be divided into 50 S and 30 S
particles.    The RNA in these ribosomes is at a different level of polymerization.    The 50 S ribosomes contain molecules of RNA of two types,
having sedimentation constants of 28 S and 18 S (molecular weights of
$1 \cdot 5 \times 10^6$ and $5 \times 10^5$) while the 30 S ribosomes only contain 18 S RNA.
However, the authors succeeded in finding some very fine 20 S ribosomes
in the cells of *E. coli* and these contained comparatively low-molecular
RNA with sedimentation constants of 4 S and 8 S (molecular weights
30,000-144,000).    The RNA with a molecular weight of 30,000 was
different from S-RNA and could be separated from it.    It is interesting
that, when dialysed against a buffer of low ionic strength, the large molecules (28 S and 18 S) of RNA broke down successively into 13 S particles
(mol. wt. 300,000), 8·8 S particles (mol. wt. 144,000) and 4·4 S particles
(mol. wt. 29,200) with a high degree of homogeneity in each of the fractions.    This breakdown thus had something of the nature of a fragmentation into sub-units.    The smallest units were the most stable.    On the
basis of these results the authors suggested that the synthesis of large
molecules of RNA may occur by aggregation in the reverse order.    The
kinetics of the incorporation of [14C]uracil into 4 S and 8 S RNA did, in
fact, suggest that they may act as precursors of 18 S and 28 S RNA.    4 S
and 8 S RNA incorporated [14C]uracil more quickly, and then the decrease
in the activity of this fraction coincided with its incorporation into larger
fragments.    The authors suggest that during the synthesis of ribosomal
RNA there is a successive increase in the size of the particles of something like 4 S → 8 S → 13 S → 18 S → 28 S with a consequent increase
in the size of the ribosomes according to the scheme 20 S → 30 S →
50 S → 70 S → 100 S.

It is interesting to note that viral RNA breaks down in successive
stages on standing into fragments constituting one third, one sixth and
one eighth of the original length of the RNA, which is evidence for the
possible existence of the sub-units of RNA which have been observed by
other authors (33).    Hall, Storck & Spiegelman (54) have recently published the results of their investigations which coincide almost exactly
with those of Aronson & McCarthy.    The "small" RNA (4 S-10 S)
contained in the ribosomes of *E. coli* was able to incorporate labelled

compounds quickly but reversibly and the "small" component was found as an admixture to the larger ones in the 30 S, 50 S and 70 S ribosomes.

The distinction between S-RNA and the low-molecular fractions of RNA with a specially rapid rate of renewal was found in liver cells as well (98). It is interesting that a polynucleotide phosphorylase has recently been found in E. coli ribosomes (119a).

In some recent, interesting and original studies by a large group of scientists (31, 49) using the method of "pulse labelling" in which $^{32}$P was administered to the cells for 10-20 sec., it was found that there was a very quick turnover of a fraction of RNA having a molecular weight of the order of 300,000 (14 S) and forming a part of the 70-100 S ribosomes. However, this actively synthesized RNA differed from the high-molecular RNA of the ribosomes in its nucleotide composition, which argues against its being used for building up the larger molecules. So far we have no certainty that this RNA is a quite special fraction which is different from the 4 S and 8 S fractions of Aronson & McCarthy. The rapidly synthesized 14 S RNA found in E. coli had a nucleotide composition related to that of the DNA of the cells. When the cells were infected with the T2 phage an analogous RNA quickly appeared in them, having a composition related to the DNA of the phage (31). The authors suggest that this RNA is the special form of RNA which acts as a "messenger" carrying information from the DNA and which is synthesized on the DNA and then combines with the ribosomes. In their opinion it is this "messenger" RNA which acts as the immediate matrix for protein synthesis while the polymeric RNA of the ribosomes carries out other, non-specific functions.

We shall discuss the hypothesis put forward by these authors as to the significance of the different forms of RNA in bringing about protein synthesis in the next chapter. Here we must emphasize that the clearly manifested functional and molecular heterogeneity of RNA makes the problem of the way in which it is synthesized in the cells considerably more complicated than one might have expected from experiments carried out *in vitro*.

## 5. The synthesis of a special form of RNA under the direct control of DNA

### a) *Theoretical considerations on the possibility of synthesis of RNA on the surface of DNA*

It would seem that the transfer of information from DNA to RNA can only take place by direct means, by the actual synthesis of a poly-ribonucleotide chain on the surface of a polydesoxyribonucleotide chain with the transfer of information based on the principle of the complementarity of the nitrogenous bases.

It is interesting to note that, even several years before the experimental discovery of the synthesis of special forms of RNA on the surface of

DNA, various possible theoretical schemes of such a synthesis had been worked out. It is interesting to compare these theoretical schemes with the facts which have later been brought to light experimentally.

The original model of a possible mechanism for the self-reproduction of DNA with the simultaneous synthesis of RNA on the surface of the DNA, was presented by Stent in his address to the Fourth International Congress of Biochemistry (116). This scheme was based primarily on evidence from the genetics and hybridization of phages. Stent departed quite radically from the ideas of Watson & Crick about complementary synthesis and tried to represent the synthesis of DNA as a two-stage, complex process including the simultaneous mutual transfer of specificity between DNA and RNA. Stent suggested that when the molecules of phage DNA entered the bacterial cell they presented themselves as a double helix, serving as matrices for the synthesis of *ribonucleoprotein* molecules composed of a protein "backbone" with a single ribbon of RNA which grew as a third polynucleotide in the deep groove of the double helix of the DNA, thus forming a three-ply helical structure (Fig. 47). The nitrogenous bases of this RNA thus interacted with the complementary pairs of bases making up the DNA, each such pair forming two subsidiary hydrogen bonds with a single, and therefore highly determinate, type of third base, such that the bond between the carbon of its ribose and the nitrogen of its base is always in the same position. The nature of these interactions was represented by Stent as follows (Fig. 48). In this scheme the ordinary pairs of bases of the Watson & Crick model determine the position of the bases of the RNA as follows: the pair guanine-cytosine fixes the position of cytosine in the RNA, the pair thymine-adenine fixes that of adenine, the pair adenine-thymine fixes that of uracil and the pair cytosine-guanine fixes that of guanine. The structural blocks for the synthesis of the third polynucleotide chain could, in Stent's opinion, be complexes of an amino acid with a nucleotide and thus the polypeptide "backbone" could be built up at the same time as the RNA. After a new molecule of RNA has separated from its matrix the whole process can be repeated. According to Stent this is the

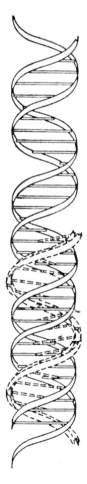

FIG. 47

The growth of a third polynucleotide-protein chain (lower part of diagram) in the deep groove of duplex DNA (upper part of diagram).

way in which genetic information is transferred from DNA to RNA. However, the author puts forward the further suggestion that the whole process may be reversible in the sense that the polynucleotide chain of the RNA, when formed, can serve as a matrix for the reproduction of the double helix of the DNA.

Although it is so hypothetical, a discussion of this scheme is of a certain theoretical interest, especially as concerns those parts of it which deal with the possible mechanism of transfer of specificity from DNA to RNA. An analogous scheme of the possible synthesis of RNA on the surface of DNA has been worked out by Zubay (126) independently of Stent. He also takes the view that the matrix for the synthesis of the RNA is the double helix of DNA and that the position of an individual nucleotide of RNA is determined by a pair of complementary bases of the DNA joined by hydrogen bonds. (There can be four such pairs: A—T, T—A, G—C, C—G.) In order to convert his purely mathematical schemes into stereochemical models of such triplets (two bases from the DNA and one from the RNA) the author has undertaken a number of stereochemical calculations and began his work with the construction of a three-dimensional model of this kind. However, even in the preliminary communication which we have mentioned, he showed that if the specificity of these

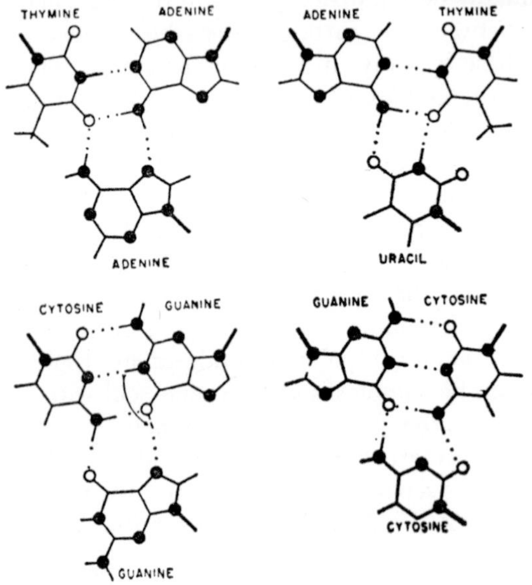

FIG. 48. Triplets of purine and pyrimidine bases. In each triplet the two upper bases represent the complementary pairs of the Watson & Crick scheme in the double helix of DNA while the lower base belongs to the polyribonucleotide chain of RNA. The nitrogen atoms are shown as black circles and the oxygen atoms as white ones. Hydrogen bonds are shown as dotted lines and the bonds joining the nitrogen atoms to carbohydrate groupings are shown as heavy lines.

triplets was to be ensured it would require a number of fairly complicated changes in the configuration of the DNA (the breaking of some hydrogen bonds between monomers, alteration of the projection of the bases, etc.).

Unlike Stent and Zubay, some other authors consider it more likely that the synthesis of RNA is under the control of a single polynucleotide chain of DNA rather than the double helix. A typical example of a hypothesis of this sort is the scheme of Lockingen & DeBusk (77). In their scheme parts of one of the chains of the double helix may become detached and a "skin" may form over the "free" area, consisting of the nucleotides of the RNA which reproduce the characteristic sequence of the DNA with the substitution of uracil for thymidine. The chain of RNA which has been formed is then removed from the DNA and itself begins to reproduce new, complementary polynucleotides.

An interesting analysis of the stereochemical possibilities for inter-action between RNA and DNA during their synthesis has been made recently by Rich (98a). This author undertook a critical review of schemes like those of Stent and Zubay, which we have described, based on the transfer of information by the double helix of DNA, so that a pair of bases in the DNA determines the position of one nucleotide of the RNA. Rich showed that experiments on the association of artificial polynucleo-tides with the formation of a 3-ply structure do not confirm the reality and selectivity of the triplets of nucleotides postulated by Stent. For example, the interaction of artificially synthesized polyadenylic and polyuridylic acids leads to the formation of a two-stranded molecule, very similar in many physico-chemical respects to DNA (99). X-ray structural analysis shows that in these molecules there are pairs of bases analogous to the thymine-adenine pairs of DNA. It has been found experimentally that such a two-stranded polynucleotide can combine with a third poly-nucleotide if this is a polyuridylic acid, but will not form complexes with polycytidylic or polyadenylic acids. X-ray structural analysis showed that in such a structure of two polyuridylic and one polyadenylic chain the third additional polyuridylic chain was only combined with the polyadenylic chain through the amino group of the adenine and the nitrogen atom in the 7 position in the purine nucleus.

Rich considers that neither Stent nor Zubay has proved the strict stereochemical selectivity of the triple combinations and he is therefore inclined to the view that it is more likely that RNA is synthesized in contact with only one of the polynucleotides of DNA. He considers it possible that hybrid double helices are formed of one chain of RNA and one of DNA, with subsequent removal of the RNA, and he suggests that experiments with artificial polynucleotides might help in finding the most probable way of transferring information from DNA to RNA.

It must be mentioned that the experimental evidence, derived from models, concerned with the possible synthesis of RNA on the surface of DNA also corresponds better with a scheme of synthesis in which the nitrogenous bases of a single-stranded polynucleotide of DNA serve as the surface on which the new polyribonucleotide chain is manufactured. In model experiments Rich (100) has recently obtained a hybrid helix in which polyriboadenylic acid was bound complementarily to polydesoxy-ribothymidylic acid with the formation of a double helix.

*b) The enzymic synthesis of an RNA complementary to a DNA which is present as a primer in the synthesizing system*

In 1959-60 workers in several laboratories discovered the process of "incorporation" of labelled nucleoside triphosphates (ATP, GTP, CTP and UTP) into RNA, which depends (both *in vitro* and *in vivo*) on the presence of DNA in the synthesizing system (61, 62, 25, 46, 90, 93, 94, 118, 121-124). This process was called the DNA-dependent synthesis of RNA. In some cases the DNA-dependent "incorporation" of nucleoside triphosphates into RNA really consisted in their incorporation into pre-existing RNA (123, 124, 94) while in other cases the process had all the characteristics of *de novo* synthesis. The "incorporation" of nucleoside triphosphates into pre-existing DNA required the presence of native, two-stranded molecules of DNA (94, 118) while *de novo* synthesis required the presence of partly denatured and untwisted molecules of DNA. The enzymic system concerned with the DNA-dependent synthesis of RNA seems to be localized in the nuclei of the cells (25).

Naturally these new results at once attracted the attention of workers in many laboratories, as they led up to the idea that this synthesis of RNA which depends on the presence of DNA also constitutes the formation of the theoretically postulated "carrier of genetic information" which translates the genetic information of the DNA into the form of active matrices for the synthesis of specific proteins.

Apart from the discovery of an enzyme for synthesizing RNA, which was only active in the presence of DNA, it was also discovered that special forms of RNA are synthesized in bacterial cells infected with phage, their nucleoside composition being related to that of the DNA of the phage (22, 31, 53, 90). In recent studies it has been found that this special form of RNA synthesized in the cells of *E. coli* (the nucleoside composition of which is related to that of the DNA of the phage or of its own DNA), when combined with the ribosomes, does in fact serve as active matrices for the synthesis of proteins and is the carrier of genetic information (messenger RNA) (49, 125). The question of the role of this RNA in the synthesis of protein will be discussed in a later chapter.

Further confirmation of the fact that the mechanisms being worked out in all these cases were different aspects of one and the same phenomenon was provided by the observations of Hall & Spiegelman (53) who found that specific complementary complexes could be formed between one polynucleotide chain of the DNA of the phage T2 and the RNA synthesized in the course of the infective process. Single polynucleotides of

DNA derived from other sources could not form complexes with this RNA in the form of a double helix. This shows that the correspondence between the nucleotide composition of the RNA and DNA ensured their complementarity which, theoretically, could only have arisen by the synthesis of the RNA in contact with the polynucleotide chain of the DNA. In a later paper from the same laboratory (114a) it was found that cells of *E. coli* infected with phage contained mixtures of hybrid (DNA-RNA) double helices similar to those which had been obtained artificially. The quantity of DNA-RNA hybrids amounted to $0 \cdot 1\%$-$1\%$ of the total phage-specific DNA. These figures agreed with the calculations of Sarkar (106), who found that when DNA from liver is purified by various methods it is only possible to remove about $1\%$ of RNA from it. Mixed DNA-RNA complexes have been observed in *Neurospora crassa* (106a) and in plants (30a). In the latter work Bonner and colleagues found that RNA, which had been synthesized enzymically from riboside triphosphates under the control of chromatin isolated from pea embryos, remained bound to the chromatin in the form of a DNA-RNA-protein complex. The RNA could be freed from the complex by heating to 60°. The RNA:DNA ratio in the complex was 1:2.

It must be assumed that the experimental observations of the incorporation of ribonucleotides into DNA in experiments *in vivo* (60, 72) are also connected with the synthesis of RNA in contact with DNA.

A particularly detailed study has been made of the reaction involved in the DNA-dependent synthesis of RNA in the recent researches of Weiss and colleagues (124a, 46a). They succeeded in establishing that this reaction is catalysed by a special enzyme which they called "DNA-dependent RNA polymerase". Huang & Bonner (59a) have called this enzyme "chromosomal RNA synthetase". This enzyme is localized in the nucleus and the RNA which it synthesizes corresponds exactly with the primer DNA. Using the method of Josse (cf. p. 290) to determine the approximate nucleotide composition the authors showed that, in regard to this characteristic, which was determined for all 16 possible pairs of nucleotides, the DNA and RNA were the same (with the substitution of thymine for uracil). The same enzyme synthesized different forms of RNA in the presence of primer DNA derived from different sources (*Micrococcus lysodeikticus*, T2 phage, calf thymus and *E. coli*). The reaction of synthesis of RNA took place in accordance with the following equation:

$$
\begin{aligned}
&n \text{ ATP} \\
&\quad + \quad \text{DNA} \\
&n \text{ GTP} \quad \rightarrow \\
&\quad + \quad \leftarrow \\
&n \text{ CTP} \\
&\quad + \\
&n \text{ UTP}
\end{aligned}
\quad
\begin{bmatrix}
\text{A} & \text{p} \\
& \downarrow \\
\text{G} & \text{p} \\
& \downarrow \\
\text{C} & \text{p} \\
& \downarrow \\
\text{U} & \text{p}
\end{bmatrix}_n
\quad + 4n \text{ pyrophosphate}
$$

According to the findings of Weiss and colleagues, the RNA formed in this system is complementary to the single-chained polynucleotides of DNA and, on heating and cooling of the mixed solutions, could easily give rise to hybrid DNA-RNA helices. These helices were highly specific and were only formed when the solution contained a particular DNA and the RNA formed on it.

According to these authors the formation of this type of RNA *in vivo* requires the presence of single-stranded DNA and it seems that the process of unwinding the DNA may be incomplete and may, according to the authors, be carried out by the enzyme itself. When this happens it may be that both of the chains of the partly unwound DNA can act as primers.

The authors suggest that the RNA formed by the DNA-dependent polymerase of the nucleus is the same as "messenger RNA". We must also mention that according to the recent results of Hayashi & Spiegelman (55a) the "messenger RNA" formed in the cells of *E. coli* and other bacteria *in vivo* has many properties in common with the RNA formed by DNA-dependent polymerase *in vitro* when the DNA of *E. coli* is used as a primer. "Messenger RNA" forms the same complementary hybrids with single-stranded homologous DNA and has a corresponding composition. The enzyme which synthesizes this RNA in *E. coli* has recently been isolated by Stevens (118a) and was only active in the presence of DNA.

The role of the primer in the DNA-dependent synthesis of RNA has been studied in two different laboratories, each group using the same approach, a comparison of the priming activity of the single-chain form of DNA of the small phage $\phi$X174 with its double-chain (reproductive form (32a, 32b, 45a)). The following data illustrate the results of one of these experiments (32a):

| DNA primer or product | Base composition (moles %) | | | |
| --- | --- | --- | --- | --- |
| | A | U or T | G | C |
| Single-chain DNA | 24·6 | 32·8 | 24·1 | 18·5 |
| RNA from single-chain DNA | 32·0 | 24·1 | 19·5 | 24·3 |
| Double-chain DNA | 28·7 | 28·7 | 21·3 | 21·3 |
| RNA from double-chain DNA | 28·8 | 29·1 | 20·9 | 20·9 |

As the newly synthesized RNA had complementary base composition to that of the DNA regardless of whether it was single-chain or double-chain, it follows that, in the presence of double-chain DNA, both of its chains are copied. DNA-dependent RNA synthesis was also studied in detail in the second paper of Chamberlin & Berg (32b). This RNA was found to have sedimentation constant 6-7·5 S. The newly synthesized RNA in these experiments exceeded in amount the primer DNA by a factor of 2-15.

It is interesting that in the work of Schulman & Bonner (106a), which has already been mentioned, there is found in *Neurospora* a DNA-RNA complex to which the authors tentatively assigned a triplet character (DNA:RNA = 2:1). This is different from the findings of other workers and, if confirmed, supports the ideas of Stent and Zubay on the possible occurrence of such triplets.

*c) Is the synthesis of DNA on polyribonucleotide matrices possible?*

The complementary synthesis of RNA on the matrix surface formed by the nitrogenous bases of a single-chain polynucleotide of DNA implies the theoretical possibility of the reverse process, i.e. the synthesis of DNA under the complementary control of RNA.

The only experimental indication yet available of the possibility that such a process may occur is the work of Gershenzon and colleagues (4, 5) on the synthesis of the DNA of the polyhedral virus of the silkworm under the influence of the RNA of the infected host (cf. Chapter VII). However, their first publication along these lines was of a preliminary nature and may be susceptible to other interpretations so that, before making any final assessment of whether such processes are possible, we must await the results of the further investigations which are being pursued intensively in that laboratory.

*d) The dependence of RNA synthesis on protein synthesis*

Gros & Gros (50) have established that the auxotrophic mutants of *E. coli*, which require methionine, tryptophan and β-phenylalanine or proline, could only synthesize RNA in the presence of whichever amino acid was required by the strain in question. The strain which required threonine and valine together only began to synthesize RNA (judging by the incorporation of [$^{14}$C]adenine) in the presence of both amino acids. The authors made a very detailed study of this dependence and found that only very small, catalytic amounts of all amino acids were required for the synthesis of RNA. Calculations showed that one molecule of amino acid is responsible for the polymerization of eight mononucleotides. It was important, from this point of view, that the amino acid should have a free carboxyl group but a free amino group was not necessary. The authors therefore suggest that the amino acids are needed for the activation of the synthesis of RNA from precursors, which is accomplished by the formation of complexes of nucleotides with amino acids. It is also interesting that the introduction of the amino acid analogue, *p*-fluorophenyl-alanine, into the medium inhibited the synthesis of RNA.

The interesting system of the interrelated syntheses of protein, RNA and DNA has been studied by Okazaki & Okazaki (96). These authors cultivated *Lactobacillus acidophilus* on media which were deficient in thymidine, uracil and amino acids, which specifically suppressed the formation of DNA, RNA and protein respectively. It was found that prevention of the synthesis of DNA impeded the synthesis of RNA slightly but had no effect on the synthesis of proteins. Prevention of RNA synthesis impeded the formation of proteins but had no effect on DNA. Deficiency of amino acids, in its turn, halted the synthesis of RNA but also had no effect on the synthesis of DNA.

Maaløe & Hanawalt (78) found recently that the synthesis of protein and RNA is necessary to start off a new cycle of duplication of DNA but that the cycle, when once begun, will complete itself even under conditions under which the synthesis of proteins and RNA is inhibited.

The nature of these interrelationships is still obscure. The formation of DNA certainly requires the parallel formation of protein bonds (73). As for RNA, the facts mentioned above and the augmentation of the synthesis of RNA by mixtures of amino acids (83, 87, 88, 73) may indicate that the system of RNA, which is functionally specialized to ensure the synthesis of proteins, is a dynamic one and is only set to work when synthesis is necessary. In the absence of protein synthesis the system does not continue to work to no purpose but curtails its metabolism, which not only economizes the resources of the cell, but also does not lead to unbalance in the composition of the cells. The dependence of the synthesis of both soluble and ribosomal RNA on the presence in both medium and cells of amino acids (even when protein synthesis has been inhibited by chloramphenicol) has been the subject of an interesting investigation by Kurland & Maaløe (73a). The correlation observed by these workers led them to the hypothesis of a mutual linkage of amino acid activation and RNA synthesis, based on the scheme of induction-repression as developed by Jacob & Monod in explanation of the induced synthesis of enzymes (cf. Chapter VIII). According to Kurland & Maaløe the amino acids play the role of inducers, while the soluble RNA acts as a repressor, inhibiting the synthesis of template or ribosomal RNA at the corresponding genetic locus. The combination of amino acids with the soluble RNA removes this repression and thus stimulates the synthesis of high-polymeric RNA. Aronson & Spiegelman (21a) showed recently that a certain amount of newly formed, specific protein combined with RNA is necessary to change RNA from the free state into an active component of the ribosomes. Not only is this protein physically bound to the RNA, but its formation occurs in parallel with that of the RNA. This, in the opinion of the authors, explains the need for amino acids in the synthesis of RNA.

## Conclusion

This chapter only contains a very short survey of the facts and theoretical concepts concerning the synthesis of nucleic acids. We have tried to select from among the extensive and rapidly growing literature on the synthesis of nucleic acids only those results which refer to their functions as matrices for protein synthesis and as self-reproducing polymers.

In comparing DNA, RNA and proteins with regard to the mechanisms by which they are synthesized, it is easily seen that the first member of this series has the simplest system of reproduction and itself plays the greatest part in the process of its own reproduction, being responsible for its specificity. The synthesis of RNA is more complicated and several enzymic systems help to bring it about. As well as the autonomous and apparently less accurate reproduction and the increase in length of the polymers, a certain part of the synthesis of RNA takes place in direct contact with DNA, thus achieving the transfer of information by reproducing in the RNA the nucleotide sequence of certain parts of the DNA. In the RNA this "information" controls the further synthesis of proteins which takes place on RNA matrices. In earlier chapters we have already made a detailed survey of the connection of various fractions of RNA with protein synthesis and the role of RNA as the matrix of protein synthesis. However, the question of the way in which the information carried by the RNA is converted into a sequence of amino acids has not yet been dealt with in this book. So far we have only referred to the facts concerned with this problem and have not dwelt on the theories and hypotheses which try to explain the mechanism of this process. They will form the subject of the next chapter.

## REFERENCES

1. BELOZERSKIĬ, A. N. (1959). *Nukleoprotidy i nukleinovye kisloty rasteniĭ i ikh biologicheskoe znachenie.* (*Bakhovskaya lektsiya*), Moscow: Izd. Akad. Nauk S.S.S.R.
2. BELOZERSKIĬ, A. N. & SPIRIN, A. S. (1956). *Uspekhi sovremennoĭ Biol.* **41**, 144.
3. BOGOYAVLENSKAYA, N. V. & TONGUR, V. S. (1959). *Uspekhi sovremennoĭ Biol.* **48**, 19.
4. GERSHENZON, S. M., KOK, I. P., VITAS, K. I., DOBROVOL'S'KA, G. M. & SKURATOVS'KA, I. N. (1960). *Dopovidi Akad. Nauk Ukr. R.S.R.*, p. 1638.
5. — (1961). *Proc. V int. Congr. Biochem., Moscow*, **9**, 150.
6. GRYUNBERG-MANAGO [GRUNBERG-MANAGO], M. [V.] (1958). *Biokhimiya*, **23**, 307.
7. GRUNBERG-MANAGO, M. (1959). *Proc. first internatl. Sympos. on the origin of life on the Earth* (ed. A. I. Oparin *et al.*), p. 344. London: Pergamon.
8. DEBOV, S. S. (1954). *Uspekhi biol. Khim.* **2**, 115.
9. EFIMOCHKINA, E. F. & POSNANSKAYA, A. A. (1957). *Voprosy med. Khim.* **3**, 243.
10. KOROTKORUCHKO, V. P. (1958). *Ukraïn. Biokhim. Zhur.* **30**, 128.
11. — (1959). *Obmen purinov v tkanyakh zdorovykh i porozhennykh opukholyami zhivotnykh.* Kiev: Izd. Akad. Nauk Ukr. S.S.R.
12. MEDVEDEV, ZH. A. (1953). *Uspekhi sovremennoĭ Biol.* **36**, 161.

13. NUZHDIN, N. I. (1958). *Agrobiologiya*, No. *1*, 3.
14. PLISHKIN, YU. M., LUCHNIK, N. V. & TALUTS, G. G. (1959). *Biofizika*, **4**, 275.
15. SALGANIK, R. I., MOROZOVA, T. M. & KIKNADZE, I. I. (1959). *Doklady Akad. Nauk S.S.S.R.* **129**, 947.
16. TONGUR, V. S. (1960). *Uspekhi sovremennoĭ Biol.* **49**, 156.
16a. KHESIN, R. B. (1961). *Zhur. vsesoyuz. khim. Obshchestva im. D. I. Mendeleeva*, **6**, 254.
17. CHARGAFF, E. (1959). *Proc. first internatl. Sympos. on the origin of life on the Earth* (ed. A. I. Oparin *et al.*), p. 297. London: Pergamon.
18. CHEPINOGA, A. P. (1956). *Nukleinovye kisloty i ikh biologicheskaya rol'*. Kiev: Izd. Akad. Nauk Ukr. S.S.R.
19. SHMERLING, ZH. G. (1954). *Uspekhi sovremennoĭ Biol.* **38**, 18.
19a. ABRAMS, R. (1961). *Ann. Rev. Biochem.* **30**, 165.
20. ADLER, J., LEHMAN, I. R., BESSMAN, M. J., SIMMS, E. S. & KORNBERG, A. (1958). *Proc. natl. Acad. Sci. U.S.* **44**, 641.
21. ARONSON, A. I. & McCARTHY, B. J. (1961). *Biophys. J.* **1**, 215.
21a. ARONSON, A. I. & SPIEGELMAN, S. (1961). *Biochim. biophys. Acta*, **53**, 70.
22. ASTRACHAN, L. & FISHER, T. N. (1961). *Proc. V int. Congr. Biochem.*, *Moscow*, **9**, 121.
23. BESSMAN, M. J., LEHMAN, I. R., ADLER, J., ZIMMERMAN, S. B., SIMMS, E. S. & KORNBERG, A. (1958). *Proc. natl. Acad. Sci. U.S.* **44**, 633.
24. BESSMAN, M. J., LEHMAN, I. R., SIMMS, E. S. & KORNBERG, A. (1958). *J. biol. Chem.* **233**, 171.
25. BISWAS, B. B. & ABRAMS, R. (1961). *Fed. Proc.* **20**, 362.
26. BLOCH, D. P. (1955). *Proc. natl. Acad. Sci. U.S.* **41**, 1058.
27. BOLLUM, F. J. (1958). *J. Amer. chem. Soc.* **80**, 1766.
28. — (1959). *Fed. Proc.* **18**, 194.
29. BOLLUM, F. J. & POTTER, V. R. (1958). *J. biol. Chem.* **233**, 478.
30. BOLLUM, F. J. (1959). *J. biol. Chem.* **234**, 2733.
30a. BONNER, J., HUANG, R.-C. C. & MAHESHWARI, N. (1961). *Proc. natl. Acad. Sci. U.S.* **47**, 1548.
31. BRENNER, S., JACOB, F. & MESELSON, M. (1961). *Nature (Lond.)*, **190**, 576.
32. CAVALIERI, L. F. & ROSENBERG, B. H. (1961). *Biophys. J.* **1**, 317, 323, 337.
32a. CHAMBERLIN, M. & BERG, P. (1962). *Fed. Proc.* **21**, 385.
32b. — (1962). *Proc. natl. Acad. Sci. U.S.* **48**, 81.
33. CHEO, P. C., FRIESEN, B. S. & SINSHEIMER, R. L. (1959). *Proc. natl. Acad. Sci. U.S.* **45**, 305 (1959).
34. COHEN, S. S. (1957). In *A Symposium on the chemical basis of heredity* (ed. W. D. McElroy & B. Glass), p. 651. Baltimore, Md.: Johns Hopkins Press.
35. — (1960). *Cancer Res.* **20**, 698.
36. DAOUST, R., LEBLOND, C. P., NADLER, N. J. & ENESCO, M. (1956). *J. biol. Chem.* **221**, 727.
37. DAVIDSON, J. N. (1960). *The biochemistry of the nucleic acids* (4th edn.). London: Methuen.
38. DAVIDSON, J. N., SMELLIE, R. M. S., KEIR, H. M. & McARDLE, A. H. (1958). *Nature (Lond.)*, **182**, 589.
39. DELBRÜCK, M. (1954). *Proc. natl. Acad. Sci. U.S.* **40**, 783.
40. DELBRÜCK, M. & STENT, G. S. (1957). In *A Symposium on the chemical basis of heredity* (ed. W. D. McElroy & B. Glass), p. 699. Baltimore, Md.: Johns Hopkins Press.
41. DINNING, J. S., SIME, J. T. & DAY, P. L. (1956). *Biochim. biophys. Acta*, **21**, 383.

42. FEINENDEGEN, L. E., BOND, V. P. & PAINTER, R. B. (1961). *Exptl. Cell Research*, **22**, 381.
43. FORRO, F. (jun.) & WERTHEIMER, S. A. (1960). *Biochim. biophys. Acta*, **40**, 9.
44. FRIEDKIN, M., TILSON, D. & ROBERTS, D. (1956). *J. biol. Chem.* **220**, 627.
45. FRIEDKIN, M. & WOOD, H. (IV) (1956). *J. biol. Chem.* **220**, 639.
45a. FURTH, J. J., HURWITZ, J. & ANDERS, M. (1962). *Fed. Proc.* **21**, 371.
46. FURTH, J. J., HURWITZ, J. & GOLDMANN, M. (1961). *Fed. Proc.* **20**, 363.
46a. GEIDUSCHEK, E. P., NAKAMOTO, T. & WEISS, S. B. (1961). *Proc. natl. Acad. Sci. U.S.* **47**, 1405.
47. GERBER, G., GERBER, GISELA & ALTMAN, K. I. (1960). *J. biol. Chem.* **235**, 1433.
48. GOLDWASSER, E. (1955). *J. Amer. chem. Soc.* **77**, 6083.
49. GROS, F., HIATT, H., GILBERT, W., KURLAND, C. G., RISEBROUGH, R. W. & WATSON, J. D. (1961). *Nature (Lond.)*, **190**, 581.
50. GROS, F. & GROS-DOULCET, F. (1958). *Exptl. Cell Research*, **14**, 104.
51. GRUNBERG-MANAGO, M., ORTIZ, P. J. & OCHOA, S. (1955). *Science*, **122**, 907.
52. — (1956). *Biochim. biophys. Acta*, **20**, 269.
53. HALL, B. D. & SPIEGELMAN, S. (1961). *Proc. natl. Acad. Sci. U.S.* **47**, 137.
54. HALL, B. D., STORCK, R. & SPIEGELMAN, S. (1961). *Fed. Proc.* **20**, 362.
55. HART, R. G. & SMITH, J. D. (1956). *Nature (Lond.)*, **178**, 739.
55a. HAYASHI, M. & SPIEGELMAN, S. (1961). *Proc. natl. Acad. Sci. U.S.* **47**, 1564.
56. HEPPEL, L. A., ORTIZ, P. J. & OCHOA, S. (1957). *J. biol. Chem.* **229**, 679, 695.
57. HEPPEL, L. A., SINGER, M. F. & HILMOE, R. J. (1959). *Ann. N.Y. Acad. Sci.* **81**, 635.
58. HERBERT, E. (1958). *J. biol. Chem.* **231**, 975.
59. HERSHEY, A. D. (1957). *Virology*, **4**, 237.
59a. HUANG, R. C. & BONNER, J. (1962). *Fed. Proc.* **21**, 384.
60. HURWITZ, J. (1959). *J. biol. Chem.* **234**, 2351.
61. HURWITZ, J., BRESLER, A. & KAYE, A. (1959). *Biochem. biophys. Research Commun.* **1**, 3.
62. HURWITZ, J., FURTH, J. J. & GOLDMANN, M. (1961). *Proc. V int. Congr. Biochem., Moscow*, **1**, 110.
63. JEHLE, H. (1957). *Proc. natl. Acad. Sci. U.S.* **43**, 847.
64. JONES, A. S., LETHAM, D. S. & STACEY, M. (1956). *J. chem. Soc.* p. 2579.
64a. JORDAN, D. O. (1960). *The chemistry of nucleic acids*. London: Butterworths.
65. JOSSE, J. (1961). *Proc. V int. Congr. Biochem., Moscow*, **1**, 29.
66. JOSSE, J., KAISER, A. D. & KORNBERG, A. (1961). *J. biol. Chem.* **236**, 864.
67. KORNBERG, A. (1957). In *A Symposium on the chemical basis of heredity* (ed. W. D. McElroy & B. Glass), p. 699. Baltimore, Md.: Johns Hopkins Press.
68. — (1957). *Advanc. Enzymol.* **18**, 191.
69. — (1960). *Science*, **131**, 1503.
70. KORNBERG, A., LEHMAN, I. R., BESSMAN, M. J. & SIMMS, E. S. (1956). *Biochim. biophys. Acta.* **21**, 197.
71. KOZINSKI, A. W. (1961). *Virology*, **13**, 124.
72. KRAKOW, J. S. & CANELLAKIS, E. S. (1961). *Fed. Proc.* **20**, 361.
73. KRAUSE, M. & PLAUT, W. (1960). *Biochim. biophys. Acta*, **42**, 179.
73a. KURLAND, C. G. & MAALØE, O. (1962). *J. mol. Biol.* **4**, 193.
74. LEHMAN, I. R. (1959). *Ann. N.Y. Acad. Sci.* **81**, 745.
75. LEHMAN, I. R., BESSMAN, M. J., SIMMS, E. S. & KORNBERG, A. (1958). *J. biol. Chem.* **233**, 163.

76. LEHMAN, I. R., ZIMMERMAN, S. B., ADLER, J., BESSMAN, M. J., SIMMS, E. S. & KORNBERG, A. (1958). *Proc. natl. Acad. Sci. U.S.* **44**, 1191.

77. LOCKINGEN, L. S. & DeBUSK, A. G. (1955). *Proc. natl. Acad. Sci. U.S.* **41**, 925.

78. MAALØE, O. & HANAWALT, P. C. (1961). *J. mol. Biol.* **3**, 144.

79. MANTSAVINOS, R. & CANELLAKIS, E. S. (1958). *Fed. Proc.* **17**, 268.

80. — (1959). *Cancer Res.* **19**, 1239.

81. — (1959). *J. biol. Chem.* **234**, 628.

82. MCCARTHY, B. J. & ARONSON, A. I. (1961). *Biophys. J.* **1**, 227.

83. MEHTA, R., WAGLE, S. R. & JOHNSON, B. C. (1960). *Biochim. biophys. Acta*, **39**, 504.

84. MESELSON, M. & STAHL, F. W. (1958). *Proc. natl. Acad. Sci. U.S.* **44**, 671.

85. MESELSON, M., STAHL, F. W. & VINOGRAD, J. (1957). *Proc. natl. Acad. Sci. U.S.* **43**, 581.

86. MOAT, A. G. & FRIEDMAN, H. (1960). *Bact. Revs.* **24**, 309.

87. MUNRO, H. N. & CLARK, C. M. (1959). *Biochim. biophys. Acta*, **33**, 551.

88. MUNRO, H. N. & MUKERJI, D. (1958). *Biochem. J.* **69**, 321.

89. NAKADA, D. & RYAN, F. J. (1961). *Nature (Lond.)*, **189**, 398.

90. NOMURA, M., HALL, B. D. & SPIEGELMAN, S. (1960). *J. mol. Biol.* **2**, 306.

91. OCHOA, S. (1956). *Fed. Proc.* **15**, 832.

92. GRUNBERG-MANAGO, M. & OCHOA, S. (1955). *J. Amer. chem. Soc.* **77**, 3165.

93. OCHOA, S., BURMA, D. P., KRÖGER, H. & WEILL, J. D. (1961). *Fed. Proc.* **20**, 362.

94. — (1961). *Proc. V int. Congr. Biochem., Moscow*, **1**, 96.

95. OCHOA, S. & HEPPEL, L. A. (1957). In *A symposium on the chemical basis of heredity* (ed. W. D. McElroy & B. Glass), p. 615. Baltimore, Md.: Johns Hopkins Press.

95a. OCHOA, S. & MII, S. (1961). *J. biol. Chem.* **236**, 3303.

96. OKAZAKI, R. & OKAZAKI, T. (1958). *Biochim. biophys. Acta*, **28**, 470.

97. ORTIZ, P. J. & OCHOA, S. (1959). *J. biol. Chem.* **234**, 1208.

97a. REICHARD, P. (1959). *J. biol. Chem.* **234**, 1244.

98. REID, E. & STEVENS, B. M. (1961). *Biochim. biophys. Acta*, **49**, 215.

98a. RICH, A. (1959). *Ann. N.Y. Acad. Sci.* **81**, 709.

99. RICH, A. & DAVIES, D. R. (1956). *J. Amer. chem. Soc.* **78**, 3548.

100. RICH, A. (1960). *Proc. natl. Acad. Sci. U.S.* **46**, 1044.

101. ROLL, P. M. (1958). *J. biol. Chem.* **231**, 183.

102. ROLL, P. M., WEINFELD, H. & CARROLL, E. (1956). *J. biol. Chem.* **220**, 455.

103. ROLL, P. M., WEINFELD, H., CARROLL, E. & BROWN, G. B. (1956). *J. biol. Chem.* **220**, 439.

104. See (102).

105. ROLL, P. M. & WELIKY, I. (1955). *J. biol. Chem.* **213**, 509.

106. SARKAR, N. K. (1961). *Fed. Proc.* **20**, 147.

106a. SCHULMAN, H. M. & BONNER, D. M. (1962). *Proc. natl. Acad. Sci. U.S.* **48**, 53.

107. SIMON, E. H. (1961). *J. mol. Biol.* **3**, 101.

108. SINGER, M. F., HEPPEL, L. A. & HILMOE, R. J. (1957). *Biochim. biophys. Acta*, **26**, 447.

109. — (1960). *J. biol. Chem.* **235**, 738.

110. SINGER, M. F., HEPPEL, L. A., HILMOE, R. J., OCHOA, S. & MII, S. (1959). *Can. Cancer Conf.* **3**, 41.

111. SINGER, M. F., HILMOE, R. J. & HEPPEL, L. A. (1960). *J. biol. Chem.* **235**, 751.

111a. SINSHEIMER, R. L., STARMAN, B., NAGLER, C. & GUTHRIE, S. (1962). *J. mol. Biol.* **4**, 142.

112. SMELLIE, R. M. S. (1960). *Biochem. J.* **77**, 15P.
112a. — (1961). *J. Chim. phys.* **58**, 965.
113. SMELLIE, R. M. S., GRAY, E. D., KEIR, H. M., RICHARDS, J., BELL, D. & DAVIDSON, J. N. (1960). *Biochim. biophys. Acta*, **37**, 243.
114. SMELLIE, R. M. S., KEIR, H. M. & DAVIDSON, J. N. (1959). *Biochim. biophys. Acta*, **35**, 389.
114a. SPIEGELMAN, S., HALL, B. D. & STORCK, R. (1961). *Proc. natl. Acad. Sci. U.S.* **47**, 1135.
115. STEINER, R. F. & BEERS, R. F., jun. (1961). *Polynucleotides: natural and synthetic nucleic acids.* Amsterdam: Elsevier.
116. STENT, G. S. (1958). *Proc. IV internatl. Congr. Biochem.*, Vienna, **7**, 200.
117. STENT, G. S., SATO, G. H. & JERNE, N. K. (1959). *J. mol. Biol.* **1**, 134.
118. STEVENS, A. (1961). *Fed. Proc.* **20**, 363.
118a. — (1961). *J. biol. Chem*, **236**, PC 43.
119. SUEOKA, N. (1960). *Proc. natl. Acad. Sci. U.S.* **46**, 83.
119a. WADE, H. E. & LOVETT, S. (1961). *Biochem. J.* **81**, 319.
120. WEINFELD, H., ROLL, P. M. & BROWN, G. B. (1955). *J. biol. Chem.* **213**, 523.
121. WEISS, S. B. (1960). *Proc. natl. Acad. Sci. U.S.* **46**, 1020.
122. — (1961). *Proc. V int. Congr. Biochem.*, Moscow, **1**, 45.
123. WEISS, S. B. & GLADSTONE, L. (1959). *J. Amer. chem. Soc.* **81**, 4118.
124. WEISS, S. B. & NAKAMOTO, T. (1961). *J. biol. Chem.* **236**, PC 18.
124a. — (1961). *Proc. natl. Acad. Sci. U.S.* **47**, 694, 1400.
125. WOESE, C. R. (1961). *Nature (Lond.)*, **189**, 920.
126. ZUBAY, G. (1958). *Nature (Lond.)*, **192**, 1290.

CHAPTER XV

# THE MOLECULAR MECHANISM FOR REPRODUCTION OF THE SEQUENCE OF AMINO ACID RESIDUES DURING PROTEIN SYNTHESIS: THE DISCOVERY OF "MESSENGER" RNA AND THE DECIPHERING OF THE NUCLEOTIDE CODE FOR AMINO ACIDS

*Introduction*

The structure of proteins is remarkable for its extraordinary variety and the almost endless possibilities for the occurrence of new alterations. It is just this capacity for almost unlimited variation which forms the basis for the evolution of the whole of living nature. At the same time the structure of any individual protein is highly specific and a change, for example, in some one amino acid residue in its chain may lead to the complete loss of its functional properties and to the disorganization of the physiological processes in which it is involved. It is interesting that the different parts of the polypeptide chains are of very different significance in this respect, especially in proteins having enzymic or hormonal functions. The slightest alteration in the disposition of the amino acids in the "active centre" of the molecule will destroy the function, while changes in other parts of the molecule may not have any substantial effect on the biological activity of the protein.

As the order in which the amino acids are arranged in the polypeptide chains is an accurate expression of the functional properties of the proteins and of the species to which they belong, so, in the continuous succession of molecular generations, this sequence must be reproduced accurately and there must exist in the cells a biochemical mechanism which determines the reproduction of the specificity of proteins.

Among all the various problems of biochemistry there seems to be none for the explanation of which such varied concepts and hypotheses have been put forward as those offered for the mechanism for the reproduction of the specificity of proteins. Many of these hypotheses are now only of historical interest, as the impetuous development of experimental studies of protein synthesis very quickly changes theoretical ideas in this field.

The historical aspects of researches in the field of protein biosynthesis have already been dealt with many times in reviews and monographs (6,

316

54, 15, 14, 79, 80) and we shall only touch on them very briefly here. We shall only mention such of the earlier hypotheses in the field as are logically connected with the brilliant discoveries which are now being made.

The sensational nature of the latest discoveries in the field of protein synthesis and the deciphering of the general code of genetic information which regulates this synthesis has meant that these discoveries have quickly become known to wide circles of the scientific public. There is nothing fortuitous in this, for the importance of the discoveries which have been made in this field are such that they rank among the greatest achievements of science. However, these discoveries did not happen by accident; the whole course of the development of biochemistry has prepared the way for them and it is our task to indicate the whole complicated sequence of discoveries which led up to the outstanding successes in this field which became so clearly revealed in the last few years.

## 1. Short historical synopsis of some hypotheses concerning protein synthesis

Although the hypotheses which will be described briefly in this section were worked out between 1951 and 1957 we may already speak of them as being historical. Although these hypotheses are no longer "working" ones we must nevertheless consider the ideas involved in them in order that the antecedents of the schemes now current may become clear.

### a) Hypotheses assuming a direct reaction between the amino acids and the matrices

The concept of the theory of information discussed above only indicates the problems which must be solved in explaining the mechanism of protein synthesis but there are several types of hypothesis which try to find a real biochemical solution to these problems. The hypothesis which we should first set apart from among all others of this sort is that which states that the proteins are formed on matrices which have specific surfaces. This hypothesis has passed through many stages of development, has been criticized, altered, disproved, reinstated and finally has become practically universally accepted, although the mechanisms whereby these matrices act were at first unknown.

The idea of the existence of some kind of biological matrices came into being in 1928-30 in immunology and in genetics and at first it was quite abstract in character because the possible biochemical substrates for such matrices had not been studied by then. There is now an almost complete unanimity on this question because only nucleic acids can satisfy the essential requirement of the matrix theory, the power of self-reproduction. The idea of the self-reproduction of proteins, put forward by several authors about 10 years ago, has now died a natural death. The priority of nucleic acids as matrices seems to be decisive. In the first schemes the explanation given for the mechanism whereby the matrix acted was that the amino acids sat directly on the surface of the matrices where there were supposed to be certain stereochemical limitations caused by combinations of bases which were responsible for the selection of amino acids in the required order. However, these schemes laboured under several serious difficulties, the chief of which was the absence of any strict stereochemical complementarity between the surfaces of polypeptides and polynucleotides. A particular polynucleotide surface has a very narrow range of possibilities for the formation of hydrogen bonds and thereby it produces complementary specificity in a new polynucleotide chain. It cannot, however, use the same characteristics for the construction of an even more specific polypeptide chain consisting of 20 different amino acids.

Most workers now disagree with the concept of direct moulding on the matrix and a critical analysis of some hypotheses of this type (54, 40, 76-78) may therefore be considered superfluous. It should also be mentioned that the critical attitude towards the matrix theory which was characteristic of certain workers in 1953-5 (15, 20, 21, 73) was in fact directed against the theory of direct moulding on a matrix, which was held to be abstract and formalistic. These objections do not amount to arguments which can lead to the demolition of the idea of matrices as such and the new variants of this theory which will be discussed in this chapter naturally cannot inherit the criticisms which were levelled at the old imperfect attempts in this field. Now, however, when all the arguments about this problem have been shifted on to a new and higher plane, we can assess the value of the matrix hypothesis and we must admit that, even in the simplified form in which it appeared in 1952-6, it played a very formative part in the development of theoretical and experimental studies and the present, more elaborate schemes of the action of matrices, which we shall discuss in this and later chapters, result from the development of this concept while some of the ideas involved in it still keep their importance at the present time.

A new and very ingenious approach to the erection of a model of the functions of matrices in protein synthesis has recently been worked out and published by Jehle (61) who has made several new models representing direct synthesis on a matrix. Being a physicist Jehle has made a considerably deeper analysis than most other workers in this field of the possible selectivity of the forces of interaction between amino acids and polynucleotides. He mentions that the complementarity of biochemical structures, which gives their interaction its specificity, can be of various kinds.

In the case of the double-stranded polynucleotide of DNA we have the complementarity of systems of hydrogen bonds. There is also steric complementarity, complementarity of van der Waals contacts, complementarity of electrostatic charges or complementarity of all these factors together.

Jehle considers that the matrix for the synthesis of a protein, by which he means the nucleoprotein on which the protein chain is combined specifically with the polynucleotide, plays the chief part in determining the sequence of amino acids. He also thinks, though he is not yet quite certain, that it is only RNA which really acts as a matrix.

Jehle has put forward stereochemical models of four possible types of nucleoprotein matrices: (1) a matrix in which the protein chain is bound to the ribose of a single-stranded polynucleotide ribbon of RNA; (2) a matrix in which the protein chain is combined with the bases of the RNA; (3) a matrix in which the protein chain is combined with the phosphate groups of the RNA; (4) a matrix in which the individual amino acids are combined with the RNA by covalent bonds.

All these matrices would, theoretically, provide for a certain amount of stereochemical selectivity in their reaction with amino acids, in which the protein component plays the active part while the role of the RNA is essentially that of a carrier. Thus, in this case we are dealing with a hypothesis which, in fact, postulates the autoreproduction of proteins.

However, such schemes come up against many objections, both factual and theoretical, especially in trying to explain the morphogenetic and ontogenetic appearance of new proteins, formed under genetic control exercised by DNA. Nevertheless we still cannot be sure that it is absolutely impossible that matrices might exist in which proteins would fulfil a specific function, though only a partial one, and which would be reproduced together with the RNA. In such matrices the RNA would act as a backbone supporting the proteins, although at present, most biochemists think the reverse.

*b) Hypotheses concerning the reproduction of the specificity of proteins
by their assembly from peptides*

As the original variant of the matrix hypothesis lacked the sober convincingness necessary for its general acceptance, many authors of this period naturally advanced other explanations of the mechanism of protein synthesis, the most widely supported of which was one based on the idea of the assembly of peptides and this has not been altogether abandoned even now (15, 4, 5, 6).

The various facts concerning the intermediate formation of peptides in protein synthesis which we have already reviewed and the fact that they can be included directly in the cycle of protein metabolism provided the basis for hypotheses of this sort.

Hypotheses of protein synthesis which base its specificity on the selective assembly of large components, like hypotheses based on matrices, are very unsatisfactory. The actual principle of assembly is quite beyond doubt. In fact, if we compare the cell with an automobile assembly plant which takes in finished components and blocks of components of a particular vehicle, then these components cannot be assembled in such a way as to produce any other vehicle. If we consider the peptides and poly-peptides to be the ready-made components for the synthesis of proteins, then the specificity of the synthesis of only a single possible model from them is "technically" quite understandable. The biochemical reality of such a synthetic assembly has been more or less demonstrated in a series of researches by Bresler and colleagues on the resynthesis under pressure of proteins from products of their incomplete enzymic breakdown in the presence of proteases (4-11).

However, unsuccessful attempts to repeat these experiments have been made in two well-known biochemical laboratories, that of the Institut Pasteur in Paris (85) and that of the University of California (51). In 1955 Bresler (7) tried to explain the lack of success of the French authors in repeating his experiments. He attributed this to the unsuitability of the proteinase preparations used, which were contaminated with other enzymes and inert proteins, the breakdown-products of which inhibited resynthesis. Bresler maintains that resynthesis can only take place in a system consisting of a crystalline enzyme and a crystalline protein.

The American workers who tried to carry out resynthesis under pressure (51) were investigating a simpler system (carboxypeptidase + in-sulin). Both the enzyme and the protein used were in crystalline form. Carboxypeptidase removes the terminal alanine from insulin far more quickly than it acts on the next residue in the chain which is asparagine. The authors therefore tried to demonstrate, by Bresler's method, the resynthesis of insulin after the removal of its terminal alanine residue, i.e. to accomplish the reincorporation of alanine into its proper place. The experiments were carried out with [$^{14}$C]alanine added to the system as a tracer. All attempts at such resynthesis were, however, unsuccessful. In spite of wide variations in the concentration of alanine, pH and time of reaction, no case of reversal of the reaction under pressure was observed.

In forming an opinion on this subject we do not want, in any way, to cast doubt on the reliability of the original results obtained in Bresler's laboratory. There can, however, be no doubt that the conditions and methods required for such a synthesis, and the universality of the principle involved require to be worked out in further detail. Only when this has been done shall we be able to draw any theoretical conclusions from the results of this work. We must, however, mention the fact that work along these lines has not been continued in Bresler's laboratory and those working there have not published any new experimental work confirming their earlier results.

The main objection to any hypothesis which presupposes the synthesis of proteins by means of some sort of spontaneous assembly of specific peptide and polypeptide fragments of itself is, in the first place, that it underestimates the specific role of nucleic acids in this process which is brought out so clearly by recent researches on the mechanism of synthesis of viral particles and adaptive enzymes as well as many other protein syntheses in all kinds of living things.  Nevertheless, a survey of the evidence obtained in earlier researches concerning the part played by peptides in the synthesis of proteins leads to the conclusion that these products, of synthetic or hydrolytic origin, in fact only take part in the intermediate reactions of protein synthesis so that, in setting out the general, theoretical considerations about the reproduction of the specificity of proteins, one must also take account of the evidence as to the part played by RNA.

It is also important to note that, among the many and varied processes of protein synthesis occurring within the organism, there are cases which conform fully with the principle of assembly from peptides.  Good examples of such a synthesis seem to be those of the formation of fibrin from fibrinogen and the phenomenon of the condensation of peptides in ripening peas.  The latter has already been described in the chapter on protein synthesis in plants.  Furthermore, selective assembly from peptides must represent a certain stage in the formation of such protein molecules as consist of a complex of selectively bound peptide chains. Taking insulin as our example, its molecules are made up of two or four parallel peptide chains bound together (in pairs) by disulphide bonds, and this secondary structure must certainly arise by the selective, spontaneous assembly of peptides which are formed on matrices, maybe even on different matrices.  However, although it is clear that such assembly occurs in the case of proteins which are not made up of a single polypeptide chain but are aggregates of like or unlike peptides, there is nothing impossible about the idea that the formation of long uninterrupted polypeptide chains might take place in accordance with the same principle but on the surfaces of matrices.

In judging the value of this group of hypotheses we must therefore take account of the fact that, although they did not reflect the true complexity of protein synthesis, nevertheless they contained ideas which are partly valid even today.  Although we are now quite justifiably attracted by the new variants of the matrix hypothesis, we cannot, for this reason, write off either the ideas or the facts concerning the selective assembly of peptides.  The working out of these ideas and the classification of the relevant facts were undoubtedly of positive value and if our present-day concepts of protein synthesis are to have objective theoretical foundations they must be based, to a certain extent, on some elements of these ideas.

*c*) *Hypotheses concerning the reproduction of the specificity of proteins as a result of kinetic features of the relationship between enzymic reactions*

The most obvious importance of the kinetic features of protein synthesis in the reproduction of specificity has been emphasized by Oparin and Pasynskiĭ (19, 20, 21).

According to Pasynskiĭ, the amino acid composition and the sequence of amino acids of a polypeptide chain are determined by "the sum total of the kinetic conditions prevailing during the biosynthesis of the protein molecule; the nature and concentrations of the amino acids, enzymes, cofactors, vitamins, nucleic acids, hormones, salts, etc. which take part in the biosynthesis".

Similar ideas had already been put forward by Oparin (19) who believes that the constancy of the substances produced in the cell "is simply a manifestation of the constancy of the order in which one chemical reaction replaces another in the long chain of chemical transformations". This idea was later worked out in more detail in Oparin's book on the origin of life (20). According to Oparin it is actually the kinetic factors, the relationships between the rates of the individual processes, the organization of protoplasm in time as well as in space, which gives to biosynthesis its great lability, as a result of which what are synthesized are not proteins in which each molecule is identical with every other one, but extensive families of proteins which are very similar to one another.

This aspect of the study of the problem of the specificity of synthesis is of considerable interest but, unfortunately, it has not yet received enough experimental investigation. This is probably due to the extreme complexity of the experimental elucidation of the kinetic picture of the synthesis of a protein from twenty different amino acids and an indefinite number of peptide and polypeptide fragments. In a paper published in 1955 (15) we have also indicated the theoretical difficulties in approaching the problems of synthesis from this angle. If we assume that the production of individual proteins is determined by a relationship between the occurrence of a particular concatenation of reactions, on the one hand, and the existence of a particular structural and biochemical organization in the protoplasm on the other, then we must also explain the production, during growth, of the specificity of that organization, i.e. the specificity of the internal structure of the organelles and the localization of the enzymes and substrates. These difficulties must not, however, prevent the working out of such ideas. The complicated and biochemically purposive structure of protoplasm is a real fact and the harmonious reproduction of this structure during the growth and multiplication of cells is also a real phenomenon and forms the basis of the sequence of biological development. There can also be no doubt that the kinetic laws of protein synthesis to which we have already referred are some of the most important features of this process from whichever angle one may look at it. We have, however, very few exact data which would enable us to give a more concrete form to these ideas in our study of protein synthesis.

The importance of the co-ordination between the rates and extents of the preparatory reactions in protein synthesis is quite obvious from analysis of the material which we have already reviewed concerning the nature of the preliminary activation of amino acids. We saw that no less than 20 enzymes take part in this activating system, activating each free amino acid individually. Each amino acid residue is then transferred to receptor RNA by means of special enzymes. The further transfer of amino acids from the RNA of the activating system to the RNA of the microsomes also takes place at a particular rate and the rates of all these processes must be accurately co-ordinated.

This extremely complicated and quickly acting biochemical system, whereby molecules which are already ready for "incorporation" are carried to the place where the final polymerization occurs, is certainly an important link in the organization of the processes whereby specificity is reproduced. All the same there must clearly be some special specific mechanism for the final completion of the very last stages of protein synthesis.

All of the complicated preliminary biochemical work produces optimum quantities of the necessary kinds of components. It creates a general pool which fulfils the requirements for the synthesis of a very large number of proteins having very varied internal structures. It is easy to explain the spontaneous synthesis of one specific protein simply in terms of different rates and combinations of preparatory reactions which comprise a system of synthesis giving rise to different products at different times and under different circumstances. However, if the same materials are used to form tens of different proteins, then the difficulties which arise expand to form an insoluble problem if we do not look for new factors directing the metabolic reactions of the general stock of materials along particular, strictly defined courses and it is the matrices which constitute this essential directing factor. This is the conclusion reached by Pasynskiĭ (22) who has stated in a recent paper that "clearly we must recognize the stage of stereometric disposition of the activated amino acids on the matrices as being the stage which limits the overall rate of synthesis".

Thus, in our evaluation of this "kinetic" group of hypotheses of protein synthesis, we must draw the conclusion that, although at first sight they seemed to stand in opposition to the matrix theories, yet the development of our ideas concerning the synthesis of proteins, which has also changed our ideas as to matrices and as to the nature of the activating process, has diminished and even, maybe, resolved the contradiction between these ideas. Although the kinetic explanation of the reproduction of the specificity of proteins cannot, in its "pure" form, explain the real complexity of the process, nevertheless the development of these ideas has been valuable and was necessary to the working out of the general ideas of protein synthesis which are now current.

*d) Hypotheses concerning adaptor-transfer amino acids on matrices and hypotheses concerning common precursors for the simultaneous synthesis of proteins and RNA*

In biology and biochemistry there are now hardly any theoretical ideas which have developed on truly virgin soil and one can almost always find direct or indirect precursors for each of them. The hypotheses which we shall discuss in the present section are also the direct precursors of ideas which are now very widely accepted and which will be discussed later.

The dissatisfaction of many biochemists with the hypothesis of direct "seizure" of amino acids by the surfaces of the matrices and the accumulation of facts which contradicted such a simplified scheme, caused many authors to look for other ways of explaining the mechanism of reproduction of the specificity of proteins.

We have already discussed some such attempts to avoid using the actual principle of the matrix. In some hypotheses, however, this principle was retained but the mechanism by which the function of the matrix was carried out was quite elaborate. At the time immediately after they had been worked out (1955-7) these hypotheses were nothing but flights of fantasy, for at that time the biochemical composition of protoplasm had already been well studied and nobody really believed that there was a special group of adaptor molecules within the cells which carried the activated amino acids to particular loci on the matrices. However, only two years had passed since the idea was put forward before it was found that the cells really do contain a complicated and heterogeneous system of adaptors in the form of small molecules of soluble ribonucleic acid.

The first person to introduce the idea of adaptor molecules which would transfer amino acids to the matrix and, in so doing, would elicit the information which they contained was Crick, who was one of the authors of the complementary model of the structure of DNA. He made his original formulation of the idea in 1955 in a publication which is very hard to come by, so we refer the reader to the recent review by Hoagland (56). In his review Hoagland gives the following quotation of Crick's original remarks on this subject. "What DNA structure does show (and probably RNA will do the same) is a specific pattern of *hydrogen bonds* and very little else. It seems to me that we should widen our thinking to embrace this obvious fact." ... "Each amino acid would combine chemically at a special enzyme with a small molecule which, having a specific hydrogen-bonding surface, would combine specifically with the nucleic acid template. This combination would also supply the energy necessary for polymerization. In its simplest form there would be 20 different kinds of adaptor molecule, one for each amino acid, and 20 different enzymes to join the amino acid to their adaptors."

In his address to a symposium on the mechanism of the synthesis of macromolecules which was held in 1957, Crick had already developed this idea in greater detail (34). He suggested that the adaptor molecules might be nucleotides according to the idea which he had already worked out as to a trinucleotide non-repeating code in the polynucleotides of RNA and therefore, he concluded, on purely theoretical grounds, that the adaptors must obviously be trinucleotides having a structure complementary to that of the trinucleotides of the information-imparting sections of the molecule of the matrix.  Crick explained the fact that the structure of the adaptors was complementary to that of particular parts of the RNA quite simply and ingeniously by postulating that they were formed by the incomplete breakdown of molecules of RNA which were already present in the cell and were therefore complementary to its matrices.

Crick considers that the order of the successive syntheses after the combination of the adaptor with the matrix is still unknown.  It is possible that new polynucleotides are first formed and then a new polypeptide chain.  Maybe these processes occur in the reverse order or both syntheses may happen at the same time.

Such ideas have also been developed by other authors, independently of Crick.

In assessing the evidence concerning the interaction between protein synthesis and RNA synthesis in his review, Spiegelman (80) put forward the idea of the possibility that there might be common amino acid-nucleotide precursors for the simultaneous synthesis of both polymers.

In his paper at a symposium on the transfer of information in biological systems, Yčas (92) also formulated such an idea.

This hypothesis was worked out in detail by Loftfield (68).  He considers that the nucleotide component of the intermediate complex determines the specificity of the deposition of the amino acid on the matrix by virtue of the complementarity of the nucleotide carrier to a particular part of the polynucleotide chain.  This supposition at once provides the reason for the stereochemical advantages of these complexes.  It requires, however, that each amino acid should have its own carrier which would only be able to combine specifically with the appropriate amino acid. Mononucleotides would not be able to provide for such selectivity and, according to Loftfield, the simultaneous synthesis of RNA and protein occurs from complexes of amino acids with trinucleotides which would, as it were, represent the free, complementary copies of the "three-lettered" symbols of the non-repeating code "written" on the surface of the RNA. The activated amino acids combined with these trinucleotides coincide exactly with particular parts of the matrix and then combine spontaneously with one another to form a polypeptide chain with a determinate sequence of amino acid residues.

It must be admitted that these hypotheses are certainly very original and ingenious and, in spite of their apparently abstract nature, they show great boldness, while at the same time taking account of the real biochemical potentialities of RNA.  These hypotheses were put forward in 1955-1957, that is to say before the discovery of soluble, adaptor RNA and their rather weaker points have now become obvious.

For example, the suggestion that there is a simultaneous twofold synthesis of protein molecules from amino acids and of new polymeric molecules of RNA from the adaptors is a very dubious one, although such suggestions are sometimes made even now (70).

These interesting hypotheses were originally developed in a completely abstract way, but the discovery of acceptor S-RNA at once gave reality to the working out of these ideas. A collection of real adaptor molecules was discovered in the form of the molecules of low-molecular acceptor RNA and the course of theoretical studies was naturally diverted in this direction.

*e) Analysis of the problem of the reproduction of the specificity of proteins on the basis of the theory of the transfer of information within biochemical systems at the time when the first hypotheses were constructed*

During the last few years there has developed within biology a new field of theoretical studies based on the application of the principles of the theory of information to biological processes. In a recent paper on the theory of information and coding Elias (41) remarks that, in its narrowest sense, the theory of information deals with problems of the generation, conservation, transfer and manifestation of any sort of information, regardless of the object. In its widest sense the concept of the theory of information coalesces with the concept of "cybernetics". One of the clearest examples of the working out of these principles in a biological context which is being carried out at present is the analysis of the mechanism of protein synthesis.

Let us accept the ideas which were put forward in earlier chapters and which are based on facts, namely that nucleic acids constitute the system responsible for the reproduction of the specificity of proteins, by which we mean the unique sequences of amino acid residues in their polypeptide chains. If we do so it means that this reproduction can only take place if the nucleic acids can concentrate in their structure some particular stock of chemical information which is a coded expression of some particular combination of amino acid residues. Furthermore the cell must contain some kind of biochemical system which can decipher the code in which this information is "written". This could be done by the polynucleotide structure of the nucleic acids. The activity of the biochemical system activating and preparing the amino acids to interact to form peptide bonds (the system of activating enzymes and adaptor RNA) must also be directed in accordance with the code.

The information which determines the potential formation of secondary bonds in pre-formed proteins and their specific, three-dimensional structure is contained in the sequence of amino acids which is determined by the matrix, i.e. it depends on the interaction of these amino acids.

The only theoretically possible biochemical way in which this information could be recorded in the molecules of nucleic acids is by using some sort of arrangement of the different nucleotides of which their polynucleotide chains are composed.

The question of the ability of nucleic acids to determine the sequence of amino acids in polypeptide chains resolves itself, in this case, into the question of how the combinations of arrangements of the four nucleotides of RNA and DNA can correspond with any possible arrangement of 20 different amino acids in polypeptide chains.

The idea that the specificity of amino acids is determined by a sequence of nucleotides was first put forward by Dounce in 1953 (39), but the mathematical realization of this idea was carried out by Gamow (46-48) who then expanded it in a number of theoretical reviews written in collaboration with others (49, 50). We shall only deal briefly with Gamow's ideas on this question, not concerning ourselves with corroboration of the calculations involved. Many ingenious schemes of coding have been suggested by Gamow, based on the assumption that the position of each amino acid is determined by a group of three nucleotides in the double complementary chain of RNA, one of which is combined directly with the amino acid while the other two act as "neighbours". For example, the sequence AGCUA should contain the coding for three amino acids, AGC for the first, GCU for the second and CUA for the third. Such a code, depending on overlapping triads, contains too much information, as 64 different triads can be constructed from the 4 nucleotides to correspond with the 20 amino acids. With a view to removing this discrepancy the proponents of the code schemes have postulated a number of geometrical and logical limitations on the recording of information in triads of nucleotides which have already provided the basis for several schemes (the rhombic code, the triangular code, the code of "oldest-youngest", the code of the successive series, codes with and without "commas") and many more can be thought up.

The possibility of encoding the sequence of amino acids by means of overlapping triads of nucleotides has been put to the test by Brenner (30) on 28 protein and polypeptide compounds, the amino acid sequence of which had already been established. As each later triad contains two nucleotides of the preceding one, any particular triad can only precede one of four others and can only be preceded by one of four others. According to Brenner the formation of the tripeptide groups of the protein structures which had already been worked out by 1957 using such triads would require 70 triads while only 64 could exist. Brenner therefore considers that any such overlapping of the units of the code is impossible but does not deny that there may be other ways in which some information regulating the synthesis of a protein is recorded in a nucleic acid.

Crick, Griffith & Orgel (37) also came to the conclusion that an overlapping type of code was not very plausible and in 1957 they, in turn, put forward their own variant of a non-overlapping code. They hold that the position of each amino acid is controlled by a group of three nucleotides, that of the next amino acid being controlled by the next three, etc.

Yčas (93) has recently put forward a new idea concerning the problem of the biological code in nucleic acids. As a result of his studies of the correlation between the nucleotide and amino acid compositions of several viruses he put forward the idea that the position of each amino acid is determined by a single nucleotide residue. For this purpose the amino acids are divided into four groups each containing amino acids related to one nucleotide link. However, Yčas thinks that some other source of information which does not exist in the RNA is required to determine which of the amino acids in the group shall occupy a particular position. This idea was supported by Woese (91) who brought forward a number of additional arguments in support of it.

It is quite obvious that all these hypothetical schemes of recording information in the structures of DNA and RNA, as well as some others which are discussed in several reviews (25, 35, 65) have a certain theoretical interest. Although these schemes, when they were first worked out, appeared to be mathematical problems without any practical application, they have turned out to be a peculiar mathematical exploration of possible biochemical relationships. The main importance of this exploration is that it has completely changed all the earlier objections, made by many biochemists, that the nucleic acids, consisting of four forms of nucleotides, are simply incapable, on theoretical grounds, of accounting for the reproduction of the whole varied range of specific proteins containing twenty different kinds of amino acids. Mathematically speaking this objection is quite invalid.

These researches have also stirred up a great interest in biology and biochemistry amongst physicists, mathematicians, chemists and cyberneticists and this convergence will be an important factor in the further development of biology through the exact sciences.

At present, however, all schemes of this kind are only of historical interest, because in 1961 biochemistry embarked on a successful experimental approach to the problem. These researches were soon marked by a series of brilliant and sensational advances, exciting to the scientific community and comparable in significance with the greatest discoveries in natural science.

## 2. Hypotheses connecting the reproduction of the specificity of proteins with special forms of reaction between the matrices and the molecules of "soluble" RNA which transfer the accepted amino acids to the organelles of the cells

In Chapters III and IV we have already discussed most of the known facts concerning the biochemical properties of the enzymic systems for the acceptance of amino acids and the evidence for the structure, metabolism and functions of the "adaptor" S-RNA which is supposed to accept the activated amino acids and transfer them to matrices in the organelles.  We have already mentioned that the incorporation of amino acids into proteins during synthesis is, in fact, observed to be associated with an interaction between molecules of soluble RNA and the RNA of the organelles.  In this respect the present theoretical section may be considered as a logical extension of Chapters IV and V.

The investigations discussed in these chapters demonstrate the extremely selective nature of the activation of amino acids, such that the terminal assembly in the organelles does not proceed from free amino acids, arranged in sequence directly by a matrix, but from amino acids the metabolic paths of which have diverged as early as the stage of their carboxyl activation and which have then been firmly accepted by smallish polynucleotides with which they react by an enzymic process which is specific both in respect of the amino acid and in respect of the acceptor RNA.  This being so, the nature and specificity of the reaction of the complex (acceptor RNA-amino acid) with the high-molecular RNA acting as the matrix is primarily determined, not by a specific relationship of the amino acid to some part of the matrix, but by a relationship between two polynucleotide structures, although there is a 50-100-fold difference between their molecular weights.

An extremely interesting communication on the nature of the reaction between soluble RNA and the matrices of protein synthesis was made by Hoagland (55) to the Fourth International Congress of Biochemistry. The author proceeds from the well-known hypothesis that the RNA of the microsomes and other organelles plays the part of a stable matrix in which the linear sequence of the bases determines the final sequence of the amino acids in the proteins which are synthesized on its surface. According to Hoagland's idea, the molecules of soluble RNA may have structures which, apart from the terminal group, are complementary to various parts of the matrices and, by forming pairs of bases with these particular parts, they play the part of adaptors, depositing the amino acids on the matrices in a definite position.  Having accomplished this the adaptor may release itself and be used again, for the same purpose.  The complementary nature of the relationship between the matrix and the adaptor has, according to the author, been established in the nucleus. Hoagland also shows that this idea gives rise to many questions and new problems which will have to be solved in the future.  For example, it is still not clear what part is played by GTP, what is the nature of the force which brings about the formation of peptide bonds and what finally sets the protein free.  We still do not know the stereochemical interaction between the components nor whether more than one amino acid can combine with an adaptor and there are many other unknown factors. Similar ideas in a more generalized form have been put forward by Fraser & Gutfreund (43) and by Lipmann et al. (67).

It must be mentioned that, even before this, in an address to a symposium on molecular biology in 1957, Hoagland and his colleagues (59) made a first attempt at creating a hypothetical model of the reaction between soluble RNA and the matrix (Fig. 49).

According to this scheme, the terminal groups of the adaptor forms of RNA hang out, as it were, from the helix of the matrix and the amino acids which are combined with them react with one another without making contact with the surface of the matrix.

The authors feel that this scheme is very tentative and its main defect lies in the arbitrariness of the structure and the distance between the amino acids. The molecules of soluble RNA are composed of 70-120 nucleotides and therefore, each turn of the spiral which brings the amino acids alongside one another must also consist of nearly 70-100 nucleotides which is certainly far more than would be required to bring about contact between the amino acids (2-2·5 Å).

Such a way of determining protein synthesis, whereby 100-150 nucleotides of RNA of the matrix carry only one amino acid and the approximation of the amino acids to one another is brought about by structural compression of the whole system, like the compression of a spring, is improbable on theoretical grounds. In discussing this in a more recent review, Hoagland (56) has made a number of changes in this concept. "However, there are two matters which cause some perplexity" writes Hoagland in discussing the original hypothesis. "One is the fact that

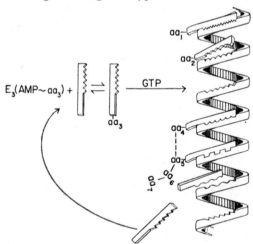

FIG. 49. Diagram of the terminal reaction of protein synthesis by interaction between the microsomal RNA and the complex of S-RNA with an amino acid. The complementarity between the molecules of S-RNA and parts of the helical matrix is indicated diagrammatically by a correspondence between the shapes of the reactive parts.

$E_3$ = the enzymic system of activation.

AMP~aa = macroergic compounds of adenosine monophosphate with an amino acid.

GTP = guanosine triphosphate, the mode of action of which has not yet been determined.

transfer RNA is a larger molecule, by a factor of at least 10, than one would require on the basis of current coding concepts. Perhaps these concepts are totally wrong. It certainly seems difficult to conceive, however, of a reason why such a large molecule should accompany each amino acid into the particle. Perhaps a 'coding piece' is split off the transfer RNA during the transfer reaction and only this piece accompanies the amino acid into the particle. The remainder of the molecule would only serve as a polymerized carrier for the end group."

Hoagland thinks that this assumption is supported by the interesting observations of Chantrenne (32) who found that, immediately after the induction of maltase formation in resting yeast cells, the pool of acid-soluble nucleotides increased rapidly, apparently by the breakdown of RNA which was already in the cells. This increase was mainly associated with material which has been identified in a preliminary way as oligonucleotides. After the enzyme had been synthesized there was a sharp decrease in the pool of nucleotides.

In his unpublished work Hoagland also shows that when soluble RNA containing amino acids is incubated with microsomes which do not contain GTP, oligonucleotide-like products are formed from the soluble RNA and these do not contain amino acids.

According to this somewhat altered idea, the specificity of proteins is supposed to be determined, not by the turns of a helix, but by considerably shorter segments of the polynucleotide chain of RNA, continuously along its length. However, the role of the terminal trinucleotide segments is primarily that they enable the amino acids to get close enough together to react with one another without their being "deposited" directly on the matrix. When such a "shortened" adaptor is "deposited" on the matrix its terminal part unbends itself somehow and acts like a "hand" to approximate the amino acids to one another.

In a recent article in an American popular-science magazine Hoagland (57) represented this mechanism in action in the form of a very striking and picturesque scheme which seems appropriate for reproduction here (Fig. 50). A similar scheme has also been put forward in a paper by Lipmann et al. (67).

However, this new variant of Hoagland's scheme also runs into several serious difficulties. In the first place such a scheme takes no account of any need for polymerization in the carrier. In the second place, when it has fulfilled its function the adaptor (coded section) has to recombine with its carrier by some reaction which is not yet known. It is hard to imagine that all carriers have the same structure but, if they are heterogeneous, then they might include other complementary units in their composition. Furthermore, it is clear that the enzyme which gives rise to the terminal CCA group which is common to all adaptors must have the additional and hardly credible ability to "recognize" whether a molecule of soluble RNA contains a coded section or whether it has not yet returned to its place after having been released from the chain.  And finally, if each adaptor enzyme, having "transferred" its amino acid to another neighbouring adaptor, were then to set itself free from its attachment to the matrix, then peptide linkages might again arise in the free places and incomplete peptides which could not form the whole chain might branch out from the matrix at different places.  And this is by no means the whole list of the difficulties which confront any attempt to imagine such a system in action.  In a recent paper Hoagland & Comly (58) showed that, when they have carried out their function of transferring amino acids, the molecules of S-RNA do not break down but remain complete.  This means that the idea of the carrier becoming detached from the S-RNA is not confirmed and it becomes necessary to look for some way of showing what is the part played by the molecule of S-RNA in its entirety in the reactions of protein synthesis.

Hoagland's ingenious hypothesis was certainly a great step forward in the theoretical working out of the problem of protein synthesis.  Although many of the assumptions made in this hypothesis are tentative and unproven it still serves as a basis for further experimental and theoretical researches in this direction and, whatever may be its fate, it will remain a noteworthy landmark in the working out of the problem of protein synthesis.

A number of suggestions in this field, based on several of the fundamental principles of this idea, have recently been set out by the author of the present book (16, 17, 18).  According to the first variant of the scheme

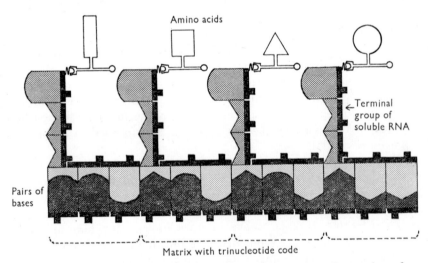

Amino acids

Terminal group of soluble RNA

Pairs of bases

Matrix with trinucleotide code

FIG. 50. Final stage of protein synthesis.  Formation of complex of "soluble" RNA, charged with amino acids, with RNA of matrix.  Each triplet in the "soluble" RNA molecule finds a complementary group on the matrix.  The terminal trinucleotide moieties, carrying the amino acids, bring them close together for the purpose of peptide-bond formation.

the molecules of S-RNA which are complementary to the polynucleotide of the matrix react temporarily with it and transfer the amino acids directly to the matrix. The terminal group of the acceptor RNA acts as an adaptor which enables the deposition of the amino acids to occur, not on the nitrogenous bases of the matrix, but somewhere to the side of them, on the hydroxyl groups of the ribose or on the phosphate bridges.

A second variant of this scheme, shown in Fig. 51, also postulates complementary interaction between S-RNA and a matrix but the amino acids are deposited on the protein stroma of the ribosomes and not directly on the matrix.

The combination of the amino acids in another plane from that in which the nucleotide fragments react with one another provides the best steric conditions for the continual growth of the polypeptide chain, while the protein underlay does not give rise to the problem of the "splitting off" of the polypeptide from the matrix because the linkage between the new molecule and the ribosome is very weak.

Peptides formed at different parts of the ribosomes would have gradually to coalesce to form a common polypeptide chain. In such a case it is not impossible that the peptides should grow in one direction only. If that were so, the nature of the peptide surface and thus that of the protein component of the nucleoprotein might play an additional "specializing" part, that is to say, it might act as a specific component of the matrix, as has been suggested also in the hypothetical models of matrices put forward by Jehle (61). As we have seen in Chapter V, the proteins combined with RNA in the ribosomes do not show great specificity and are not identical with the specialized proteins manufactured in the ribosomes, so their importance cannot be decisive, as Jehle suggests, but yet they may carry out a more complicated function and not just serve as a support for the matrix.

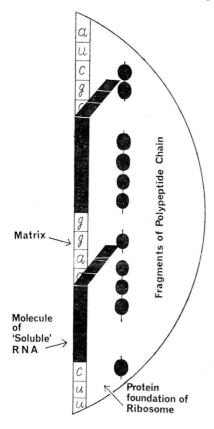

Matrix

Molecule of 'Soluble' RNA

Fragments of Polypeptide Chain

Protein foundation of Ribosome

FIG. 51

Possible mode of interaction of "soluble" RNA with matrix on surface of ribosome, with formation of polypeptide chain on surface of protein stroma of ribosome.

This scheme also gives rise to many questions which require a logical explanation. In the first place it requires the existence of a very large assortment of molecules of soluble RNA which, taken as a whole, would have species and tissue specificity. The question as to the presence or absence of species specificity among soluble RNA's in this stage of reaction with the matrix has still only been very poorly studied. Some preliminary evidence for the existence of such specificity has been obtained recently by Rendi & Campbell (74). In a new, short, preliminary communication Rendi & Ochoa (75) put forward new evidence as to the specificity of the enzymes activating the amino acids. In their experiments the leucine-activating enzyme of yeast activated [$^{14}$C]leucine and transferred it to S-RNA from yeast and from liver. The corresponding enzyme of liver could transfer [$^{14}$C]leucine to the S-RNA of yeast. However, neither of these enzymes could transfer [$^{14}$C]leucine to the S-RNA of *Escherichia coli* or other bacteria. The reverse was also true, that the activating enzyme of *E. coli* could not react with the S-RNA of yeast or liver. The authors also found a similar specificity for the reactions catalysed by the transport enzyme of S-RNA.

Interesting facts have been discovered by Kruh, Dreyfus and their colleagues (63) studying the synthesis of haemoglobin *in vitro* by the biochemical systems of reticulocytes. When microsomes from the reticulocytes were incubated with a fraction which is soluble at pH 5 (the activating enzymes + S-RNA) *in vitro* haemoglobin was synthesized. However, the synthesis of haemoglobin required the presence of the soluble fraction from reticulocytes and did not take place if the activating enzymes + S-RNA from other organs were present. If the microsomes from the reticulocytes of one species of animal are incubated with the soluble fraction from another species the haemoglobins of both species are synthesized. At first the authors supposed that the S-RNA of these fractions was specific and carried information. However, in later work with reticulocytes, Lamfrom (64) showed that the synthesis of the haemoglobin of two species on the "foundation" of the ribosomes of one species was not determined by the S-RNA but by some accessory factor which Lamfrom believed to be messenger RNA which was also present to some extent in the soluble fraction. It has also been shown that the ribosomes of liver can react with RNA from *E. coli* (71). Species specificity of the transfer enzymes was also observed in the experiments of some Japanese workers (84).

Finally we must mention the evidence which points to the heterogeneity of S-RNA as seen in leucine- and threonine-carrying S-RNA after separation from other forms of S-RNA by countercurrent distribution (38).

Thus, we cannot exclude the possibility that the heterogeneity of fractions of S-RNA is greater than that required by the scheme of the non-overlapping deposition of the adaptors on the matrix.

Great difficulties also arise in the analysis of the features of the reaction between the activating enzymes and the molecules of soluble RNA. If there is a very large assortment of molecules of soluble RNA it is hard to imagine how the activating enzyme "knows" with which molecule it should combine leucine, with which alanine and so forth. The question of the nature of the selective interaction between the activating enzyme and the matrix also takes a leading place in Hoagland's scheme. However, his scheme only presupposes the existence of twenty adaptor molecules and therefore he suggests that in the various activating enzymes there are sections, at the places with which the soluble RNA combines, which are complementary to the code-bearing trinucleotide sections of the molecule of soluble RNA and that this is the way in which they "recognize" the appropriate molecule of adaptor with which to combine any particular amino acid. This "distinguishing" region must be situated at a distance from the activated amino acid of at least the three nucleotide lengths represented by the terminal group of the soluble RNA. Hoagland (57) represented the whole cycle of the reaction of activation of the amino acids and their interaction with the molecules of S-RNA in the form of a very picturesque and striking scheme (Fig. 17, p. 60).

In our scheme for the explanation of this cycle we have made a somewhat different assumption. We have supposed that what the activating enzyme "recognizes" by the form of its surface is, not the nucleotide sequence, but the size or shape of the whole molecule of "soluble" RNA and that the variation in the lengths of the polynucleotides of this fraction is an expression of the occurrence of twenty different types of length, each specifically associated with a particular amino acid.

This suggestion was not, however, supported by the recent work of Klee & Cantoni (62). These authors made a special study of the question as to whether the acceptor specificity of the molecules of S-RNA is associated with the length of their polynucleotide chains. Study of different fractions of the S-RNA of liver in the ultracentrifuge showed that the difference in the specificity of S-RNA for the acceptance of valine, leucine and proline was not connected with the size of the molecules. The sedimentation coefficient of this RNA had a sharp peak at 4·25 S which was considerably sharper than that of ordinary RNA. Thus, what is important for the specificity of molecules of "soluble" RNA is, apparently, the sequence of the nucleotides or the secondary or tertiary structure determined by it.

A great difficulty about the adaptor hypothesis is that of explaining how the molecules of "soluble" RNA are formed so as to be of varied sorts, complementary to the matrix and having dimensions adapted to correspond with three different types of specificity in respect of the amino acid, of the enzyme and of the position of the amino acid in the polypeptide chain which is expressed in the structure of the matrix. Hoagland and his colleagues (59) have suggested that this complementarity between the

S-RNA and the matrix is established in the nucleus and that it is a reflection of the complementarity of the polynucleotides of the double helix of DNA. This supposition is very logical. This sort of suggestion has already been fully discussed in a recent paper of Stanley & Bock (81). These authors consider that if each nucleotide of the double helix of DNA carried information, then the "reflected" complementary information of the second polynucleotide can only have any meaning if one polynucleotide of the DNA produces the matrices while the other produces the molecules of S-RNA which are complementary to them. In this case the "information" of the complementary chains of DNA culminates in the synthesis

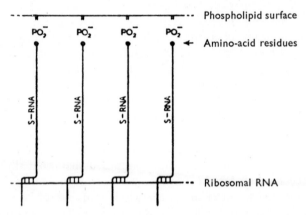

Fig. 52. Outline of structure of proposed "template" for protein biosynthesis.

of one particular protein. We have put forward a somewhat different idea (16, 17, 18). We have assumed that autonomous, complementary synthesis of the RNA of the cytoplasm is possible. In this case one of the polynucleotides of the RNA serves as a matrix in the ribosomes while the complementary molecule of RNA may break down partially into smaller fragments which function as acceptors. In this case there are two complementary chains of polymeric matrices, each of which, individually, possesses the same information which is only made manifest by the complementary adaptor. In such a case the same "half" store of information is also contained in each complementary molecule of DNA.

Hunter & Godson (60) have recently suggested yet another new variant of the reaction between the S-RNA and the matrix. Basing their argument on certain evidence as to the importance of phospholipid membranes in the synthesis of protein (cf. Chapter XII) and on the triplet code of Crick, Griffith and Orgel, these authors have suggested the following model of protein synthesis (Fig. 52). The authors pick out the following postulates derived from the scheme: (1) All chains of S-RNA are of the same length from the "triplet" region to the amino acid; (2) the phospholipid membrane is situated at a strictly determinate distance from the matrix; (3) each amino group of an amino acid only reacts with one phosphate group. We must, however, mention that there is, as yet, very little evidence in favour of this scheme. Lingrel & Webster (66) have recently shown that the ribosomes of the liver, when freed of all lipid membranes, can bring about a net synthesis of albumin. The authors consider that the membranous material is not necessary for the synthesis.

In a recent paper Hendler (54b) has published yet another variant scheme involving obligate participation of lipid membranes in protein synthesis. In his opinion, ribosomes have only about one hundredth of the synthetic activity of native microsomes, and this can be attributed to contamination of the ribosomal preparations by microsomes. Hendler considers that protein formation takes place in the narrow space between the RNA-matrix and the lipoprotein membrane, and that, besides amino acid-S-RNA complexes, transport of amino acids to the matrices is mediated by lipid-amino acid complexes. However, Hendler's views on the inactivity of ribosomes amount to assertions which are not supported by detailed experimental evidence.

A very complicated problem over which all the schemes fall down is that of how the molecules of S-RNA find the complementary parts of the matrices. Until the discovery of the transfer enzymes of S-RNA it was supposed that the interaction between the matrix and the molecule of S-RNA took place as a result of chance encounters at a molecular level. Having carried out a number of mathematical calculations, Ts'o & Lubell (87) reached the conclusion that even if $1:100,000$ of the encounters between the matrix and S-RNA were effective from the point of view of complementarity this would be enough to account for the rate of protein synthesis normally observed. But what determines the effectiveness of the collisions and is even this safety factor of 100,000 enough? In the first place, at any given moment not all of the molecules of S-RNA can be combined with amino acids. Loftfield (69) found that activated valine was present in the reticulocytes at a concentration of $3 \times 10^{-10}$ moles/ml. If all of the activated amino acids were present in the same concentration, then the sum of their concentrations in the cell would be $6 \times 10^{-6}$ M (taking into account that contained in the proteins). This is ten times less than the molar concentration of S-RNA. If this were correct one effective collision might occur for every 10,000-15,000 ineffective ones. In the second place, 20 amino acids take part in the synthesis of proteins and it is therefore obvious that, if a ribosome needs serine to add on to the protein it is making, its collision with S-RNA-glycine will not be effective. This diminishes the number of collisions among which one may be effective by another 20 times and it becomes about 500-700. However, 500-700 is still a large reserve which should guarantee the necessary effectiveness if there really are only 20 forms of S-RNA, corresponding to the 20 amino acids.

Thus, even random collision would provide for the synthesis of proteins with ample reserves. Obviously the presence in the cytoplasm of special transfer enzymes would provide conditions for the occurrence of these reactions which would differ from those provided by chance encounters but the mechanism of action of these enzymes is still not known.

The matter becomes rather more complicated if we assume that the synthesis of all the proteins of the cell takes place from a single pool of S-RNA in the cell sap, for the number of proteins synthesized by the cell is very great. In fact, however, not only do many organelles have their own internal systems of matrices, but they also have autonomous complexes of S-RNA, and the synthesis of various proteins takes place at different places within the cell and within the organelles. It is this spatial differentiation which also leads to the localization of enzymes in the cellular organelles which is indispensible for the carrying out of complicated and co-ordinated reactions. There are some observations which are a bit out of line with the hypotheses which have already been discussed but they can only be considered after they have been verified in greater detail. Bloemendal, Bosch & Sluyser (29) have shown that when S-RNA reacts with the matrix covalent bonds are formed. There is a great difference between S-RNA and the RNA of the ribosomes in respect of their content of unusual bases (cf. Chapter IV). There is, however, some preliminary evidence that the RNA which is synthesized by the system which depends on DNA also contains a fairly large amount of the unusual nucleotide pseudouridylic acid.

Chapeville, Lipmann and their colleagues (32a) have obtained direct evidence that the placing of an amino acid in the protein chain during synthesis depends solely on interaction between S-RNA and the matrix and not on the nature of the amino acid. In molecules of cysteine-specific RNA the cysteine associated with the polynucleotide was reduced to alanine by a special reagent. These "hybrid" forms were then introduced into a synthesizing system containing artificially synthesized uridylic-guanylic polynucleotides. These usually stimulate the incorporation into the protein of cysteine, but not alanine. The alanine which was combined with the S-RNA was, however, incorporated during the synthesis, thus supporting the hypothesis of adaptors.

An original hypothetical scheme of the interaction between S-RNA and the matrix was published in November 1962 by Zamecnik (93b). He suggested that the matrix RNA takes on a secondary structure in the form of a double helix. According to his ideas this helix can quickly untwist itself to form pulsating waves of the length of three pairs of nucleotides. The double helix would then always have zones of a length of three nucleotides which were single-stranded and which would have free hydrogen bonds with which the coding triplets of S-RNA would combine.

## 3. New lines of approach to the mechanism whereby the RNA matrix takes part in the reproduction of the specificity of proteins

The hypothesis of Hoagland which has already been described and its modifications only represent the initial stages of the theoretical working out of the problem of the reproduction of specificity and therefore it is only natural that they should suffer from very many defects. They are not yet concepts which can explain the essential laws of any group of facts but, rather, they are an attempt to suggest how the process might conceivably be brought about if we make a model based on the evidence we now have about the matrices and S-RNA and on the theory of the nature of the information contained in the polynucleotides of RNA. Nevertheless, the hypothesis represents a definite step forward in comparison with such hypotheses as a direct moulding on matrices, assembly of free peptides, etc. In these earlier hypotheses discussion was confined to the postulation that some unknown hypothetical process might occur (selective adsorption, synthesis in two layers, selective assembly, ideal co-ordination of the rates of reactions) and it was not possible to get as far as considering any explanation of the actual mechanism of synthesis. In Hoagland's hypothesis the occurrence of interactions between the molecules of S-RNA and the nucleoproteins of the ribosomes is based on facts, the theorizing is directed towards the explanation of the interaction of the two forms of RNA which accompanies the incorporation of amino acids into the protein being synthesized.

This does not mean that Hoagland's theory is now the only possible one and we have put forward some other suggestions in connection with the problem under review. It does, however, mean that any new hypotheses about protein synthesis must start from the facts as they are now known and that, when trying to reach new explanations for the reproduction of specificity in the course of protein synthesis, it is inappropriate to return to a level of knowledge which has already been superseded by scientific advances. This must be emphasized particularly because attempts to set up such simplified schemes of protein synthesis without taking into account the most recent factual discoveries have very recently been made (13, 24).

The collection of facts in this and allied fields is going on so quickly that even now, although it is only a short time since the publication of the schemes already described, we may expect the appearance of other hypotheses, worked out in more detail, which will explain a greater number of the features of this synthesis.

The commoner schemes for describing the way in which the information of the RNA of the matrices is put into effect assume that the transfer of information for the synthesis of protein is brought about as a result of the complementary interaction of the nucleotides of some particular parts of the adaptors and matrices. But what is the position in regard to the putting into effect of the information in the RNA of the ribosomes if the molecules of this RNA acquire a secondary structure in parts of their chains and twist themselves up so that the complementary hydrogen bonds are "locked up" (cf. Chapter V)?

In this connection we must take a look at a completely new treatment of the question of the nature of the information in the RNA of the matrices which was set out in recent papers by Fresco, Alberts & Doty (44, 45).

The authors consider that the secondary structure of the single polynucleotide of RNA which occurs in the ribosomes is made by a special type of complementary folding. According to the authors this folding is brought about by complementary interaction between adenine and uracil on the one hand and cytosine and guanine on the other.

However, not all of the bases find themselves pairs in the course of this folding so that the extra, "unpaired" bases form curious "loops" or knots along the double helix and can form structural bonds with the ribosomal proteins. A diagram of the folding of such forms of the poly-

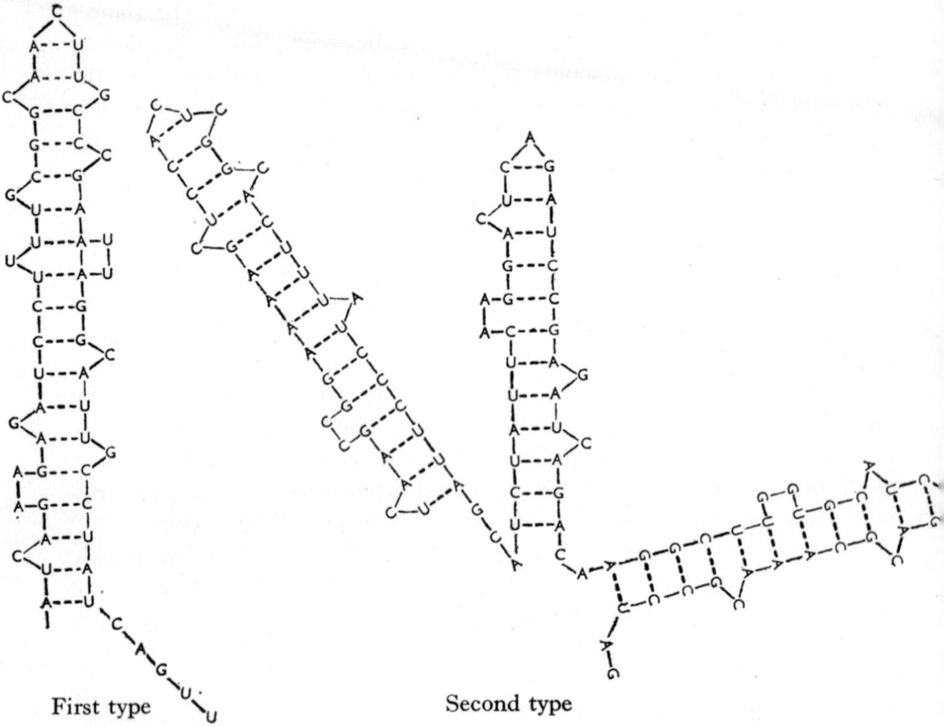

First type          Second type

FIG. 53. Possible modes of folding of single polynucleotide chain of RNA (44, 45). Complementary pairs form hydrogen bonds; "extra" noncomplementary nucleotides form "loops". The authors consider the second type the more probable.

nucleotide of RNA is given in Fig. 53. The unpaired nucleotides are stable and this gives new possibilities of coding through special sequences of these "loops".

If we take account of the fact that the molecules of "soluble" low-molecular RNA also form some sort of fairly stable secondary structure of the nature of a double helix (33, 94) we may suppose that the complementarity between the molecules of S-RNA and the matrices is brought about by the selectivity of their reaction, which is determined solely by the interactions of the unpaired nucleotides. This explains the necessity for fairly large adaptors having 80-140 nucleotides.

This scheme also faces several difficulties, the chief of which is concerned with the great length of the sections of the matrices corresponding to one amino acid. However, if it is really established that RNA is folded in this way it would certainly lead to corrections being introduced into the schemes postulated earlier as to the nature of the reactions between S-RNA and the matrices.

A new idea concerning the nature of the code in the matrices has been put forward recently by Blyumenfel'd & Benderskiĭ (2). Blyumenfel'd and colleagues (3) have recently been studying the paramagnetic resonance of organic compounds. In doing so they discovered some most unusual magnetic properties of nucleic acids and nucleoproteins which are caused by the presence in these compounds of unusually large numbers of unpaired electrons.

These facts indicate that the molecules of RNA in the nucleoproteins act as something in the way of small but strong magnets. The authors of this study have not yet been able to find any definite connection between the magnetic and biological properties of RNA but they took the view that some connection of this kind must exist. In their opinion a possible way in which such an association might exist would be if the biological specificity of nucleic acids were associated, not so much with the sequence of the nucleotides in the chain of RNA or DNA, as with the sizes and mutual arrangements of the oriented areas which have electronic and magnetic anomalies. The cloud of unpaired electrons might then play some part in the special biological properties of these compounds. Samoĭlova & Blyumenfel'd (23) found that there is a change in the magnetic properties of the cytoplasm of yeast at the beginning of active growth.

In setting up schemes of protein synthesis it will apparently be necessary in future to study the evidence concerning the mechanisms and peculiarities of the transfer of energy along polypeptide and polynucleotide chains (82, 12, 83, 42).

## 4. The discovery of a special "messenger" form of RNA transferring genetic information from DNA and its function as a matrix in protein synthesis

In the last chapter we have already touched on the recent interesting studies of a number of laboratories which led to the discovery of a new form of RNA in the cells, synthesized on DNA and resembling DNA in its nucleotide composition and in the arrangement of the nucleotides in the chain. As we saw in Chapter XIII, the existence of such a special form of RNA was foretold by Belozerskiĭ & Spirin (1, 28) from their analysis of the extensive comparative material on the nucleotide composition of RNA and DNA of various species of bacteria. This analysis showed that, while there is great variability in the composition of DNA as between one species and another, there is only a small fraction of the RNA which shows a correlated variability, while the bulk of the RNA shows only slight variability according to species.

The form of RNA which is correlated with the DNA was later actually found in bacteria after phage DNA had been introduced into them and it turned out to be a special form of RNA with unusual biochemical properties (72, 53, 26, 27, 88-90, 52, 31, 93a). This RNA is distinguished by having an extremely rapid rate of turnover, which meant that it could be discovered in the cells by the method of "pulse labelling". The characteristic feature of this method is that the analysis of the various forms of RNA takes place at very short intervals (5-10 sec.) after the introduction into the medium of compounds of $^{32}P$ which has a very high specific activity. When this is done it is found that the greatest activity is present in the RNA which has a composition correlated with that of the DNA and which is separable by ultracentrifugation from the polymeric RNA of the organelles (molecular weight 1,500,000) and from the soluble RNA (molecular weight 30,000-40,000). The molecular weight of the most "active" RNA was found to be different from that of the other forms of RNA which were already known; it is about 300,000 and can therefore fairly simply be separated from the other forms, although compared with them it is only present in extremely small amounts in the cells. The role of this RNA has recently been studied in the interesting researches of two groups of biochemists (31, 52) who have put forward a new concept as to the synthetic activity of the ribosomes. In experiments on the infection of *Escherichia coli* with phage (31) it has been shown that the DNA of the phage is not synthesized during the first seven minutes after the introduction of the phage into the culture but a small amount of a special phage-specific RNA with a constitution corresponding to that of the phage DNA is synthesized. This phage-specific RNA then combines with the ribosomes of the infected cell (72). As a result of this combination the ribosomes of the *E. coli turn over to the synthesis of the proteins of the phage without any breakdown of their own RNA.*   No new phage ribosomes are formed, protein synthesis is based upon the ribosomes of *E. coli* in conjunction with this "messenger RNA".

The same process has also been found to occur in normal cells of *E. coli* but in this case the composition of the messenger RNA was correlated with that of the DNA of the bacterium itself (52). The authors suggest that it is precisely this form of RNA, emanating from the nucleus, which acts as the direct information-carrying matrix for protein synthesis, while the polymeric RNA of the ribosomes (mol. wt. $1 \cdot 5 \times 10^6$) does not carry any genetic information and only acts as a stable surface on which the S-RNA, carrying its amino acids to the messenger RNA, is deposited. The authors take the view that the participation of the RNA in protein synthesis brings about its disintegration and, according to them, this explains the rapid turnover of this form of RNA.   In their discussion of the experiments in which it was established that there are "active" and "inactive" 70 S ribosomes in *E. coli* (86) (cf. Chapter V), the authors suggest that "active" ribosomes are ribosomes combined with messenger RNA.

The study of a new form of RNA so closely associated with protein synthesis soon attracted the attention of many laboratories.   In an earlier chapter we have already remarked on the presence within the nucleus of a special polymerase which brings about the synthesis of RNA, but only in the presence of DNA acting as a primer.   The composition of the RNA synthesized in these experiments (mainly *in vitro*) was completely correlated with that of the DNA matrix.   In recent work Weiss and colleagues (50a) have shown that this DNA-dependent RNA is complementary to the DNA of the matrices and is formed in contact with one of the polynucleotides of the DNA, being liberated after formation.   In this connection the authors put forward the suggestion that the RNA synthesized *in vitro* in the presence of DNA and the messenger RNA found *in vivo* may be one and the same compound.

It is interesting, that when this form of RNA, which is complementary to DNA, is heated with DNA (unwinding the double helices), subsequent cooling leads to the formation of mixed DNA-RNA double helices identical with the DNA-RNA "hybrids" which were discovered in Spiegelman's laboratory (80a).   The formation of such "hybrids" was very specific.   Hayashi & Spiegelman (54a) have recently made a detailed study of the properties of the messenger RNA formed in the cells of *E. coli* when it is infected with T2 phage.   This RNA was more heterogeneous than the S-RNA or the high-polymeric RNA of the ribosomes (23 S and 16 S).   *In vitro* it formed good double helices with homologous single-chained DNA from T2 phage but could not form helices with DNA isolated from other sources.

Notwithstanding the very short time which has passed since methods for the demonstration and study of messenger RNA were developed, evidence on this subject has rapidly piled up, and already the first reviews of this field have appeared (23a, 60a, b, 87a). The existence of messenger RNA has also been demonstrated in plant material (29a), in fractions of the RNA of thymus nuclei (78a), in liver cells (54c) and other objects. Great interest attaches to the DNA-dependent synthesis of messenger RNA *in vitro*, and also in systems containing all the other factors required for protein synthesis (90a). Under these conditions such systems demonstrate intensified protein synthesis. A very interesting method has now been developed for the isolation of this RNA (27b), which uses DNA-cellulose columns, which adsorb specifically this complementary form of RNA. This should contribute to further progress in this field.

It must be noted that this form of RNA has not yet acquired a generally recognized name in the literature and a number of synonyms are to be met with, e.g. messenger RNA (mRNA), informative RNA (iRNA), complementary RNA (cRNA), intermediary RNA, mediator RNA, etc.

The discovery of the special form of messenger RNA which carries genetic information is certainly a great achievement. It was really only made in 1961, so many of the features of the reactions which take place are not yet elucidated. It has, however, begun to fill up the gaps in our knowledge of the way in which the information carried by DNA is translated into material form, which used to be filled only with a series of hypotheses and assumptions.

## 5. Experimental synthesis of polypeptides on synthetic polynucleotides and the discovery of the nature of the nucleotide code for amino acids

At the end of 1961 and the beginning of 1962 one of the greatest discoveries of science was made. It was the partial solving of the nature of the code in the polynucleotides in the RNA matrices of protein synthesis and the first steps towards the identification of the genetic code of the DNA which controls individual development.

We shall discuss the question of the coding of the information contained in DNA in the chapter specially concerned with heredity. Here we shall only deal shortly with the experimental solution of the problem of the coding of amino acids on the RNA which is the immediate matrix for protein synthesis.

Biochemists have approached the experimental solution of the problem of coding along three lines. These are the correlation of changes in the nucleotides of tobacco mosaic virus with the corresponding changes in the protein of the TMV; theoretical analysis of punctate mutations of phages; and the use of model matrices in the form of synthetic polynucleotides, in place of the messenger RNA. They led biochemists almost simultaneously to the beginning of a solution of this problem, a problem which until recently was thought to be a matter for the distant future.

The third line led to the most sensational and productive results. The first communication concerning the possibility of using a synthetic polynucleotide (polyuridylic acid) as an analogue for a matrix for protein synthesis was made by Nirenberg at the Fifth International Congress of Biochemistry in Moscow in August 1961. This work was fully published by Nirenberg and Matthaei in October of the same year (71a). The authors naturally drew the conclusion that the nucleotide represented by . . . pUpUpUpU . . . contains the information for the synthesis of the polypeptide represented by . . . Phe.Phe.Phe.Phe.Phe. . . . This result was immediately used to solve the problem of the code, because all that was required for this purpose was to vary the composition of the synthetic polynucleotides and to study the nature of the amino acids used for the synthesis of protein in each case. Another laboratory, besides that of Nirenberg, was involved in these studies; it was the laboratory of Ochoa, in which an extensive study was made in order to work out methods of synthesis of polynucleotides. Ochoa and his colleagues were the first to use mixed polynucleotides in the study of these phenomena and this enabled them quite quickly to work out the code for each of the twenty amino acids or, more accurately, to find out the "meaning" of most of those triplets of nucleotides which contain uridylic acid as a necessary component (65a, 65b, 79a, 79b). The work in Nirenberg's laboratory afterwards followed the same line (70a).

There are many questions connected with this problem which still remain obscure, the foremost being the question of the sequence of nucleotides in the code-bearing triplets. However, the solution of these problems is clearly a task for the immediate future.

We must, however, spend a short time on the experimental methods by which these questions are being answered and give consideration to the peculiarities of the amino acid-determining code of the polynucleotides. In a later chapter on heredity we shall return to the subject of the genetic code, its universality, etc. Here we shall only deal, in the meanwhile, with the biochemical nature of that information inherent in the composition of the active matrices of protein synthesis which is directly involved in protein synthesis. The question of the genetic information in the chromosomes is more complicated and includes the problem of the control of the specialization of cells, the control of individual development, the reproduction of information, the formation of repressors and other questions whose analysis must be left till Chapters XVII-XVIII.

The method of working out the amino acid-determining code by the use of mixed synthetic polynucleotides has been described in detail by Ochoa and colleagues in their first paper in this series (65a). In the experiments of these authors the synthetic system consisted of ribosomes isolated from the bacterium *Escherichia coli* during the period of logarithmic growth and from the liver of rats. Having been cleaned up in the usual way the ribosomes were suspended in a solution at pH 7·9 ("tris-HCl" buffer, 0·01 M-MgCl$_2$ and 0·02 M-KCl). The concentration of the ribosomes in the solution was such that there was 10 mg. protein/ml. of suspension. In this state the ribosomes could be preserved unchanged by freezing to $-18°C$. The polynucleotides were synthesized by means of a polynucleotide phosphorylase isolated from *Azotobacter vinelandii*. The appropriate nucleotide diphosphates served as substrates for the reaction.

Before the beginning of the experiment the ribosomes underwent a so-called "preincubation" in a solution containing optimal concentrations of all the components required for protein synthesis in an unlabelled form (tris-HCl buffer at pH 7·9, Lubrol, KCl, MgCl$_2$, mercaptoethylamine, twenty unlabelled amino acids, ATP, GTP, creatine phosphate, creatine kinase, the supernatant fraction containing S-RNA and the necessary transfer and activation enzymes). "Preincubation" in this way decreased the likelihood of later "blank" incorporation of amino acids into proteins by non-enzymic reactions and residues of messenger RNA on the ribosomes. A fresh medium was then added to the "preincubated" ribosomes as well as synthetic polynucleotides (0·04 M calculated as mononucleotide), S-RNA and [$^{14}$C] amino acids with a high specific activity (1-10μc/μmole).

It is interesting to notice that the sedimentation coefficients of the synthetic polynucleotides varied between about 8 and 19 S, that is to say, they were of about the same length as messenger RNA molecules.

In these experiments the effect of synthetic polynucleotides on the nature of the incorporation of amino acids into proteins (in the acid-precipitated fraction) was also studied.

The question as to the nature of the sequence of polynucleotides determining the incorporation of phenylalanine in Nirenberg's first experiment was easily solved, it could only have been a mono-, di-, tri- or polyuridylic group. Ochoa and colleagues repeated Nirenberg's experiments and found that the molecular weight of the polypeptide obtained was related to that of the polynucleotide used as a matrix. They showed that one mole of amino acid incorporated into peptide corresponds with 3·25 moles of uridylic acid. This might be evidence in favour of the triplet, non-overlapping code (3U:1 Phe) but, to prove this on the basis of the relationship mentioned, it would be necessary to show that all of the polyuridylic chains in the medium were acting as matrices throughout the whole of their length simultaneously.

When mixed synthetic polynucleotides are used as matrices the analysis of the code depends on indirect calculations. If, for example, the polynucleotide is synthesized from uridylic and cytidylic components in a ratio of 5:1 (U:C = 5:1), then it would be theoretically possible to calculate by statistical methods the probability of the occurrence of different types of triplets in the chain. Thus, if the ratio U:C = 5:1 the ratios of the triplets UUU:UCC, UUU:CUC and UUU:CCU should theoretically be 25:1. This means that the group UUU will be found in the chain 25 times as often as the groups CCU, CUC and UCC. A similar calculation shows that the groups $U_2C$ (UUC, UCU, CUU) will occur five times less frequently in the chain than the group UUU.

As it is known that the group UUU is responsible for the incorporation of phenylalanine in the protein, the quantitative relationship between the amino acid being studied and phenylalanine in the protein formed indicates which are the triplets most likely to represent this amino acid in the polynucleotide code. Thus, using a polynucleotide of uridylic and cytidylic acids in a ratio of 5:1 the ratio between the incorporation of phenylalanine and serine was 4·4:1. It was therefore concluded that the triplet code for serine must be $U_2C$ (UUC, CUU or UCU), etc. In this way, without establishing the sequence and without any absolute proof of the triplet basis of the code, a calculation has been made which is still only provisional but which gives combinations of nucleotides containing uracil which correspond with all of the 20 amino acids. In this way it was shown that the code is "degenerate", i.e. each amino acid can correspond with several triplets (theoretically 4 nucleotides can give rise to 64 different triplets for 20 amino acids). It would seem that each theoretically possible triplet has a meaning.

Table 13 gives the composition of the code-bearing triplets as established by the first work of Ochoa's and Nirenberg's laboratories to which

TABLE 13

| Amino acid | Coding triplet, sequence unknown |
|---|---|
| Alanine | 1U, 1C, 1G |
| Arginine | 1U, 1C, 1G |
| Asparagine | 1U, 2A (1U, 1A, 1C) |
| Aspartic acid | 1U, 1A, 1G |
| Cysteine | 2U, 1G |
| Glutamic acid | 1U, 1A, 1G (UGC)* |
| Glutamine | 1U, 1C, 1G |
| Glycine | 1U, 2G |
| Histidine | 1U, 1A, 1C |
| Isoleucine | 2U, 1A |
| Leucine | 2U, 1C (2U, 1A) (2U, 1G) |
| Lysine | 1U, 2A |
| Methionine | 1U, 1A, 1G (UG)* |
| Proline | 1U, 2C |
| Phenylalanine | 3U |
| Serine | 2U, 1C (U, G, C)* |
| Threonine | 1U, 1A, 1C (1U, 2C) |
| Tryptophan | 1U, 2G |
| Tyrosine | 2U, 1A |
| Valine | 2U, 1G |

* As found by Nirenberg and colleagues (70a) who found different triplets. In the other cases the findings of Nirenberg's and Ochoa's laboratories agreed.

we have already referred. The code triplet for glutamine given here was not determined by Ochoa by the method just described but as a result of a theoretical analysis of the biochemical study of mutants of the tobacco mosaic virus carried out in another laboratory (Fraenkel-Conrat & Tsugita).

It is very important to take note of the fact that in the recent work of Nirenberg and colleagues under discussion (71b) the formation of poly-phenylalanine on a polyuridylic matrix is carried out in the usual bio-chemical system in which the participation of activating enzymes and soluble RNA as well as enzymes accepting the activated phenylalanine residues and transferring them to the matrix were all obligatory. This suggests that the composition of the S-RNA can only include some small code-bearing group which is responsible for combination with the matrix. Logically, it must also be assumed that there are as many different types of S-RNA as there are altogether of code-bearing triplets for the 20 amino acids.

In continuing this series of investigations, it was possible for workers in the laboratories of Nirenberg and Ochoa not only to study in more detail the nature of the coding triplets but also to attack some other complicated problems connected with protein biosynthesis. Nirenberg and colleagues (70b, 70c, 71a, b) found that, in the incorporation of phen-ylalanine into protein which is stimulated by polyuridylic acid, an increase in length of the polyuridylic chain also increases its functional capability, incorporation becoming more vigorous. At the same time they observed that, contrary to the view of some authors, the matrix does not seem to break down after synthesizing a single protein molecule, inasmuch as one mole of polyuridylic acid effected the synthesis of several moles of poly-phenylalanine. They also concluded that some types of nucleotide linkage are lacking in "sense", i.e. they do not code for amino acids and do not stimulate the synthesis of protein (poly A, poly AG). In their opinion such inactive zones may also exist in biological matrices, so as to set limits to the polypeptide chain; they would, as it were, mark the locations of the end-groups of the protein. The necessity for such end-group markers is also recognized in the work of Goldstein (50b). Ochoa and colleagues (27a) in a recent communication observed changes in the coding properties of polynucleotides when the bases of these synthetic polymers had been treated with nitrous acid (conversion of adenine to hypoxanthine, guanine to xanthine and cytosine to uracil). The authors thus realized "model mutations", where the changes effected in the proteins corresponded with those predicted on theoretical grounds.

As we can see, the working out of the amino acid code of the RNA-matrix of protein synthesis has only just begun, but it represents the beginning of a new stage in the development of the problem of protein synthesis and heredity. We shall come back to a discussion of this question in later chapters.

## 6. Conclusion. General hypothetical scheme of protein synthesis accounting for reactions between all of the components of the synthetic system discovered by the end of 1962

So far the following components of the system of protein synthesis have been found to be obligatory: 70 S "inactive" ribosomes made up of 30 S and 50 S ribonucleoprotein particles of which the protein and RNA components are relatively non-specific (p. 79), 70 S "active" ribosomes

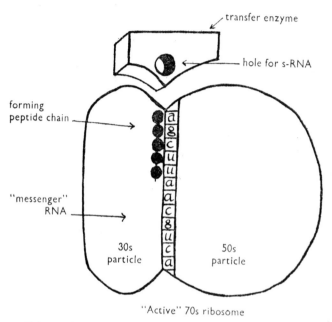

FIG. 54. Scheme for possible functioning of transfer enzyme, 70 S ribosome and "messenger" RNA in protein biosynthesis.

(70 S ribosomes + the messenger RNA carrying genetic information from DNA), enzymes activating the amino acids (p. 40), a heterogeneous mixture of molecules of low-molecular acceptor-RNA containing comparatively large amounts of unusual nucleotides (S-RNA) (p. 49) and enzymes transferring the molecules of S-RNA to the 70 S ribosomes (p. 66). ATP, GTP and Mg++ are co-factors of this system. It has also been found that there is a connection between the sequence of nucleotides and the determination of the sequence of amino acid residues.

Consideration of these recently discovered aspects of the process of synthesis has led us to work out a new hypothetical scheme of the processes going on in the ribosomes which proposes a rather different solution to the problem of the arrangement of information on the interacting polynucleotide chains. This scheme is set out in Figs. 54 and 55 (18a). According to this model of protein synthesis the enzyme transferring the S-RNA is sterically adapted to react with the 70 S ribosome and has a canal or groove for carrying the S-RNA. It is known that the molecule of S-RNA carrying different amino acids may have chains of the same length. However, we take the view that it is the pseudouridine in the chain of the S-RNA which can mark off the limits of the information-carrying zone. We also suppose that the positions of the amino acids are

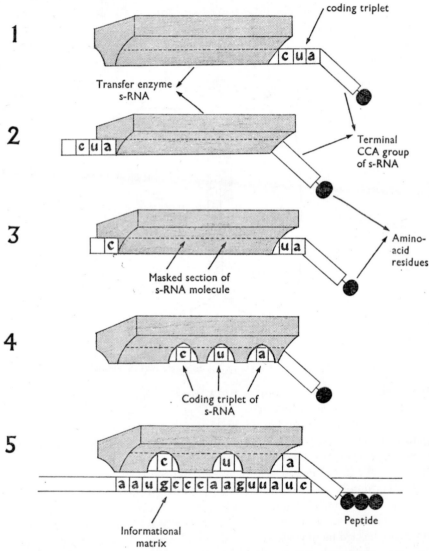

FIG. 55. Scheme for possible modes of interaction of matrix-RNA with the transfer enzyme—S-RNA complex in protein biosynthesis.

"coded" by a combination of three nucleotides but it is not impossible that this "coding" group may be divided up by the transfer enzyme into two or three sections $(2+1)$ $(1+1+1)$. The number of possible combinations in this case would be 64, that is to say, greater than the number of amino acids, but it is possible that each amino acid may correspond not to one, but to two, three or four types of combination.

The length of the molecule of messenger-RNA in bacteria is 1000 nucleotides. The average length of the molecule of S-RNA is 100 nucleotides. It is now known that there are some continuous polypeptide chains (e.g. that of serum albumin) which consist of 500-600 amino acid residues. In this case the site of "deposition" on the molecules of S-RNA must be overlapping and even the trinucleotide code with breaks must overlap. In this scheme, as before, we assume that the "assembly" of the polypeptide chain does not take place on the matrix itself but alongside it on the protein stroma of the ribosomes.

The idea that the code in the polynucleotide chains does not overlap is now accepted by almost everyone. There is, however, only a limited amount of evidence that this is so. It is based on the idea that the coding sequence of the nucleotides is uninterrupted. If this is true, then it does, in fact, follow from the factual evidence that the code cannot overlap. The main argument in favour of the non-overlapping nature of the code is found in the nature of the mutational and evolutionary changes in the proteins which often involve substitution of a single amino acid. This argument is also the main foundation for the non-overlapping types of code structures suggested in the new and extremely interesting work of Crick and colleagues (36). The essential feature of this idea is quite clear; if the code overlapped the alteration of one nucleotide would lead to the alteration of three amino acid residues, as each nucleotide in the polynucleotide chain would, in this case, be a component of three triplets. For example, in the sequence GCAGUC the nucleotide A participates in three triplets, GCA, CAG and AGU. Alteration at A, by substitution of G for example, would alter the meaning of three triplets. As each triplet "codes" a single amino acid, the alteration of one nucleotide must lead to the alteration of three amino acids. In most cases, both in experiments on the chemical mutation of viruses and in the evolutionary and pathological variability of proteins (haemoglobin) there is substitution of a single amino acid in the polypeptide chain.

This phenomenon, however, is not direct proof that the code does not overlap, as it is susceptible to several other explanations. For example, if we take it that the code is degenerate, then each amino acid will correspond with several triplets and it may be supposed that the substitution of the nucleotide G for nucleotide A in the nucleotide sequence given above would lead to an alteration of meaning in the case of the formation of the new triplet GCG but not in the cases of CGG or GGU. This is theoretically possible and in this case the alteration of only one nucleotide in the RNA would lead to an alteration in only one amino acid residue in the polypeptide chain.

Another possibility, which we incline to think of as a probability, is that of an overlapping code with breaks. Such a code would explain all the facts on which the assumption of non-overlapping codes is based. It also explains the need for a comparatively long chain in the soluble RNA and would ensure a considerably more economical type of recording of information in the make-up of the polynucleotides of the RNA. It also makes the equality of the distance between adjacent nucleotides and that between adjacent amino acids comprehensible. Clearly the length of the breaks need not be great. If we imagine a trinucleotide code with breaks:

```
A  G U C  A G C U U G  A A U C G      matrix
U — — — — — G — — — — U — — — — —⎫    coding
— — A — — — — — A — — — — — — G —⎪    triplets
                                  ⎬    of
— — U — — — — — — — — — U — — — A ⎭    S-RNA
```

then it becomes clear that the alteration of a single nucleotide in the matrix may lead to one of several consequences. 1. To an alteration in the position of a single amino acid residue (if the code is degenerate or if the nucleotide concerned only forms part of one triplet on account of the extensiveness or constriction of the next triplet). It must also be considered that, in the overwhelming majority of cases the length of the polypeptide chain is less than one third of that of the polynucleotide of S-RNA and, where the breaks are long enough, hardly any of the code-bearing nucleotides of the RNA need to take part in more than one triplet. 2. It may lead to alteration in two or three amino acid residues in different parts of the protein molecule (a very common occurrence). Furthermore, in such a case the insertion or removal of one or two nucleotides in the chain will not lead to a wrong interpretation of the code along the whole length of the chain but only to a comparatively localized disturbance (punctate mutation) limited to a part corresponding in length with the code-bearing triplet.

An interesting model of a partially overlapping code has recently been worked out by Wall (87b), who adduces a number of tentative data on the possibility of an overlapping code in nucleic acids. This is a novel "compromise" of partially overlapping codes, in which neighbouring coding groups are demarcated by a single nucleotide in common.

As for the part played by high-molecular RNA ($0.6 \times 10^6$ and $1.5 \times 10^6$), it may take part either by determining the particular morphology of the ribosomes or it may provide the matrices for the synthesis of the proteins of the ribosomes themselves at the time when the ribosomes are being formed in the nucleus or cytoplasm. In this model the pseudouridine plays a hypothetical part in fixing the position of the soluble RNA in the canal of the transfer enzyme. The possibility of fixing a polynucleotide chain in a "canal" formed by a special form of protein is not unexpected on theoretical grounds. In fact almost all forms of plant virus particles show just such a type of interrelationship between the protein sub-units and the molecule of RNA.

Of course, the model suggested here is hypothetical but the setting up of such models must, to some extent, make it easier to work out experimentally the nature of that complicated system of reactions which, in the long run, leads up to protein synthesis.

## REFERENCES

1. BELOZERSKIĬ, A. N. & SPIRIN, A. S. (1960). *Izvest. Akad. Nauk S.S.S.R.* **25**, No. *1*, 64.
2. BLYUMENFEL'D, L. A. (1960). *Doklady Akad. Nauk S.S.S.R.* **130**, 887.
3. BLYUMENFEL'D, L. A. & BENDERSKIĬ, V. A. (1960). *Doklady Akad. Nauk S.S.S.R.* **133**, 1451.
4. BRESLER, S. E. (1950). *Uspekhi sovremennoĭ Biol.* **30**, 90.
5. — (1951). *Voprosy Filosofii*, no. **3**, 82.
6. — (1954). *Uspekhi biol. Khim.* **2**, 66.
7. — (1955). *Biokhimiya*, **20**, 463.
8. BRESLER, S. E., GLIKINA, M. V., KONIKOV, A. P., SELEZNEVA, N. A. & FINO-GENOV, P. A. (1949). *Izvest. Akad. Nauk S.S.S.R., Ser. fiz.*, **13**, 392.
9. BRESLER, S. E., GLIKINA, M. V., SELEZNEVA, N. A. & FINOGENOV, P. A. (1952). *Biokhimiya*, **17**, 44.
10. BRESLER, S. E., GLIKINA, M. V. & FRENKEL', S. YA. (1954). *Doklady Akad. Nauk S.S.S.R.*, **96**, 565.
11. BRESLER, S. E., KONIKOV, A. P. & SELEZNEVA, N. A. (1949). *Doklady Akad. Nauk S.S.S.R.*, **65**, 521.
12. VLADIMIROV, YU. A. & KONEV, S. V. (1959). *Biofizika*, **4**, 533.
13. KRITSMAN, M. G. & KONIKOVA, A. S. (1959). *Uspekhi sovremennoĭ Biol.* **48**, 136.
14. LESTROVAYA, N. N. (1958). *Uspekhi biol. Khim.* **3**, 97.
15. MEDVEDEV, ZH. A. (1955). *Uspekhi sovremennoĭ Biol.* **40**, 159.
16. — (1959). *Doklady Moskov. sel'skokhoz. Akad. im. K. A. Timiryazeva*, No. 47, p. 77.
17. — (1960). *Uspekhi sovremennoĭ Biol.* **50**, 121.
18. — (1960). *Izvest. Timiryazev. sel'skokhoz. Akad.* No. *1*, p. 103.
18a. — (1962). *Nature (Lond.)*, **195**, 38.
19. OPARIN, A. I. (1948). *Soveshchanie po Belku, Akad. Nauk S.S.S.R. (5-ya Konferents. vysokomolekulyarnym Soedineniyam)* (ed. A. I. Oparin & A. G. Pasynskiĭ), p. 5. Moscow: Izd. A.N. S.S.S.R.
20. OPARIN, A. I. (1957). *The Origin of life on the Earth* (3rd edn., trans. A. Synge). Edinburgh: Oliver & Boyd.
21. PASYNSKIĬ, A. G. (1951). *Doklady Akad. Nauk S.S.S.R.* **77**, 863.
22. — (1960). *Biofizika*, **5**, 16.
23. SAMOĬLOVA, O. P. & BLYUMENFEL'D, L. A. (1961). *Biofizika*, **6**, 15.
23a. SPIRIN, A. S. & SMIRNOV, V. N. (1962). *Izv. Akad. Nauk S.S.S.R. Ser. biol.* **27**, 477.
24. TONGUR, V. S. (1960). *Uspekhi sovremennoĭ Biol.* **49**, 156.
25. CHAVCHANIDZE, V. V. (1958). *Biofizika*, **3**, 391.
26. ASTRACHAN, L. & FISHER, T. N. (1961). *Proc. V int. Congr. Biochem., Moscow*, **9**, 121.

27. ASTRACHAN, L. & VOLKIN, E. (1958). *Biochim. biophys. Acta*, **29**, 536.
27a. BASILIO, C., WAHBA, A. J., LENGYEL, P., SPEYER, J. F. & OCHOA, S. (1962). *Proc. natl. Acad. Sci. U.S.* **48**, 613.
27b. BAUTZ, E. & HALL, B. D. (1962). *Proc. natl. Acad. Sci. U.S.* **48**, 400.
28. BELOZERSKY [BELOZERSKIĬ], A. N. & SPIRIN, A. S. (1958). *Nature (Lond.)*, **182**, 111.
29. BLOEMENDAL, H., BOSCH, L. & SLUYSER, M. (1960). *Biochim. biophys. Acta*, **41**, 454.
29a. BONNER, J., HUANG, R.-C. C. & MAHESHWARI, N. (1961). *Proc. natl. Acad. Sci. U.S.* **47**, 1548.
30. BRENNER, S. (1957). *Proc. natl. Acad. Sci. U.S.* **43**, 687.
31. BRENNER, S., JACOB, F. & MESELSON, M. (1961). *Nature (Lond.)*, **190**, 576.
32. CHANTRENNE, H. (1958). *Rec. Trav. chim.* **77**, 586.
32a. CHAPEVILLE, F., LIPMANN, F., EHRENSTEIN, G. VON, WEISBLUM, G., RAY, W. J. & BENZER, S. (1962). *Proc. natl. Acad. Sci. U.S.* **48**, 1086.
33. COX, R. A. & LITTAUER, U. Z. (1960). *J. mol. Biol.* **2**, 166.
34. CRICK, F. H. C. (1958). *Symp. Soc. exp. Biol.* **12**, 138.
35. — (1959). In *Brookhaven Symposia in Biology*, No. *12. Structure and function of genetic elements* (ed. M. Levine *et al.*), p. 35. Upton, N.Y.: Biology Department, Brookhaven National Laboratory.
36. CRICK, F. H. C., BARNETT, L., BRENNER, S. & WATTS-TOBIN, R. J. (1961). *Nature (Lond.)*, **192**, 1227.
37. CRICK, F. H. C., GRIFFITH, J. S. & ORGEL, L. E. (1957). *Proc. natl. Acad. Sci. U.S.* **43**, 416.
38. DOCTOR, B. P., APGAR, J. & HOLLEY, R. W. (1961). *J. biol. Chem.* **236**, 1117.
39. DOUNCE, A. L. (1953). *Enzymologia*, **15**, 251.
40. — (1953). *Nature (Lond.)*, **172**, 541.
41. ELIAS, P. (1959). *Revs. modern Phys.* **31**, 221.
42. FÖRSTER, T. (1960). *Radiation Res. Suppl.* No. *2*, 326.
43. FRASER, M. J. & GUTFREUND, H. (1958). *Proc. roy. Soc.* **B149**, 392.
44. FRESCO, J. R. & ALBERTS, B. M. (1960). *Proc. natl. Acad. Sci. U.S.* **46**, 311.
45. FRESCO, J. R., ALBERTS, B. M. & DOTY, P. (1960). *Nature (Lond.)*, **188**, 98.
46. GAMOW, G. (1954). *Kgl. Danske videnskap. Selskab. biol. Medd.* **22**, No. *3*, 1.
47. — (1955). *Kgl. Danske videnskap. Selskab. biol. Medd.* **22**, No. *8*, 1.
48. GAMOW, G. & YČAS, M. (1955). *Proc. natl. Acad. Sci. U.S.* **41**, 1011.
49. GAMOW, G., RICH, A. & YČAS, M. (1956). *Adv. biol. med. Phys.* **4**, 23.
50. GAMOW, G. & YČAS, M. (1958). *Symposium Inform. Theory Biol.*, Gatlinburg, Tenn. 1956 (ed. H. P. Yockey *et al.*), p. 63. New York: Pergamon.
50a. GEIDUSCHEK, E. P., NAKAMOTO, T. & WEISS, S. B. (1961). *Proc. natl. Acad. Sci. U.S.* **47**, 1405.
50b. GOLDSTEIN, A. (1962). *J. mol. Biol.* **4**, 121.
51. GRODSKY, G. M. & TARVER, H. (1957). *Arch. Biochem. Biophys.* **68**, 215.
52. GROS, F., HIATT, H., GILBERT, W., KURLAND, C. G., RISEBROUGH, R. W. & WATSON, J. D. (1961). *Nature (Lond.)*, **190**, 581.
53. HALL, B. D. & SPIEGELMAN, S. (1961). *Proc. natl. Acad. Sci. U.S.* **47**, 137.
54. HAUROWITZ, F. (1950). *Chemistry and biology of proteins.* New York: Academic Press.
54a. HAYASHI, M. & SPIEGELMAN, S. (1961). *Proc. natl. Acad. Sci. U.S.* **47**, 1564.
54b. HENDLER, R. W. (1962). *Nature (Lond.)*, **193**, 821.
54c. HIATT, H. H. (1962). *Fed. Proc.* **21**, 381.
55. HOAGLAND, M. B. (1958). *Proc. IV intern. Congr. Biochem.*, Vienna, **8**, 199.
56. — (1960). In *Nucleic acids* (ed. E. Chargaff & J. N. Davidson), **3**, 349. New York: Academic Press.
57. — (1959). *Scient. Amer.* **201**, No. *6*, p. 55.

58. HOAGLAND, M. B. & COMLY, L. T. (1960). *Proc. natl. Acad. Sci. U.S.* **46**, 1554.

59. HOAGLAND, M. B., ZAMECNIK, P. C. & STEPHENSON, M. L. (1959). In *A Symposium on molecular Biology* (ed. R. E. Zirkle), p. 105. Chicago: Univ. of Chicago Press.

60. HUNTER, G. D. & GODSON, G. N. (1961). *Nature (Lond.)*, **189**, 140.

60a. HURWITZ, J. & FURTH, J. J. (1962). *Scient. Amer.* **206**, No. *2*, p. 41.

60b. JACOB, F. & MONOD, J. (1961). *J. mol. Biol.* **3**, 318.

61. JEHLE, H. (1959). *Proc. natl. Acad. Sci. U.S.* **45**, 1360.

62. KLEE, W. A. & CANTONI, G. L. (1960). *Proc. natl. Acad. Sci. U.S.* **46**, 322.

63. KRUH, J., ROSA, J., DREYFUS, J.-C. & SCHAPIRO, G. (1961). *Proc. V int. Congr. Biochem., Moscow*, **9**, 95.

64. LAMFROM, H. (1961). *J. mol. Biol.* **3**, 241.

65. LEVINTHAL, C. (1959). *Revs. modern Phys.* **31**, 227, 249.

65a. LENGYEL, P., SPEYER, J. F. & OCHOA, S. (1961). *Proc. natl. Acad. Sci. U.S.* **47**, 1936.

65b. LENGYEL, P., SPEYER, J. F., BASILIO, C. & OCHOA, S. (1962). *Proc. natl. Acad. Sci. U.S.* **48**, 282.

66. LINGREL, J. B. & WEBSTER, G. [C.] (1961). *Biochem. biophys. Res. Commun.* **5**, 57.

67. LIPMANN, F., HÜLSMANN, W. C., HARTMANN, G., BOMAN, H. G. & ACS, G. (1959). *J. cell. comp. Physiol.* **79**, Suppl. 1, p. 75.

68. LOFTFIELD, R. B. (1957). *Progr. Biophys. biophys. Chem.* **8**, 347.

69. LOFTFIELD, R. B., EIGNER, E. A. & HECHT, L. I. (1958). *Proc. IV int. Congr. Biochem., Vienna*, **8**, 222.

70. MANDEL, P., WEILL, J. D., LEDIG, M. & BUSCH, S. (1959). *Nature (Lond.)*, **183**, 1114.

70a. MARTIN, R. G., MATTHAEI, J. H., JONES, O. W. & NIRENBERG, M. W. (1962). *Biochem. biophys. Res. Commun.* **6**, 410.

70b. MATTHAEI, J. H. & NIRENBERG, M. W. (1962). *Fed. Proc.* **21**, 415.

70c. MATTHAEI, J. H., JONES, O. W., MARTIN, R. G. & NIRENBERG, M. W. (1962). *Proc. natl. Acad. Sci. U.S.* **48**, 666.

71. NATHANS, D. (1960). *Ann. N.Y. Acad. Sci.* **88**, 718.

71a. NIRENBERG, M. W. & MATTHAEI, J. H. (1961). *Proc. natl. Acad. Sci. U.S.* **47**, 1588.

71b. NIRENBERG, M. W., MATTHAEI, J. H. & JONES, O. W. (1962). *Proc. natl. Acad. Sci. U.S.* **48**, 104.

72. NOMURA, M., HALL, B. D. & SPIEGELMAN, S. (1960). *J. mol. Biol.* **2**, 306.

73. NOVELLI, G. D. & DEMOSS, J. A. (1957). *J. cell. comp. Physiol.*, **50**, Suppl. 1, 173.

74. RENDI, R. & CAMPBELL, P. N. (1959). *Biochem. J.* **72**, 435.

75. RENDI, R. & OCHOA, S. (1961). *Science*, **133**, 1367.

76. SCHWARTZ, D. (1955). *Proc. natl. Acad. Sci. U.S.* **41**, 300.

77. — (1958). *Nature (Lond.)*, **181**, 769.

78. — (1959). *Nature (Lond.)*, **183**, 464.

78a. SHIBATANI, A., KLOET, S. R. DE, ALLFREY, V. G. & MIRSKY, A. E. (1962). *Proc. natl. Acad. Sci. U.S.* **48**, 471.

79. SIMKIN, J. L. & WORK, T. S. (1958). *Symposia Soc. exptl. Biol.* **12**, 164.

79a. SPEYER, J. F., LENGYEL, P., BASILIO, C. & OCHOA, S. (1962). *Proc. natl. Acad. Sci. U.S.* **48**, 63.

79b. — (1962). *Proc. natl. Acad. Sci. U.S.* **48**, 441.

80. SPIEGELMAN, S. (1957). In *A Symposium on the chemical basis of heredity* (ed. W. D. McElroy & B. Glass), p. 232. Baltimore, Md.: Johns Hopkins Press.

80a. SPIEGELMAN, S., HALL, B. D., & STORCK, R.   (1961).   *Proc. natl. Acad. Sci. U.S.* **47**, 1135.

81. STANLEY, W. M. (jun.) & BOCK, R. M.   (1961).   *Nature (Lond.)*, **190**, 299.

82. STEELE, R. H. & SZENT-GYÖRGYI, A.   (1958).   *Proc. natl. Acad. Sci. U.S.* **44**, 540.

83. STRYER, L.   (1959).   *Biochim. biophys. Acta*, **35**, 242.

84. SUZUKI, I. & SHIMURA, K.   (1960).   *J. Biochem. (Tokyo)*, **47**, 551.

85. TALWAR, G. P. & MACHEBOEUF, M.   (1954).   *Ann. Inst. Pasteur*, **86**, 169.

86. TISSIÈRES, A., SCHLESSINGER, D. & GROS, F.   (1960).   *Proc. natl. Acad. Sci. U.S.* **46**, 1450.

87. Ts'O, P. O. P. & LUBELL, A.   (1960).   *Arch. Biochem. Biophys.* **86**, 19.

87a. VOLKIN, E.   (1962).   *Fed. Proc.* **21**, 112.

87b. WALL, R.   (1962).   *Nature (Lond.)*, **193**, 1268.

88. WEISS, S. B.   (1961).   *Science*, **133**, 1370.

89. — (1961).   *Fed. Proc.* **20**, 362.

90. — (1960).   *Proc. natl. Acad. Sci. U.S.* **46**, 1020.

90a. WOOD, W. B. & BERG, P.   (1962).   *Proc. natl. Acad. Sci. U.S.* **48**, 94.

91. WOESE, C. R.   (1961).   *Nature (Lond.)*, **190**, 697.

92. YČAS, M.   (1958).   *Symposium on information theory in biology* (ed. H. Yockey *et al.*), p. 70.   London: Pergamon.

93. — (1960).   *Nature (Lond.)*, **188**, 209.

93a. YČAS, M. & VINCENT, W. S.   (1960).   *Proc. natl. Acad. Sci. U.S.* **46**, 804.

93b. ZAMECNIK, P. C.   (1962).   *Biochem. J.* **85**, 257.

94. ZUBAY, G. & WILKINS, M. H. F.   (1960).   *J. mol. Biol.* **2**, 105.

# CHAPTER XVI

# ERRORS IN THE REPRODUCTION OF NUCLEIC ACIDS AND PROTEINS AND THEIR BIOLOGICAL SIGNIFICANCE

## *Introduction*

Theoretically it is hard to imagine a system of reproduction of the specificity of polymers which would always work absolutely accurately and would never make a mistake. This rule also holds for the synthesis of proteins and nucleic acids. It would seem that there must always be a certain variability in any synthesis, especially when there are changes in both internal and external conditions, and this variability forms the main basis for the natural selection by means of which living systems are improved.

An understanding of the causes of these "errors" of synthesis, and, in particular, of the synthetic systems which make these "errors" possible, is not only of great theoretical but also of practical interest. For example, the use of analogues and antimetabolites in medicine, agriculture and breeding is based on the ability of synthetic systems to "err" and incorporate altered products in the cycles of their biochemical reactions.

One of the main causes of "errors" in any synthesis must surely be the variability of the internal and external environment (chemical, physical, physiological, etc.), the background against which the biochemical processes take place (the correlation and nature of the substrates, activity of the enzymes, etc.). All of these factors are continually changing and this constitutes the essential cause of every variation.

However, if these "elemental" forces were allowed to act in an uncontrolled way in the organism, then, of course, no development would be possible. Living nature gives rise to complicated systems of reproduction of specificity in order to counteract the levelling out effect of the "elemental" forces of variability and maintain the purposive achievements of evolution which have been selected by natural selection.

It is therefore clear that the second main reason for errors of synthesis is the imperfection of the controls of the system responsible for the reproduction of specificity, i.e. their inability to ensure the reproduction of the necessary sequence of amino acids or nucleotides in the macromolecules of the living body always and under any conditions.

The limits of the accuracy in the working of these systems depend on many factors but they too have been subjected to a preliminary survey. The attempts so far made to calculate the "project accuracy" of synthetic systems such as that for protein synthesis are still very imperfect and take no account of the stepwise character of the reproduction of specificity which is determined in the first place by the selectivity of the enzymes, then by the selectivity of molecules of S-RNA and finally by the selectivity of the matrices.

In this chapter we cannot consider all the questions involved in the problem of "errors of reproduction". The question of the biochemical nature of mutation will be partly discussed in the next chapter. The nature of the external and internal factors which promote the occurrence and especially the accumulation of errors over a period will be the subject of special discussion in the chapter on molecular ageing. In the present chapter we shall concentrate on pointing out some of the imperfections in the system for the biochemical reproduction of proteins, RNA and DNA and the nature of possible "errors". Furthermore, we should like to indicate that the induction of experimental errors of synthesis of proteins and amino acids is now one of the most widely used methods for studying the mechanism of protein synthesis and that the results obtained in this way are of great practical and theoretical significance.

## 1. Errors in the reproduction of DNA and their biological consequences

### a) Spontaneous "errors" in the synthesis of DNA

Errors in the reproduction of DNA are, as a rule, the basis for genetic mutations if they are localized in reproductive cells, and for somatic mutations if they are localized in the cells of somatic tissues at the time of morphogenesis. One must evidently, therefore, distinguish especially a group of "hidden" somatic mutations in specialized, already formed tissues, because these mutations, having a random, chaotic character, do not lead to any particular alteration of the tissue, but resemble the accumulation of "noise" in cybernetic systems, and lead to the gradual inactivation or "wearing out" of the biological system as a whole. In particular cases they lead to the appearance of malignant neoplasms; this happens when the aggregate of changes brings the cell outside the control of its regulatory systems (11a).

Mutations of an individual gene in reproductive cells arise on the average with a frequency of 1 in $10^5$ to $10^6$ per generation. The number of genes in the human being is also of the order of $10^5$. The diploid array of human chromosomes numbers 46, while in each chromosome there are not less than 5000-6000 molecules of DNA, having molecular weight 5-10 millions. Thus each human cell contains not less than 300,000 molecules of DNA. If we assume that each "mistake" in the reproduction of DNA leads to a mutation, this would mean that at each cell division one molecule of DNA is altered. In fact, however, the number of "mistakes" in the reproduction of DNA is higher, since not all mutations come to be noticed, and also because a particular amino acid may correspond with various combinations of nucleotides, so that not every change in nucleotide sequence leads to mutation.

Thus, for purposes of argument, we may assume that not more than one "mistake" arises during the reproduction of 20,000-30,000 molecules of DNA; this is a very high accuracy of reproduction, considering that the molecular weight of each molecule of DNA is about six million, which corresponds to 20,000 nucleotides in double helix, or 10,000 nucleotide pairs. Inasmuch as the reproduction of molecules of DNA consists of the reproduction of nucleotide pairs, a rough calculation shows that a single change in sequence (punctate mutation) arises in the course of $3 \times 10^8$ complementary reproductions of nucleotide pairs. Such a high accuracy of reproduction of DNA is to be expected, since it is this system which conserves the hereditary information through millions of successive generations. Of course selection has a stabilizing role in conserving information during the passage of the generations, but even selection would be powerless to preserve a species unchanged through many generations if the frequency of "mistakes" in the reproduction of DNA were to rise above a certain level. The factors responsible for spontaneous changes in DNA are various (formation of excursions (loops) in the complementary chain, oxidation of nucleotides, substitution, action of destructive agents such as free radicals, spontaneously arising in biological systems, etc.). The spontaneous formation of "the wrong" pairs, e.g. guanine-thymine, etc., is less probable. However, not long ago Donohue & Trueblood (32) sought to show that, besides the usual base pairs in DNA (A—T, T—A, G—C, C—G) other pairs, such as A—G, could occur. However, the hypotheses of these authors have a somewhat abstract character, and they did not base their arguments on any experimental or factual material.

Thus we see that the system of the synthesis of DNA is powerfully protected from spontaneous variations, just because it has to preserve and translate the hereditary information from generation to generation. While "resting", DNA exists in the cell in the form of a double helix, in which the active groups are bound to one another in accordance with their complementary interaction. Metabolism of DNA, in the absence of mitosis, proceeds tens of times more slowly than that of RNA and protein, with correspondingly lower opportunity for chance variations than exists, say, for RNA. One should also take into account that, when a change of sequence of nucleotides takes place in some molecule of DNA, perhaps under the influence of some artificial factor such as increased radiation, then such a change will be included in the information transmitted to RNA, and so on to proteins, whereas the reverse reaction, to possible changes arising in these latter, will not exist, as far as DNA is concerned.

*b) Experimentally induced "errors" in the reproduction of DNA*

There are many experimental means of inducing changes in DNA but we shall confine ourselves to a discussion of one, namely the effect of analogues of nucleotides on DNA. The formation of analogues and their incorporation into DNA may occur to a very slight extent under natural conditions and this is evidence that the discriminatory systems are not perfect. In experimental studies further evidence of imperfection can be provided by increasing the variety and concentration of endogenous analogues millions of times. In many experiments carried out in several laboratories it has been shown that various analogues (bromouracil, iodouracil, chlorouracil and others) can be incorporated into DNA (13, 37, 69a, 97, 98, 99). In the bacterium *E. coli* it has been found possible to substitute bromouracil for nearly 28% of the thymine of DNA (97). At the same time thiouracil and thiothymine could not be incorporated into DNA. Further cultivation of the bacteria on an ordinary medium led to restoration of the normal composition of the DNA as the analogues did not bring about any marked change in the genetic constitution of the cells (although certain changes did take place). It was found possible to substitute bromouracil for thymine to the extent of nearly 80% by artificial means in mammalian cells grown in culture (55). In many cases, however, analogues have a pronounced genetic effect and disturb the function of the DNA (50). According to Shapiro & Chargaff (89) bromouracil causes quite a significant alteration in the sequence of nucleotides in bacterial DNA. It is also possible that analogues may have secondary effects, for example by altering the relationship of DNA containing them to a number of factors in the external environment. Thus, according to Zamenhof (97), cells containing 5-bromouracil in their DNA are much more sensitive to ultraviolet radiations than ordinary cells.

The mutation-producing effect of analogues of nucleotides is certainly associated with their incorporation into DNA. In spite of the experimental nature of this sort of variation it cannot be denied that it may play a certain part in the development of some of the spontaneous mutations occurring in living things. It is also possible that occasional errors may take place in the process of synthesis of nucleotides and that therefore a minute quantity of analogues of nucleotides may be continually being formed endogenously. We have also mentioned already that RNA and DNA isolated from natural sources have been found to contain very small quantities of unusual nucleotides and it is not yet known whether these have any biochemical function or whether they arise as a result of the incorporation of endogenous analogues.

In this connection, study of the mechanism of the mutagenic effect of analogues of nucleotides might lead to the discovery of the mechanism of one of the factors affecting the molecular variability of DNA. The closest approach to an understanding of the molecular mechanism of the mutagenic effect of analogues of nucleotides has been made in the study of phages, which are the simplest of living things containing DNA.

Several years ago Litman & Pardee (70) showed that 5-bromouracil, which is an analogue of thymine, can replace thymine in T2 phage and that this substitution is very productive of mutants in the phage.

In view of these findings Benzer & Freese (20) compared mutants of T4 phage induced by 5-bromouracil with mutants of the same phage arising spontaneously, by analysing the detailed genetic structure of the mutants. They found that the localization of the mutations caused by 5-bromouracil differed from that of the mutations arising spontaneously. Analysis of the spontaneous mutations showed that they are not random (as regards their localization) but are most often found in particular parts of the genetic system (the so-called "hot spots"). The mutations induced by bromouracil were also localized in certain limited parts of the genetic system. The discovery of these parts where the mutability is great and the possibility of studying the mutation of the simplest living organisms in terms of their DNA have attracted a lot of attention and further studies have been made in these laboratories (44-47, 71) in which the authors have tried to unravel the mechanism of the mutagenic effect of 5-bromouracil and some other analogues.

By using a very wide range of analogues of nucleotides for his experiments Freese (46-48) found that the most effective ones in producing mutations in T4 phage were 5-bromouracil and 5-bromodesoxyuridine (analogues of thymine), 2-aminopurine (an analogue of adenine) and 2, 6-diaminopurine.

The mutagenic effect of analogues of nucleotides was very closely correlated with the concentration of the natural substrates for the synthesis of DNA. A significant mutagenic effect was usually only observed when the formation of the normal metabolites was suppressed. The nature of the mutations and their classification, reproducibility and stability and the possibility of their "reversion" have been studied in detail by Freese. By means of genetic maps it has been established that the mutations induced by bromouracil and 2-aminopurine are localized in an area which is not large and is the same in both cases. As most of these mutations are reversible the author suggests that they consist in alterations in individual pairs of bases in the DNA. Having established a number of interesting features of the mutagenic process Freese tried to work out the mechanism of the mutagenic effect of analogues on the basis of theoretical "structural" considerations. It is known that the reproduction of the sequence of nucleotides in a DNA depends on the structural complementarity of the bases and the formation of pairs of adenine with thymine and guanine with cytosine (or hydroxymethylcytosine in the case of phages of the T2 and T4 groups) which are bound together by contacts between the active groups and the formation of hydrogen bonds (cf. Chapter II). Structural analysis of the possible complementary reactions of 5-bromouracil and

2-aminopurine showed that these analogues are complementary to adenine and thymine respectively but their complementarity is not so strict as that of the normal nucleotides. If it were simply a question of the replacement of one nucleotide by its analogue the genetic properties of the phages would not be altered because, in a new generation of polynucleotide chains, the original "maternal" sequence of nucleotides would be reproduced in its entirety. In fact this is what did happen in 90% of descendants of phages in which thymine was quantitatively replaced by 5-bromouracil. They grew normally and no mutations were found. Mutations arose as a result of errors in the sequence of nucleotides of the daughter polymers and the presence of analogues in the mother chains determined, to some extent, the percentage of these errors.

In looking for possible structural causes for the increased percentage of errors, Freese took note of the mention by Watson & Crick (92) of a rather uncommon possibility for the occurrence of errors in daughter chains when adenine is replaced by its tautomer having the imino form, thus becoming complementary to cytosine. When this happens the position in the daughter chain which would normally be intended for thymine may instead be occupied by cytosine and this gives rise to a stable change in the nucleotide sequence of the daughter chains and of subsequent generations of chains.

According to Freese something analogous happens when 5-bromouracil and 2-aminopurine are exerting their mutagenic effects. The strong electronegative properties of bromine (compared with the 5-methyl group of thymine) increase the likelihood of the formation of the tautomeric form and also, the loss of the hydrogen atom in the 1-position by the bromouracil may cause errors in the formation of pairs of hydrogen bonds. For this reason 5-bromouracil, as well as being complementary to adenine, may also form a pair with guanine, thus causing an alteration in the sequence of nucleotides as, in the next generation of polynucleotides the guanine will determine the position of a cytosine (or 5-hydroxymethylcytosine residue).

As for 2-aminopurine, although it is normally complementary to thymine, it can react with 5-hydroxymethylcytosine and thus it too can induce a stable alteration in the sequence of nucleotides. Fig. 56 represents a chemical model of such errors in the formation of complementary pairs. Freese suggests that we should, apparently, distinguish between two types of error in the synthesis of DNA, i.e. "errors of incorporation" and "errors of reproduction". In cases where there is an error of incorporation, some analogue or nucleotide is incorporated into a wrong position in an earlier chain and disturbs the sequence of the complementary section of any copies. Errors of reproduction are determined by the peculiarities of the process we have just been describing.

Litman & Pardee (71) have studied the mechanism of the mutagenic effect of 5-bromouracil on T2 phages quite independently of Freese.

They also tried to give a picture of the mechanism of the mutagenic action of this analogue of thymine based on its dual complementarity (to adenine and to guanine).   In our present discussion of the significance of the dual complementarity of the structure of DNA we cannot avoid mentioning that, according to recent calculations (83), the pair guanine-

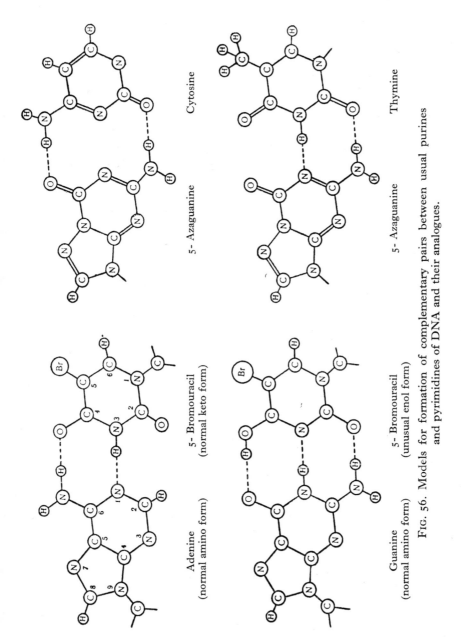

FIG. 56. Models for formation of complementary pairs between usual purines and pyrimidines of DNA and their analogues.

cytosine is more close-fitting and therefore, presumably, more accurately reproduced than the pair thymine-adenine.

The effects of analogues, whether endogenous or artificial is, of course, not the only factor causing errors in the reproduction of DNA. The mutagenic effects of ionizing and ultraviolet radiation and several other factors acting on the cells are well known. It has been found that free radicals also have a mutagenic effect (4).

Until recent times the study of the mechanism, frequency and nature of changes in DNA has been mainly the concern of geneticists. This phenomenon is, however, of general biological interest and can also be used, as we shall see later, in the study of certain features of the development of individual organisms.

## 2. "Errors" in the reproduction of RNA and their biological consequences

### a) Spontaneous "errors" in the reproduction of RNA

Variation in RNA as a result of purely spontaneous errors of synthesis is certainly commoner than variation in DNA. In the first place RNA obtains its information from DNA and therefore, even if there were not other forms of variation of RNA, there would still be a summation of the errors derived from repetition of all the mistakes in the synthesis of DNA as well as the additional ones which arise in the system DNA → RNA.

It must also be reckoned that the synthesis and metabolism of RNA in non-dividing nuclei takes place 50-100 times more actively than that of DNA. RNA functions in the cell as a single polynucleotide and some fragments of it (especially the "loops" of the secondary structure) are more susceptible than others to the action of chemical factors. The enzymic synthesis of RNA *in vitro* depends on the relationship between the substrates in the medium. This does not imply that a similar variation, connected with the relationship of the substrates, also occurs *in vivo*, but still it is clear that the limiting factors in the synthesis of RNA are not as accurate as they are for DNA. There is evidence suggesting the occurrence of prolongation (growth) of molecules of RNA *in vivo*. In the cells, RNA performs very intensive functional work which is associated with the movement of its molecules from place to place (from the chromosomes to the nucleolus, from the nucleolus to the cytoplasm), with temporary interactions with S-RNA and with more frequent contact with ribonuclease. Furthermore, the cytoplasm is the first part of the cell to be affected by different physical and chemical factors from the environment and it protects the nucleus, functioning as an active buffer. Finally we must take account of the fact that a considerably greater number and variety of processes take place in the cytoplasm than in the chromosomes and this leads to an increased likelihood of side reactions and non-enzymic reactions in the metabolism of RNA.

It is hard to assess the degree to which these factors affect the spontaneous variation of RNA but it is clear that this must be an order of 50 greater than the variations in DNA (during the same time).

Nevertheless, however great the difference may be between the frequency of errors of synthesis and alterations in existing molecules of RNA and DNA it is considerably harder to detect errors in the structure of RNA. Not having hedged round the synthesis of RNA with such rigid limitations as that of DNA, nature has "taken care" that alterations in RNA should not be hereditary. Errors in the synthesis of RNA are more frequent but more transient in their effects, only a very few of them can be handed on to future generations and each new individual starts its development from a zygote which contains a stock of information which has not been significantly affected by variations in RNA occurring during the ontogenesis of previous generations. In other words, *the number of errors in the synthesis of the RNA formed in the nucleus under the direct control of DNA is always proportional to the number of errors which have become fixed in the DNA but the increase of errors in RNA under the influence of the factors which have already been discussed mainly takes place during the autonomous reproduction of RNA (in the cytoplasm and nucleolus) and their accumulation is limited by individual development.* We wish to lay special emphasis on this because it is very important for the analysis of some problems of individual development.

Alterations in DNA (in the reproductive cycle) easily become fixed because, as a rule, they are expressed in hereditary characteristics of the individual. It is also comparatively easy to discover alterations in proteins as these change, destroy or distort some biochemical function. Alterations in RNA, however, especially when they are of a random nature, are hidden and their existence is only revealed by indirect indications.

### b) Experimentally induced "errors" of synthesis of RNA

An obvious indicator of possible variation in RNA is provided by the many facts concerning the incorporation into RNA of various analogues of nucleotides both *in vivo* and *in vitro*. In several cases (in experiments with bacteria) the replacement of the ordinary nucleotides of RNA with analogues (e.g. 8-azaguanine or fluorouracil) amounted to 25-40% (73, 75, 61). The literature dealing with the incorporation of dozens of different analogues during the synthesis of RNA is extremely extensive, as many of these substances are used in the treatment of malignant and viral diseases. It is, of course, unnecessary to give a detailed review of such work in this book.

It is only of interest to note that there can be a very considerable incorporation of 5-fluorouracil into the RNA of tobacco mosaic virus, in the course of which 28-47% of the uracil occurring in this RNA is replaced by the analogue. This, however, does not deprive the virus of its infectivity or upset the relationship between the bases in the descendants of the viruses (53). The replacement of uracil by fluorouracil did not either have any noticeable mutagenic effect. The infectivity of the viral RNA was also retained when the uracil was partly replaced by 2-thio-uracil (43).

It has been mentioned that various conditions in the cultivation of the bacterium *E. coli* bring about changes in the ratios of the bases in its RNA (87). We must also mention that observations have been made which indicate that atypical, functionally inactive RNA may be synthesized in cells treated with chloramphenicol, which is an inhibitor of protein synthesis (3, 42).

It may seem that our conclusions as to the relatively great variation in RNA apparently run counter to the numerous results from Belozerskiĭ's laboratory (1, 2) which indicate that the differences between different species are considerably more pronounced in the nucleotide composition of their DNA than in that of their RNA. However, this contradiction is only apparent. In the comparison of different species the DNA fraction must show greater variability because random variation of the autonomous synthesis of RNA is not hereditary. Furthermore, the total amount of RNA formed under the direct control of DNA only comprises a small part of the total stock of RNA of the cells. The greatest part of this consists of ribosomal RNA, which is not very heterogeneous.

Great importance attaches to the study of the systems for discrimination against analogues of nucleotides, either those arising spontaneously or those entering the organism from outside. The study of these is only just beginning. The work of Kempner (66a) is worthy of attention. Kempner compared the utilization by cells of a culture of *Candida utilis* of uracil with that of its analogue fluorouracil at various stages of metabolism. The accumulation of both bases by the cells was competitive. However, maximal accumulation of the normal component was attained more rapidly. As far as the intracellular stock of nucleotides was concerned, the normal metabolite was used 9-10 times more actively. This was because the enzymic systems converting the base to nucleotide possessed marked selectivity. In the final phase, incorporation of nucleotides into RNA, there was little selective effect; the proportion of fluorouracil incorporated into RNA corresponded closely with that in the free nucleotides. In a different laboratory (52a), some selectivity was observed in the incorporation into RNA of pseudouridine triphosphate, compared with that of uridine triphosphate.

### 3. The presence in RNA and DNA of small amounts of "unusual" additional nucleotides and their possible biological role

The classical theory of self-reproduction of nucleic acids is based on the theory of the complementarity of nucleotide pairs, which forms the basis for the DNA model of Watson & Crick (92). The idea of complementarity is also applicable, in theory, to the case of self-reproduction of RNA and to the phenomenon of the synthesis of RNA at the surface of DNA. In its original form the principle of complementarity was applied to polynucleotide chains constructed from four nucleotides (with, in the case of RNA, replacement of thymine by uracil). However, in recent years there have been detected in several laboratories small quantities of "additional" nucleotides in RNA and DNA, differing in their structure from the ribonucleotides and desoxyribonucleotides based on adenine, guanine, cytosine, uracil and thymine, from which the greatest part of these polymeric molecules is constructed. Sometimes such an "additional" base may completely replace a common one, as does hydroxymethylcytosine in the DNA of some bacteriophages of the T group, where it presumably fulfils the functions normally carried out by cytosine. However, in other cases both the quantity and nature of the additional bases suggests that they may have a special significance for the realization of the biochemical functions of nucleic acids. The content of additional nucleotides in nucleic acids seems to follow a definite rule, namely that the greatest variety of these components is to be found in the soluble, low-molecular S-RNA, which performs the function of transferring activated amino acids to the matrices of protein synthesis. Much smaller variety, and smaller proportions of the additional bases, are observed in the RNA of ribosomes and microsomes, while in the RNA of viruses still less are found. Only in very rare cases have any unusual components been found in DNA.

Several of the first descriptions of the presence in total RNA of a number of additional nucleotides (30, 5, 31a, 31b, 72, 14) were later explained as due to the high content of these bases in the "soluble" RNA. Here were observed relatively large amounts of methylcytosine, pseudouridylic acid, 5-ribosyluracil, 6-methylaminopurine, 1-methyladenine, 1-methylguanine and thymine ribonucleotide (33, 34, 35, 38, 38a, 81a, 81b, 91a). We have given already in Chapter IV some figures for these.

However, some of these components have also been observed in high-polymeric forms of RNA from organelles. Littlefield & Dunn (72) observed small amounts of 6-methylaminopurine and 6-dimethylaminopurine in the RNA of liver microsomes. In further work from the same laboratory (38a), the microsomal fraction of the bacterium *Escherichia coli* was subjected to further fractionation and the RNA of the 70 S ribosomes, which play a very active role in protein synthesis, was analysed. It was found that the 70 S particles were almost free of additional nucleotides, containing only 0·8 moles of pseudouridylic acid for every 100 moles of uridylic acid. However, ribosomes from *Neurospora* cells seemed to have a considerably higher content of supplementary nucleotides (35).

Finally, we should note that in DNA significant amounts of unusual nucleotides have only rarely been observed, apart from 6-methylaminopurine (36, 91b), which is present in small amounts in the DNA of some *E. coli* mutants and of some other organisms. It should also be noted, however, that in DNA of several origins, especially from plants, 5-methylcytosine is frequently present, and may amount to 0·2-6·0% of the total nucleotides of the DNA.

An interesting observation in this connection has been made by Mandel & Borek (76, 77). When cells of *E. coli* become deficient in methionine, there results an accumulation of an unusual RNA which cannot participate in protein synthesis when the cells are returned to a normal nutritive medium. This RNA does not differ in composition from the normal in respect of the main components, but the supplementary unusual bases 2-methyladenine, 6-*N*-methyladenine and thymine, which are present normally, were absent under the conditions of methionine deficiency.

More recently the same workers (21a) have observed that these differences in composition of RNA are due in the first place to the formation of S-RNA free from methylated base analogues. In their opinion, methylation of these nucleotides is effected at the expense of methionine.

Hypotheses concerning any special role for the additional nucleotides have been few and cautious. Littlefield & Dunn (72) have suggested that, when present in the matrices, the "unusual" nucleotides could code for the positions of the rarer amino acids such as tryptophan or cysteine, or mark a limit for polymerization, necessary for the termination of the polypeptide chain. Markham (77a) considers that the methylated components spontaneously replace the usual ones simply because they happen to be present in the cell, having been formed there for some quite other purpose. Crick (31) supposes that the incorporation of methylated bases should not influence the character of the code contained in DNA, since the methylated base forms a pair with the same base as goes with the "usual" component.

It would be quite natural for there to be no general explanation for the role of the additional nucleotides, since each might have its own specific function. However, we think it possible to put forward a preliminary hypothesis which to some extent permits assessment of the role of compounds of this kind. In view of the mutagenic effect of analogues of nucleotides and the induction of changes in the nucleotide sequence (changes in information) when they are incorporated into a nucleic acid it may be suggested that the "unusual" nucleotides and the systems by which they are synthesized constitute a *built-in factor in the organism for the generation of endogenous mutations*. The phylogenetic significance of mutations is generally understood but the accumulation of somatic mutations may also regulate several developmental processes of the individual. This possibility will be discussed in Chapter XIX.

## 4. "Errors" in the synthesis of proteins and their biological consequences

### a) Spontaneous "errors" in protein synthesis

On superficial examination it would seem that "errors" in protein synthesis must, in any case, be more frequent than those in the synthesis of RNA because protein synthesis takes place on matrices and therefore it must reflect all changes occurring in the matrices. There may also be additional sources of variation in protein synthesis associated with inaccurate activation, the incorrect combination of S-RNA with the matrix, etc. and these will add extra mistakes in protein synthesis to the number caused by alterations in the matrices. This, however, is only one aspect of the possible reactions between the variability of RNA and that of protein. The other aspect of this reaction is the counteracting, stabilizing tendency which makes it very much harder to solve the problem of the relative frequency of changes in RNA and protein and may even give rise to conditions which will reduce the number of possible alterations in proteins to less than that for RNA.

*The chief of these counteracting influences seems to be that molecules of RNA which are altered may, at the same time, lose their ability to act as matrices for protein synthesis.* The numerous observations, discussed in earlier chapters, that the incorporation of analogues such as 8-azaguanine into RNA inhibits protein synthesis, suggest that this factor is a genuine one. As the incorporation of amino acids during protein synthesis is determined by an interaction between two forms of RNA (S-RNA and the RNA of the organelles) which would seem to be complementary, it may be that molecules of S-RNA can "recognize" the altered parts of the matrices and not react with them.

It is therefore very hard to decide *a priori* whether variations occur more often in proteins or in the matrices on which they are synthesized. The complementary synthesis of a new polynucleotide on the surface of an old one is like the taking of a simple impression, any change in the stamp causes an equivalent change in the impression. The synthesis of a protein may be compared to an automatic assembly line directed by a programme. If any change occurs in a working part the automatic control can simply halt the whole process.

However, even if proteins do not repeat in their synthesis all the "errors" of their matrices and adaptors, other additional "errors" may occur during their synthesis (incorrect reaction with activating enzymes, incorrect transfer to RNA, etc.). These errors are certainly of a random, scattered nature and as the protein molecules are not capable of auto-synthesis there cannot, theoretically, be an accumulation of errors to such an extent that they would disturb some biochemical function. In practice, however, such accumulation does take place during the development of the individual and some of the reasons for the autonomous accumulation of variations in the structure and synthesis of proteins will be discussed in the chapter on ageing.

Petrenko & Karasikova (9) have recently described another similar case of non-equivalent substitution in the proteins of muscles of the Baltic herring. In the maturing of the sexual products of fish arginine-rich proteins are formed. In the females this arginine is derived from the muscles, in which the arginine content falls from 9·9% to 4·5%. In the males, on the other hand, the arginine content of both the muscles and the gonads increases as the fish becomes mature (from 3·7% to 7·5%). Some changes in the concentrations of other amino acids have also been observed.

Kishi & Takei (68) have recently made an interesting study in which they showed that the content of glycine, tyrosine and threonine in such a specialized protein as the fibroin of silk depended on the age of the mulberry leaves on which the silkworms were fed. If they ate young leaves the content of glycine and tyrosine was higher while that of threonine was lower than if they ate old leaves. The addition of glycine to the food of the silkworm also increased the glycine content of the silk (69).

Many observations have been made which show that, in plants too, the amino acid composition of proteins may vary within definite limits depending on nutritional factors, but, in these cases, the analysis has generally been done on the total proteins, which makes it impossible to judge whether these changes are associated with alterations in the sequence of the amino acids. Recently, however, Sisakyan & Markosyan (11) have found clearly-defined alterations in the amino acid composition of individual proteins of wheat grains (gliadin, albumin and globulin) depending on the conditions of cultivation. See also (7, 51).

Different syntheses of proteins would seem to have different degrees of accuracy of reproduction and some of them show a large number of errors. Bawden (19) and Matthews (78) think that a large amount of the viral protein which remains in infected cells and a considerable proportion of the non-infective particles among the "descendants" of a virus are the results of "errors" in viral synthesis. There is a very close steric correspondence between the protein particles and the RNA of viruses but the use of a "foreign" synthetic system will obviously not ensure a very high output of correctly formed molecules.

Spontaneous "errors" in protein synthesis may apparently be divided into three groups: hereditary (involving the whole cycle DNA → RNA → protein), ontogenetically stable (RNA → protein) and reversible, depending on inaccurate operation of the synthetic production line carrying the amino acids to the matrices.

Ontogenetically stable and hereditary changes in protein synthesis are now receiving intensive study, not only because some of them provide the material for natural selection, but also because some of these alterations cause pathological changes in man, and medicine today is very often confronted with the phenomenon of molecular pathology. Nowadays the study of many diseases at a molecular level is one of the most promising lines of work in medical chemistry. Naturally we cannot undertake anything like a complete review of all the facts which are known in this field and shall limit ourselves to a single example.

Ingram and colleagues (21, 62, 63, 65, 66) have established that so-called "sickle-cell" and "C-disease" haemoglobins are associated with a stable alteration in the position of some of the amino acid residues in the terminal part of the haemoglobin. Fig. 57 shows the sequence of amino acids in the terminal group of the haemoglobin in these pathological conditions.

$$
\text{Haemoglobins}
\begin{cases}
\text{A} & \text{Val His Leu Thr Pro } \mathit{Glu} \text{ Glu Lys} \\
\text{S} & \text{Val His Leu Thr Pro } \mathit{Val} \text{ Glu Lys} \\
\text{C} & \text{Val His Leu Thr Pro } \mathit{Lys} \text{ Glu Lys}
\end{cases}
$$

A = Normal

S = From the cells of a case of sickle-celled anaemia

C = From the cells of a case of C-disease

FIG. 57. Sequence of amino acids in the N-terminal fragment of three forms of human haemoglobin.

In these pathological conditions a particular amino acid link in the large molecule of haemoglobin has undergone a stable change but, when the normal haemoglobin is completely replaced by S-haemoglobin, the organism dies. It is interesting to note that Hill & Schwartz (59, 60) have succeeded in isolating from the blood of a single individual a special G-haemoglobin in which the same glutamic acid residue is replaced by glycine. Such variability of one particular link in the chain of haemoglobin is certainly not accidental and it would seem to be connected with instability in some determining part of the genetic matrices.

The work of Ingram and colleagues has attracted much attention to the molecular variation of haemoglobin. During a relatively short period several other hereditary, pathological varieties of this protein have been discovered. In a paper by Braunitzer and colleagues (23) comparison of the N-terminal sequence of amino acids of normal and S-haemoglobin was carried out over a length of 28-30 amino acid residues. In this way it was found that there was a substitution in the 26 position which was also occupied by a glutamic acid residue in the normal protein. In a later paper by Hunt & Ingram (64) the authors compared normal haemoglobin with the pathological E-form found in South Asia. Only one difference was found among the peptides of a tryptic hydrolysate and here again it took the form of replacement of a glutamic acid residue by lysine. These facts taken together indicate that the glutamic determinant is unstable.

Study of the variability of haemoglobins is now going on over a wide front (11a, 12, 16, 17, 52) and also the elucidation of the amino acid sequence of the whole molecule is being completed and being converted into a three-dimensional model (cf. Chapter I). All this gives us the hope that we shall soon have a complete picture of the main molecular variations of this protein.

The case of haemoglobin is particularly interesting because here a study has been made of the connection between the pathological changes and the amino acid sequences.

In dozens of other cases it has simply been established that in the course of various pathological processes (usually in the blood) certain abnormal proteins with several specific features (electrophoretic properties, molecular weight, amino acid constitution) appear in the organism.

We may suppose that certain forms of tumours also arise as a result of such errors of synthesis.

Individual, random errors of protein synthesis may arise as a result of the uneven working of all the stages of protein synthesis. Theoretically these are reversible and it is hardly likely that there will be any marked accumulation of such errors during the development of the individual. There is, however, no doubt that the tendency to develop such errors is strengthened or weakened by many internal and external factors. A typical case of such an error is, for example, the ability of some specific activating enzyme to react with some amino acid other than its own. In recent work by Wong & Moldave (93), for example, it has been shown that purified tryptophan-activating enzyme may react to a very small extent with glycine adenylate and transfer its glycine residue to S-RNA. Such errors may be "corrected" at a later stage of the synthesis but obviously they may go on unhindered and lead to random alterations in the amino acid sequence of some part of the protein molecules.

In Chapters III and IV we have already discussed the question of possible inaccuracies in the work of activating enzymes and S-RNA. An interesting case of alteration of a protein has been described recently by Schapira and colleagues (88). These authors observed the conversion of glycine into serine and of phenylalanine into tyrosine after their incorporation into globin. They suggest that this conversion may occur by alteration of the amino acids *in situ*. In a recently published detailed study (88a) these authors presented a closer analysis of the phenomenon. This was based on observations of variations in the specific activity of amino acids in the haemoglobin of mature erythrocytes. Thus, after injection of labelled glycine, the specific activities of haem and of globin remained constant during the course of 33 days. However, this constancy of radioactivity was based on mutually opposed changes in the specific activity of serine and glycine as components of the haemoglobin. The activity of the glycine fell, while that of the serine rose, in a strictly equivalent way. Similar results have been obtained in experiments with myosin. Detailed study of the phenomenon excluded the possibility that conversion of glycine into serine or of phenylalanine into tyrosine took place in the non-protein fraction. Finally we must mention the possibility of spontaneous formation of D amino acids during metabolism (22).

At present we know of only two attempts at a theoretical analysis of the accuracy of the working of the system of protein synthesis and both of these analyse a single-stage synthesis (amino acid + matrix) and do not reckon with alterations in the matrices themselves. The probability of the occurrence of errors is therefore naturally magnified on the one hand while on the other it is diminished and can therefore obviously only refer to the first stage of synthesis (amino acid + activating enzyme). The first attempt is that of Pauling (10) and the second that of Pasynskiĭ (8).

*b) Experimentally-induced errors in protein synthesis*

There are many descriptions in the literature of cases in which the D-forms of amino acids have been incorporated into proteins in place of the ordinary L-isomers (15, 25, 40, 41, 49).

There is a particularly wide and interesting collection of material on the error of the synthetic systems which is obtained by the use of various analogues of amino acids (ethionine, thienylalanine, selenomethionine, fluorophenylalanine, 5-methyltryptophan, azatryptophan and many others). In many cases such analogues inhibit the synthesis of proteins, but in several they are incorporated into the protein being synthesized and thereby alter its molecule (18, 28, 29, 27, 39, 54, 58, 67, 79, 84, 91, 85, 90, 82, 94, 86).

In many of these researches, proteins containing the analogues were isolated and studied. It is very interesting that in many cases this substitution was not accompanied by a loss of the biological activity of the protein. For example, when part of the methionine of the amylase of *Bacillus subtilis* was replaced by ethionine and part of its phenylalanine by fluorophenylalanine the enzyme did not lose its specific properties (94, 95, 96). It was thus shown that ethionine and fluorophenylalanine really did replace methionine and phenylalanine in the polypeptide chain. Munier (79) found almost total replacement of phenylalanine by fluorophenylalanine in the proteins of *E. coli* while the bacteria still remained able to grow, though more slowly than usual. In another paper Munier (80) described the complete replacement of phenylalanine by fluorophenylalanine or β-2-thienylalanine in the alkaline phosphatase of this bacterium. A new and interesting line of work on the experimental alteration of proteins has recently developed as a result of a series of studies in three laboratories (26, 56, 57, 81, 24). These experiments were designed to study the effect of altering RNA by means of analogues of nucleotides (thiouracil, azaguanine, etc.) on the proteins synthesized by the affected cells. In these experiments the effects on protein synthesis in the cell were studied in relation to changes in the RNA caused by nucleotide analogues (thiouracil, azaguanine). In a number of such cases the altered RNA led to the production of enzymic proteins of markedly changed activity ("defective" proteins). Naono & Gros (81) have recently shown that when fluorouracil is incorporated into RNA the incorporation of proline and tyrosine into the phosphatase molecule is hindered.

## 5. Some considerations on the possibility of increasing the selectivity of the action of antimetabolites and analogues on the synthesis of proteins

Many methods of treatment of neoplastic and viral diseases are nowadays based on the ability of the synthesizing systems which produce nucleic acids and proteins to make mistakes.

In the last few years antimetabolites to the metabolism of nucleic acids and proteins have also been used very extensively in experimental work and for the hindrance of pathological processes.

Various firms and undertakings dealing in biochemical preparations have carried out the synthesis and production of several hundreds of analogues of amino acids and nucleotides for this purpose. An even larger collection of analogous inhibitors has been synthesized and tested under laboratory conditions. A detailed review of the latest uses of analogues of amino acids and nucleotides in the treatment of neoplasms has been published by Mandel (74).

The theoretical basis for the use of such antimetabolite-inhibitors is quite clear. Antimetabolites usually become involved in a system of reactions characteristic of some normal metabolite and, until some particular stage in the process, they behave in a similar way to it. At some stage in the metabolic process, however, the biochemical systems "discover their mistake", so to speak, and the cycle of transformations is halted so that the process is brought to a standstill. Some analogues inhibit the earlier, preparatory stages of a process while others bring the cycle of transformation to a standstill at some intermediate stage while, finally, there are quite a number of such compounds (azaguanine, thiouracil, ethionine, etc.) which pass through the whole cycle of reactions and are incorporated into RNA, DNA or proteins alongside the usual substrates, although the polymers synthesized with them do not have the required degree of biological activity. They constitute, as it were, rejected components, which, if used for building up more complicated molecular complexes, prevent these complexes from becoming involved in the normal processes of living.

Such antimetabolites have recently received great attention in the treatment of viral and neoplastic diseases in plants and animals. Their use in such cases is based on the fact that the synthesis of proteins and nucleic acids in tumours and during the reproduction of viruses takes place considerably more actively than it does in the surrounding infected tissues. It is therefore possible to suppress these pathological processes by concentrations of antimetabolites which would not have a harmful effect on ordinary tissues. In this case, however, the course of the pathological process is only slowed, not halted, because the concentrations required to stop it altogether are such that they would have a bad effect on the surrounding tissues.

The reason for this defect in therapy with antimetabolites, which still has not been overcome, is that their effect is not strictly selective, affecting only the group of proteins and nucleic acids, the production of which must be suppressed. Antimetabolites usually affect all syntheses of a particular type and it is only the intensiveness of the synthesis which determines the extent of the inhibition.

It is quite clear that the discovery of a way of increasing the selectivity of the action of antimetabolites on proteins and nucleic acids having a particular specific structure, regardless of their rate of synthesis, would open up unlimited opportunities for antimetabolite treatment in stopping pathological processes. One of the theoretically possible means of increasing the specificity has been discussed in part by the present author (6) but he has not yet given experimental proof of its efficacy.

We suggested that one of the most practically hopeful methods of solving this problem would be the production of inhibiting forms of the complex of S-RNA with an amino acid, that is to say, by using analogues of amino acids, for example, not in the free state, but in the form of compounds with S-RNA which would only be complementary to particular matrices and would, therefore, only carry out the transfer and incorporation of the amino acids during the synthesis of strictly determinate proteins. It might be supposed that the inhibitor-analogues of amino acids, when accepted by particular fractions of S-RNA, say that from the neoplastic cells, if introduced into the tissues would, in the first instance, block just those matrices which synthesize the proteins of the neoplastic cells. However, present evidence gives no support for the existence of tissue specificity in "soluble" RNA and this makes it less easy to solve the problem in this way.

It is very likely that other intermediate products of the synthesis of proteins and nucleic acids (such as peptides, polypeptides and oligonucleotides) containing analogues of amino acids or nucleotides might exhibit considerably greater specificity in their incorporation into strictly determinate synthetic processes.

Ardry and colleagues (14a) have injected solutions of labelled RNA and DNA from various organs (liver, pancreas) and from yeast intramuscularly into rats and observed the way in which they were broken down and assimilated in the organism. During the four days after the beginning of the experiment, when labelled nucleic acids were being injected daily, there was a clearly marked tropism of the breakdown products of the RNA and DNA which reflected the source from which they were derived, those from the liver and pancreas being incorporated into the nucleic acids of the analogous tissues of the recipient. Such an effect might result from the reutilization of products of incomplete depolymerization which still retained some tissue specificity. The nucleic acids from yeast were used non-specifically. The selective accumulation of injected nucleic acids (or of the products of their partial breakdown, as polymeric molecules can only pass through the cell membrane with difficulty) in homologous organs is certainly connected with the phenomenon of tissue specificity of the sequences of nucleotides in RNA and DNA and with the phenomenon of the complementarity of these polynucleotides and the products of their partial breakdown to the other polynucleotides in some particular tissue. Such complementarity determines the selective accumulation of circulating polynucleotides by a strictly determinate tissue. When certain analogues of nucleotides (bromine, iodine and chlorine derivatives) are incorporated into RNA and DNA they increase the sensitivity of the tissues to irradiation hundreds of times over. Thus, if we could find out how to increase the selective assimilation of these analogues by neoplastic cells we might bring about a marked increase in the effectiveness and selectivity of the treatment of cancers by irradiation.

*These results have led us to suggest that, as concerns the artificial synthesis of tissue-specific nucleic acids (DNA or RNA) saturated with inhibitor analogues, their introduction into the organism in partly depolymerized form (to avoid immune reactions) should lead to selective accumulation of the analogues in the corresponding tissue (in neoplasms, for example) and to the selective suppression of growth in that tissue without a harmful effect on the other tissues.* It would be possible to synthesize tissue-specific RNA by using a DNA-dependent enzymic system (cf. Chapter XV).

The carrying out of these suggestions would certainly require a very large amount of experimental work but it seems to us that the importance of the vistas which such work would open up render it necessary for an extensive research along these lines to be undertaken.

## Conclusion

The study of errors of synthesis of proteins and nucleic acids is a new but very fruitful field of biochemistry. In practice, this is the field in which biochemistry begins to study the molecular nature of various forms of variation, whether mutational, pathological or developmental, and it is clear that the discovery of the underlying nature of this variation opens up some very promising lines of work on the development and regulation of various organisms. This class of biochemical phenomenon is already used extensively in the application of chemical mutagens. Tests of substances for the treatment of viral and cancerous conditions are also directly connected with the study of the limits of accuracy of the systems for the reproduction of proteins and nucleic acids. Work in this field will also provide a firm basis for many fields of molecular pathology. All this indicates that the greatest possible stimulus must be given to the development of these studies, which form the most promising borderland between biochemistry, genetics and medicine.

REFERENCES

1. BELOZERSKIĬ, A. N. & SPIRIN, A. S. (1956). *Uspekhi sovremennoĭ Biol.* **41**, 144.
2. — (1960). *Izvest. Akad. Nauk S.S.S.R., Ser. biol.* **25**, No. *1*, 64.
3. HABERMANN, V. (1961). *Proc. V int. Congr. Biochem., Moscow,* **9**, 123.
4. DUBININ, N. P., SIDOROV, B. N. & SOKOLOV, N. N. (1959). *Doklady Akad. Nauk S.S.S.R.* **128**, 172.
5. EVREINOVA, T. N., DAVYDOVA, I. M., SUKOVER, A. P. & GORYUNOVA, S. V. (1961). *Doklady Akad. Nauk S.S.S.R.* **137**, 213.
6. MEDVEDEV, ZH. A. (1959). *Doklady Moskov. sel'skokhoz. Akad. im. K. A. Timiryazeva,* No. *47*, p. 77.
7. OREKHOVICH, V. N. (1956). *Priroda,* No. *5*, 35.
8. PASYNSKIĬ, A. G. (1960). *Biofizika,* **5**, 16.
9. PETRENKO, I. N. & KARASIKOVA, A. A. (1958). *Doklady Akad. Nauk S.S.S.R.* **122**, 1071 (1958).
10. PAULING, L. (1959). In *Proceedings of the first International Symposium on the Origin of Life on the Earth* (ed. A. I. Oparin *et al.*), p. 215. London: Pergamon.
11. SISAKYAN, N. M. & MARKOSYAN, L. S. (1959). *Biokhimiya,* **24**, 1094.
11a. EFROIMSON, V. P. (1961). *Problemy Kibernetiki* (ed. A. A. Lyapunov), No. *6*. Moscow: Izd. Fiz. Mat. Lit.
12. *Abnormal haemoglobins* (ed. J. H. P. Jonxis & J. F. Delafresnaye). Oxford: Blackwell (1959).
13. ADLER, J., LEHMAN, I. R., BESSMAN, M. J., SIMMS, E. S. & KORNBERG, A. (1958). *Proc. natl. Acad. Sci. U.S.* **44**, 641.
14. ADLER, M., WEISSMAN, B. & GUTMAN, A. B. (1958). *J. biol. Chem.* **230**, 717.
14a. ARDRY, R., KIRSCH, F. & BRUX, J. DE (1956). *Thérapie,* **11**, 658.
15. WRETLIND, K. A. J. & ROSE, W. C. (1950). *J. biol. Chem.* **187**, 697.
16. BAGLIONI, C. & INGRAM, V. M. (1961). *Biochim. biophys. Acta,* **48**, 253.
17. — (1961). *Nature (Lond.),* **189**, 465.
18. BAKER, R. S., JOHNSON, J. E. & FOX, S. W. (1958). *Biochim. biophys. Acta,* **28**, 318.
19. BAWDEN, F. C. (1959). *Proc. roy. Soc.* **151B**, 157.
20. BENZER, S. & FREESE, E. (1958). *Proc. natl. Acad. Sci. U.S.* **44**, 112.
21. BENZER, S., INGRAM, V. M. & LEHMANN, H. (1958). *Nature (Lond.),* **182**, 852.
21a. BOREK, E., MANDEL, L. R. & FLEISSNER, E. (1962). *Fed. Proc.* **21**, 379.
22. BOULANGER, P. & OSTEUX, R. (1960). *Acta Unio. intern. contra Cancrum,* **16**, 1044.
23. BRAUNITZER, G., HILSCHMANN, N. & MÜLLER, R. (1960). *Hoppe-Seyl. Z.* **318**, 284.
24. BUSSARD, A., NAONO, S., GROS, F. & MONOD, J. (1960). *Compt. rend. Acad. Sci., Paris,* **250**, 4049.
25. CAMIEN, M. N. & DUNN, M. S. (1955). *J. biol. Chem.* **217**, 125.
26. CHANTRENNE, H. (1959). *Biochem. Pharmacol.* **1**, 233.

27. COHEN, G. N. & COWIE, D. B. (1957). *Compt. rend. Acad. Sci., Paris,* **244,** 680.
28. COHEN, G. N. & MUNIER, R. (1959). *Biochim. biophys. Acta,* **31,** 347.
29. COWIE, D. B., COHEN, G. N., BOLTON, E. T. & ROBICHON-SZULMAJSTER, H. DE (1959). *Biochim. biophys. Acta,* **34,** 39.
30. COHN, W. E. (1960). *J. biol. Chem.* **235,** 1488.
31. CRICK, F. H. C. (1939). In *Brookhaven Symposia in Biology, No. 12, Structure and Function of genetic Elements* (ed. M. Levine *et al.*), p. 35. Upton, N.Y.: Biology Dept., Brookhaven National Laboratory.
31a. DAVIS, F. F. & ALLEN, F. W. (1957). *J. biol. Chem.* **227,** 907.
31b. DAVIS, F. F., CARLUCCI, A. F. & ROUBEIN, I. F. (1959). *J. biol. Chem.* **234,** 1525.
32. DONOHUE, J. & TRUEBLOOD, K. N. (1960). *J. mol. Biol.* **2,** 363.
33. DUNN, D. B. (1959). *Biochim. biophys. Acta,* **34,** 286.
34. — (1961). *Biochim. biophys. Acta,* **46,** 198.
35. — (1961). *Proc. V int. Congr. Biochem., Moscow,* **9,** 126.
36. DUNN, D. B. & SMITH, J. D. (1958). *Biochem. J.* **68,** 627.
37. — (1954). *Nature (Lond.),* **174,** 305.
38. DUNN, D. B., SMITH, J. D. & SPAHR, P. F. (1960). *J. mol. Biol.* **2,** 113.
38a. DUNN, D. B., SMITH, J. D. & SIMPSON, M. V. (1960). *Biochem. J.* **76,** 24P.
39. DUNN, M. S. & MURPHY, E. A. (1958). *Cancer Res.* **18,** 569.
40. EL-SHISHINY, E. D. H. & NOSSEIR, M. A. (1957). *Plant Physiol.* **32,** 360.
41. EL-SHISHINY, E. D. H. & NOSSEIR, M. A. (1957). *Plant Physiol.* **32,** 639.
42. EZEKIEL, D. H. (1961). *J. Bact.* **81,** 319.
43. FRANCKI, R. I. B. (1960). *Virology,* **10,** 374.
44. FREESE, E. (1959). *J. mol. Biol.* **1,** 87.
45. — (1959). *Brookhaven Symposia Biol.* **12,** 63.
46. — (1958). *Cold Spring Harbor Symp. quant. Biol.* **23,** 13.
47. — (1959). *Proc. natl. Acad. Sci. U.S.* **45,** 622.
48. — (1961). *Proc. V int. Congr. Biochem., Moscow,* **1,** 204.
49. FRIEDBERG, F. (1953). *Experientia,* **9,** 425.
50. FRISCH, D. M. & VISSER, D. W. (1960). *Biochim. biophys. Acta,* **43,** 546.
51. GALE, E. F. & FOLKES, J. P. (1953). *Biochem. J.* **55,** 721.
52. GAMMACK, D. B., HUEHNS, E. R., LEHMANN, H. & SHOOTER, E. M. (1961). *Acta genet. et statist. Med.* **11,** 1.
52a. GOLDBERG, I. H. & RABINOWITZ, M. (1961). *Biochem. biophys. Res. Commun.* **6,** 394.
53. GORDON, M. P. & STAEHELIN, M. (1959). *Biochim. biophys. Acta,* **36,** 351.
54. GROSS, D. & TARVER, H. (1955). *J. biol. Chem.* **217,** 169.
55. HAKALA, M. T. (1959). *J. biol. Chem.* **234,** 3072.
56. HAMERS, R. & HAMERS-CASTERMAN, C. (1959). *Biochim. biophys. Acta,* **33,** 269.
57. HAMERS, R. & HAMERS-CASTERMAN, C. (1961). *J. mol. Biol.* **3,** 166.
58. HANCOCK, R. (1960). *Biochim. biophys. Acta,* **37,** 47.
59. HILL, R. L. & SCHWARTZ, H. C. (1959). *Nature (Lond.),* **184,** 641.
60. HILL, R. L., SWENSON, R. T. & SCHWARTZ, H. C. (1960). *J. biol. Chem.* **235,** 3182.
61. HOROWITZ, J. & CHARGAFF, E. (1959). *Nature (Lond.),* **184,** 1213.
62. HUNT, J. A. & INGRAM, V. M. (1959). *Nature (Lond.),* **184,** 640.
63. — (1960). *Biochim. biophys. Acta,* **42,** 409.
64. — (1961). *Biochim. biophys. Acta,* **49,** 520.
65. INGRAM, V. M. (1958). *Scient. Amer.* **198,** No. 1, p. 68.
66. HUNT, J. A. & INGRAM, V. M. (1958). *Nature (Lond.),* **181,** 1062.
66a. KEMPNER, E. S. (1961). *Biochim. biophys. Acta,* **53,** 111.

67. KENDREW, J. C. (1959). *Fed. Proc.* **18**, 740.
68. KISHI, Y. & TAKEI, T. (1957). *Nippon Nôgei-Kagaku Kaishi*, **31**, 504.
69. KISHI, Y. & TAKEUCHI, T. (1959). Through *Ref. Zh., Khim., Biol. Khim.* Ref. 5C780 (1961).
69a. KRISS, J. P. & RÉVÉSZ, L. (1961). *Cancer Res.* **21**, 1141.
70. LITMAN, R. M. & PARDEE, A. B. (1956). *Nature (Lond.)*, **178**, 529.
71. — (1960). *Biochim. biophys. Acta*, **42**, 117.
72. LITTLEFIELD, J. W. & DUNN, D. B. (1958). *Biochem. J.* **70**, 642.
73. MANDEL, H. G. (1957). *J. biol. Chem.* **225**, 137.
74. — (1959). *Pharmacol. Revs.* **11**, 744.
75. MANDEL, H. G. & ALTMAN, R. L. (1960). *J. biol. Chem.* **235**, 2029.
76. MANDEL, L. R. & BOREK, E. (1961). *Fed. Proc.* **20**, 357.
77. — (1961). *Biochim. biophys. Res. Commun.* **4**, 14.
77a. MARKHAM, R. (1958). *Symposium Soc. gen. Microbiol., 8th, London* (ed. S. T. Cowan & E. Rowatt), p. 163.
78. MATTHEWS, R. E. F. (1958). *Virology*, **5**, 192.
79. MUNIER, R. L. (1959). *Compt. rend. Acad. Sci., Paris*, **248**, 1870.
80. — (1960). *Compt. rend. Acad. Sci., Paris*, **250**, 3524.
81. NAONO, S. & GROS, F. (1960). *Compt. rend. Acad. Sci., Paris*, **250**, 3889.
81a. OSAWA, S. (1960). *Biochim. biophys. Acta*, **42**, 244; **43**, 110.
81b. OTAKA, E., HOTTA, Y. & OSAWA, S. (1959). *Biochim. biophys. Acta*, **35**, 266.
82. PARDEE, A. B. & PRESTIDGE, L. S. (1958). *Biochim. biophys. Acta*, **27**, 330.
83. PULLMAN, B. & PULLMAN, A. (1959). *Biochim. biophys. Acta*, **36**, 343.
84. RABINOVITZ, M. & MCGRATH, H. (1959). *J. biol. Chem.* **234**, 2091.
85. RABINOVITZ, M., OLSON, M. E. & GREENBERG, D. M. (1957). *J. biol. Chem.* **227**, 217.
86. RICHMOND, M. H. (1959). *Biochem. J.* **73**, 261.
87. SANTER, M., TELLER, D. C. & ANDREWS, W. (1960). *J. mol. Biol.* **2**, 273.
88. SCHAPIRA, G., DREYFUS, J. C., KRUH, J. & LABIE, D. (1961). *Proc. V int. Congr. Biochem., Moscow*, **9**, 116.
88a. SCHAPIRA, G., DREYFUS, J. C. & KRUH, J. (1962). *Biochem. J.* **82**, 290.
89. SHAPIRO, H. S. & CHARGAFF, E. (1960). *Nature (Lond.)*, **188**, 62.
90. SIDRANSKY, H. & FARBER, E. (1956). *J. biol. Chem.* **219**, 231.
91. SIMPSON, M. V., FARBER, E. & TARVER, H. (1950). *J. biol. Chem.* **182**, 81.
91a. SINGER, M. F. & CANTONI, G. L. (1960). *Biochim. biophys. Acta*, **39**, 182.
91b. THEIL, E. C. & ZAMENHOF, S. (1962). *Fed. Proc.* **21**, 372.
92. WATSON, J. D. & CRICK, F. H. C. (1953). *Nature (Lond.)*, **171**, 737.
93. WONG, K. K. & MOLDAVE, K. (1960). *J. biol. Chem.* **235**, 694.
94. YOSHIDA A. (1958). *Biochim. biophys. Acta*, **29**, 213.
95. — (1960). *Biochim. biophys. Acta*, **41**, 98.
96. YOSHIDA, A. & YAMASAKI, M. (1959). *Biochim. biophys. Acta*, **34**, 158.
97. ZAMENHOF, S. (1959). *Ann. N.Y. Acad. Sci.* **81**, 784.
98. ZAMENHOF, S., GIOVANNI, R. DE & RICH, K. (1956). *J. Bact.* **71**, 60.
99. ZAMENHOF, S., RICH, K. & GIOVANNI, R. DE (1959). *J. biol. Chem.* **234**, 2960.

# GENERAL CONCLUSION TO PART III

In Part III of this book we have been mainly concerned with theoretical concepts, trying to explain the hitherto unsolved riddle of the synthesis of proteins and nucleic acids. We also took note of the presence in cells of more complicated systems which regulate the reproduction of the specificity of the polymeric components of living matter and discussed the evidence which indicates that this system always works perfectly accurately.

Theoretical analysis of the possible mechanisms of the synthesis of proteins on matrices is very important. The various hypotheses in this field are important, not so much because they give a unique explanation of the available facts, but because they provide a temporary filling for the gaps in problems which cannot yet be worked out by direct experiment. In erecting these hypotheses the authors rely mainly on their own logic, which is often considerably simpler than the real logic of nature or natural selection. We discussed a series of different hypotheses and it is still too early to choose from among them the one which corresponds most closely with existing facts. Time and further advances in experimental biochemistry are necessary before this can be done.

The evidence which we have already discussed makes it quite clear that the determination of the specificity of protein is a process which takes place over many stages and is regulated from two quarters, by the system including DNA (DNA → RNA → proteins) and by the cytoplasmic system (activating enzymes → S-RNA → proteins).

Incontrovertible proof of all these complicated interconnections is still a matter for the future, but it is already in order to ask the question: What is the evolutionary and biological point of such complicated correlations? A provisional answer may be given in terms of considerations which will be developed in greater detail in later chapters. The DNA of the nucleus is obviously a copious arsenal of information which can determine the direction of the development and differentiation of the many biochemical systems of the species concerned. The nucleotide chains of DNA cannot, therefore, be the direct producers of proteins because not all the various proteins which might be produced on the basis of this information are required at the same moment; nor can it form the basis for the functioning of the individual cell. There thus arises the need for an intermediary which will take up only a part of this information and carry it over to protein synthesis. The messenger RNA synthesized in the chromosomes acts as such an intermediary. However, this chromosomal RNA obviously cannot fulfil all the requirements of the cytoplasm for matrices for protein synthesis and so it would seem to be used mainly as a pattern for the synthesis of RNA in the nucleolus and for the synthesis by the cytoplasm of its own RNA.

The possibility that a large store of "superfluous" information may be accumulated in the composition of DNA, capable of not merely controlling current syntheses, but also of determining the processes of differentiation during ontogenesis, seems to be fundamental for evolutionary development. The storage and transfer of this information is by no means always carried out by purely biochemical means. Evolution and selection supplement the biochemical processes by creating many complicated intracellular structures such as the chromosomes which still continue to be the elements of the reactions themselves, ensuring in many cases the orientation of a biochemical process in one direction or another. These intracellular structures represent, as it were, the second "level" of biochemistry and there are many such "levels" in the process of the evolution of living matter. In this book we cannot analyse the problem of protein synthesis through all these "levels" but one of them, namely the special part played by the nucleus in bringing about the synthesis of proteins and nucleic acids and in the control of their specificity, must not remain without discussion as this control is the connecting link between the synthesis of protein and the phenomena of heredity.

On analysis the hidden properties of DNA localized in the nucleus are revealed as the biochemical determinants for the phenomena of heredity and the development of the individual. Such an approach to the evaluation of the hidden potentialities of DNA which might cover the current requirement of the cells has been called "preformism" by some biologists. In this concept they express their critical attitude to the possibility of such a relationship existing between a cell and its DNA. However, such a critical attitude is too unconvincing and it arises from an unwillingness to work out the real laws of living nature. The proponents of this view talk, on the one hand, about the qualitative peculiarity of living material but, on the other hand, they cannot detect this qualitative peculiarity in the actual biochemical processes carried out in the cell. Nevertheless it is just these biochemical processes which display for us the special qualitative peculiarities of living matter and the fundamental qualitative property of biological cellular structures is their ability, not only to carry on a flowing functional and biochemical activity, but also their ability to form within themselves specialized structures which can fix the whole course of the evolution of living matter in a particular way, while at the same time providing for the actual succession of millions of generations. Structure and function determine one another and it is therefore natural that by no means every structure of the cells can determine heredity.

It must be mentioned that the application of the principles of cybernetics (i.e. the theory of information) to biological processes has been extremely fruitful. In essence, the study of cybernetics, as a special subject, arose mainly out of progress in the study of the mechanisms for the regulation, co-ordination and organization of complicated biological processes.. However, the new and *ad hoc* science of cybernetics which had arisen in this way began to exert a considerable influence in the reverse direction and stimulated the development of a number of new lines of research within biology, giving rise to new concepts, new methods and a new outlook. Owing to the theory of information biologists acquired the ability to analyse in an accurate and strictly mathematical way practically all forms of biological purposiveness, the reproduction of the specificity of biological structures at various levels of organization, the rules governing the regulation of processes of response to internal and external impulses and many other phenomena. Cybernetics, on the other hand, was able to use technical means for the artificial imitation and creation of qualities and properties which had hitherto only been known in living nature. This gave rise to the possibility of not merely recreating biological properties such as the power of memory, but also of increasing these properties several million times.

It seems to us that analysis of biological phenomena in terms of the principles of coding, the transfer and reproduction of information is the very road along which biology may attain most quickly to the ranks of the exact sciences. If we consider the problem on the evolutionary plane it becomes obvious that even if we meet with the phenomena of metabolism, interacting processes, catalysis, etc. in non-living nature, the phenomenon of the transfer and codification of information has only arisen phylogenetically in living systems and provides the essential basis for the biological form of the motion of matter.

It would seem that it is now no longer enough to say that metabolism and self-renewal are the main indicators of living matter. We must add that *the essential characteristics of life are metabolism and self-renewal based on the codification, transfer, reproduction and alteration of information reflecting the purposiveness and co-ordination of biochemical and physiological processes.*

# PART IV

# THE BIOSYNTHESIS OF PROTEINS AND NUCLEIC ACIDS AND PROBLEMS OF INDIVIDUAL DEVELOPMENT

# INTRODUCTION

The various extremely complicated and purposive morphological, physiological and biochemical features of animals and plants do not come into being all of a sudden, but as a result of variation and selection acting on a particular system for tens and hundreds of millions of years to adapt it more accurately to its purpose. The system may then pass on, unchanged, through hundreds if not thousands of generations if it is perfectly adapted to meet some particular requirement of the organism or it may be changed, completed, developed or replaced as changes in the environment require.

The same phylogenetic path is followed during the ontogenesis of living systems in an extremely abbreviated though accurate way. In this case the development is not determined by selection, nor by external factors, but the formative process finds the necessary forms straight away and synthesis leads to the production of the necessary compounds at once. By reacting with one another these compounds at once give rise to the necessary, almost perfectly worked out process which may have taken tens of millions of years to develop during phylogenesis by the trying out and replacing of hundreds and thousands of variants.

There is only one main factor which exerts a decisive effect on this process and that is the factor of heredity, which is the ability of the organism to fix and record the whole evolutionary path of development in the form of a specific biochemical "code" and to decipher and reproduce this record over a short period in the form of a morphological ontogenesis.

In the earlier chapters we have surveyed the basic rules governing the synthesis of proteins and nucleic acids and guaranteeing the reproduction of their specificity both in ontogenesis (by means of the RNA system) and in phylogenesis (by means of DNA). However, this ability of biological structures to reproduce is only one aspect of biological phenomena as they actually occur. The other aspect is represented by the constant occurrence of variation of these structures, which is characteristic of all biological processes when considered in a long-term way.

There are many different types of variability of proteins and nucleic acids: evolutionary, embryonic, morphological, developmental, pathological, functional, ecological and other forms. We cannot examine all the evidence which exists concerning this matter. It is very extensive, especially as concerns evolutionary and functional variability.

One of the main ways in which evolutionary variability manifests itself is in what is known as species specificity, which has been discussed already in a general way in several sections of this book.

The causes of evolutionary variations of nucleic acids and proteins are

384

very diverse (hybridization, external circumstances, nutritional factors, mutational variability, natural selection, etc.) and there is now a large amount of experimental material concerning the effect of these factors on the amino acid and nucleotide compositions of particular biological objects. Although the experimental studies of the influence of various factors on the structures of proteins and, especially, of nucleic acids are naturally, on the whole, very insignificant and fragmentary, yet they nevertheless show that the structure of proteins and the system whereby they are synthesized (the matrices) are not invariable but undergo an orderly process of evolutionary development. In this connection a stock of "information" concerning heredity is developed and makes its appearance.

In its individual development living matter covers the complicated course of its phylogenetic development in a very short space of time and naturally the organism cannot, while doing this, reproduce all the external and internal factors which, with the help of variability and selection, have given rise, in the course of millions of years, to any particular complicated enzymic system and to the specific matrices for its synthesis; but it can quickly reproduce the last link in this chain, namely the synthesis of the proteins of a given enzymic system on matrices which already exist and which are handed on to the next generation. The presence of this hereditary biochemical basis for ontogenesis is the most important factor determining its course. The rules of ontogenesis are not, however, delimited by this determination alone. Ontogenesis is an integrated development of composition, structure and function. On the one hand, this development is regulated by the hereditary biochemical apparatus of the cell while, on the other hand, it is a genuine development, with the appearance of new properties determined by new forms of interaction between proteins and other structures, taking place under control of the mechanism of heredity. For example, the synthesis of a particular enzymic system, even if it is brought about by means of matrices, is, from the ontogenetic point of view, a genuine development because owing to this synthesis new properties appear in the organization; in particular, enzymic processes develop, which could not have been brought about by the matrices themselves and which, by interacting with other enzymic processes, can, so to speak, give rise to biochemical function *de novo*. In controlling the synthesis of proteins the matrices at the same time give rise to forces which they can no longer control and which begin to "lead a new life of their own", obeying new rules. The complexity and mechanism of all these interactions is still far from being worked out and the study of them is only just beginning now.

In this connection we must mention the tremendous importance of those branches of biochemistry and embryology concerned with the morphogenetic variability of proteins, that is to say, with the increase in complexity of protein structure during the period of active morphogenesis, the appearance at various stages of embryonic development of certain proteins which are "new" for the individual, and the manifestation of their functional activity. We have only begun, during the last few years, to accumulate evidence along these lines.

The morphogenetic processes lead to the creation of a system which is more or less well-adapted biologically and which can live an actively independent life and can reproduce itself in a similar way by some means of multiplication. The rapid carrying out of the hereditary "programme" of development gradually slacks off and at a certain time it is completed by the creation of an adult individual, capable of reproduction. The organism, however, continues to change even after completion of the morphogenetic "programme", but these changes now are purely concerned with ageing and usually lead to destruction and deterioration of the bio-chemical and physiological harmony which had been attained on com-pletion of the active process of morphogenesis. Are the characteristic features of ageing essentially the same as those of morphogenesis? Does it occur in accordance with a "programme" which is also recorded in the form of hereditary information in special centres for the genetic control of development? We are inclined to answer this question in the negative, although the continuation of life does show features which are hereditary and which are characteristic for particular species. The relationship between spontaneous ageing and morphogenesis is not the same for all representatives of the living world. In many plants, especially annual ones, morphogenesis is dominant and overshadows the process of ageing and, for example, the vegetative organs, which could go on living for a long time under favourable conditions, quickly grow old and die shortly after flowering or with the onset of winter, owing to the outflow of plastic substances from them into the organs concerned with reproduction or storage. This subordination of the life cycle to the morphogenetic and reproductive processes also occurs in many of the lower animals such as the insects, which are sometimes incapable of feeding after they have completed the sexual process and die of starvation, not as a result of slow ageing with the gradual development of degenerative phenomena. If, however, we isolate a leaf of an annual plant, which is itself unable to regenerate, and cultivate it in a rooted condition (considerable numbers of such experiments have been done, even during last century), then it can survive and grow for many years, carrying out all the characteristic functions of leaves, such as photosynthesis, protein synthesis, etc. How-ever, its life in isolation from any plant is still not everlasting and after a certain time has elapsed it still dies owing to gradual ageing.

This example enables us to consider the nature of ageing in its "pure" form, independent of morphogenesis. As a result of an active formative process there arises a complicated biological system which is, in its way, fully adapted to the carrying out of metabolism. If this system, by means of its metabolism and in spite of constant interaction with its environment, were to remain an accurate copy of that which corresponded to the completion of the "formative programme" of development, then the life of the system would be practically everlasting. However, the properties of biological systems do not remain so constant in time and, owing to their metabolism, spontaneous changes accumulate continually in them and these gradually but regularly destroy the harmony which had been attained. They are reflected in deviations from the optimal conditions for the accomplishment of biochemical, physiological and functional processes and lead to a great variety of secondary changes associated with the presence within the organism of reactions at various levels, molecular and intermolecular, intracellular and intercellular, as well as reactions between different tissues or different organs, etc.

The changes which disturb the accurate adjustment of the biological system after it has reached maturity may be of many different kinds and, obviously, may come about in many different ways. The actual variable factor which makes itself felt depends on the order of complexity of the system which is being studied.

In this book we are only concerned with the primary mechanism of ontogenesis, namely that connected with the changes in proteins and nucleic acids on the molecular plane.

# CHAPTER XVII

# THE BIOSYNTHESIS OF PROTEINS AND NUCLEIC ACIDS AND THE PROBLEM OF HEREDITY

## *Introduction*

At present genetics is one of the most complicated, accurate and important of the biological sciences and its development is having a tremendous effect on many fields of biology, especially biochemistry. In this chapter we shall only deal with the very limited aspect of the subject known as biochemical genetics. There are two main branches of biochemical genetics, namely, the study of the material, biochemical basis of the phenomena of heredity and the study of the nature of the inheritance of biochemical characteristics and their connection with morphological and biochemical variations. We shall scarcely touch on the second of these aspects because it lies outside the scope of this book. The fact that the recent monograph of Wagner & Mitchell (148) gives a detailed survey of many of the problems of biochemical genetics on this same molecular plane makes it even more superfluous for us to deal with them here.

Excellent reviews of the contemporary literature on biochemical genetics have been published recently in the monographs of Strauss (141), Anfinsen (56), Chantrenne (72) and Dubinin (11a).

Among all the numerous and varied problems of modern genetics (the study of the laws of cleavage, cytogenetics, problems of cytoplasmic inheritance, mutational variability, immunogenetics, the chromosomes and karyogenesis, evolutionary genetics, polyploidy, etc.), each of which has its own connection with the problems of the synthesis of proteins and nucleic acids, we shall only deal here with those aspects which are logically connected with the subject of the present monograph.

Many of the questions concerning the link between heredity and the synthesis of proteins and of RNA and DNA have been reviewed recently in a number of theoretical papers (6, 25, 26, 33, 36, 48, 50, 109, 10, 79a, 125, 104a, 144), and have also been touched on in earlier chapters of this book in our survey of the factual material concerning the biological role and the synthesis of RNA, DNA and proteins in the living cell. A large number of comprehensive reviews of this problem have also been published in the proceedings of a recent symposium on the chemical basis of heredity (153). In this chapter, therefore, we shall not try to give an exhaustive review of the factual material relating to the problem, which is already largely known to a wide circle of biologists and biochemists, but shall, for the most part, bring out a number of ideas concerned with those controversial problems of biochemical genetics, the evaluation of which gives rise to the most discussion.

It must be pointed out that in no branch of biology has there been such continuous and all-embracing discussion over the past 30 years as that which is continually arising out of the assessment of genetic problems. It seems to us that the reason for this is not only that it is a field on which there can be a clash between two views of the world, the idealistic and the materialistic. The reason for the controversy in the field of genetics is rather, perhaps, that *this is the very field in which we encounter the group of phenomena and laws which manifest the greatest qualitative differentiation between living material and non-living nature, which are most peculiar to the living cell and living material in general and which constitute the necessary condition for the sequence of phylogenesis and the orderliness and accuracy of ontogenesis*. Furthermore, the mechanism by which these phenomena are brought into being, as well as their nature, are not only the most complicated and extraordinary of all those with which we meet in biology (with the possible exception of the mechanisms of thought), but they also operate in a region which is the meeting point for many of the other practical and theoretical problems of biology. Too many people have wanted to take possession of this key point in biology but, unfortunately, only very few have, in fact, been well enough equipped to find the correct position from which to attack the problem.

Any scientific controversy must be looked at historically. Many disputes have become irrelevant in the light of the latest achievements of science. The idea of matrices, the idea of self-reproduction of DNA, the idea of the extreme constancy of DNA, all of these were subjected to the most embittered criticism 10-12 years ago. Now they are no longer ideas but facts, the truth of which cannot be doubted. The possibility that viral particles might be infective solely by virtue of their RNA might have been held to be quite absurd 10-12 years ago. Now this possibility has been demonstrated convincingly. It is no longer possible to deny these facts. There is no need to keep going over the discussion of problems which science has already solved. The rapid development of science quickly consigns the many incorrect ideas to their fate and therefore, in discussing the biochemical aspects of heredity now we shall not consider who was right and who wrong some years ago about problems which are already solved. Now there are new problems and new mistakes will be made in solving them, and new controversies will arise. We cannot skirt round these controversies without setting out our own opinions. It is quite natural that mistakes should be made in science for, of all the dozens of roads open, only one will lead to the intended goal. Perhaps there is only one thing which is not in doubt at present and that is that the key to the problem of the mechanism of heredity is in the hands of the biochemists, and they have already practically solved the problem of the means by which heredity is controlled.

## 1. Some considerations as to the connection between heredity and ontogenetic morphogenesis

In the large number of controversial papers on heredity which have been published during the past few years (9, 10, 11, 12, 13, 14, 16, 18, 19, 27, 30-32, 35, 37, 38, 29, 43, 45, 47) one may find very different approaches to the evaluation of facts, as they are now known, concerning the importance of the specificity of molecular polymers (especially RNA and DNA) in the production and regulation of the phenomena of inheritance. The aspects surveyed in these papers are, however, by no means the only important aspects of this connection and the theoretically important question of the connection between the specificity of syntheses and ontogenesis seems to have received very little attention.

Many of these papers are published in the journal *Voprosy Filosofii* and the nature of the journal determined, to some extent, the approach adopted in the papers, each author trying to give a particular philosophical colouring to his statements and each going to great lengths to show that his point of view is materialistic, dialectical and the only correct one. Naturally we cannot here take up all the arguments of all the participants in this controversy and we shall only give our opinion on a few of the most important questions.

In the earlier chapters we have already discussed the fundamental laws of protein synthesis. The material brought out in these chapters showed clearly that the main characteristic of the biochemical system for the synthesis of proteins is its ability to reproduce the specificity of their structure and that this same mechanism for the reproduction of specificity would appear to be associated with the mechanisms of heredity. When viral particles reproduce themselves within the living cell the synthesis of a specific protein by means of RNA constitutes, in essence, the realization of the hereditary properties of the virus, i.e. its multiplication. The simple synthesis of proteins by means of nucleic acids and the self-reproduction of the latter is largely the basis for the multiplication of phage particles and bacteria. However, the picture of the relationship between syntheses and ontogenesis becomes markedly different when we pass on from unicellular to multicellular organisms, the reproduction of which takes place through intensive morphogenesis, beginning with the division of one fertilized (or unfertilized) oocyte. In the course of this morphogenesis we observe, alongside the ordinary "growth" syntheses of proteins and nucleic acids, what appear to be exceptions to the principle of the reproduction of specificity as the process of morphogenesis occurs by means of the continual production of new forms of enzymic and structural proteins which are not to be found in the oocyte nor in the earlier stages of embryogenesis. If we look at ontogenesis without reference to the evolution of the species in question, then the morphogenetic variability of proteins and the appearance during embryogenesis of a colossal variety of new forms of sequences of amino acids in polypeptide chains and new types of macromolecular structures will appear to be a true new formation and not a simple reproduction by means of previously existing matrices. In this case the process of new formation would have to be flexible, easily controlled and dependent on the widest variety of factors in the external environment and the results of such morphogenesis could never be worked out in advance.

However, the process of ontogenesis is really brought about in a rather different way. During morphogenesis there arise new proteins and structures which were absent during the first phases of development but which approximately or, in most cases, absolutely accurately, reproduce the analogous proteins and structures of the parents and of many hundreds and thousands of preceding generations of the species. The ability to reproduce in this way through a number of generations also constitutes heredity, which is the most important property of living things. *Heredity is that property of living material which ensures the succession of generations, creates the material possibility for phylogenetic evolution and predetermines the character of ontogenesis, also determining the character and specificity of the syntheses at various stages of embryonic and postembryonic development.*

Thus, if we regard ontogenesis as a part of the general evolution of life, it does not appear as an autonomous and haphazard development and it may, in some degree, be compared with the construction of a building in accordance with drawings which have already been prepared.

What is the biochemical nature of these drawings? Does the oocyte as a whole correspond with the drawings or are they recorded by means of a definite code in special structures intended for this very purpose? This controversial question of genetics can only be answered on the biochemical plane and modern biochemical geneticists have accumulated enough accurate evidence for it to be answered in principle although many of the details are still quite obscure.

The proteins of the cytoplasm of the oocyte are reproduced during ontogenesis and exert a definite influence on its character. This is also true of a number of self-reproducing cytoplasmic structures (chondriosomes, plastids, etc.). Ontogenesis, however, does not merely involve the reproduction of the structures of the oocyte, it consists primarily in the origin of multitudes of new proteins and structures, the synthesis of which cannot be determined by the proteins of the cytoplasm, which have a different structure and strictly determinate biochemical functions connected with the vital activities of the cell concerned. In order to regulate the synthesis of new proteins at the different stages of ontogenesis, there must exist within the cells a self-reproducing, specialized biochemical system which can concentrate within itself a store of hereditary information far greater than that possessed by the self-reproducing cytoplasm. This biochemical system is also the main material carrier of heredity, which manifests its activity in connection with the general metabolism of the cells and primarily by determining and regulating the synthesis of proteins.

Thus, the main biochemical properties of this system must be: the ability to reproduce itself, the ability to concentrate a vast, almost unlimited, stock of information and the ability to determine the synthesis of proteins and other compounds which possess specificity. The only system within the cell which fulfils all these conditions is the system of nucleic acids and, in the first place, DNA, a fact which follows clearly from all present-day evidence as to the part played by DNA within the organism.

The recognition of the existence of a specialized biochemical mechanism of heredity cannot be regarded as idealistic. This approach is materialistic, in that the actual relationships are worked out by consideration of phylogenesis and ontogenesis, and in that it gives rise to the possibility of presenting the evolutionary development of living material as a single process with a definite material connection between the separate cycles which, in this case, are taken as the real sequence of events which constitute phylogenesis and ontogenesis. It also provides a straightforward approach to the experimental verification of particular hypotheses. At the same time the proponents of such an approach cannot deny that ontogenesis is development, even if of a specifically controlled type.

The existence of morphogenetic syntheses of a strictly determinate nature does not rule out the possibility of extensive adaptive phenomena or even the possibility that adaptability might itself be an inheritable, evolutionarily favourable characteristic. Thus, for example, under the influence of photoperiodic induction or prolonged cooling (vernalization), a definite shift takes place in the protein metabolism of plants, which has a morphogenetic significance. It is quite clear that it is not possible to control in advance the character of the changes in either the lengths of the days and nights or the fluctuations of temperature, but on the other hand, these changes in metabolism, by controlling development, are very important in the ecological adaptation of the plant. In this case the "mechanism of inheritance" has had to *ensure variability of response because evolution and selection were obliged to place these changes directly under the control of the environment, making them depend on it.* It is most likely that it is just in morphogenetic reactions of this sort that there is the most real possibility of the environment having a direct effect on the character of the development of the organism. In this case the plasticity of the organism is itself a hereditary characteristic.

This peculiar feature of ecological morphogenesis cannot, however, be the rule for all morphogenetic processes. It seems to us that it is the absence of limits of a qualitative kind on the different morphogenetic reactions and their different degrees of dependence on the environment which have also, to a certain extent, given rise to the controversy on this question.

If we reject the idea that there is a specialized material basis for heredity and regard the morphogenetic processes of individual development as being no more than processes of development which are not controlled by the material carriers of heredity but only associated with the cell as a whole, then we shall be bound to recognize that the material formative mechanisms of phylogenesis and ontogenesis are of the same sort, for the effect of each of these forms of development is to produce an organism of the same kind. It is, however, impossible to adopt this view, because the formative mechanism of phylogenesis is unrepeatable, like the conditions of phylogenetic development.

What are the peculiar features of that specific controlled form of development which constitutes ontogenesis, and is it possible, having answered this question, to find the key to the morphogenetic variability of proteins? Are all characteristics arising during ontogenesis predetermined by the structure of the hereditary bases or are only some so determined while others develop by interaction with them? What is the mechanism which enables hereditary characteristics to take the form of morphogenetic processes? In order to find out how far from or near to a solution these questions are, it is necessary to have a short discussion of the way in which genetic systems have arisen during evolution and about present-day ideas as to the biochemical basis of heredity.

## 2. The evolutionary origin of different forms of heredity

There are very many definitions of heredity as a biological phenomenon and, although we have not tried to define this characteristic, it remains certain that heredity manifests itself in the ability of living matter to retain its specificity, its own peculiar properties, in spite of continual exchanges with a continually changing environment and in spite of the processes whereby generations succeed one another. It is quite clear that this property can manifest itself at various levels of complexity, arising out of the development of living nature. The fact that heredity is a property of living matter does not in the least mean that there are no definite mechanisms by which it occurs. The idea that this property is something inherent in a particular body and not requiring any explanation is quite naïve. Every perceptible property has a definite explanation: the weight of a body is explained by the gravitation of the Earth, colour by a particular absorption and reflection of light, etc. To say that heredity is a property inherent in living matter in general and in every structure of the cell, and that therefore there cannot be any special mechanism of heredity developing by evolution, for example no biochemical mechanism, is not to take into account the true essence and nature of the processes going on in the cell. The single property of heredity postulated by several authors as being diffusely distributed throughout living nature does not exist, and heredity, in its contemporary form, is determined by real biochemical systems in strict accordance with the nature of the character-istic features which must be conserved or reproduced in the course of time.

Protein molecules as such, taken in isolation from the biochemical systems by which they are synthesized, even though they may retain their specific biological functions (e.g. enzymic functions), do not possess the power of self-reproduction. They do, of course, retain their specific properties for a time, but only in the same way as a stone lying in the road, that is to say, only until such a time as they are broken down as a result of some external circumstance. Given favourable conditions a protein molecule can, of course, be conserved indefinitely, but that is not biological heredity, it is merely stability. After certain changes, such as the uncurling of the globule, proteins may regain their structure, but that is not heredity either, it is reconstruction. Heredity is necessarily asso-ciated with reproduction, with multiplication. It is quite natural that the property of heredity should have developed and become more compli-cated in parallel with the development and increase in complexity of biological systems.

Only the primaeval protein coacervates, the phylogenetic evolution of which in fact merged with their ontogenesis, were able to do without a specialized system to ensure heredity, and their multiplication could take place by simple, unequal division, after which each part went its own way, which might have been quite different from that of the other part, and was almost entirely dependent on external circumstances. At that stage in the development of nature there were no differentiated bio-logical species, there were only individual lumps of organic polymeric material of endless heterogeneity and variety, which could still not be described as really being alive. If there was to arise, during the selection and evolution of these lumps, the ability to reproduce particular proteins while preserving their specificity, then a system for this reproduction would have to arise first, a system such as that of a self-reproducing matrix, in fact a prototype of RNA. Clearly it was just at this point that true biological evolution came into being. It must be pointed out that the formation of purines and pyrimidines could take place under the con-ditions which obtained before life existed, just as amino acids could be formed at that time (122). According to the theory of the origin of life put forward in Oparin's book (34), the appearance of living material must have coincided with the appearance of processes for the reproduction of this material, not just with the appearance of metabolism but with the appearance of that form of metabolism which ensures self-reproduction.

Even during this phase of the development of life, however, each onto-genesis, in fact, merged into phylogenesis and each half of a lump which had divided had all the features characteristic of that form of living material. Under these conditions the matrices must have contained just those items of "information" which were needed for current syntheses of proteins and the self-reproducing prototype of RNA would seem to have completely fulfilled these conditions. Can one refer to the RNA of even these "primitive biological systems" as the "special substance of heredity", the existence of which is denied by some authors? It seems to us that the whole concept of a "substance of heredity" is very un-fortunate and it is mainly used by those biologists who criticize it. There can, however, be no doubt that, in this case, RNA was the main com-ponent of the specialized biochemical system reproducing the specificity of the proteins, just as it serves this purpose now. The existence of a specialized system for the reproduction of proteins is a fact which cannot be questioned. In such primitive systems the RNA, by ensuring the reproduction of the specificity of the proteins, also ensured the repro-duction of the whole "cell". RNA still retains this function in the case of the multiplication of viruses. Certainly the synthesis of RNA depends on the proteins themselves as well, but this dependence is not of the same kind.

The synthesis of proteins by means of RNA only merged with the reproduction of hereditary characteristics until a certain stage of evolution was reached, that is to say so long as each ontogenesis could carry on evolutionary development without any significant reorganization or re-newal of the system during the acts of multiplication. However, the evo-lution of the primaeval "cells" gradually made this less and less possible from a biochemical, physiological and morphological standpoint. There arose within the cells more complicated and specialized systems, whose existence and functions required a very high specificity of proteins. Such systems could not be transferred from one generation to another in the finished state by the simple division of existing entities. They had to be reproduced anew in each fresh generation. In this way, true ontogenesis came into existence and phylogenesis became intermittent.

It is quite clear that if, in any cell, in addition to the protein com-ponents $A$, $B$, $C$, $D$ and $E$, which had been present for a long time, and the matrices $a$, $b$, $c$, $d$ and $e$ for their production, the process of evolution and selection, occurring over thousands of years, were to lead to the development of a new and more complicated system of proteins $X$-$Y$-$Z$ which could not be transferred to the daughter cells on account of its complexity but had to be formed anew in these cells (while the molecules of the proteins $A$, $B$, $C$, $D$ and $E$ and their matrices could simply be divided into equal amounts among the daughter cells) then the transfer to the daughter cells of components $A$, $B$, $C$, $D$ and $E$ and their matrices would simply be a return to an earlier state. In such a case thousands of years and conditions the same as those to which the original cell was subject when $X$-$Y$-$Z$ first arose would be required to produce this component in the daughter cell. It is clear that such "multiplication" could not serve as the basis for the development of living material. It is *quite* obvious that if the system $X$-$Y$-$Z$ cannot be transferred ready-made from one generation to the next, then at least the specific ability to reproduce the system accurately and quickly must be so transferred, that is to say the matrices for synthesis of the proteins $X$, $Y$ and $Z$ and matrices containing the information which leads to the formation of the complex $X$-$Y$-$Z$, unless this forms spontaneously. As proteins are continually being renewed and there are matrices in the cells for synthesizing them, the transfer from one cell to the next of only one matrix, which would then give rise, in the new generation, to everything which had been present in the old, represented a great economy, extended the possibilities for the development of living systems, and was an extremely advantageous characteristic. However, the further development of the organization of living systems made even such a means of transfer of hereditary characteristics insufficient. We may explain this by extending the same scheme. Let us assume that, during the process of development of the cell consisting of the proteins $A$, $B$, $C$, $D$ and $E$, the system $X$-$Y$-$Z$ and their matrices, there arose a new complex system $P$-$Q$-$R$ which would normally not function all the time, but only for some short period, for example, at the end of ontogenesis, and which was therefore only formed in the cells for a short while, for example, at the end of their lives. In the earlier period of their lifetime, shortly after the beginning of ontogenesis, although the proteins $A$, $B$, $C$, $D$ and $E$ and the system $X$-$Y$-$Z$ were present, the system $P$-$Q$-$R$ could not and should not exist. Naturally, in such a situation, which is very common in alternating generations, the mother cells cannot transmit to the daughter cells either the active matrices for the formation of the components $P$, $Q$ and $R$ or the finished system $P$-$Q$-$R$. In such a case the simple transfer of the active matrices would not fulfil the requirements of the organism, because the system would need to be formed at the end of ontogenesis, so that it would be necessary to transfer, in some form, the ability to form this system; for repetition of the whole path of evolution would be impossible under changing circumstances. To meet the requirements of such cases, the ordinary factors controlling evolution (variability and selection) have given rise to special, self-reproducing, periodically inactive matrices which can induce the synthesis of active matrices and of the proteins $P$, $Q$ and $R$, exactly at some predetermined period in the ontogenesis of the cell. At present the inactive molecules of DNA serve as such periodically inactive matrices. They are, in essence, similar to the polynucleotides of RNA, the active, protein-synthesizing surfaces of which "interlock" in the form of complementary pairs of bases.

The development of such a system of remote transmission of characteristics signified a new level in the development of living matter and opened up unlimited possibilities for the development of living systems. When multicellular organisms came into being, a large proportion of their characteristics could be handed on from generation to generation by means of this sort of remote "concealed" transfer of the ability to form some particular system which would act, for the most part, during post-embryonic life.

The complicated organism of our times cannot transfer its brain, heart or liver directly to its offspring. It cannot, however, get along without transferring to them, in a preformed, hidden, encoded state, the ability to form these systems. This is known as a programme of development and consists of exact, although coded, information which will enable a particular system to be formed at the right moment and in the right amount during the new ontogenesis, almost exactly copying the original model. The appearance of a copy quite out of the blue does not take place either in theory or in practice.

We have made a point of going over these ideas, which have already been discussed in a rather different form long ago and have been widely accepted, in order to emphasize that the appearance of the self-reproducing system of DNA was an evolutionary necessity. It was an important stage in the development of living material on our planet. In this connection it must be recognized that the fact that information is transferred along the path DNA → RNA → protein does not in any way imply that it could follow the reverse path or give rise to a cycle after the pattern

$$DNA \rightarrow RNA$$
$$\nwarrow \qquad \swarrow$$
$$protein$$

Protein systems certainly do have an influence on DNA, but the ability of DNA to reproduce itself does away with the need for the development of synthetic systems with a reverse transfer of information.

## 3. On the importance of nucleic acids as carriers of genetic information

The hypothesis put forward more than 15 years ago by Beadle and Tatum in the form "one gene—one enzyme" or in its more general form "one gene—one protein" has been worked out on the basis of the requirements of mutants of neurospora for various nutrients. This hypothesis has played a great part in the development of modern genetical concepts and in bringing together genetics and biochemistry. This formulation has received factual confirmation in recent years from the study of the genetic functions of DNA and the synthesis of special informational RNA, which carries the information of particular discrete genetic loci.

The formation of this recently-discovered informational RNA is essentially a manifestation of the activity of genes in the form of temporary complementary copies. The information in the DNA acting as a gene exists in the "negative" form. The formation of informational RNA produces a "positive" print in accordance with the same principles which lead to the appearance of a positive desoxyribonucleotide chain during the self-reproduction of DNA.

These complementary copies serve as matrices for the synthesis of specific proteins. As the various types of cell form different proteins in different proportions, the sets of informational RNA synthesized in the cells should also reflect the nature of their specialization. They should show up differences between cells in respect of the assortment of functionally active genes which they contain, since genes which are not working do not form informational RNA.

The problem of informational RNA is, thus, the problem of the functionally active form of the gene. The length of the polynucleotide of the individual informational RNA must be an indication of the length of the individual gene. The discovery of the mechanism by which this form of RNA takes part in the synthesis of proteins is also a step towards clearing up the mechanism of the genes. Informational RNA can be regarded as the channel through which the genetic information hidden in DNA is made manifest in metabolism.

The problem of the genetic functions of DNA has been very considerably illuminated during recent years by means of a large series of theoretical reviews (25, 26, 33a, 51, 55, 59, 71, 79a, 91, 103, 106, 101, 102, 127, 138) and we do not intend to bring forward here all the proofs of the connection between DNA and the phenomenon of heredity. This may be taken as a proven fact, as may the fact that all enzymes are protein bodies. We shall only deal with a few aspects of the problem.

Our present concepts of the role of nucleic acids as a specialized biochemical system for inheritance leads us to take up the following position. As our main proposition concerning the genetic significance of nucleic acids, let us assume that it is the actual sequence of nucleotides along the polynucleotide chains of desoxyribonucleic and ribonucleic acids which constitutes the biochemical record of the hereditary information which controls the synthesis of the whole colossal variety of enzymic, supporting, structural, contractile and reserve proteins which are formed during the development of the individual and, in their turn, organize the multifarious actual physiological and biochemical processes. The occurrence of metabolism in the living organism depends mainly on proteins, because it is primarily carried out by means of enzymic reactions. It is therefore these very proteins which constitute the intermediate link between the specialized system of heredity and any particular complicated features of the organism. However, if they are to regulate the reproduction of the specificity of many thousands of proteins, not only at a given moment, but also through all stages of ontogenesis, the polynucleotide chains of DNA and RNA must possess the following properties:

1. They must have almost unlimited possibilities for variation, thus reflecting the almost unlimited variation in the structures of protein (species, tissue, intracellular, etc. specificity).
2. They must be able to ensure the reproduction of their specificity, i.e. to serve as matrices (patterns) for their own reproduction.
3. They must have the ability to transform, in some way, a special sequence of the four types of nucleotide into a unique sequence of the twenty amino acids of which proteins are made up. In other words, there must exist within the cell some special biochemical system which can decipher the information "recorded" in the form of a sequence of nucleotides and carry this information over into protein synthesis, translating it into a sequence of amino acids.
4. They must respond to the effects of all those external physical, chemical and biological factors which are known to lead to the development of hereditary changes (hybridization, mutagenic substances, ionizing and other radiation, etc.).
5. As they contain the total stock of information referring to the species, the nucleic acids must also contain a definite programme for the successive release of this information in such a way as to ensure differentiation of the cells of the organism as it develops in space as well as in time.

The evidence which we have surveyed in the earlier chapters indicates quite clearly that the system of nucleic acids has the first four of these properties in full measure, and this alone should be enough to focus the attention of geneticists on a most serious study of the part played by these polymers in the phenomena of inheritance. As for the fifth condition for the theoretically postulated specialized system for inheritance, it is of special interest, just from the point of view of the control of development. The first four properties of the system of DNA and RNA, which point to the genetic importance of nucleic acids, mainly reflect the *ability of the system to maintain the constancy of biological structures and to ensure and preserve variations of an evolutionary character.* The fifth property of the system of inheritance must ensure its *ability to maintain and control the ontogenetic development and differentiation of the organism by bringing the different genetic loci into action at the appropriate time and place.*

The first four genetic properties of nucleic acids are ensured by their own structure. The fifth function pertains to the genetic system as a whole and would seem to be brought about by the organization of the nucleic acids in the complicated structure of the chromosomes. In the present chapter we shall only take a look at the question of the mechanisms which maintain constancy and subserve evolutionary variability. The problem of the genetic control of ontogenetic development will be dealt with separately in the next chapter.

Since it was established, several decades ago, that the chromatin, of which the chromosomes are made and which is associated with heredity, belongs to the biochemical category of nucleoprotein, a curious competition developed between proteins and DNA. Scientists tried hard to determine which of these components is the main, specific carrier of hereditary information. At first pre-eminence was naturally given to protein bodies, but this pre-eminence was not based on fact but simply on a traditional faith in proteins as the essential component of living matter. Later this pre-eminence was gradually transferred to nucleic acids and this reassessment took place rather quickly under the violent impact of a tremendous number of convincing facts providing both direct and indirect evidence as to the primary importance of nucleic acids, especially DNA, as a specialized system for the collection, storage, reproduction and realization of hereditary information.

A great many of these facts and the theoretical concepts arising out of them have already been discussed in detail in earlier chapters concerned with various aspects of protein synthesis. This was inevitable, as the synthesis of proteins is, in essence, the most important link in the chain of realization of hereditary information, and the study of the mechanism of protein synthesis is also the study of the system of heredity in action. However, in discussing the function of nucleic acids from the point of view of their participation in protein synthesis and from the point of view of the mechanism whereby they reproduce themselves, we did not touch on certain other extremely important aspects of this problem, concerned with the proof of the genetic significance of nucleic acids. This is, in the first place, the question of the localization of genetic information, the question of the phenomena of transformation under the influence of DNA and the question of the biochemical nature of mutations. These questions must, however, be touched on briefly, not only because they are of great interest from the point of view of an understanding of the mechanisms of heredity, but also because they reveal the place, role and significance of protein synthesis in realizing the relationship between hereditary information and the processes of ontogenetic morphogenesis.

## 4. The sequence of nucleotides and genetic information

In Chapter XV we have already taken a look at the problem of the nature of the functional link between the sequence of nucleotides in the matrices of protein synthesis and the sequence of amino acids in the proteins synthesized on them (the nucleotide code for amino acids). Then, however, we only examined one aspect of the problem of the code, namely, that which shows up in experimental studies of protein synthesis.

However, the problem of the coding of the information which regulates the synthesis of proteins and the processes of morphogenesis (genetic information) is broader and scientists approach the study of it from several angles. Although, until quite recently, the methods used for the study of this problem were purely abstract (74, 127, 131), direct experimental approaches to its solution have now been found. One typical line of approach to the solution of the problem was discussed in Chapter XV.

Another very profitable approach to the experimental solution of the problem of coding is the study of the connection between changes in the RNA and proteins of viruses, especially tobacco mosaic virus (TMV). In the part of this book concerned with mutations we have already discussed the principles of these experiments. They involve the gentle oxidation of the individual nucleotides of the RNA of TMV by means of nitrous acid and the analysis of the changes in the sequence of amino acids in the protein which corresponds with a particular change in RNA. Evidence from this source is available for a thorough analysis. Ochoa and his colleagues (138a) have made such a study with a view to comparing the code of viral RNA with that established in their laboratory by means of experiments with synthetic polynucleotides. A large amount of evidence concerning the code of viral RNA has been obtained in the laboratories of Wittmann and Fraenkel-Conrat but the scheme which we give below has been taken from a paper by Ochoa which the author very kindly showed to us in manuscript (120a).

*Amino acid substitutions in nitrous acid mutants of TMV*

| Substitution | No. of observations | Changes in coding nucleotides | Changes in base caused by $HNO_2$ | Correspondence with code from synthetic polynucleotides |
|---|---|---|---|---|
| Asp' → Ala | 6 | CUA → CUG | A → G | Yes |
| Asp' → Gly | 2 | GUA → GUG | ,, | ,, |
| Glu → Gly | 1 | AUG → GUG | ,, | ,, |
| Glu' → Gly | 2 | AUG → GUG | ,, | ,, |
| Ileu → Val | 1 | UUA → UUG | ,, | ,, |
| Leu → Phe | 1 | CUU → UUU | C → U | ,, |
| Pro → Leu | 2 | CUC → CUU | ,, | ,, |
| Ser → Phe | 3 | UCU(UUC) → UUU | ,, | ,, |
| Thr → Ser | 1 | UCC → UCU(UUC) | ,, | ,, |
| Pro → Ser | 3 | CUC → UUC | ,, | ,, |
| Thr → Ileu | 7 | UCA → UUA | ,, | ,, |
| Asp' → Ser | 4 | UGA or UAA(CUA) → UUC(UCU) | ? | No |
| Arg → Gly | 5 | GUC → GUG | ,, | ,, |
| Ser → Leu | 1 | UUC(UCU) → CUU | ,, | ,, |
| Thr → Met | 3 | UCA(UCC) → UGA | ,, | ,, |
| Tyr → Phe | 1 | AUU → UUU | ,, | ,, |

A = adenine   G = guanine   C = cytosine   U = uridine
Asp' = aspartic acid or asparagine
Glu' = glutamic acid or glutamine

By comparing their results with the code based on experiments with artificial polynucleotides the authors came to the conclusion that there might be a universal code. The cases in which there was no correspondence could be accounted for even if there were a universal code by the fact that there is not yet evidence as to the "meaning" of every one of the 64 triplets. It is not yet possible, however, to draw any conclusion as to the universality of the code. To do this it would be necessary to establish the "meaning" of each of the 64 theoretically possible triplets and compare them in systems differing widely from one another in their evolutionary origins.

The recent work of Crick and his colleagues (73a) is of great interest in connection with the problem of coding. They have worked out a very complicated method for experimental calculation of the code of DNA which, though indirect, is certainly important. They used the genetic method of studying the localization of point mutations in $T_4$ phages brought about by mutagens (acridines) which give rise to changes in the nucleotide pairs of the DNA of the phage. The results obtained were studied by making special genetic maps of the localization of the mutations by the method worked out by Benzer (64). These authors did not study the proteins of the mutant strains of the phage and their method is therefore only circumstantial but the conclusions they draw are of the nature of more or less well founded guesses. Furthermore, it must be noted that the underlying assumption of the authors that the code is not overlapping is not based on their own researches but on experiments with mutant strains of TMV which showed that, for the most part, one amino acid was substituted for another when there was a transposition of nucleotides, and also on studies on the variability of haemoglobins. If we accept the authors' assumptions, then their conclusions on the nature of the code of DNA lead to the establishment of the following rules:

a. One amino acid is represented by a group of three (or, less probably, a multiple of three) bases.
b. The code is not overlapping.
c. The sequence of bases is "enumerated" from a fixed starting-point. This determines the correct "calculation" of the information in the form of triplets. In the polynucleotides there are no special markers showing how to select the correct triplets. If the starting-point is altered by one base the numbering of the triplets is also altered and becomes incorrect.
d. The code seems to be a degenerate one in that each amino acid is represented by one of several possible triplets of bases.

Such a coding of genetic information as may be postulated from the work of Crick and his colleagues would give rise to a number of consequences which could be verified experimentally. Such verification will certainly be attempted very soon and the following lines of research could be pursued for this purpose:

1. If the code is of this type the synthesis of proteins can only proceed linearly, by a lengthening of the chain from the amino or carboxyl end, not by coalescence of peptides which have been formed on different parts of the matrix. In Chapter VI we saw that even if the linear extension of haemoglobin and bacterial amylase may be considered as typical, such confirmation is not forthcoming in the case of serum albumin.

2. The incorporation into the chain of RNA or DNA of an extra nucleotide or two or the loss of a nucleotide will, in fact, lead to inactivation of the gene and production of a protein in which all the amino acids are changed (a completely foreign protein). The occurrence of such proteins in mutants has not been established. If the code is of the nature suggested, mutations arising by the addition or removal of a single nucleotide could scarcely serve as material for evolution, although they would be very likely to occur.

An interesting attempt was recently made by Yčas (151a) to ascertain the manner in which sequences of amino acids are coded by sequences of bases in polynucleotides. Yčas brought together a large volume of work relating to sequences of nucleotides and amino acids in the RNA and proteins of various mutants and strains of viruses; he also considered changes in proteins related to species differences. He arrived at the tentative conclusion that sequence of amino acid residues is determined in two stages, in the first of which one nucleotide determines one amino acid. In this case several amino acids correspond to each nucleotide, and he suggests the following possible relationships: adenylic acid (Glu, Gly, Leu, Phe, Try); uridylic acid (Asp, Ileu, His, Ser); guanylic acid (Ala, Arg, Tyr, Val); cytidylic acid (Cys, Lys, Met, Pro, Thr).

Yčas considers that viral RNA carries only a part of the information necessary to determine the position of each amino acid. The additional information is provided by some structure or other in the cells of the host. The nature of this determinant of the second stage of protein biosynthesis is as yet uncertain.

Woese (150b) has elaborated a number of new coding schemes.

Great interest also attaches to the work of Sueoka (142a), who has studied the correlation between the base composition of the total DNA of a number of bacteria and the amino acid composition of the proteins of these bacteria. In the 11 species taken for comparison, the nucleotide composition (content of the pair guanine-cytosine) varied from 35% to 72%. The author observed a marked correlation between these changes and changes in the content of certain amino acids. Alanine, arginine, glycine and proline were correlated positively with the (guanine + cytosine) content of the DNA, some amino acids were indifferent and others exhibited a negative correlation. He regarded the presence of such a correlation as providing evidence of the "universality" of the code.

A very serious attempt to analyse the correspondence between particular coding schemes for the transfer of information from DNA to RNA on the basis of the existing factual evidence as to the nucleotide sequences of different forms of DNA and RNA has been undertaken recently by Leslie (103). We shall not go into all the interesting comparisons which the author made. We shall merely remark that this work revealed a considerable difference between the correlation of the compositions of DNA and RNA in bacteria and in multicellular organisms. In this connection Leslie suggested that there is a different system of coding for the exchange of genetic information between DNA and RNA in bacteria from that used in multicellular organisms (animals and plants).

Detailed reviews of the various hypotheses and of the factual evidence concerning the genetic code have recently been published by Dvorkin (8a) and Gavrilov & Zograf (6a). The number of publications in this field is growing rapidly (97a, 130, 154) and we may expect that many interesting communications will be made in the very near future.

An original approach to an analysis of the information in DNA was used in the work of Morowitz & Cleverdon (117). They decided to compare the most extensive possible amount of information in the DNA with the most extensive possible amount of information in the proteins in the very simplest bacteria. For this analysis the authors chose the smallest known forms of independent living systems, the "pleuro-pneumonia-like" microbe Avian 5969 (*Mucoplasma gallisepticum*). The "total" molecular weight of the DNA in a single cell of this microbe was 70 million. Assuming the validity of the trinucleotide code of Crick, Griffith & Orgel and reckoning that the average molecular weight of a nucleotide is three times that of an amino acid, the authors calculated that the whole mass of the DNA could provide the information for building proteins with a combined molecular weight of 7 million. If we take the mean molecular weight of a protein to be 50,000, then a "total" molecular weight of 7,000,000 would represent 140 different proteins. If we reckon that DNA consists of a double helix, the number of types of protein which can be determined in accordance with the trinucleotide system must be reduced to 70.

Having made this calculation the authors concluded that either these bacteria could manage with only 70 proteins (which is not very likely) or the DNA contained a shortened type of code which was, perhaps, one third more economical. According to the calculations of the authors the cells of these microbes each contained not less than 100 different proteins and 10,000 protein molecules and approximately 300 ribosomes. In another paper Morowitz (117a) describes the occurrence of cells of this bacterium in which the "total" molecular weight of the DNA in the cell is 45 million. The author believes that the information for one protein can be recorded on a polynucleotide with a molecular weight of 1 million. However, the author's calculations obviously give too high a result as most proteins with molecular weights of the order of 50,000 are aggregated and consist of single or conjugated peptide chains with molecular weights of 3-6 thousand.

These interesting calculations by Morowitz and Cleverdon, together with several other similar calculations, bring to the forefront the suggestion that in many lower organisms the total amount of specificity which has to be reproduced is beyond the capacity of the DNA present in them. However, there is no need for all the necessary information to be recorded on the DNA. In discussing the evolutionary origin of the genetic systems (pp. 393-7) we saw that DNA would appear to have arisen at a later stage of evolution than RNA in order to provide for the reproduction during ontogenesis of characteristics which only appear at particular times. It would seem, however, that the synthesis of some proteins, especially in bacteria, might depend on RNA alone. This suggestion provides further explanation for the evidence of Belozerskiĭ & Spirin (4, 61) which we have already quoted, to the effect that in bacteria the nucleotide composition of the DNA varies markedly between species while the composition of the RNA only varies slightly from one species to the next, apart from that small fraction of RNA, the composition of which is correlated with that of the DNA. We have already tried to explain this phenomenon in an earlier chapter. We should like here to put forward an additional suggestion. It seems to us that a certain part of the pool of RNA in bacterial cells comprises, as it were, a partly autonomous and phylogenetically more ancient genetic system which hands on, from one generation to the next, characteristics which are phylogenetically stable, acting throughout life and common to many species of bacteria. The synthesis of this RNA is autonomous and the DNA only carries the genes which regulate its activity. As for the DNA of bacteria, this concentrates in itself the information needed to provide for the appearance of phylogenetically more recent and periodically active systems (such as adaptive enzymes). The differences in these systems from one species to the next are quite sharp and therefore the differences in the composition of the DNA as between one species of bacteria and the next is also just as considerable.

In considering the evidence of Belozerskiĭ and Spirin as to the wide differences in the composition of the DNA in different species of bacteria, Sinsheimer (138) mentions that, in this case, the composition of the DNA varies so much that, in the most divergent species, there can be no two molecules of DNA of identical composition if we are to judge by the discrepancy of the nucleotide compositions, although the compositions of the RNA and proteins in the different species did not show the same marked differentiation. Assuming that the synthesis of the whole mass of the RNA in the bacteria is dependent on the DNA, Sinsheimer tried to explain how this paradoxical situation could have come about, by suggesting that bacteria have a special binary code for the transfer of information from DNA to RNA (137) according to which a single nucleotide in RNA is determined not by one but by two pairs of nucleotides in the double helix of DNA and, for example, the pair C—G may be replaced by the pair A—T without changing the sequence of bases of the derived RNA.  This code would thus enable the composition of the DNA to be varied without changing the information contained in it.  It seems to us, however, that Sinsheimer's suggestion is rather an artificial construction and that an evolutionary approach to the explanation of the phenomenon would be more likely to give results.  The system of DNA arose at a later stage of evolution than that of RNA.  Having come into being to meet a particular evolutionary requirement, it cannot at once take control of all the evolutionary characteristics which were being kept in existence by heredity.  As there exist in nature biological species which carry out their reproduction solely by means of RNA, it seems to us very probable that there should exist species in which DNA exerts only a partial control over the reproduction of specificity.

The DNA content of the cells of mammals is very considerable and amounts to a "total" molecular weight of $2\text{-}4 \times 10^{11}$ which could provide the information for the synthesis of some millions of forms of protein (using a non-overlapping code) if the system were working to capacity.  However, it is hardly likely that more than a few thousand different proteins enter into the composition of even a complicated organism, so the information could, theoretically, be recorded in the pool of DNA available even if it were only used to 10-15% of its capacity.  Using an overlapping code in which the number of nucleotides in the chain of the matrix is the same as that of the amino acid residues in the protein, the capacity of the DNA to carry information would be three times greater still.

In considering the form in which information is recorded in the composition of DNA the view is sometimes put forward that, if a single-chained means of recording information is being used, only one of the polynucleotides of the DNA can act as the matrix.  The second component is only a reflection of the first and must therefore be blocked (79).  It seems to us that this is not so.  If one of the nucleotides of RNA fixes the matrix for protein synthesis in the right position, then the other can fix the adaptor RNA complementary to the matrix (cf. Chap. XV).  Furthermore, if the idea of the complementarity of adaptors and matrices were proven, then even high-molecular polynucleotides which were complementary to one another could possess identical information for protein synthesis if the activating enzymes which attach the amino acids to the adaptors have, as it were, a double zone of contact which can react in exactly the same way with complementary sequence of nucleotides and could attach one and the same amino acid to either of them.  In this case the synthesis of the complementary polynucleotide, whether it were RNA or DNA, would not involve the modification or conservation of information but would merely take the form of its reproduction.

## 5. A direct connection between genetic mutations and alterations of nucleic acids

The problem of the connection between certain mutations and a disturbance in the reproduction of RNA and DNA has already been discussed to some extent in an earlier chapter. Then, however, we only dealt with the mutagenic effect of analogues of nucleic acids which were incorporated into nucleic acids. In fact a considerably wider range of mutagenic factors is associated with changes in nucleic acids.

The problem of mutation and mutagenesis now appears to be a very broad one. It is only in the last few years that any extensive information has been collected and we are not in a position to go into it here at all thoroughly. Detailed reviews of the biochemical nature of mutation have been published recently by a number of authors (104, 108, 82, 83, 92, 93, 1, 70, 76, 81, 84, 140, 151) and we shall therefore only touch upon the subject very shortly although it is in this field that the most direct evidence has been obtained indicating the central role of nucleic acids in ensuring heredity.

In the first place we shall deal with the results obtained in a series of researches by Stent, Pardee and their colleagues (115, 124, 86). In their experiments these workers used a very interesting method to determine the effect of the breakdown of $^{32}P$ on particular characters. When an atom of $^{32}P$ serving as a bridge in a polynucleotide breaks down, the corresponding internucleotide bond is also broken. If the isotope of phosphorus is administered in small, so-called indicator doses, such breakdown will not give rise to any noticeable effects because the number of molecules of $^{32}P$ is only a negligible fraction of the number of molecules of the stable isotope. In some bacteria, which have a very high degree of stability to irradiation, the concentration of $^{32}P$ may be increased so far that its breakdown begins to exert a great physiological effect on its own account, not on account of the irradiation. This effect may be increased by means of a simple device used in the experiments of Stent and Pardee. $^{32}P$ of high specific activity is incorporated into the nucleic acids of *Escherichia coli*, after which the metabolism of the bacteria is halted by freezing at $-196°C$. The breakdown of the $^{32}P$ continues in the meantime and by using different lengths of freezing time it is possible to obtain variants of the bacteria in which there have been different amounts of breakdown of $^{32}P$ and to study the effects of the different degrees of breakdown on the metabolism of the cells and, in particular, on their ability to produce enzymes and RNA. These experiments showed that the first thing which the cells lost as a result of the breakdown was the ability to synthesize adaptive enzymes. These syntheses seemed to be more dependent on the integrity of the phosphorus-containing polymers than protein synthesis in general. In later experiments it was shown that it was, in fact, the disruption of DNA which caused the loss of the ability to synthesize adaptive enzymes. These experiments were based on the use of various methods for the incorporation of different relative amounts of $^{32}$P into the RNA and DNA of the bacteria and the evidence obtained indicated that it was DNA which was the responsible factor. The incorporation of different proportions of $^{32}$P into RNA and DNA was brought about by varying the amount of thymine in the nutritive medium. The absence of this substance paralyses the synthesis of DNA. It was strikingly demonstrated in this way that the integrity of the molecules of DNA, or of parts of them, is necessary for the synthesis of adaptive enzymes, that is to say for the realization of a definite hereditary capability.

Quantitative measurements showed that 50 disintegrations per nucleus were, on the average, enough to stop the synthesis of a number of enzymes (e.g. β-galactosidase and L-threonine dehydratase). The same method was used to bring about genetic changes in bacteriophages. The authors consider that the breakdown of $^{32}$P might entail not only the breakdown of the polynucleotide but also changes of some sort in the integrity of the chromosomes and that this might, therefore, be the cause of the lethal effect.

In publishing these results Pardee (124) reaches the conclusion that it is the presence of the polymeric form of DNA which is needed for the synthesis of enzymes.

Another line of research into mutagenesis which is being intensively developed and which gives a striking demonstration of the genetic specialization of nucleic acids is that of submitting the organism to the action of chemical agents which have a selective effect on DNA and, in several cases, RNA (in experiments on plant viruses). In an earlier chapter we have already discussed the genetic effects which are evoked, in many cases, by the alteration of only one pair of bases in DNA as a result of the disturbance in the sequence of nucleotides caused by analogues of thymine and other nucleotides.

Another commonly used method of producing a local alteration in the nucleotides of RNA and DNA is to submit certain objects (mainly viruses) to the action of solutions of nitrous acid.

In the first set of experiments in this series done by Mundry & Gierer (118), RNA isolated from tobacco mosaic virus (TMV) by the phenol method was exposed directly to the action of weak solutions of nitrous acid. This treatment led to the oxidation of parts of the nitrogenous bases; adenine was converted to hypoxanthine, guanine to xanthine and cytosine to uracil. Under the conditions described, such modifications did not lead to depolymerization of the nucleic acid but brought about, not only a partial loss of infectivity but also a marked increase in the number of mutants, the number of which was estimated by the appearance of necrotic lesions on the leaves of the variety of tobacco known as "Java" (normal RNA and intact TMV only produce chlorotic lesions on the leaves of this sort of tobacco). The number of necrotic lesions on one leaf of this tobacco after infection with chemically modified RNA was more than 20 times greater than that in the untreated controls which developed as the result of spontaneous mutations, while the infectivity of the RNA was only reduced to a half by the treatment. The number of mutants observed after infection with modified RNA was proportional to the duration of the treatment. The quantitative expression of this result enabled the authors to draw the conclusion that the oxidation of one of the 3000 nucleotides which go to make up the RNA will give rise to a mutant. The development of a mutation took place independently of the protein component of the virus. As well as the mutation described, which produced necroses in the leaves, a number of other mutants were also observed.

The connection of the mutagenic action of nitrous acid with the deamination of a number of bases has also been confirmed in experiments with phages (83, 146). These studies have been extended in many other laboratories and some very interesting results have now been obtained in this field (cf. the reviews 140, 135).

A specially interesting feature of this work is the possibility of establishing a connection between changes in the sequence of nitrogenous bases in RNA when the virus is caused to mutate by the action of nitrous acid, and changes in the sequence of the amino acids in the viral proteins synthesized under the influence of this RNA. The first evidence in this field has been obtained by Tsugita & Fraenkel-Conrat (145). In their experiment, the protein of the mutant obtained by treatment of the viral RNA alone showed a definite change in its amino acid composition. In it residues of proline, aspartic acid and threonine were replaced by residues of leucine, alanine and serine.

This line of work has been extended by the work of Wittmann (150a) which was communicated by him to the Fifth International Congress of Biochemistry. Using the action of nitrites and nitrous acid on the RNA of TMV to produce his mutations, the author decided to try to find out whether the sequence of amino acids in the protein molecules of the mutants is different from that in the normal virus and thus to establish a direct relationship between changes in the sequence of nucleotides and changes in the sequence of amino acids.

Wittmann did, in fact, succeed in finding changes in the proteins in about half the mutants. He tried to explain the absence of any detectable change in the proteins of some of the mutants by the suggestion that only part of the RNA of TMV contains information relative to the synthesis of the protein sub-units. Wittmann puts forward the idea that an alteration in the viral protein can only occur if an alteration in the nucleotide, whether spontaneous or chemically induced, takes place in that part of the molecule of RNA which contains information for the synthesis of protein. If the alteration in the nucleotide takes place outside the "gene", then a change will occur which will only affect the external characteristics of the infection but the viral protein will remain unchanged. Thus the "gene" is not made up of all the 6500 nucleotides which constitute the thread of RNA. Wittmann considers that it is possible to calculate the size of this "genic" region from the number of transformations of bases brought about in a single molecule of RNA by treatment with nitrite (which can be determined by experiment) and the number of alterations in the amino acids of the corresponding mutant. When the size of this region is known it will be possible to determine the average number of nucleotides which determine a single amino acid in the protein of TMV.

Recently Tsugita & Fraenkel-Conrat have published a detailed study along these lines (145a). Twenty-nine TMV mutants were studied for alterations in the structure of their proteins. In half of these mutants, no changes of protein structure were detected, while the remainder exhibited a very wide range of changes. In 12 mutants there were observed from one to three replacements of amino acids, while in three of the mutants, replacements were distributed over 16-17 positions. Sometimes two neighbouring amino acids were both replaced.

Analogous studies have been made with material other than viruses. Helinski & Yanofsky (89a) recently determined changes in the amino acid make-up of bacterial tryptophan synthetase (replacement of glutamic acid by glycine and of glycine by arginine) occasioned by mutations of the A-gene.

Very interesting researches into the process of mutation in phages and bacteria carried out by the groups associated with Benzer and Freese (64, 83, 132), Jacob (97), Wagner (149) and Hotchkiss (100) demonstrated clearly, not only the association of the mutations arising as a result of the action of ultraviolet irradiation and chemical treatment on DNA, but also the linearity of the information in DNA. This drew attention to the analogy between the development of mutations and the alterations in pairs of nucleotides.

The association of the mutagenic effect of ultraviolet irradiation with the metabolism of nucleic acids, and especially of DNA, has been demonstrated clearly in the striking studies of Haas and Doudney (89, 78) in which they used E. coli. The mutagenic effect of formaldehyde on Drosophila has also been found to be involved with the metabolism of nucleic acids (53).

A clear connection between mutational changes and DNA has also been shown to exist in phages in the very original researches of Pratt & Stent (126). By confining the incorporation of analogues to one of the polynucleotides of a double helix of DNA, the authors showed that in the descendants of this phage there was a separation of mutants from unchanged particles, corresponding to the separation of the two polynucleotides of the DNA.

An interesting scheme of the origin of point mutations, based on changes in the groups of nucleotides in DNA, has recently been put forward by Fresco & Alberts (85). These authors point out that errors in the reproduction of polynucleotides of DNA can be of three types, namely the substitution, removal or addition of a nucleotide. When such altered chains form a twin helix special loops must be formed at certain links (Fig. 58) but in later reproductions these loops disappear owing to the synthesis of complementary chains. As a result of such errors the chain of DNA may become longer or shorter.

In recent work in the same laboratory Doty and his colleagues (77) carried out a very interesting approximation to hybridization on a molecular level. These workers have devised a method for artificially separating the double helices of DNA into single threads by changes in temperature and then reconstructing the original molecules without any appreciable loss of biological activity. This method also enabled the authors to recombine polynucleotides of DNA from different bacterial sources.

Following up these experiments, two other workers from the same laboratory, Marmur & Schildkraut (110) gave a paper to the Fifth International Congress of Biochemistry describing results which enable this method to be used for the solution of some genetic problems. They found that "hybrid" molecules of DNA are only formed if the original specimens of DNA have been isolated from organisms which are closely

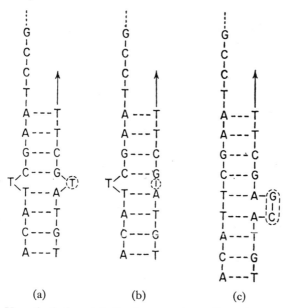

(a)        (b)        (c)

Fig. 58. Hypothetical model of point mutations. In each case the polynucleotide acting as the matrix is on the left. It may be noticed that, although the matrices are identical, the sequence of the newly formed chain is different in each case according to the type of "mistake" arising during replication. The "mistakes" in the growing chain are ringed with dotted lines.

related genetically and taxonomically and have DNA of similar nucleotide composition.

A very wide field of studies of the relationship between the genetic system of DNA and hereditary characteristics is the determination of the relationship between mutations and changes in particular enzymes. It is as a result of just such studies that a firm foundation has been laid for the concept of "one gene—one enzyme" which gives the concept of the "gene" reality in terms of the accurate findings of biochemistry.

In recent years many papers have been published, establishing a direct connection between an alteration in chromosomal DNA and an inheritable alteration in some enzyme or another (54, 58, 90, 87, 116, 73, 99, 143, 105, 100, 123, 129, 147, 136, 151). It is interesting to note that the synthesis of adaptive galactosidase in the cells of *E. coli* requires the presence of a special, specific form of DNA which is absent in cultures which are not capable of adaptation (120). As a result of these numerous studies a great many genetic effects have now been successfully associated with the actual properties of linear molecules of DNA (65). That linear extent of the DNA molecule which contains the information for a single protein has been named a "cistron". One molecule of DNA may consist of several cistrons.

Thus, even this short review shows quite clearly the special genetic part played by DNA, which is mediated by the presence of biological information in this polymer, correlated with the sequence of nucleotides in the polynucleotide chains.

## 6. The genetic role of DNA in the phenomena of transformation of bacteria

The processes of transformation of bacteria under the influence of purified preparations of DNA are the clearest and most direct illustration of the genetic functions of DNA. By now an extensive collection of facts, admitting of no doubts, has already been gathered together concerning the transforming effect of DNA and published in a number of reviews (46, 48, 91-95, 139, 153).

This being so, we shall only very briefly recall the main results of these researches. The transforming effect of DNA was first demonstrated in 1944 in the classical work of Avery, MacLeod & McCarty (57). These authors found that DNA extracted from capsulated penumococci of type S transformed non-capsulated cells of type R into capsulated ones, with a corresponding alteration in their immunological properties. The most significant features of this transformation were as follows:

1. The capsule of the newly altered cells always consisted of a polysaccharide of the same serologically specific type as that of the type from which the DNA had been isolated.

2. These transformations affected pneumococci of type R, in which spontaneous mutation into type S, or any other type, has never been observed.

3. The "smooth" cells which developed could be multiplied indefinitely, always giving rise only to cells of type S even without any further treatment with DNA; furthermore they, themselves, proceeded to form unlimited quantities of DNA with the same properties.

Thus, in this case, not only did the DNA lead to the production of a new property, but also to reproduction of itself. The transforming factor was later shown to be identical with the pure DNA itself in a very large series of studies which demonstrated clearly that small amounts of other contaminants such as peptides or polysaccharides had no activity.

The discovery of Avery et al. stimulated a number of attempts to bring about the transfer of other characteristics by means of DNA. Many of these attempts were unsuccessful, mainly because of the difficulty of introducing such polymers into undamaged cells, but there were a few successes. Many of the accessory features of the capsule were successfully transferred from organisms of one type to organisms of another in E. coli, Haemophilus influenzae and pneumococci. DNA could also transform colonies which were susceptible to a particular antibiotic into resistant colonies, and, what was specially important, it could endow the transformed cells with the ability to synthesize a number of new enzymes.

Genetical analysis of the transformational changes showed that they were identical with mutational variations and demonstrated the parallelism between the transforming agents and the hypothetical genes. It must, of course, be noted that DNA does not have any direct effect on the synthesis of the capsular polysaccharides but determines the formation within the cell of those enzyme systems which determine their synthesis. In other cases the transforming action of DNA was also associated with the formation of new enzymes or with an increase in the functioning of some enzymic system.

Many of the studies of transformation have been devoted to the question of finding out the minimum molecular weight of DNA required for the retention of transforming activity. As it usually exists, DNA has a molecular weight of between 1 and 6 million. However, by means of calculations, Guild & Defilippes (88) have managed to determine the transforming activity in preparations having a molecular weight of 300,000. However, in later work by Doty and his colleagues (107), on the same cases of transformation of the characteristics of resistance to antibiotics, the critical molecular weight below which there was a loss of activity was raised to 1,000,000.

The question of the minimum size of the genetic unit is important for the solution of a number of problems. According to the calculations of

Zamenhof (152) one bacterial cell only contains 250 polymeric molecules of DNA and many consider that this pool is too small for it to be justifiable to assume that each polymeric molecule of DNA determines one characteristic (or one protein synthesis). This has led to the development of a tendency to believe it possible that the information required for the production of several characteristics may be "recorded" on a single molecule of DNA. It has now been proved that this is perfectly possible. However, we must not forget that, in bacteria, many characteristics may be transferred from cell to cell, not only by means of DNA, but also by means of RNA as we have already discussed on pp. 313-7.

Among the recent achievements in the study of transformation we must mention the transfer of mutations by means of DNA (128) and the transformation of cells of E. coli as a result of the action of DNA isolated from bacteriophage (98).

In the last few years there have appeared papers dealing with the possibility of transferring certain genetic characteristics by means of DNA, not only in bacteria but also in highly organized creatures, namely ducks (49, 62, 63). When the experiments on ducks have been repeated, however, it has not been possible to confirm the facts of transformation (5, 142). Attempts to produce genetic transformation in Drosophila, the silkworm and the rat by the injection of DNA have also proved unsuccessful (2, 7). The method of carrying out these experiments is still not well enough worked out and it is hardly likely that a simple injection of DNA into the blood would bring about a genetic substitution in a particular locus.

## 7. Molecular mechanisms of cytoplasmic inheritance

The facts concerning cytoplasmic inheritance have now been constituted into a separate branch of genetics which does not contradict nuclear inheritance but fills out the general picture of the factors involved in the occurrence of the phenomenon of inheritance in living nature. Thorough analyses of the facts of cytoplasmic inheritance have recently been made in a number of papers (42, 52, 121, 66, 119). An interesting communication on the part played by cytoplasmic inheritance has been made by Danielli (75). He thinks that it is possible to distinguish two levels of genetic control. In the first case it is the synthesis of specific macromolecules which is controlled, and this synthesis is regulated by chromosomal inheritance. In the second case it is the organization of the macromolecules into physiological units which is controlled and this process may occur under the influence of cytoplasmic inheritance. Such a division of the transfer of information is, of course, quite possible. We have already seen, in the first chapter, that although proteins are not capable of self-reproduction, they can still serve as carriers of the biochemical information which regulates many processes. In fact it is just the particular sequence of amino acids which determines the development of the secondary structure of proteins and their ability to form particular complexes such as nucleoproteins and lipoproteins on which, in its turn, the organization of the intracellular structures depends. If we approach this phenomenon in the light of the theory of information, it becomes obvious that, although the sequence of amino acids has a directly functional significance in many cases, in others it is acting as a special method of recording information which will control the reproduction of the specific properties of considerably more complicated formations and complexes.

It is easy to include cytoplasmic inheritance, as a biological pheno-menon, in the scheme of the role of RNA and DNA in the phylogenetic development of life (p. 390). We usually run up against the facts of this form of inheritance when studying the transfer from one cell to another of systems which function continuously. The leucoplasts of plants, which are present in all plant cells, often multiply in the cells as auto-nomous systems and their transfer from generation to generation some-times affords an example of cytoplasmic inheritance. Many examples of cytoplasmic inheritance are to be met with among the infusoria and in a number of algae, the classical cases in this respect being *Chlamydomonas* and *Acetabularia*. Non-chromosomal inheritance in *Chlamydomonas* has recently been studied in detail by Sager (133). According to the results of Brawerman & Chargaff (69), however, the cells of unicellular algae contain an autonomous, specialized, genetic system controlling the de-velopment of chloroplasts from the colourless to the green state. Further-more, there is evidence that the chloroplasts of unicellular algae contain their "own" DNA (96).

It seems to us that the fundamental principle of contemporary ideas on inheritance does not consist in the allocation to DNA of exclusive governing influence over all genetic phenomena, but in the consideration of genetic phenomena from the point of view of the storage, reproduction and manifestation of information. Nucleic acids constitute the system best adapted to this function and the synthesis of proteins is impossible without nucleic acids. As for the localization of nucleic acids, both RNA and DNA, which function genetically, their organization into one or several systems, their actions in concert or individually, there is very great variety in living nature in this respect, and this fact does not in any way minimize the significance of DNA as a polymer which is specially "designed" for the *storage, reproduction and manifestation of the genetic information controlling individual development*.

An interesting attempt to classify the different ways in which in-formation is transferred in the organism has recently been made by Salser (134). By "information" he means the phenomenon of orderliness and he distinguishes a number of forms of information within the cell, hereditary and non-hereditary information, nuclear and extranuclear information, information associated with DNA and information not associated with DNA, constructional and non-constructional information etc.

The occurrence of cytoplasmic inheritance is in no sense a proof of the diffuse distribution of the "property" of heredity throughout the cell or the non-existence of the special genetic functions of DNA or anything of that sort. All the facts connected with it require to be examined in their own right and judged in terms of the generally accepted concepts of heredity. When there is cellular organization it would seem that, in many respects, chromosomal heredity is always of the same type. It is, however, an important characteristic of cytoplasmic heredity that it is different as between one species or form of organism and another. Furthermore cytoplasmic inheritance is displayed by certain more or less individual traits while chromosomal inheritance provides for the passing on as a whole of the comprehensive complex of features characteristic of the species.

In weighing up the facts concerning cytoplasmic heredity we must also take account of evolutionary mechanisms. It has already been mentioned that, from a genetic point of view, leucoplasts and chloroplasts are largely autonomous, having their own DNA. This is apparently not fortuitous and indeed there are many indications which lead some authors to suggest a symbiotic origin for green cells, the chloroplasts having originally evolved from some independent biological systems which were capable of self-reproduction.

The work of Brawerman & Chargaff (69) on the genetic autonomy of the chloroplasts of algae has recently been extended in a series of papers from their laboratory on the peculiarities of the synthesis of RNA in the chloroplasts. We have already discussed this work in Chapters V and XV. The authors have succeeded in showing that, in this case, the genetic autonomy of the chloroplasts is assured, not only by their having their own DNA, but also by their synthesis of their own informational RNA which is situated on the surface of the microsomes inside the plastids and there carries out the synthesis of the proteins specific to the plastids. Thus, in this case, the mechanisms of nuclear and cytoplasmic inheritance are identical.

Another case of cytoplasmic inheritance has been studied in detail by Rabinovitch & Plaut (127a) in amoebae, which they found to have special granules in their cytoplasm. These could be spun down in a centrifuge at 10,000 $g$ and contain DNA. In this case the presence of DNA was established by microradioautography after incorporation into the cytoplasm of thymidine labelled with tritium and by the sensitivity of the process of incorporation to the presence of desoxyribonuclease. This extranuclear DNA confers a certain autonomy on the cytoplasmic particles. The same cytoplasmic particles also contain RNA. It has been found that these particles can duplicate themselves and that the number of DNA-containing particles is correlated with the rate of growth of young amoebae. In young cells the number was about 54,000 per cell, but by the time they were ready to divide this number had fallen to 11,000. It was observed that after the removal of the nucleus from the amoebae there was a considerable increase in the number of such particles. It has been shown that the DNA-containing granules are genuine intracellular formations and not bacteria multiplying within the cells or the remains of bacteria which have been ingested and are being digested within their vacuoles. It was important to prove this point as the cytoplasm of amoebae usually contains organisms which "infect" them and this could give a false impression of the localization of DNA in the cytoplasm.

The part which the cytoplasmic granules discovered by Rabinovitch & Plaut play in the metabolism of the cell is still not clear but it would seem that they might be responsible for some cases of cytoplasmic inheritance in these unicellular creatures, in accordance with the same molecular mechanisms which perform the genetic work of the chromosomes of the nucleus. When all is said and done, even the chromosomes, that is to say all the particles located within the nucleus, in fact become cytoplasmic structures during mitosis when they "float" in the cytoplasm. The evolution of forms is also associated with the evolution of the number of the chromosomes. This is brought about by a number of processes such as the formation of polyploid forms, reduction, and the breaking up or linkage of chromosomes. It is, however, possible in some cases that the first rudimentary small chromosomes may develop from some sort of cytoplasmic DNA-containing particles while, on the other side, "fragments" of chromosomes may become adapted to cytoplasmic multiplication.

Extranuclear DNA is often found as a normal component of the cells of other objects and is commonly associated with some sort of autonomy or partial autonomy of the cytoplasmic organelles.

All cases of cytoplasmic inheritance must involve the existence of structures which can reproduce themselves and control the synthesis of proteins. So far there are only two types of such structures known to biochemistry, namely DNA and RNA, and there can therefore be no doubt that cytoplasmic and nuclear inheritance are both based on one and the same molecular systems.

A systematic exposition of the phenomena of cytoplasmic inheritance has recently appeared in a book by R. Hagemann (89b). In a preface to the Russian translation of this book the well-known Soviet geneticist B. L. Astaurov rightly remarks that inheritance and heredity in the general biological sense can only be realized through an intimate interaction between the nucleus and the cytoplasm. When this occurs, the cytoplasm plays a part in the transfer of hereditary properties. However, according to Astaurov, this does not mean that the parts played by the nucleus and the cytoplasm are of equal importance. The nucleus is a specialized system for the transfer of genetic information and serves as the chief factor controlling individual development, while the organization of the cytoplasm is directed towards the carrying out of other functions, namely metabolism, growth, interaction with the environment and regeneration of cytoplasmic structure, which is characterized by replication of individual components, and these are the functions it performs. It is typical of chromosomes that they have a unique structure and that they do not undergo effective regeneration when damaged.

## 8. On the constancy of the quantity of DNA in cells

Those wishing to demonstrate that DNA is not required for the storage of hereditary information very often make use of results which were first obtained in 1953 by Marshak, concerning the disappearance from the nuclei of the egg-cells of certain echinoderms at particular phases of development of the Feulgen reaction for desoxyribose which is characteristic of DNA (112, 113, 114). Similar results have also been obtained by other authors and have been interpreted as the "disappearance" of DNA (23, 30, 44). It should be noted that some stages of oogenesis are exceptional in this respect, as extensive results obtained on other objects show that the DNA content remains fairly constant whether calculated for the nucleus or for the chromosomes (cf. reviews 67, 68). Continuing his experiments on the "disappearance" of DNA from the egg-cells of echinoderms Marshak (111, 112) used more direct, quantitative analyses and in this way he found that it might not be a case of the "disappearance" of DNA but only of a diminution of the amount present, which was sometimes very considerable.

The question of whether the "disappearance" of DNA from the egg-cell could be proved solely by the absence of the Feulgen reaction has since been evaluated critically. In his review of the cytochemical studies of nucleic acids (143a) the well-known cytochemist Swift shows, from an analysis of the evidence obtained by means of the Feulgen reaction, that "... one should never conclude, on the basis of a negative Feulgen reaction, that DNA is 'absent' from a nucleus. Even where all steps have been properly followed, no perceptible colour can be seen in some nuclei, for example, in many mature oocytes. The DNA in such nuclei is obviously too dilute to produce a visible reaction." The work of a number of authors on the "disappearance" of DNA from the nuclei of egg-cells has been repeated by Ficq & Brachet (81a) using the same material and the same Feulgen reaction but it has not been confirmed. In making a detailed analysis of the reasons for these contradictions in his monograph, Brachet (67) criticized the methods of study used by Marshak and those others who confirmed his conclusions. In the microphotographs which Brachet adduced it was quite clear that what is changed in the markedly hypertrophic nucleus of the egg-cell is only the localization of the DNA, which is distributed around the periphery as though it were adjacent to the nuclear membrane and it is therefore hard to find. In this connection we must understand that, in general, owing to its enormously increased size (in the absence of any increase in the amount of DNA) and to its being laden with various sorts of storage inclusions, the egg-cell is the least favourable object in which to study the localization and amount of DNA, the more so if cytochemical methods are being used.

It must also be mentioned that, having made a detailed analysis of the numerous papers concerned with the DNA content of cells, Brachet came to the conclusion that it remains fairly constant (when calculated on the basis of the chromosomes). In the course of this he noted that this constancy is not absolute, that it only represents a tendency and that the facts that DNA shows a certain dynamism and that DNA is slowly renewed and resynthesized do not in any way preclude the possibility that this compound may have genetic functions.

The possibility that there may be some changes in the amount of DNA in the cells does not, in itself, mean that it has no genetic role. It is still unknown whether the whole of the stock of DNA in the nucleus carries hereditary information or whether there is a certain "reserve" of non-specific DNA in the nucleus.

It is quite obvious that far-reaching conclusions about the "disappearance" of DNA should not be drawn without a full check being made on the causes for the weak or negative Feulgen reaction and without the necessary chemical analyses and spectroscopic studies.

In this connection we must draw attention to the fact that "disappearance" of DNA may simply mean that it has changed its localization and gone out of the nucleus into the cytoplasm. It must be pointed out that Finamore & Volkin (80) actually found DNA in the cytoplasm of the eggs of amphibians.

Bell (60) has recently made a particularly detailed study of this possibility by following up the interactions between the nucleus and the cytoplasm during oogenesis in bracken (*Pteridium aquilinum* L.). In this case also the egg-cell lost its power of giving the colour reaction of the Feulgen test for a certain period. In order to study the fate of the DNA, this author used the considerably more objective method of autoradiography to record the incorporation of labelled thymidine into DNA. Quantitative estimation of the [³H]thymidine being incorporated into DNA in the cells, before and after the "disappearance" of the Feulgen reaction, showed quite clearly that no diminution in the amount of DNA took place. And what is more, it was found that the amount of DNA in the cell doubled. The material "incorporating" the thymidine was identified as DNA by means of desoxyribonuclease. It was, however, observed that, in the later stages of maturation of the ovum, the DNA migrated into the cytoplasm in polymeric form. The author calculated that this involved a 25-fold dilution of the DNA with cytoplasm and that the Feulgen reaction was therefore not sensitive enough. It is significant that the nuclei of the oocytes are also enlarged to 25 times the size of those of the somatic nuclei.

These results show quite clearly that far-reaching conclusions about the "disappearance" of DNA without any influence on heredity and about the later resynthesis of DNA with no direct connection with the pre-existing stock of genetic information are no more probable than, for example, the "disappearance" of a printing press and its subsequent assembly from the newspapers and magazines which it had printed. In reaching this conclusion, however, we cannot omit to mention an attempt to give a different explanation of the "disappearance" of DNA which has been made quite recently (in 1961) by the Leningrad cytochemist Makarov in the proceedings of a conference on genetics.

"Studies carried out on the developing ova of the *Ascaris* of horses and on the oocytes of amphibia", writes Professor Makarov, "lead to the conclusion that DNA is dispersed during the synthesis of the acid nuclear proteins and that these latter, in their turn, are used in the reproduction of DNA. If this conclusion is confirmed by later researches it follows that the hereditary substrate of the cell is not either unitary or stable. The reproduction of DNA occurs, not only by self-reproduction, but by more complicated means, with the participation of the intracellular proteins. The hereditary information must be handed on, like the Olympic torch, from DNA to protein and from protein to DNA, corresponding with their molecular organization and with the distribution of particular atomic groupings along the chain of the DNA or the protein. Alterations from without may be reflected in the course of the synthetic processes which we are discussing and thus lead to a disturbance of hereditary characteristics" (24).

Makarov takes the view that the "disappearance" of nucleic acids is reversible because the proteins can recreate the DNA with all its unique sequence of nucleotides, corresponding to its own information. However, this is quite without foundation.

At present there is much controversy about the DNA-content of cells. This is associated with differences in methods for measuring the DNA in the tissues. Only one form of determination is correct from a genetic point of view and that is the DNA content of the chromosome. The DNA content calculated for the nucleus as a whole cannot be of any value for the solution of genetic problems as it can vary on account of the polyploidy of somatic cells, which is very common in animal tissues. In a later chapter we shall discuss several papers in which it has been shown that the formation of tetraploid, octoploid and multinuclear cells in various tissues is a typical feature of increasing age.

From the genetic point of view the determination of DNA on the basis of the dry or wet weight of tissues, on single cells or on organs, etc. are all unsatisfactory.

It may also be supposed that the migration of a portion of DNA into the cytoplasm need not necessarily mean a disturbance of the unique structure of the chromomere in which the particular sequence of genes is also a self-reproducing product of evolutionary development.

## 9. The importance of studies of molecular mechanisms of heredity in the development of the theoretical and practical problems of science

The discovery of the molecular mechanisms of heredity, the genetic code in the form of different combinations of nucleotides in a molecule of DNA, the linkage between the problem of the synthesis of proteins and that of the synthesis of amino acids and, finally, the artificial synthesis of DNA, RNA and proteins have been the most important scientific achievements of the past decade. They have brought biology to the forefront of scientific progress and have created the necessary conditions for the formation of well-founded concepts of the mechanism of variation and for its control for the benefit of mankind.

The material discussed in this book, along with much other material, indicated that the problem of the origin of the specific biological polymers cannot be understood, studied or explained without the use of the principles of the theory of information and without recognizing that cells contain a particular biochemical system which is specialized for the retention, accumulation and reproduction of the information. The development of living nature cannot be understood in materialist terms without recognizing the need to preserve all the "achievements" of earlier species to ensure the supremacy of the greater complexity of biological individuals in spite of mortality and of reproduction through single cells. The development of life without such a system for the concentration of such preformed information would have been just as impossible as the development of mankind without any means whereby one generation could hand on the results of its mental and manual work to the next. In human society the most usual means for this transfer of information is by the written word, in books where all the evidence is spelled out by means of the letters of the alphabet. The existence of books does not prevent information from being passed on from one generation to the next in other ways (direct transfer of experience by discussion or demonstration, by samples or by tools, etc.). The same is true for living nature. Here the main vehicle for the transfer of hereditary information is the system of molecules of DNA organized into chromosomes in which each item of information is "spelled out" in the four-lettered nucleotide alphabet. However, this does not exclude other means of passing information from one generation to the next, such as extranuclear inheritance, the autonomous reproduction of RNA, etc.

It is not every scientist who recognizes the general biological significance and some persist in interpreting the problem of heredity in other ways (3, 8, 12, 13, 14, 14a, 15, 15a, 17, 28, 19, 20, 21, 22, 39, 40, 41, 43a), assigning the main role in the maintenance of species specificity, not to a molecular, internal, specialized system for the reproduction of the characteristic traits, but to external conditions. Nevertheless, the importance of the changes which have been introduced into biology is becoming more and more evident. The tremendous significance of changes in the external environment for the origin of variability and in determining the direction of evolutionary selection admits no question but that the maintenance of the dynamic constancy of complicated intracellular structures is the function of a special biological system. We have already pointed out that the genetic system ensures the constancy of the organism in a special sense; it ensures the constancy, within certain limits, of the form of individual development. It is the controlling system for the individual development of the individuals of the successive generations of living matter. The existence of such a system is a prerequisite for evolution. In its reproduction any complicated individual only passes on to its progeny an insignificantly small part of its body in the form of the ovum or sperm. Under exactly the same environmental conditions completely different individuals are formed from thousands or hundreds of thousands of ova while at the same time often reproducing the smallest details of the parent individuals, even under different conditions. It is thus clear that some definite structures in the sex cells must be responsible for such a strictly determinate type of development and that they must contain a structural "programme of development". The discovery of the genetic code of DNA has made clear the nature of this programme and the mechanism of its realization.

According to the rough calculations of L. Pauling the number of different proteins in the human body must amount to at least 100,000 polypeptides each having, on the average, 150-200 amino acid residues. The nucleus of the human cell contains 46 chromosomes, 23 from each parent. In each chromosome there are several thousand molecules of DNA and in each molecule of DNA several tens of thousands of nucleotides. If this is correct the total length of all the molecules of DNA of a haploid sex cell will be no less than 5-6,000,000,000 nucleotides which, using a trinucleotide genetic code, is enough, with a lot to spare, to record the information for the synthesis of all the proteins of the living body.

It must not be thought, however, that our modern ideas of heredity and ontogenesis are simply the result of discoveries which have been made in the past few years. On the contrary, the discoveries of the past few years have brilliantly confirmed and particularized conclusions which geneticists had arrived at far earlier in the form of hypotheses which were logically derived from the classical experiments which laid the foundations of their science.

As is well known, the very concept of genes as discrete carriers of hereditary traits arose without any reference to biochemistry, as a theoretical explanation of the laws governing the distribution of parental characteristics on crossing, in the course of the so-called segregation in hybrids which was first studied about a hundred years ago in experiments on the hybridization of plants performed by the Czech scientist G. Mendel. The characteristics of the parent varieties, such as red or white flowers, were mixed in the hybrids to give pink flowers. In the second generation they appeared again. The hybrids were segregated and white and red flowers appeared again in this generation and they did so in a strict numerical ratio to one another, namely 1:1, while the number of red and white flowers together was equal to the number of pink ones. A similar sort of segregation was found to occur in respect of many other characteristics of plants. In many cases in several species of plants the hybrids showed the characteristics of only one of the parents, for example the red colour, which was dominant. In such a case, when segregation occurred, the ratio of red to white flowers was, on the average, 3:1. These rules were later confirmed in thousands of experiments on almost all those living objects which reproduce sexually. These facts show, with inescapable logic, that the external characteristics correspond with some factors present in the sex cells. When the sex cells unite these factors do not vanish, just as atoms do not vanish with the formation of molecules. When the sex cells of the hybrid are formed they again pass into different sex cells but in such a way that the factors, later called genes, move independently of one another, and this corresponds with an independent distribution of the external characteristics. Such an explanation of the facts was, of course, only a hypothesis, but a hypothesis with every chance of future confirmation. Three important discoveries made about 50 years ago made it possible to develop the hypothesis. These were the establishment that, for any given species, the number and form of the chromosomes is constant; the observation of the exact parallelism between the distribution of characteristics during the segregation of hybrids and the distribution of the chromosomes between the cells in the so-called reduction division of the sex cells and the discovery that there is a difference in regard to one chromosome as between male and female individuals and the fact that there is an equal distribution of sexes in the offspring of a cross corresponding with the distribution of the sex chromosomes. It was later found that the characteristics of the parents are not distributed quite independently among their offspring but go in groups and the number of these groups in any species (groups of linked characteristics) is exactly the same as the number of chromosomes in the species concerned. These facts gave rise to the chromosomal theory of heredity which postulated that the factors controlling heredity—the genes—were localized in the chromosomes. At that time nobody had seen the genes, nobody had yet isolated them, but the chromosomal theory of inheritance which postulated their existence, based as it was on an abstract analysis of these and many other factors, was just as incontrovertible as, say, the supposition of the existence of atoms and molecules even before physicists had found the means of observing them, or the discovery of enzymes long before biochemists had succeeded in isolating them in the pure state or studying their structure.

In spite of this, the mechanism of action of the genes remained the subject of speculations and hypotheses, though this aspect of the problem was the most important one. The insolubility of the problem of the mechanism of action of the genes formed the basis for continual discussions, for the course of the investigations pursued by the geneticists was very complicated and the temptation to find some very simple solution to the problem of heredity was great.

At just this period attempts to find simplified explanations of heredity were beginning to be made. Sometimes these had not even the least connection with the solution of the problem of the material succession of generations, that of the mechanism of morphogenesis and so on, but dealt exclusively with the influence of uniform (heredity) and varied (variation) factors in the external environment.

The discovery, in 1953, of the complementary double-helical structure of DNA and the explanation of the mechanism of its self-reproduction were very important and caused a rapid change in this situation. This provided an explanation of the process of self-reproduction, the occurrence of which was an essential theoretical requirement for the existence of genetically significant structures (150).

The second decisive factor in the development of modern genetic concepts was the application of the principles of the theory of information to biological phenomena. This was first undertaken by Schrödinger in 1946. At first this development was purely abstract but after the discovery of the mechanism of the self-reproduction of DNA the ideas were carried over to nucleic acids which formed a mathematical model, fulfilling quite satisfactorily all the requirements which the theory demanded of any possible carrier of genetic information. By this time the work of E. Chargaff and A. N. Belozerskiĭ had established the species specificity of DNA and this provided the basis for the suggestion that the sequence of the nucleotides might be the biochemical means of recording hereditary information. Thus there arose the problem of the biochemical genetic code which, as we saw in Chapter XV, was first worked out abstractly and then received an extensive experimental basis from studies of the mechanisms of the biosynthesis of proteins.

We have already seen that the study of the synthesis of proteins laid emphasis on the principle of the use of matrices and it was established that it was ribonucleic acids which acted as the matrices, the linear sequence of the nucleotides in these compounds determining the linear sequence of the amino acids in the protein molecules.

The synthesis of RNA under the control of DNA was the last stage to remain hypothetical, though it had been presupposed for a long time: there was much indirect evidence for the occurrence of this process but it was not till 1960-1 that direct methods and approaches to its demonstration were devised. All of these achievements have been described fully enough in various chapters of this book and we shall now turn to a general discussion of them in order to emphasize that, in the long run, studies of extremely varied problems and different phenomena have led to the solution of the problem of the mechanism of heredity and have given rise to a concept of heredity which has ceased to be the province of genetics alone and has become a central concept of many of the biological sciences. Even 10-12 years ago the proof of genetic ideas was complicated, involving the study of the splitting off of characteristics by segregation, genetic mapping, the study of the nature of segregation and so forth. The understanding of the material and a competent scientific, theoretical analysis of it was possible for only the very narrow circle of genetic specialists who had at their disposal the rather bulky apparatus required for cytogenetics. As the scope of genetics has expanded through its merging with biochemistry in the field of the synthesis of proteins, the incorporation of the theory of information, the study of the problems of heredity in the simplest systems of viruses, phages and bacteria, the methods and indicators have been extended rapidly and it has become possible to understand the problems of genetics in terms of biochemistry, biophysics, radiobiology, embryology and so on. These subjects have thus become integrated in the process of establishing the genetic significance of the nucleic acids and especially DNA.

By this time the great practical importance of genetics had already become apparent. The discovery of the intimate mechanism for the transfer of "hereditary information" could not but affect the solution of many of the practical problems. To take only the medical aspects of genetics, it must be said that the solution of the problem of the role of DNA in heredity has had a tremendous effect on many of the everyday problems of medicine. We will only mention a few of them here. The science of virology has been almost entirely reconstructed in the last few years as a result of the advances in genetics. The process of controlling the variation of viruses by means of their RNA or DNA has been firmly established as an achievement of science. The study of the genetic nature of many illnesses has developed and methods have arisen for establishing the accurate correlation of particular pathological syndromes with chromosomal anomalies of the sexual or somatic cells. The exact genetic and chromosomal mechanism of dozens of very common pathological syndromes has been established and methods have been devised for their diagnosis. The problem of the development and regression of tumours has been placed on a new footing by the idea of the genetic control of the synthesis of proteins. The question of damage to the system of DNA is an important one in assessing the biological effect of irradiation. The whole of the manufacture of antibiotics is based on the production of special mutant strains of bacteria or moulds in which there is a hereditary change in the cycle DNA → RNA → enzyme, and so on. This is only a beginning. The future prospects for this genetic control are very extensive and very fruitful.

## 10. Different forms of molecular variation (hereditary and non-hereditary variation)

We have already given a short discussion of two different types of genetic molecular variation, namely mutation and transformation. Each of these is associated in some way with DNA, in the first case with its alteration and in the second with its transfer from one species into another. In nature, however, there is a far greater variety of forms of molecular alterations and it would be useful to classify them and investigate their biological significance. We shall return to this matter in a later chapter in connection with the relative importance of different forms of variation in the generation and accumulation of changes in the living body as it grows older.

In the first place, variation should be classified in two main categories, variations in the sex cells and variation in the somatic cells, because the genetic significance is quite different in the two cases. This difference applies to any form of variation. Changes affecting the DNA and the chromosomes, i.e. mutations, when they occur in the sex cells, are transferred to the next generation which will develop from these cells, whereas changes in the somatic cells are not hereditary except in the case of vegetative reproduction (bud mutations, mutations of the seeds of plants, etc.). In many cases, however, the mutation of somatic cells cannot be differentiated from the mutation of sex cells because there is a period in the life cycle of almost all species during which the sex cells which will form the next generation do not yet exist and all the cells of the body are somatic. In animals these periods usually occur in the very earliest stages of embryogenesis while in plants there is often a very long period of vegetative growth which is greater than the period of reproductive development and sexual differentiation begins at the growing points and the reproductive cells are formed there. In such a case any mutation of the somatic cells of the growing point may turn out to be one of the hereditary group if the affected cells take part in the formation of the reproductive organs. As to the mass of the remaining somatic cells, changes in them cannot be directly incorporated into the material genetic chain which links together the different generations. Mutations of the sex cells are, then, new, acquired characteristics. This category also includes mutations of somatic cells from which the sex cells are later derived by differentiation. However, mutations of specialized cells, or any other changes in them, cannot be included in the genetic reproductive cycle. They are only of local significance. This is quite natural, as in the living bodies of complicated organisms the somatic cells may be millions or thousands of millions of times as many as the reproductive ones, especially in females, but if there were any channel for conveying genetic information from them to the sex cells, then if there was a fairly high mutation rate among the somatic cells, no conservation of species specificity would be possible because all the characteristics and all the genes of the sex cells would be changed over a very short period of time.

The difference between somatic and genetic mutation also extends to their frequency, which is two to three orders greater for somatic than for genetic ones. In the specialized cells only a very small part of the genome is active and only a small fraction of one per cent of all the molecules carries out the synthesis of informational RNA. For this reason many of the changes in their DNA, both in its finer and its coarser structure, cannot affect the carrying out of the specific functions of the cell. In the reproductive cells any such change is of genetic importance and alters some process or characteristic. Furthermore, the frequency of genetic mutations cannot exceed a certain definite level at which a certain number of the sex cells do not contain any changes so that each haploid genome is the same as the genomes of the parental individual.

An increase in the frequency of mutation of the somatic cells leads to a shortening of the life span. An increase in the frequency of mutation of the sex cells above a certain limit may lead to the gradual extinction of the whole species and the extinction of some species in the history of the flora and fauna of our planet must certainly have been due to this phenomenon. It is a very simple law that operates here. The frequency of mutation of a single gene is about $10^{-6}$ in each generation (the figure for man is not known). Let us assume that the individual has $10^5$ genes. In this case one in ten of the sex cells has a new mutation which it passes on to the offspring. Medical geneticists believe that this is a reasonable figure for man. However, mutations are disorderly and the vast majority of them are harmful. Only one mutation in 1000 or 10,000 can be considered as useful, increasing the adaptation of the individual to its environment or improving it. If, then, the frequency of mutation is increased twentyfold each sex cell will contain two altered genes. Although these changes might be insignificant so long as they were concealed by the heterogeneity of the affected individual, progressive selection would cease under these circumstances because the great majority of favourable mutations would not occur alone but would be associated with harmful ones. With the passage of the generations the species as a whole would begin to accumulate harmful mutations and this would lead to the ageing of the species as such and finally to its extinction. This is a very important aspect of the problem of variation and is of vital significance for mankind because it draws attention to the need for constant control over all mutagenic factors, both internal and environmental.

PB 2F

One of the main mutagenic factors acting on all living things is radiation. Any increase in the external or internal background radiation increases the rate of mutation. This fact was known to geneticists more than 30 years ago. The first artificial mutations were, in fact, obtained by means of radiation. Different forms of radiation have different mutagenic effects and different species of animals and plants have different sensitivities to irradiation. According to the calculations of N. P. Dubinin as set out in his book *Molecular genetics and the effect of radiation on heredity*\* the dose which would double the mutation rate in man is 10 roentgen over the whole lifetime. By the age of 30 a person living in the northern hemisphere has received 3 roentgen from natural sources. If the background radiation were to be raised to, say, 50 or more times its present level and kept there by some means, this might lead to the gradual extinction of the human race as a result of hereditary anomalies. However, even a very slight increase in the background radiation cannot occur without serious consequences. According to Dubinin's calculations an increase in the background radiation of only 30-40% of its natural value will lead to an increase of 800,000 hereditary anomalies for every 200 million born in each succeeding generation. For mankind as a whole this would mean about 10 million people in each generation would be handicapped by hereditary anomalies, not to speak of the increase in the incidence of those diseases which are associated with irradiation, such as cancer and leukaemia.

These simple and irrefutable calculations show how great is the importance of the study of the mechanism of molecular genetic variation for the future of all mankind and how important is action aimed at the stoppage of the testing of atomic weapons and their complete abolition. It is also certain that even the peaceful use of atomic energy cannot be undertaken without genetic safeguards and checks.

As we have seen, mutations of the somatic and sex cells arising under the influence of the same factors and having the same nature are, nevertheless of different genetic significance. The analysis of other aspects of molecular variation demands just as separate treatment. From the genetic point of view, it is very important to classify forms of molecular variation according to the factors causing them.

The commonest way of classifying these factors is into external, exogenous or environmental and internal or endogenous. Let us take the case of variations in DNA as an example, as they are the most important from a genetic point of view. Each of the factors inducing changes in the information-carrying DNA can occur both externally and internally. Radiation, the effects of which are of obvious genetic importance, acts on the organism as a factor from the external environment as ionizing radiation of the type of cosmic rays or as ultraviolet radiation. It can also

---

\* *Molekulyarnaya genetika i deĭstvie izluchenii na nasledstvennost'*. Moscow: Gos. Izd. Lit. v Oblasti Atomn. Nauki i Tekhn. (1963).

be endogenous in the form of micro-emissions from biochemical reactions. Analogues of nucleotides, which cause disturbances of the codes of nucleic acids, can enter the organism in the food and can also arise endogenously as a result of errors in enzymic reactions. The latter is constantly happening.

Keeping the organism under strictly constant conditions does not stop its variation and does not put an end to evolution, for it is not only the environment but also the individual itself which gives rise to selection, especially that directed towards progressive development. Changes in the external environment (temperature, lighting, food, etc.) affect the direction of selection because they give rise to alteration in the relative values of particular molecular changes. Variation can exist on its own, even when environmental conditions are kept constant because it is not only changes in the environment relative to the organism which are important in inducing variation, but also changes in the organism relative to the environment. Any endogenous change in the organism, any spontaneous mutation, will alter the relation between the organism and the environment and create the conditions for selection leading to evolution; the possibilities of the ensuing changes in the conditions of interaction with the environment are of endless variety and very stable.

Ordinary changes in the environment are, as a rule, episodic, transient or rhythmic. Adaptation to the conditions of the environment usually amounts to the manifestation of biological rhythms, daily, seasonal or annual, for most of the processes carried out in the organism. Changes relative to the environment occurring as a result of mutation are stable and it may be that they play a predominant part in the formation of progressive variation.

No less important is the classification of changes according to the structures affected, for this usually determines the biological significance of the change. As an indication of what we mean we may point to the following kinds of genetically significant forms of change: non-migration of chromosomes, the breaking of nucleotide chains, point mutations (changes in the sequence of nucleotides), cytoplasmic mutations (mutation of the plastids or other autonomous structures), transformation and transduction. Non-genetic molecular changes are considerably more widely distributed and their classification and characterization requires the analysis of an extensive range of facts. Some of these changes which are irreversible will be discussed in the last chapter of this book.

Non-migration of the chromosomes as a result of some disturbance of mitosis is a very common type of change in both somatic and sex cells. It has been calculated that about 4-5% of all human embryos die at various stages of embryogenesis as a result of such an anomaly of the somatic cells having occurred during the first mitoses of the zygote. However, non-migration of the sex chromosomes is generally not lethal and is associated with a very stable number of sexual anomalies in man which are associated with an increase in the number of chromosomes in all the cells of the body ($XXY = 47$ chromosomes, $XXXY = 48$ chromosomes, etc.). Such anomalies usually lead to sterility and are eliminated by selection.

Polyploidy is very common in nature, especially among plants, and a considerable percentage of contemporary species of plants are polyploids in which all the cells have, not 2 elementary genomes but 4, 6, 8, 10 or 12 and even more; historically speaking, some of the genomes belong to different species of plants and have become merged in the general genome through polyploidy.

In animals, polyploidy of the sex cells usually leads to the production of non-viable individuals, but in populations of somatic cells of many organs the accumulation of polyploid forms occurs commonly.

Rupture and reassembly of chromosomes is usually associated with anomalies but this form of change is of evolutionary importance as a way in which the sequence of genes in the chromosomes may be changed and may give rise to such phenomena as the reduction of characteristics, which are very important for evolution. Quite a large number of hereditary anomalies associated with rupture of the chromosomes have now been studied.

The commonest and most genetically significant changes, however, are point mutations associated with changes in the nucleotide sequence. In this case the change in the old information leads to the appearance of new information and the proportion of useful changes must be greater in this form of mutation. Furthermore, this is the group of mutations in which useful mutations, however rare they may be, can be isolated and this makes them very valuable for the process of evolution.

The identification of such mutations in animals is still difficult but recently, in connection with the study of the microheterogeneity of proteins and the so-called isozymes (which are hereditarily stable alternative forms of a single type of enzyme), biochemists have elucidated the heredity of this kind of variation, showing that any individual is not a stable type of organization but a transitional form. This also applies to cytoplasmic mutations. In this case we meet with the problem of endogenous selection, which is very widespread in living systems. In the case of cytoplasmic mutations selection may proceed at the level of the organelles, all of which, together, constitute the population.

Changes of the nature of transformation have already been discussed above and it is obvious that they are mainly of experimental origin. However, the extremely interesting genetic phenomenon of transduction may be one of the normal means of reconstitution and completion of the genome. The phenomenon of transduction is at present being studied on systems of bacterial cells infected with a virus (phage) but it is not impossible that these processes may also occur in more highly organized things. Transduction is what happens when bacterial viruses (phages) multiply within the cells of the bacteria, forming particles which incorporate part of the DNA of the host cells and then carry these over into other strains, incorporating them in their genomes. Transduction is usually accompanied by the phenomenon of lysogenesis, in which the DNA of the phage is incorporated in the bacterial chromosome and is reproduced in it synchronously with the bacterial DNA by the system DNA → DNA → DNA, that is to say, without the formation of the informational RNA and protein of the phage. In this case the DNA of the phage becomes, as it were, a genetic element of the bacterial cell. It becomes combined with the DNA of the bacterial cell and passes into the inactive state along with the other inactive cistrons of the cell, such as those which subserve the formation of adaptive enzymes. Such cistrons are normally reproduced according to the scheme DNA → DNA → DNA and some sort of inducer is needed to direct their reproduction along the functional pathway DNA → RNA → protein. This is also true for lysogenes. Phage DNA (prophage) can regain its infectivity and bring about the formation of structured particles and the destruction of the cell when subjected to the influence of ultraviolet irradiation.

Such phages which have been formed after prolonged reproduction with the genome of the host can absorb part of the DNA of the host and transmit it to the new cells which they infect. If the phage again reverts to the latent form after the new infection the fragments of DNA from the original cells transferred by it become part of the active genome of the new host and bring about its transformation.

Theoretically the phenomenon of transduction cannot be excluded in other types of virus infection; phenomena similar to lysogeny in bacterial cells are known to occur in virus infections of more highly organized systems. For example, something similar takes place in the case of viral forms of neoplasms. Cases are known in which viruses are transferred from one generation to the next via the reproductive cells.

We have only examined all these questions of genetic variation very shortly and schematically here because a more exhaustive and systematic presentation of the evidence would have been outside the scope of this book. It was, however, necessary to bring them out in a general way because they are important for an understanding of certain aspects of the problem of protein synthesis and especially the problem of ontogenesis and ageing. In recent years many comprehensive reviews and monographs on molecular genetics have been published in the Russian language alone and those readers who want a more detailed acquaintance with molecular and biochemical genetics can supplement the book by Dubinin which we have already mentioned by using the excellent monograph by S. E. Bresler entitled *Introduction to molecular biology.**

It has now become certain that molecular genetics will, to a great extent, give mankind such control over the processes of living as will

* *vredenie v molekulyarniyu biologiyu.* Moscow: Izd. AN SSSR (1963).

enable him to cause the life of the organism to progress in any direction. This does not only mean that he will be able to create new forms and species, but also that he will be able to direct functional processes, because even the most complicated of them such as memory is, according to present evidence, associated with the system of nucleic acids.

## Conclusion

This chapter has contained too much in the way of general discussion and too little factual material. It must, however, be remembered that a large part of the factual evidence relating to the connection between the synthesis of proteins and nucleic acids and the phenomenon of heredity has already been discussed in earlier chapters. All we wanted to do in this chapter was to state our own opinion about some controversial questions in genetics, which, on account of its rapid development, have descended to the molecular level and have impinged upon the problem of the reproduction of the specificity of proteins and nucleic acids.

In concluding our short survey of some present-day problems in biochemical genetics we should like to make some observations of a methodological nature.

In the first place we must state that the methodological approach to the theoretical analysis of any particular problem is an extremely important factor. The dialectical-materialist analysis of natural phenomena is a tremendous achievement of Soviet science and when this approach is used for the analysis of particular biological phenomena it represents, as it were, the common essence of the qualities inherent in matter in its various stages of development with which particular theoretical concepts should correspond, in a philosophical sense, regardless of their endless variety, corresponding with the endless variety of natural phenomena.

Our understanding of a particular phenomenon depends on the level of development of science, and the replacement of theoretical concepts is primarily an indication of development in science. It is a characteristic feature of all theories, including the theory of the gene. The fact that we now reject earlier and rather oversimplified hypotheses about genes does not mean that there is a crisis in genetics. It is a natural process caused by development, primarily that of biochemistry, and is an indication of the progress of genetics. Such occurrences are a characteristic feature of any science and very many of the theoretical concepts of 20-30 years ago are now only of historical interest to scientists. This is especially true of the theory of heredity, which is the fundamental property ensuring the continuity of the development of living matter.

It seems to us quite obvious that a really scientific theory of heredity, corresponding with the level of development of our knowledge of nature, has only started to be built up in recent years and that this slow process is taking place as a result of movement forward on a very wide front. It is based, in the first place, on the extremely fast development which is now taking place in the chemistry and biochemistry of proteins and nucleic acids, on the remarkable progress in virology and bacteriology, on the differentiation of genetic problems and on the study of variability. We have only taken the first steps towards an understanding of the biochemical nature of heredity and, clearly, much of that which seems certain at present will later be subject to criticism. The extremely complicated question of the determination of the morphogenetic reactions, which seems to be an inseparable part of the theory of heredity, and that of the mechanism of the mutual dependence of the phenomena of development and "hereditary information", as well as many others, will remain, for a long time, the subjects of theoretical discussions and experimental researches. This does not mean that we cannot, even now, use the achievements of genetics and biology for practical purposes, but, on the other hand, we are not justified in asserting that science is already in possession of all the secrets of development and can easily and simply alter it in the required direction. The practical achievements of modern biology are undoubted and are largely due to the increasing use of theory, as opposed to pure empiricism. This becomes greater as theoretical problems develop. However, these achievements must not be allowed to slow down the all-round development of theoretical biology, which will certainly show us how simplified, imperfect and unproductive are our present-day concepts, which may now seem to us to represent the limit beyond which it is impossible to progress.

Biology now stands on the same plane as the exact sciences, that is to say, it is beginning to have to discover the mechanisms of biological phenomena, not only within the realm of general biology, but also in regard to the precise clearing up of the physical and chemical laws of the processes of life. The discovery of these mechanisms is an extremely complicated task but it is the very basis of the solution of the problem of causing the development of organisms to proceed in any desired direction.

## REFERENCES

1. ALIKHANYAN, S. I. (1961). *Zhur. vsesoyuz. khim. Obshchestva im. D.I. Mendeleeva*, **6**, 285.
2. ASTAUROV, B. L., BEDNYAKOVA, T. A., GINTSBURG, G. I., ZBARSKIĬ, I. B. & RAMENSKAYA, G. P. (1960). *Doklady Akad. Nauk S.S.S.R.* **134**, 449.
3. BELEN'KIĬ, N. G. (1948). In *O polozhenii v biologicheskoĭ nauke: Doklady sessii VASKhNIL* (ed. V. N. Stoletov *et al.*), p. 73. Moscow: Sel'-khozgiz.
4. BELOZERSKIĬ, A. N. & SPIRIN, A. S. (1960). *Izvest. Akad. Nauk S.S.S.R.*, Ser. biol., **25**, No. *1*, p. 64.
5. BOGATYREVA, S. A., ZNAMENSKAYA, M. P., KUSHNER, KH.F., MOISEEVA, I. G. & TOLOKONNIKOVA, E. V. (1961). *Doklady Akad. Nauk S.S.S.R.* **136**, 1213.

6. VOL'KENSHTEĬN, M. V. (1958). *Izvest. Akad. Nauk S.S.S.R.*, *Ser. biol.* 23, No. *1*, p. 3.

6a. GAVRILOV, V. YU & ZOGRAF, YU. N. (1962). *Uspekhi fiz. Nauk*, 77, 597.

7. GERSHENZON, S. M., KOK, I. P., SAMOSH, L. V., TURKEVICH, N. M. & FEDOROVA, N. YA. (1960). *Zhur. obshcheĭ Biol.* 21, 387.

8. GLUSHCHENKO, I. E. (1948). In *O polozhenii v biologicheskoĭ nauke : Doklady sessii VASKhNIL* (ed. V. N. Stoletov *et al.*), p. 181. Moscow: Sel'-khozgiz.

8a. DVORKIN, G. A. (1962). *Zhur. obshcheĭ Biol.* 23, 216.

9. DVORYANKIN, F. A. (1959). *Voprosy Filosofii*, No. *12*, p. 128.

10. DUBININ, N. P. (1956). *Biofizika*, 1, 677.

11. — (1957). *Byul. Mosk. Obshchestva Ispytateleĭ Prirody*, *Otd. biol.* 62, No. *2*, p. 5.

11a. — (1964). *Problems of radiation genetics* (trans. G. H. Beale). Edinburgh: Oliver & Boyd.

12. ZHUKOV-VEREZHNIKOV, N. N. & PEKHOV, A. P. (1958). *Voprosy Filosofii*, No. *6*, p. 127.

13. ZHUKOV-VEREZHNIKOV, N. N., PEKHOV, A. P. & LYSOGOROV, N. V. (1958). *Uspekhi sovremennoĭ Biol.* 45, 234.

14, IVANOV, N. D. (1960). *Darvinizm i teoriya nasledstvennosti*. Moscow: Izd. AN SSSR.

14a. KAGANOV, V. M. (1962). *Zhur. obshcheĭ Biol.* 23, No. 3.

15. KOSTRYUKOVA, K. YU. (1948). In *O polozhenii v biologicheskoĭ nauke : Doklady sessii VASKhNIL* (ed. V. N. Stoletov *et al.*), p. 269. Moscow: Sel'khozgiz.

15a. — (1962). *Nauchnye Doklady vyssheĭ Shkoly, Seriya filosovskaya*, No. *1*.

16. KREMYANSKIĬ, V. I. (1960). *Voprosy Filosofii*, No. *9*, p. 132.

17. LYSENKO, T. D. (1948). In *O polozhenii v biologicheskoĭ nauke : Doklady sessii VASKhNIL* (ed. V. N. Stoletov *et al.*), p. 7. Moscow: Sel'khozgiz.

18. — (1958). *Voprosy Filosofii*, No. *2*, p. 102.

19. — (1959). *Voprosy Filosofii*, No. *10*, p. 103.

20. — (1959). *Agrobiologiya*, No. *4*, p. 484.

21. — (1960). *Agrobiologiya*, No. *3*, p. 323.

22. — (1961). *Agrobiologiya*, No. *1*, p. 4.

23. MAKAROV, P. V. (1956). *Uspekhi sovremennoĭ Biol.* 41, 3.

24. — (1961). *Tezisy dokladov konferentsii po eksperimental'noĭ genetike* (ed. M. E. Lobashov), p. 104. Leningrad: Izd. Leningradsk. Univ.

25. MEDVEDEV, ZH. A. (1960). *Izvest. Timiryazev. sel'skokhoz. Akad.* No. *1*, p. 103.

26. — (1961). *Zhur. vsesoyuz. khim. Obshchestva im. D.I. Mendeleeva*, 6, 268.

27. MENITSKIĬ, D. P. (1957). *Biofizika*, 2, 2.

28. MITIN, M. B. (1948). In *O polozhenii v biologicheskoĭ nauke : Doklady sessii VASKhNIL* (ed. V. N. Stoletov *et al.*), p. 221. Moscow: Sel'khozgiz.

29. NIKOL'SKII, G. V. (1958). *Voprosy Filosofii*, No. 7, p. 113.

30. NUZHDIN, N. I. (1958). *Voprosy Filosofii*, No. *8*, p. 82.

31. — (1958). *Agrobiologiya*, No. *1*, p. 3.

32. — (1959). *Uspekhi sovremennoĭ Biol.* 48, 245.

33. OLENOV, YU. M. (1959). *Tsitologiya*, 1, 527.

33a. — (1961). *Nekotorye problemy evolyutsionnoĭ genetiki i Darvinizma*. Moscow: Izd. AN SSSR.

34. OPARIN, A. I. (1957). *The origin of life on the Earth* (trans. A. Synge). Edinburgh: Oliver & Boyd.

35. PANCHEV, N. (1959). *Voprosy Filosofii*, No. *10*, p. 106.

36. PASYNSKIĬ, A. G. (1958). *Izvest. Akad. Nauk S.S.S.R., Ser. biol.*, No. *6*, p. 641.
37. PETROV, D. F. (1958). *Voprosy Filosofii*, No. 7, p. 102.
38. PLATONOV, G. V. (1958). *Voprosy Filosofii*, No. *11*, p. 129.
39. PREZENT, I. I. (1948). In *O polozhenii v biologicheskoĭ nauke: Doklady sessii VASKhNIL* (ed. V. N. Stoletov *et al.*), p. 486. Moscow: Sel'-khozgiz.
40. — (1959). *Agrobiologiya*, No. *4*, p. 489.
41. — (1960). *Agrobiologiya*, No. *4*, p. 624.
42. RYZHKOV, V. L. (1959). *Zhur. obshcheĭ Biol.* **20**, 16.
43. SEMENOV, N. N. (1959). *Voprosy Filosofii*, No. *10*, p. 95.
43a. SISAKYAN, N. M. (1954). *Biokhimiya obmena veshchestv.* Moscow: Izd. AN SSSR.
44. SKVIRSKAYA, E. B. & CHEPINOGA, O. P. (1953). *Ukrain. biokhim. Zhur.* **25**, 117.
45. SUKHOV, K. S. (1957). *Agrobiologiya*, No. *1*, p. 36.
46. TIMAKOV, V. D. & SKAVRONSKAYA, A. G. (1958). *Vestnik Akad. med. Nauk S.S.S.R.* No. *4*, p. 12.
47. TURBIN, N. V. (1958). *Voprosy Filosofii*, No. *2*, p. 112.
48. KHESIN, R. B. (1958). *Uspekhi sovremennoĭ Biol.* **46**, 113.
49. CHEPINOGA, O. P., NOVIKOVA, B. G., LYUBARSKAYA, M. A. & KHILOBOK, I. Yu. (1960). *Tezisy Dokladov 3-go vsesoyuznogo Soveshchaniya Embriologov*, p. 179. Moscow: Izd. AN SSSR.
50. SHMAL'GAUZEN, I. I. (1958). *Doklady Akad. Nauk S.S.S.R.* **120**, 187.
51. ENGEL'GARDT, V. A. (1961). *Zhur. vsesoyuz. khim. Obshchestva im. D.I. Mendeleeva*, **6**, 244.
52. EFRUSSI, B. S. (1959). *Izvest. Akad. Nauk S.S.S.R., Ser. biol.* No. *3*, 359.
53. ALDERSON, T. (1960). *Nature (Lond.)*, **185**, 904.
54. ALEXANDER, P. & LETT, J. T. (1960). *Biochem. Pharmacol.* **4**, 34.
55. ALFERT, M. (1957). In *A symposium on the chemical basis of heredity* (ed. W. D. McElroy & B. Glass), p. 186. Baltimore, Md.: Johns Hopkins Press.
56. ANFINSEN, C. B. (1959). *The molecular basis of evolution.* New York: Wiley.
57. AVERY, O. T., MACLEOD, C. M. & MCCARTY, M. (1944). *J. exptl. Med.* **79**, 137.
58. BAGLIONI, C. (1960). *Heredity*, **15**, 87.
59. BEADLE, G. W. (1957). In *A symposium on the chemical basis of heredity* (ed. W. D. McElroy & B. Glass), p. 3. Baltimore, Md.: Johns Hopkins Press.
60. BELL, P. R. (1961). *Proc. roy. Soc.* **153B**, 421
61. BELOZERSKIĬ, A. N. & SPIRIN, A. S. (1960). In *Nucleic acids* (ed. E. Chargaff & J. N. Davidson), Vol. **3**, p. 147. New York: Academic Press.
62. BENOIT, J., LEROY, P., VENDRELY, C. & VENDRELY, R. (1957). *Compt. rend. Acad. Sci., Paris*, **244**, 2320.
63. BENOIT, J., LEROY, P., VENDRELY, R. & VENDRELY, C. (1960). *Biochem. Pharmacol.* **4**, 181.
64. BENZER, S. (1957). In *A symposium on the chemical basis of heredity* (ed. W. D. McElroy & B. Glass), p. 70. Baltimore, Md.: Johns Hopkins Press.
65. — (1961). *Proc. natl. Acad. Sci. U.S.* **47**, 403.
66. BOYD, J. S. K. (1956). *Biol. Revs. Cambridge phil. Soc.* **31**, 71.
67. BRACHET, J. (1957). *Biochemical cytology.* New York: Academic Press.
68. — (1960). *The biochemistry of development.* New York: Pergamon.

69. BRAWERMAN, G. & CHARGAFF, E. (1960). *Biochim. biophys. Acta*, **37**, 221.
70. BRENNER, S., BARNETT, L., CRICK, F. H. C. & ORGEL, A. (1961). *J. mol. Biol.* **3**, 121.
71. BUTLER, J. A. V. & DAVISON, P. F. (1957). *Advanc. Enzymol.* **18**, 161.
72. CHANTRENNE, H. (1961). *The biosynthesis of proteins.* Oxford: Pergamon Press.
73. CRAWFORD, I. P. & YANOFSKY, C. (1959). *Proc. natl. Acad. Sci. U.S.* **45**, 1280.
73a. CRICK, F. H. C., BARNETT, L., BRENNER, S. & WATTS-TOBIN, R. J. (1961). *Nature (Lond.)*, **192**, 1227.
74. CRICK, F. H. C., GRIFFITH, J. S. & ORGEL, L. E. (1957). *Proc. natl. Acad. Sci. U.S.* **43**, 416.
75. DANIELLI, J. F. (1956). *Nature (Lond.)*, **178**, 214.
76. DEMEREC, M., LAHR, E. L., BALBINDER, E., MIYAKE, T., ISHIDSU, J., MIZOBUCHI, K. & MAHLER, B. (1960). *Carnegie Inst. Washington Yearbook*, **59**, 426.
77. DOTY, P., MARMUR, J., EIGNER, J. & SCHILDKRAUT, C. (1960). *Proc. natl. Acad. Sci. U.S.* **46**, 461.
78. DOUDNEY, C. O. & HAAS, F. L. (1959). *Proc. natl. Acad. Sci. U.S.* **45**, 709.
79. DOUNCE, A. L. (1959). *Ann. N.Y. Acad. Sci.* **81**, 794.
79a. DRYSDALE, R. B. & PEACOCKE, A. R. (1961). *Biol. Rev. Cambridge phil. Soc.* **36**, 537.
80. FINAMORE, F. J. & VOLKIN, E. (1958). *Exptl. Cell Research*, **15**, 405.
81. FINCHAM, J. R. S. (1960). *Advanc. Enzymol.* **22**, 1.
81a. FICQ, A. & BRACHET, J. (1956). *Exptl. Cell Research*, **11**, 135.
82. FREESE, E. (1959). *J. mol. Biol.* **1**, 87.
83. — (1959). *Brookhaven Symposia in Biol.* No. *12*, p. 63.
84. — (1961). *Proc. V int. Congr. Biochem., Moscow*, **1**, 204.
85. FRESCO, J. R. & ALBERTS, B. M. (1960). *Proc. natl. Acad. Sci. U.S.* **46**, 311.
86. FUERST, C. R. & STENT, G. S. (1956). *J. gen. Physiol.* **39**, 687.
87. GLANVILLE, E. V. & DEMEREC, M. (1960). *Genetics*, **45**, 1359.
88. GUILD, W. R. & DEFILIPPES, F. M. (1957). *Biochim. biophys. Acta*, **26**, 241.
89. HAAS, F. L. & DOUDNEY, C. O. (1960). *Nature (Lond.)*, **185**, 637.
89a. HELINSKI, D. R. & YANOFSKY, C. (1962). *Proc. natl. Acad. Sci. U.S.* **48**, 173.
89b. HAGEMANN, R. (1959). *Plasmatische Vererbung.* Wittenberg-Lutherstadt: A. Ziemsen Verlag.
90. HOROWITZ, N. H., FLING, M., MACLEOD, H. L. & SUEOKA, N. (1960). *J. mol. Biol.* **2**, 96.
91. HOTCHKISS, R. D. (1957). In *A symposium on the chemical basis of heredity* (ed. W. D. McElroy & B. Glass), p. 321. Baltimore, Md.: Johns Hopkins Press.
92. — (1959). *Can. Cancer Conf.* **3**, 3.
93. HOTCHKISS, R. D. & EVANS, A. H. (1958). *Cold Spring Harbor Symposia quant. Biol.* **23**, 85.
94. HOTCHKISS, R. D. & OTTOLENGHI, E. (1961). *Proc. V int. Congr. Biochem., Moscow*, **3**, 215.
95. HUTCHINSON, F. (1961). *Science*, **134**, 533.
96. IWAMURA, T. (1960). *Biochim. biophys. Acta*, **42**, 161.
97. JACOB, F. & WOLLMAN, E. L. (1957). In *A symposium on the chemical basis of heredity* (ed. W. D. McElroy & B. Glass), p. 468. Baltimore, Md.: Johns Hopkins Press.
97a. JUKES, T. H. (1962). *Biochem. biophys. Research Commun.* **7**, 281.
98. KAISER, A. D. & HOGNESS, D. S. (1960). *J. mol. Biol.* **2**, 392.

99. KRIMM, S. (1960). *J. mol. Biol.* **2**, 247.
100. LACKS, S. & HOTCHKISS, R. D. (1960). *Biochim. biophys. Acta*, **39**, 508.
101. LANNI, F. (1960). *Perspectives in Biol. Med.* **3**, 418.
102. LEDERBERG, J. (1959). *Angew. Chem.* **71**, 473.
103. LESLIE, I. (1961). *Nature (Lond.)*, **189**, 260.
104. LEVINTHAL, C. (1959). *Revs. modern Phys.* **31**, 249.
104a. LEVINTHAL, C. & DAVISON, P. F. (1961). *Ann. Rev. Biochem.* **30**, 641.
105. LEVINTHAL, C., GAREN, A. & ROTHMAN, F. (1961). *Proc. V int. Congr. Biochem., Moscow*, **9**, 116.
106. LINDEGREN, C. C. (1961). *Nature (Lond.)*, **189**, 959.
107. LITT, M., MARMUR, J., EPHRUSSI-TAYLOR, H. & DOTY, P. (1958). *Proc. natl. Acad. Sci. U.S.* **44**, 144.
108. LOVELESS, A. (1960). *Biochem. Pharmacol.* **4**, 29.
109. MANDELSTAM, J. (1960). *Bact. Revs.* **24**, 289.
110. MARMUR, J. & SCHILDKRAUT, C. L. (1961). *Proc. V int. Congr. Biochem., Moscow*, **1**, 232.
111. MARSHAK, A. (1958). *Symposia Soc. exptl. Biol.* **12**, 205.
112. MARSHAK, A. & MARSHAK, C. (1953). *Exptl. Cell Res.* **5**, 288.
113. — (1954). *Nature (Lond.)*, **174**, 919.
114. — (1955). *Exptl. Cell Res.* **8**, 126.
115. McFALL, E., PARDEE, A. B. & STENT, G. S. (1958). *Biochim. biophys. Acta*, **27**, 282.
116. MOHLER, W. C. & SUSKIND, S. R. (1960). *Biochim. biophys. Acta*, **43**, 288.
117. MOROWITZ, H. J. & CLEVERDON, R. C. (1959). *Biochim. biophys. Acta*, **34**, 579.
117a. MOROWITZ, H. J. & TOURTELLOTTE, M. E. (1962). *Scient. Amer.* Vol. **206**, No. *3*, p. 117.
118. MUNDRY, K. W. & GIERER, A. (1958). *Z. Vererbungslehre*, **89**, 614.
119. NANNEY, D. L. (1957). In *A symposium on the chemical basis of heredity* (ed. W. D. McElroy & B. Glass), p. 134. Baltimore, Md.: Johns Hopkins Press.
120. NOVELLI, G. D., EISENSTADT, J. M. & KAMEYAMA, T. (1961). *Science*, **133**, 1369.
120a. OCHOA, S. (1963). *Fed. Proc.* **22**, 62.
121. OKAZAKI, T. & OKAZAKI, R. (1959). *Biochim. biophys. Acta*, **35**, 434.
122. ORÓ, J. (1961). *Fed. Proc.* **20**, 352.
123. PAIGEN, K. (1960). *Acta Unio intern. contra Cancrum*, **16**, 1032.
124. PARDEE, A. B. (1959). *Exptl. Cell Res.* Suppl. **6**, p. 142.
125. PERUTZ, M. F. (1958). *Endeavour*, **17**, 190.
126. PRATT, D. & STENT, G. S. (1959). *Proc. natl. Acad. Sci. U.S.* **45**, 1507.
127. QUASTLER, H. (1959). *Lab. Invest.* **8**, 480.
127a. RABINOVITCH, M. & PLAUT, W. (1962). *J. Cell Biol.* **15**, 525, 535.
128. RAVIN, A. W. (1960). *Genetics*, **45**, 1386.
129. RILEY, M., PARDEE, A. B., JACOB, F. & MONOD, J. (1960). *J. mol. Biol.* **2**, 216.
130. ROBERTS, R. (1962). *Proc. natl. Acad. Sci. U.S.* **48**, 1245.
131. ROZEN, R. (1959). *Bull. math. Biophys.* **21**, No. *1*, p. 71.
132. RUDNER, R. (1960). *Biochem. biophys. Research Commun.* **3**, 275.
133. SAGER, R. (1960). *Science*, **132**, 1459.
134. SALSER, W. (1961). *Perspectives Biol. Med.* **4**, 177.
135. SCHUSTER, H., GIERER, A. & MUNDRY, K. W. (1960). *Abhandl. deut. Akad. Wiss. Berlin, Kl. Med.* p. 76.
136. SCHWARTZ, D. (1960). *Proc. natl. Acad. Sci. U.S.* **46**, 1210.
137. SINSHEIMER, R. L. (1959). *J. mol. Biol.* **1**, 218.

138. SINSHEIMER, R. L. (1960). *Ann. Rev. Biochem.* **29**, 503.

138a. SPEYER, J. F., LENGYEL, P., BASILIO, C. & OCHOA, S. (1962). *Proc. natl. Acad. Sci. U.S.* **48**, 441.

139. SPIZIZEN, J. (1959). *Fed. Proc.* **18**, 957.

140. STAEHELIN, M. (1959). *Experientia*, **15**, 413.

141. STRAUSS, B. S. (1960). *An outline of chemical genetics.* Philadelphia: W. B. Saunders Co.

142. SVOBODA, J. & HAŠKOVÁ, V. (1959). *Folia biol. (Prague)*, **5**, 402.

142a. SUEOKA, N. (1961). *Proc. natl. Acad. Sci. U.S.* **47**, 1141.

143. SUSKIND, S. R. & BONNER, D. M. (1960). *Biochim. biophys. Acta*, **43**, 173.

143a. SWIFT, H. (1955). In *The nucleic acids* (ed. E. Chargaff & J. N. Davidson), Vol. *2*, p. 51. New York: Academic Press.

144. TATUM, E. L. (1961). *Proc. V int. Congr. Biochem., Moscow*, **3**, 178.

145. TSUGITA, A. & FRAENKEL-CONRAT, H. (1960). *Proc. natl. Acad. Sci. U.S.* **46**, 636.

145a. — (1962). *J. mol. Biol.* **4**, 73.

146. VIELMETTER, W. & WIEDER, C. M. (1959). *Z. Naturforsch.* **14B**, 312.

147. WAGNER, R. P. & BERGQUIST, A. (1960). *Genetics*, **45**, 1375.

148. WAGNER, R. P. & MITCHELL, H. K. (1955). *Genetics and metabolism.* New York: Wiley.

149. WAGNER, R. P., SOMERS, C. E. & BERGQUIST, A. (1960). *Proc. natl. Acad. Sci. U.S.* **46**, 708.

150. WATSON, J. D. & CRICK, F. H. C. (1953). *Nature (Lond.)*, **171**, 737.

150a. WITTMANN, H. G. (1961). *Proc. V int. Congr. Biochem., Moscow*, **1**, 240.

150b. WOESE, C. R. (1961). *Biochem. biophys. Research Commun.* **5**, 88.

151. YANOFSKY, C. (1960). *Bact. Revs.* **24**, 221.

151a. YČAS, M. (1961). *J. theoret. Biol.* **1**, 244.

152. ZAMENHOF, S. (1956). *Progr. Biophys. biophys. Chem.* **6**, 85.

153. — (1957). In *A symposium on the chemical basis of heredity* (ed. W. D. McElroy & B. Glass), p. 351. Baltimore, Md.: Johns Hopkins Press.

154. ZUBAY, G. & QUASTLER, H. (1962). *Proc. natl. Acad. Sci. U.S.* **48**, 461.

# CHAPTER XVIII

# MORPHOGENETIC CHANGES IN THE BIOCHEMICAL SYSTEMS FOR THE SYNTHESIS OF PROTEINS IN THE EARLIEST STAGES OF ONTOGENESIS AND THE MECHANISM FOR CARRYING OUT THE GENETIC "PROGRAMME OF DEVELOPMENT"

*Introduction*

Surveys of the literature on the morphogenetic variability in a number of proteins have already been made in a review by Dorfman (8) and in Brachet's new book *The biochemistry of development*, which gives a comprehensive survey of the numerous questions involved in the biochemistry of embryonic development (53; see also 25a). For this reason our own review of the subject will be confined to those findings which are most closely associated logically with the problems to which this book is devoted and which reveal fragments of the biochemical mechanism underlying morphogenetic processes.

Among morphogenetic changes in proteins we include changes in these compounds which occur in the first active period of development of the organism (before sexual maturity) and which differ qualitatively from the simple changes of growth. The nature of morphogenetic changes may easily be made clear by the following example. The fertilized ovum does not contain haemoglobin or actin. The formation of these proteins during later development is a "morphogenetic" process which alters the protein composition of the developing system. This is brought about by the appearance, at some particular time, of new synthetic systems which are, however, formed under the control of the mechanisms of inheritance (see previous chapter). Once these systems have been formed, however, they do not remain unchanged but later undergo spontaneous alteration associated with growth. We shall discuss the phenomenon of the "origin" of proteins during ontogenesis and the phenomenon of their further alteration during growth separately, as the nature of these phenomena, the laws governing them and our biological knowledge of them are very different in the two cases.

## 1. Morphogenetic changes in proteins in plants

The individual development of higher animals follows, in almost all cases, the same sequence of phases, so that active embryonic and post-embryonic morphogenesis is succeeded by a fairly prolonged period of ageing of the organism which has been formed and completed. The morphogenetic process differs in its nature from the process of ageing in that it is essentially a process of manifestation of the hereditary potentialities of the organism, which are specifically determined in the "programme of development" of the species and are fixed in the form of the particular biochemical information carried in the structure of the nucleic acids (DNA and RNA). Although the process of ageing constantly accompanies that of individual development, nevertheless, during the period of active morphogenesis, it is, more or less, masked by the complex of morphological, physiological and biochemical changes which are forming the organism as a single integrated system. It is only the ending of the period of morphogenesis which unmasks the process of ageing, which is essentially destructive in nature, and enables the research worker to study it in its "pure form".

The relationship between morphogenetic development and ageing in higher plants is considerably more complicated and varied and this makes it more difficult to analyse the factual results which have been obtained from the study of different objects.

In many plants (especially annual (monocarpal) species) the morphogenetic process dominates that of ageing and the end of the morphogenetic reactions (the fertilization and ripening of the seed) is usually accompanied by the death of all the vegetative organs, the products of their breakdown being mobilized for the growth and development of the seed. A different picture is seen among perennial (polycarpal) species and especially woody plants, which undergo a gradual ageing process which has a definite general effect on the morphogenetic reactions (the reproductive process) as they occur at different stages of growth.

Finally, there is yet another extremely important specific peculiarity of plants which is the considerable degree of regional autonomy (asynchronism) of the ageing and morphogenesis of the different organs, the leaves, roots, stems, fruits, etc. and the ease with which they may be replaced so that one and the same plant almost always has dying, ageing, mature, young and embryonic organs and buds in process of formation. This means that it is often possible to study morphogenetic changes of organs, not only in plants of different ages but in plants of the same age.

In this section we shall deal with the general concepts of ontogenesis in plants, and with the analysis of many facts concerned with the biochemical alteration of different organs of plants in connection with their phase of development. All these matters have been dealt with fairly fully in several reviews and monographs (4, 5, 20, 21, 13, 32, 42). We shall only give a few examples which indicate that during the ontogenesis of plants there are definite reconstitutions of the proteins which form the basis of their protoplasm.

The morphogenetic alterations of the proteins of plants taking place during ontogenesis have not been studied nearly enough. If we do not count the indirect evidence of the work of Pashevich (31), who observed a change in the amino acid composition of wheat during vernalization, or that of Konarev (18) who has followed up the connection between changes in the isoelectric point of proteins and the stage of differentiation of the part of the growing point of the plant from which they were obtained, there only remains the work of Sisakyan and colleagues (35, 36) in which there is an attempt to study the ontogenetic changes in the composition of the proteins of the plastids by electrophoresis. The results which they obtained indicate that the protein structures of the plastids are dynamic, but very little is yet known as to the properties and the functional role of the protein components which they identified on their electropherograms. Several results (33, 37) indicate that during ontogenesis in plants various oxidative systems may be substituted for one another.

An interesting observation on the connection between the process of differentiation of plant cells and alteration of the composition of their proteins has recently been made by Wright (123). By using very sensitive immunological methods he found definite changes in the proteins of the coleoptiles of wheat at the time when they passed from the meristematic to the differentiated state.

There can be no doubt that the reasons for the morphogenetic alteration of the proteins are primarily connected with some sort of changes in the conditions under which they are synthesized, such as a redistribution of the rates of synthesis of the various proteins which function in the tissues of the plant.

An interesting explanation of the reasons for the redistribution of the ability to synthesize different kinds of proteins has recently been put forward by Nichiporovich and his colleagues (1, 25, 96). Assuming that photoperiodic induction is based on some qualitative reorganization of the proteins during the induction period, he analysed the possible causes of such a reorganization. In doing so he based himself on the observations made in his own laboratory, which showed that $^{14}C$ is incorporated into the various amino acids in a different way both qualitatively and quantitatively in the dark and in the light. On this basis Nichiporovich suggested that the actual synthesis of proteins by the incorporation of amino acids occurs in different ways under different lighting conditions.

It must be noted that when synthesis occurs in the dark it occurs in a different place, for in the dark the activity of the plastids is less than that of the microsomes.

Changes in the content of the free amino acids connected with photoperiodic induction have also been found by Metzner (90) and

Madan (85).  The former associates a characteristic change in the amino acid composition of the apical leaves with this induction.

It may be suggested that changes of this sort in the proteins and also, one must suppose, in the systems which synthesize them, may take place without any direct intervention by the specialized system of inheritance and may be a "concession" by this system which provides for a better ecological adaptation of the plant to its environment.

Some results have recently been published concerning the connection between RNA and the processes of morphogenesis in plants (19, 38) but the authors of these papers have only studied the quantitative changes in the RNA and have not concerned themselves with its nucleotide composition.

Some very interesting results, concerning changes in the microsomes of plant cells occurring at the time when they are going over from the meristematic to the differentiated state, have recently been published in the form of a preliminary communication by Loening (84).  He studied the microsomal fractions of successive segments of the tips of the roots of peas, these fractions being at different stages of differentiation.  Preparative centrifugation of the microsomal fraction at 140,000 $g$ led to its separation into a "heavy" and a "light" layer.  These microsomes were biochemically different.  After they had been treated with RNA-ase the "heavy" microsomes were found to contain two RNA components while the "light" ones contained five.  The "light" particles were also found to contain a greater amount of proteins which were not soluble in phosphate buffer.  During the processes of cellular differentiation (passing upwards from the point of growth) there was a regular change in the proportions of "heavy" and "light" microsomes.  While the "light" microsomes constituted only 10-20% of all the microsomes in the meristem at the tip of the root, in the more differentiated cells they increased to form from 75-90% of all the microsomes.  The author believes that these changes result from changes in protein synthesis occurring during differentiation.

## 2. Morphogenetic changes in the proteins of animals

Brachet (53) has devoted a special section of his monograph to the question of the synthesis of new proteins in the course of embryonic differentiation.  There is, therefore, no special need for this book to deal with all the relevant facts.  It will be enough to restrict ourselves to a few characteristic examples and to the most recently discovered facts in order to give a general idea of the nature of this process.

Although the material concerned with morphogenetic changes in proteins in the animal organism is more extensive, it too is very fragmentary.  The proteins of muscle have been studied most from the point of view of their ontogenetic origins (26). Interesting results have been obtained, especially from the study of the synthesis of myosin in embryonic muscles. The literature contains an indication that myosin has a complex structure, made up of two components (41, 48, 67). In view of this it is possible that it is only the structure of the components which is genetically determined while the formation of the complex molecule is a secondary process. In this connection we must take note of the work of Robinson (103) who compared the appearance of adenosine triphosphatase activity of the myosin of the chick embryo with the kinetics of this activity in the sarcoplasm of the muscle. In this way he found that the growth of adenosine triphosphatase activity in the myosin fibrils is closely correlated with a lowering of the adenosine triphosphatase of the sarcoplasm. Robinson therefore suggests that embryonic muscle contains a water-soluble precursor of myosin which can aggregate and thus turn into a particulate protofibrillary structure. It is possible that this does not happen by simple aggregation but is a more complicated process, comprising the formation of a complex between the protein which shows ATP-ase activity and several other proteins. In this connection it is interesting to note that, according to Kasavina (14, 15), in some animals, such as the chicken and the guinea-pig, the contractile reaction of actomyosin involving ATP occurs suddenly during embryogenesis.

It is also quite possible that the proteins of the muscle fibres are differently constituted at different stages of embryogenesis and that the final composition of the protein is arrived at as a result of several fairly complicated transformations. This is suggested by the fact that a special protein called metamyosin has been observed in the muscles of embryonic rabbits and sheep. Using electrophoresis, Ivanov and his colleagues (11, 12) have also found that the fractional composition of the proteins of the proactomyosin complex of the embryo rabbit differs from that in the adult in containing an additional fraction. The actual nature of the adenosine triphosphatase also changes during embryonic development. This has been discovered by comparison of the action of the cations $Ca^{++}$ and $Mg^{++}$ on the adenosine triphosphatase of muscles at various stages of embryogenesis in mice (118).

Although it is incomplete, this evidence, nevertheless, suggests that the formation of the muscle protein which is characteristic of any given species consists of a complex of complicated biochemical transformations. This development results in the production of the complex muscular proteins which are characteristic of the species in question and, in particular, of the parents of the individual. This inevitably suggests that there must be some sort of strict determination of the development.

We now have a lot of material concerning the way in which haemoglobins develop during ontogenesis and the characteristic changes in their composition during this process (70, 68, 71a, 72, 104 and others). There is, however, only a limited amount of evidence concerning the mechanism whereby the haemoglobin originates in the first place or even the way in which the embryonic haemoglobin is replaced by the postembryonic haemoglobin. These processes seem to be associated with the morphogenetic processes in the blood-forming organs and are under genetic control. The sequences of the amino acid residues in some chains of foetal and adult haemoglobins were also different (109a).

During embryogenesis there is also a change in the form of the myoglobin (from the foetal to the adult form), but both of these forms have a similar amino acid composition (117).

Evidence is now being obtained to show that regular changes in the composition and immunological properties of blood proteins occur in embryos and during the processes of metamorphosis (6, 43, 71, 102, 110, 113).

A good example of the formation of new proteins during ontogenesis is the formation of collagen from procollagen, which has been studied in detail by Orekhovich and his colleagues (27, 28, 29). From this work it has been concluded that collagen, which occurs in large amounts in skin and connective tissue, is formed in the organism from a special precursor, procollagen, which is similar in general type to collagen but has a different amino acid composition and a different rate of self-renewal. In the course of a detailed investigation of the proteins of the skin several protein fractions have been isolated which are intermediate between procollagen and collagen in their rate of self-renewal and amino acid composition. Orekhovich suggests that similar processes may be at work in the formation of other extracellular proteins such as elastin, keratin, etc.

The formation of collagen in embryonic life certainly takes place by several stages, for it has now been established that the structure of this protein is extremely complicated. Tustanovskiĭ and colleagues (30, 40) have followed the formation of collagen during embryogenesis and have concluded that the collagen fibrils are made of several components (including mucopolysaccharides as well as proteins) and, according to their observations, procollagen is a component of collagen. Several other workers have also shown that collagen has several components, including polysaccharides (7, 9, 47, 97). Recently Edds (60) has shown that typical collagen only appears at a certain stage of the development of the frog embryo. Lowther and colleagues (84a) have recently published a detailed study of embryonic histogenesis of collagen.

In recent years evidence has been obtained suggesting that the biochemical genetic cycle of collagen synthesis may really be even more complicated and may include the formation of another fibrillary protein, elastin, from the collagen (54, 75, 76). In this connection the ideas of Gulyĭ are of considerable interest (2, 3). He believes that the manifestation of the biological properties of many proteins is due to some secondary changes in the type of complex formation which happen to them during ontogenesis after they have been formed. In particular, he brings forward experimental evidence to show that the enzymic phosphohexokinase activity of muscles does not belong to any one individual but to a complex between two protein fractions.

Pelc (99) has studied the development of keratin, which is another metaplasmatic protein. He concluded that the histone of desoxynucleoproteins plays a direct part in the formation of this protein.

The liver is an organ which has a great variety of biochemical functions and its protein composition undergoes particularly marked variation during morphogenesis. This can easily be demonstrated qualitatively by immunoelectrophoresis (56, 57). Recent studies of this kind on extracts of the livers of developing chick embryos using antiserum to the proteins of adult fowls showed that, as early as the sixth day, six different antigens had appeared, giving special precipitation bands. The number of bands increases, reaching 16 in the adult fowl. The enzymic system of the liver also changes substantially during morphogenesis. Nemeth (93, 95) has shown that glucose-6-phosphatase and tryptophan peroxidase, which are present in the liver of adult animals, are practically absent at birth and their activity only begins to increase sharply after birth. This sort of morphogenetic appearance of enzymes has been demonstrated in the case of phenylalanine hydroxylase, tyrosine oxidase and uridinediphosphoglucurone transferase (82a, 77).

In recent work Nemeth (94) has tried to show the mechanism controlling the change in the tryptophan peroxidase activity of the developing liver of mammals. In doing this he was testing two hypotheses, the adaptive and the genetic. According to the adaptive hypothesis, a particular enzymic activity may originate during embryogenesis as a result of induction by the substrate. Until the substrate of the particular enzyme appears there is no enzyme, but with the appearance of the substrate, e.g. tryptophan, the synthesis of the enzyme is induced by a mechanism which is similar in principle to that which controls the synthesis of adaptive enzymes (cf. Chap. VIII). Indirect evidence for the existence of such a mechanism is provided by the work of Knox (81), who showed that the injection of L-tryptophan into adult rats leads to a marked increase in tryptophan peroxidase activity.

Nemeth decided to carry out such injections at different periods in the development of guinea-pigs and to compare the amounts of induction in embryonic, new-born and adult animals. In this way he found that the injection of L-tryptophan did not have any effect on the tryptophan peroxidase activity of the livers of embryos or new-born animals but that induction only began to occur one day after birth and appeared very suddenly. The effect of the adrenocortical hormones on the activity of this enzyme also only appeared one day after birth.

Thus, this work shows that not only do particular enzymes appear, or get synthesized at a particular stage of development, *but the actual ability to induce the appearance of an enzyme itself also arises at particular stages.*

It may be supposed that many of the conditions required for adaptive synthesis do not exist in the ova of mammals but come into being or manifest themselves as a result of morphogenetic development. Endogenous induction is, however, undoubtedly one of the most important factors in morphogenetic development (86, 124).

In summing up our short review of the morphogenetic variability of proteins we must admit that it would be certainly very interesting to compare them with the morphogenetic variations of the nucleic acids. A series of investigations has been made in which DNA and RNA were studied during ontogenesis, but, in almost every case, the studies were made on the basis of analyses of these compounds as a whole and not of analyses of fractions, and this makes it difficult to demonstrate any changes (44, 65, 87, 100). Only recently Leïkina, Tongur et al. (22) compared the nucleoproteins of the DNA in normal and regenerating rabbit livers using a special so-called "temperature-fractionated nucleoprotein". Regenerating liver may provisionally be regarded as being like embryonic liver. In one of the fractions there was a clear-cut difference in the nucleotide composition, the ratio $(A+T)/(G+C)$ being 1·20 in the normal liver and 1·36 in regenerating liver.

Studies of the morphogenetic variation of proteins and nucleic acids during ontogenesis represent a field of biology which is very rich in both theoretical and practical possibilities. Perhaps we have not yet got enough equipment in methods and theory to tackle this problem on an extensive scale. It is, however, certain that the preparations for a broad attack along these lines are already becoming complete.

## 3. A possible biochemical mechanism for the genetic control of morphogenetic processes during individual development: the translation into facts of the "developmental information" carried by DNA

According to our modern ideas the fertilized cell, which will have a multitude of cellular offspring from which a differentiated organism will be formed, will hand on to each of these cells the whole stock of hereditary information characteristic of the species, which is mainly concentrated in the pool of DNA. However, even though they have essentially the same stock of genetic information, different cells develop differently, muscular tissue is formed from some, nervous tissue from others, while from a third group liver is formed and so on. At the same time, the formation of these tissues during ontogenesis is certainly under genetic control. What causes morphogenetic differentiation when the cells all have the same genetic equipment?

One way of presenting the problem makes it quite clear that during morphogenesis different groups of cells receive different genetic stimuli and are governed by different loci in the general stock of information. The central problem of general biological importance is therefore, naturally, that of the mechanism of this differentiated manifestation of hereditary information.

### a) Is the differentiation of tissues accompanied by a concurrent alteration in RNA and DNA?

The possibility that there might be a parallelism between cellular differentiation and the composition of nucleic acids was, of course, the first suggestion to be made when biochemists began to study this problem. This parallelism has not, however, been clearly shown to exist.

As well as a large number of papers reporting no success in finding tissue or organ specificity of DNA (17, 55) the literature contains reports of the occurrence of slight differences in the fractional composition (50) and in certain physical properties (106) of DNA as between one tissue and another. It is, however, quite possible that these variations are connected with the later physiological activity of the organs, affecting their composition. It must also be pointed out that when Kondo & Osawa (82) tried recently to repeat one of these experiments they could not confirm the findings of Bendich and his colleagues on the tissue specificity of the fractional composition of DNA. According to them, even tissues having such widely different functions as the kidneys and the brain gave results which were virtually identical in respect of fractions separated by chromatography.

Recent work by Kit (78-80) has confirmed the absence of tissue-specific differences by the study of six organs of rats and mice and the tumours derived from them. The nature of the distribution of the subfractions and the kinetics of the denaturation of subfractions of DNA were practically the same for all the organs studied. Kok (17) studied the nucleotide composition of DNA from various organs of carp, cattle, rats and mice. He did not find any noteworthy difference in the nucleotide composition of the total DNA derived from the different organs or tissues of any one species.

In general no definite differences were found in the nucleotide composition of the total RNA derived from various organs and tissues either (79, 120). In the second of these researches it was further shown that when the mammary gland passes from the resting state to intensive lactation (and this is associated with the production of new proteins), this is not accompanied by perceptible changes in the nucleotide composition of the RNA as a whole or of that of the microsomes.

Recently there have been published two new studies containing important information obtained by new methods. Schurin & Marmur (110a) have studied the state of the DNA of larval and mature forms of *Drosophila virilis* in respect of its physical properties, heterogeneity and base composition. They used the procedure of ultracentrifugation in a density gradient of caesium chloride solution. With the larval material they obtained only one ultraviolet-absorption band, while the mature material yielded an additional ("satellite") band, with a somewhat different base composition. Finamore (65a) observed a number of differences in composition and metabolism on comparing the RNA of amphibian eggs with that of tissues of the same species.

In Chapters XIV and XV we discussed certain facts concerning a new fraction of RNA in microorganisms, which has been provisionally associated with the transfer of information from DNA to the ribosomes. It is clear that the study of the tissue and organ specificity of this RNA, which will certainly be started, should be the most promising line of approach to the study of the way in which the information manifests itself as differentiation.

A beginning in this direction has been made by Scholtissek (109b). This author studied the rapidly-turned-over nuclear RNA (corresponding to the informational-RNA fraction in bacteria). This turned out to be different in properties on isolating it from various different organs (spleen, kidneys, liver). These differences were characterized as differences of distribution of labelled phosphorus between oligonucleotides and nucleotides resulting from partial degradation of the RNA after injecting the animals with radioactive phosphate. The author considers that the nuclei of different kinds of cells transfer to the cytoplasm different kinds of RNA as "messengers". All the same it is not in doubt that differentiation is not accompanied by "loss" of the total potential fund of genetic information; there must therefore exist some mechanism for making manifest a part of the total fund of information in accordance with the "development schedule". Another study along these lines is that of Dingman & Sporn (58a), who showed that in rats the microsomal RNA of brain and liver is of a single type, while that of the nuclei of these organs is different. These differences become greater as development proceeds.

*b) Facts and ideas concerning the molecular mechanisms for the selective transfer of parts of the information of DNA determining the differentiation and morphogenesis of various organs and tissues*

When we look at the evidence for the completeness of the genetic endowment of cells of different degrees of differentiation, the suggestion naturally arises that the ability of DNA to pass the information which it contains along the line DNA → RNA → protein may be blocked and that, in the various tissues and groups of cells, different parts of the genetic system are unblocked, namely those which determine the formation of the specific tissue in question.

*The role of the chromosomal proteins (protamines and histones).* If the genetic system could not carry out such partial and differentiated activity it could not function as the determinant of ontogenetic development. What could be the mechanism of such specialized activity by DNA? In explanations of the control of the order in which items of genetic information manifest themselves, special attention has been paid to the part played by the specific proteins combined with DNA, the protamines and histones. The idea that these proteins might regulate the "flow" of information passed on by DNA was formulated ten years ago by Stedman (114) and Danielli (58). During the years that followed, many other authors arrived at the idea that the differential synthetic activity of one and the same "code" in the DNA in different cells of the organism could be explained by differences in interaction between the DNA and the proteins in the different parts of the organism (23, 121, 59, 86, 45, 91, 119). The peculiar structure of the desoxyribonucleoproteins, in which the polypeptide chain is arranged in the gap between the two strands of the polynucleotides (39, 121) makes such a suggestion very plausible (101).

Dounce (59) has recently put forward an interesting suggestion. He thinks that one of the two polynucleotide chains of DNA is distinguished by being the carrier of information and that it must therefore be blocked. In his opinion also, the blocking is brought about by the protein linked to the DNA, which thus allows only one of the chains of the DNA to carry out its genetic function.

If, however, we suggest that one of the chains of the DNA creates the matrices in the cytoplasm while the other helix, which is complementary to it, creates the adaptor forms of S-RNA, which are also complementary to the matrices, then we would not have to regard one of the chains of DNA as being superfluous.

Protamine is the simplest of the basic proteins which are usually associated with DNA, and the main features of the way in which it combines with DNA have already been worked out (61, 62, 122). The polypeptide chain of the nucleoprotamine is, as it were, wound around the molecule of DNA in the bottom of the "shallow" groove (Fig. 59). The side-chains stick out from the polypeptide chain almost at right angles to it so that the basic groups of the arginine and other basic amino acids can combine with the phosphate groups of the DNA. One third of all the amino acid residues of protamine consists of other amino acids and these residues, according to the hypothesis, occur in the folds of the polypeptide chain so that all the basic groups can combine with the phosphate groups. One residue alone cannot form a corner but two together can. This suggests that such residues will not be found singly but in pairs and there is some evidence which, in fact, suggests that this expectation is fulfilled (61).

However, nucleoprotamines are only found in the spermatozoa of some animals, and not in the nuclei of the cells of tissues and organs.

Wilkins (121) has suggested that a structure of this sort might correspond to the inert state of the genetic material before it begins to fulfil its genetic functions in the fertilized ovum. According to Wilkins, the structure of the desoxyribonucleoproteins in cells which are growing, differentiating and synthesizing proteins may be more directly related to the dynamic interaction between DNA, RNA and proteins. In these cells the DNA is combined with histones but other, acid proteins are also present in the nuclei, and in many cases these form components of the nuclear nucleo-proteins (cf. Chap. XIII).

Wilkins (121) has obtained clear X-ray evidence that histone combines with DNA in a different way from protamine. His results show that histone occupies both of the grooves in the DNA spiral, the shallow and the deep (Fig. 60).

Vendrely and his colleagues (118) also regard the change from nucleo-protamines to nucleohistones as being a change from the resting to the genetically active state.

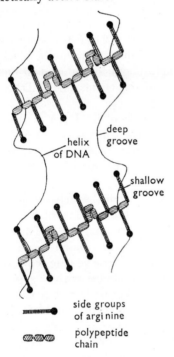

side groups
of arginine

polypeptide
chain

FIG. 59. Diagram to show how pro-tamine "winds around" the molecule of DNA. The phosphate groups are arranged alongside the black circles and coincide with the basic arginine side chains of the polypeptides. Non-basic residues are shown in pairs in folds of the chain.

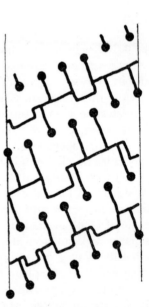

FIG. 60. Diagram showing a possible way in which a molecule of histone could "wind around" a molecule of DNA. The black dots show the posi-tion of the phosphate groups of the DNA spiral. The connecting lines represent the polypeptide chain with folds and the side groups of lysine and arginine.

A comprehensive study of the nature of the transition from protamine to histone and back in the nuclear structures of *Helix aspersa* has recently been completed by Bloch & Hew (51, 52).

In the first paper these authors describe experiments in which they used various methods to follow the transition from the "somatic" histone to protamine during spermatogenesis in this animal. This transition is accompanied by a substantial change in amino acid composition, for while arginine is the predominant amino acid in the histone, in the protamine this position is held by lysine. The authors point out that the transition from histone to protamine is, in fact, associated with the synthesis of a new protein and this synthesis is not related to duplication of DNA, i.e. it is independent. The second paper deals with the opposite process, the transition from protamine to histone in the early stage of embryonic development.

Further interesting results are those of Mauritzen & Stedman (88) who found small but definite differences from tissue to tissue in the amino acid composition of the nuclear histones isolated from different organs. These differences only affected certain amino acids (aspartic and glutamic acids, alanine, leucine, isoleucine and valine). The authors suggest that histones may be cell-specific proteins. Another group of authors (92a) have also demonstrated the specificity and great heterogeneity of the histones of various organs in fowls. After zone electrophoresis on a starch gel the histones of the liver and spleen both formed 18 zones, but the distributions of the zones were different from one another. The histones of the erythrocytes showed a smaller collection of components.

A detailed study of the fractional composition and also of the species and tissue differences of histones has recently been carried out in Butler's laboratory (70a, 74b). The authors also observed definite heterogeneity in histones isolated from a single tissue. For example, the histone associated with the DNA of the salivary gland of the calf has been fractionated into three components: f1 which is rich in lysine, f2 which contains less lysine and f3 in which arginine predominates. The authors note that, although the fractional composition of the histones varies considerably according to the species and organ studied, the amino acid composition of each fraction remains more or less the same whatever its source. In later work from the same laboratory (100a) it was established that the arrangement of arginine and lysine along the chain of a histone is irregular and therefore the arrangement of the phosphate links in the DNA interacting with the side-groups of the lysine and arginine is also not regular.

It is also interesting to note that there is a parallelism between the kinetics of DNA and the kinetics of histones in the cell during mitosis (87a). Furthermore, during the development of the pollen of *Lilium longiflorum* the appearance of desoxyribonucleosides in the cytoplasm coincides with the appearance there of arginine during the phase of duplication of the chromosomes. The lysine content, however, falls during this period (92b). The significance of these changes is still uncertain.

However, with regard to the suggestion that the transition from protamines to histones may represent the transition from the inactive to the genetically active state, it must be remembered that the nature of the linkage between histones and DNA seems far more like a blockage of information encoded in the form of a sequence of nucleotides and it may be that only particular parts of the DNA which have no such "blockade" are active. The blocking power of both protamines and histones has been indirectly confirmed by several other workers, who have shown that when protamines and histones act on cells and when they are introduced into cells they combine with nucleic acids there and at the same time inhibit the growth of the cells and the synthesis of proteins in them (10, 49, 115). Direct evidence of the ability of histones to suppress the DNA-dependent synthesis of informational RNA has been obtained recently by Huang & Bonner (70c).

It is interesting to note that neither protamine nor histone has been found in the bacterium Escherichia coli (122, 125) and that in the cells of this organism part of the DNA ($\sim$30%) exists in the cells in a protein-free form, the rest of the DNA being combined with a protein containing only half the proportion of lysine and arginine found in ordinary histones. In the cells of E. coli ontogenesis is certainly very abbreviated and this may be why the ratio of active (free) DNA to desoxyribonucleoprotein (a temporarily blocked reserve of information) is so great (1:2) in these bacteria. It is also interesting to note that the proteins of the nucleus and nucleolus show definite differences (99a).

If these hypotheses as to the possible blocking of particular genetic zones of DNA by means of proteins should really be verified, then it is very likely that this blockade is used for differentiation of the activity of the nucleus. The disorderly nature of this blockade would, however, need to be regulated in strict accord with the nature of the development. What carries out this regulation? What brings into action strictly determinate zones of the chromosomes for strictly determinate periods of time in relation to the differentiation of the tissues? This is a very complicated and also a very important theoretical question for the whole problem of ontogenesis.

*Differentiated activity of the chromosomes and genetic systems during specialization.* It has recently been demonstrated by Ficq and Brachet (63, 64) that the activity of the different parts of the chromosomes may be very highly different. They injected [$^3$H]thymidine into the larvae of *Rhynchosciara angelae* and made a microradioautographic study of its incorporation into the chromosomes of the salivary glands. It was found that at certain stages of development certain parts of these chromosomes were particularly active in incorporating thymidine. This shows very

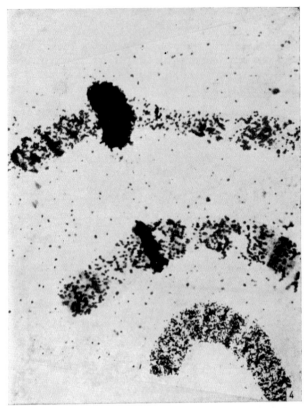

Fig. 61. Radioautograph of salivary-gland chromosomes. Distal segment of a chromosome at three different stages after beginning of incorporation of [³H]thymidine

clearly in Fig. 61, which is taken from the work of these authors. By using [14C]adenine and [14C]phenylalanine it was found that the same parts also show an unusually great activity in the synthesis of proteins and RNA. The authors suggest that this increased local activity of the chromosomes is connected with the physiology and differentiation of the cells. Similar evidence for differences in metabolic activity along the length of the chromosomes has also been obtained recently in other laboratories (16, 89, 98, 111, 112).

The last of these papers (112) gave specially clear indications in this direction. The authors studied the synthesis of proteins by parts of the giant chromosomes of drosophila, using [35S]methionine and [3H]leucine. Only certain of the loci on these chromosomes synthesize proteins. The authors suggest that the localization is determined by the freeing of the matrices by the rapid removal of the proteins which have been newly synthesized, this being controlled by the whole chromosome. The authors also point to their preliminary findings which show that the actual process of replication of the DNA in the chromosomes is not uniform. Other biochemical methods have also been used to show that the synthesis of DNA is not uniform (66).

The question of the sequence of the replication of DNA in different parts of the chromosome has recently been the object of a considerable number of studies, embracing both plant and animal material. In all cases, asynchrony of this replication has been observed. Taylor has used autoradiography to study the duplication of chromosomes in cells from a tissue culture derived from Chinese hamster. He found the portions of DNA engaged simultaneously in replication to vary from part to part of the chromosome, while each portion, in the course of a single cycle of division, replicated only once (116a). In some segments of the chromosome the retardation was very marked. Variations in the periodicity of duplication along the length of the chromosome have been observed in broad-bean shoots (*Vicia faba*) (71b), in human tissue-culture cells (83a, 91a) and elsewhere.

It is interesting that, even in homologous pairs of chromosomes, duplication may begin at different times.

Goldstein & Brown (68a), who studied the physical state of DNA during the period of duplication in *Escherichia coli*, reached the conclusion that the process of replication of the genome (DNA) is completed in asynchronous fashion. These authors produced some evidence indicating that the synthesis of "messenger" RNA took place on the DNA at a time when it was not engaged in replication and ceased when replication was in progress.

Another specially interesting observation concerns the formation of swellings on the giant chromosomes of the larvae of certain insects (53a). At particular periods in the development of the larvae, certain bands in the chromosomes become exceptionally long and form swellings known as "puffs". After a while these go down again. At other times "puffs" are formed from other bands. It seems that the formation of "puffs" is a manifestation of the activity of genes. The chromosomal bands from which "puffs" are formed are always the same for any particular type of cell but in cells of other types they are formed from different bands. The number of bands which will form "puffs" in cells of any particular type in the course of time only represents a small percentage of the total number of bands.

It is interesting that the number of such local expansions of chromosomes ("puffs") is greater in chromosomes seen in cells during the earlier stages of development of the larvae than during the later stages (65a). Autoradiographic experiments showed that the regions of "puff" formation corresponded with the regions of active synthesis of RNA and protein. "Puff" formation undoubtedly expresses an active state of some particular group of genes. It is specifically related not only to the stage of development but is also determined by the differentiation. Thus, salivary glands have two types of secretory cells, and each type has its own peculiarities in the manner of "puff" formation. The observations of Clever (55a) give a detailed picture which bears witness to the genetic importance of these formations. He succeeded in inducing "puff" formation in strictly determinate zones of the chromosomes by the action of hormones (e.g. ecdysone) which had a parallel effect on the process of metamorphosis of the larvae.

Mirsky & Allfrey (91) gave an interesting paper on the role of the nucleus in ontogenesis at a symposium on the biochemical basis of development. These authors maintain that, although the genetic potentiality of the nuclei is the same in all tissues, their development is determined, to a considerable extent, by the state of the surrounding cytoplasm. According to their ideas, the results obtained by Sax (107) in his study of the pollen grains of *Tradescantia* constitute an example of the effect of the cytoplasm on differentiation. Sax found that during the first mitosis, which leads to the formation of the pollen cells, the daughter nuclei migrated into different regions of the cytoplasm and this determined which of the two nuclei would become reproductive and which would remain vegetative.

Mirsky & Allfrey also adduce several more examples to show that an increase in the synthetic activity of the cytoplasm, as in secretory cells, will lead to a rapid increase in the metabolism of histone, RNA and DNA of the nucleus, and also that nuclei isolated from different tissues show certain differences in the composition of their enzymes. The authors of this paper did not, however, deal with the fundamental question of the reasons why the nucleus and chromosomes respond to different biochemical stimuli in different parts of the organism and at different stages of ontogenesis. In later papers (45, 90a) the same authors showed that positively charged histones mask the negatively charged groups of DNA and they suggested that this masking might be the means by which they affect the functions of the chromosomes.

A recent paper of Edström & Beermann is particularly interesting (60a). By a procedure for ultramicroanalysis of nucleotides they determined the base composition of RNA from fragments of the nuclear apparatus of *Chironomus*. In this respect chromosome no. 1 proved different from chromosome no. 4. Various parts of chromosome no. 4 likewise showed differences in the composition of their RNA. The ratio RNA/DNA varied along the length of the chromosomes and was markedly higher in the region of the "puffs".

According to Mirsky & Allfrey it must be supposed that a large amount of the magnesium in the nucleus is combined with those phosphate groups of DNA which are not occupied by histone. The metabolic processes associated with nucleotides usually require the presence of magnesium. These observations indicate that the places along the DNA chain where the phosphate groups are masked by histone are inactive while those where they are combined with magnesium (in the nuclei of the thymus only a very few) are active. The fact that, in the cells, histones form a heterogeneous group of basic proteins accords well with the idea that they are factors regulating the activity of DNA.

In studying the problem of the way in which differentiation affects the fund of genetic information in the nuclei, an interesting approach has been to transplant nuclei into cells at earlier stages of development and to observe the morphogenetic effect of such transplantations. Good reviews of experiments along these lines have recently been published (65b, 54a). Such work with amphibians has shown that nuclei from ova in the earliest stages of cell division possess so called "total potentiality" and, on transplantation into a denucleated oocyte, can promote normal development. However, nuclei from cells at later stages of differentiation evoke numerous and varied changes in development. The later the stage at which the nuclei are taken for transplantation, the wider the range of variations in embryogenesis that are observed. At the same time the morphology of the chromosomes (not counting "puffs") remains practically unchanged in all these different cells. All this strongly suggests that during individual development the general pool of genetic information has imposed on it a regulatory system which itself undergoes changes both in space and in time.

It seems that the common pool of DNA may be compared to a gigantic scoreboard with lamps alight in different parts of it, different combinations of lamps corresponding to differentiation of particular cells along different lines which, as it were, "read" the lighted instructions. But how is this scoreboard set up? What determines in advance the nature of the instructions which will appear on it? On ordinary scoreboards this operation is carried out by an automatic apparatus which is regulated and directed by some person. But what regulates the purposive and well-ordered work of the genetic pool? We have not yet got the exact facts which might enable us to give the correct answer to this question and we therefore require some hypothesis, even if only a provisional one, to fill in temporarily the considerable gap in our ideas on the genetic control of ontogenesis.

Certain suggestions have been put forward to explain this phenomenon but they are still only very indefinite. One characteristic example of such a provisional hypothesis is that put forward by Schmitt (109) who takes the view that, in addition to the information recorded in the genetic code, which must be conserved as a whole in the reproductive cells so that it may be handed on to the next generation, there is another coded programme which occurs in every cell in the body and directs the chemical and structural differentiation of the various tissues and organs during their development.

A similar idea, though in a more developed form, has recently been put forward by Jacob & Monod (73, 74), who base it on evidence as to the way in which adaptive enzymes are synthesized. They take the view that, in the synthesis of most proteins, two genes are active. The first is the structural gene which controls the way in which the protein molecule is synthesized. The other is the regulator or operator gene which regulates the activity of the first gene (or of a group of such genes) by means of some sort of repressor.

In their two contributions to a recent symposium on quantitative biology Jacob & Monod (74a) tried to adapt this scheme, which had been worked out for the synthesis of adaptive enzymes, to the explanation of cellular differentiation. Theoretically it is, of course, possible that there might be a manifestation of previously concealed synthetic potentialities of the genetic system of the cell during differentiation as much as during adaptation.

The scheme of Jacob & Monod explains the formation of adaptive enzymes by the hypothesis that, as well as the structural genes on which the matrix RNA is formed, there are also groups of regulating genes, repressors and operators, and it has recently received strong experimental support from the study of mutations.

Mutations corresponding to changes in the DNA in the region of the structural gene lead to alteration in the protein synthesized. A series of such mutations was studied in detail. According to this scheme the regulator gene forms some sort of repressor (provisionally regarded as some form of protein or RNA) which, under ordinary conditions, blocks and inactivates the structural gene. By interacting with the repressor the inducer (substrate of the reaction) removes its repressive action and converts the structural gene to its active state. The occurrence of such a regulator gene suggests the possibility that there may be mutations which do not affect the structural gene but interfere with the action of the regulator gene. As a result of this the formation of the repressor is suppressed and this increases the synthesis of the adaptive enzyme, even in the absence of the substrate. The synthesis of the enzyme becomes constitutive instead of adaptive. A series of such mutations has, in fact, been studied. In this way it has been shown that operator genes could exist and bring into action or throw out of action a group of structural genes according to the presence or absence of a repressor.

In the phenomena of differentiation there may occur, according to Jacob & Monod, a similar type of regulation of the activity of genes, but the "permissive" effect, which brings the group of genes into action, does not, in this case, depend on the substrate of the reaction, but on some other products associated with embryonic development, possibly with the action of embryonic inducers. It is not impossible that they are the same as the so-called "ontogenins" discovered by Rusev (34) which, he believed, could block the synthesis of specific proteins at a particular moment.

Encouraging evidence has also been obtained recently in regard to the chemical nature of the embryonic inducers. We have in mind two recent publications (70b, 83) which give an account of the first isolation of an embryonic inducer in pure form. It has been known for a long time that the ventral half of the embryonic spinal cord produces some substance which, on entering into the adjacent somite cells, induces the formation of vertebral cartilage. The notochord also brings about induction of this sort. The nature of the chondrogenic factor has also been studied. The production of this factor is confined to a particular period of ontogenesis, in fact to the first few hours of contact between the tissues. To isolate the active compounds the authors had to fractionate tissues from 70,000 embryos. The active factor showed up in the free-nucleotide fraction and, after electrophoretic and chromatographic separation, it was provisionally identified as a nucleotide-peptide compound (containing cytidine and guanosine monophosphates and several amino acids: aspartic and glutamic acids, glycine, alanine, valine and serine). The compound also contained hexosamine.

An interesting idea was set out recently in a paper by the well-known geneticist Nanney (92). He suggested that the genes or the genetically significant parts of DNA are inactivated by interaction with proteins or, as he thinks more likely, with RNA. One of the main features of differentiation is the occurrence of mutual exclusion, that is to say the synthesis of one functional type of protein often precludes the formation of another. Nanney suggests that the protein or RNA being formed at a particular genetic locus may form a complex with and block other loci and that this determines the character of the differentiation of the cell.

None of these hypotheses are particularly attractive, in as much as they only deal with the way in which blocking takes place and do not touch the most important question of the determination of the place and time at which the blocking occurs, that is to say, the question of the nature of the information that leads to the automatic switching on of the gigantic (as molecules go) genetic scoreboard.

*c)  The hypothesis of the hereditary "programming" of the*
*processes of differentiation of cells in space and time*

A survey of the role of DNA in heredity on the one hand, and the nature of the morphogenetic processes in ontogenesis on the other, shows clearly that the activity of DNA in morphogenesis is differentiated in both time and space.  In different parts of the body and at different times the very same pool of DNA will determine the development of different characteristics.  It is also important to note that there is a certain correspondence between ontogenesis and phylogenesis as regards the sequence of the morphogenetic reactions.

In the preceding chapter we arrived at the conclusion that DNA, as a basis of heredity, arose during evolution in order to ensure the handing on from one generation to the next of characteristics which only appear and function at some particular period in the life of the individual.  This must be why there also arose by evolution in this system a new quality, namely the ability to accumulate information regulating the conditions under which the system varied its collection of active genes in time and space.

The morphogenetic importance of this information incorporated in DNA must surely consist, not only in the fact that it contains the code for the reproduction of the specificity of proteins which are not present in some of the earlier stages of ontogenesis, but also in the fact that this information begins to operate in a particular place at a particular time and under particular conditions, both external and internal.

However, if the information which determines the sequence of amino acids in polypeptide chains is recorded in the form of a particular sequence of nucleotides, how is the information concerned with the time and place of the synthesis recorded?  The only feature of the composition of DNA which can be altered is the sequence of nucleotides.  Having been baffled at this point we would be right to refrain from regarding DNA exclusively as a coding for amino acids.

It seems beyond doubt that the macromolecular structure of DNA may contain information dependent on other features of its structure which may control the phenomena of specificity (rhythm and speed of unwinding of the double helix, presence of loops, organization of DNA in chromosomes, etc.).  These would also co-ordinate the various genetic activities.

In their studies of protein synthesis research workers ran into difficulties with the system (matrices of the organelles + S-RNA) which can translate the "text" of the information determining the amino acid sequence.  In studying the mechanism of ontogenesis we must discover the system which can translate the "text" regulating the time, place and other conditions for individual syntheses.

Thus, from the purely theoretical point of view, we may postulate that DNA can possess information determining, not merely the specificity of the proteins synthesized, but also the successive accomplishment of different syntheses at different times and under different conditions. We must find out how this information is encoded, as we are now finding out how the sequence of amino acids is encoded. This is a task for the future, not only for biochemists, but also for embryologists, histologists, cytologists and geneticists. Another task is the more intensive study of the theoretical aspects of the problem of regeneration.

The question of the nature of the information controlling the sequence of amino acids in proteins is being studied actively by biochemists, while the question of the nature of the information which determines the time and position of the differential activity of genetic loci is still only beginning to be looked at.

Very many biochemists and geneticists have remarked on the theoretical importance of solving this problem (24, 69, 91, 92, 109, 116) but so far only a few have tried to put forward any hypotheses. As we saw in the preceding section these hypotheses are either remarkably general in nature or very guarded.

As the system of DNA arose in evolution to supplement the system of RNA in assuring the hereditary transfer of particular characteristics, which only function for limited periods, and the conditions under which they appear, it is natural to pay attention, in the first place, to the differences between RNA and DNA. These are, primarily, the substitution of desoxyribose for ribose and thymine for uracil, the occurrence of the double helix and the far higher molecular weight of DNA. The substitution of desoxyribose for ribose gives DNA great stability but the significance of the substitution of thymine for uracil is not yet clear, though it is possible that thymine has a better complementarity to adenine, thus ensuring more accurate reproduction. The double helix accounts for the periodicity of synthesis, the possibility of combining with histones and protamines, the stability of the structure, the slightness of the metabolism and so forth. Finally, the large molecular weight may be associated with the fact that, as well as information determining the synthesis of a particular protein, the molecule of DNA also contains information which controls the turning on and off of the system under strictly determinate conditions.

Sadron, Pouyet & Vendrely (105) have recently tabulated the molecular weights of DNA from various sources on the basis of their own findings and those reported in the literature. In the 28 samples with which the table is concerned the molecular weight varied between 5,800,000 and 16,500,000. The DNA with this colossal molecular weight existed in the form of an unbranched chain wound up into the characteristic helix. Assuming that the molecular weight of the continuous polypeptide chain of many proteins lies between 6000 and 15,000, it is not hard to imagine that, if the macromolecular structure of DNA were to serve as the chemical basis for recording information which was then passed on in the synthesis

PB 2H

of a protein, then, theoretically, one such matrix could serve for the inscription of the structure of dozens of proteins. The first question which arises in this connection is whether it is possible to observe any repetition or periodicity of a particular piece of information along the length of a chain of DNA. It must be mentioned that several authors have admitted the possibility that the polymers of DNA may have an aggregated structure (39, 46, 108) but such ideas have not yet been confirmed.

The molecular weight of the specific RNA which carries information to the DNA in the ribosomes concerning the synthesis of proteins (and RNA) is about 300,000. The non-specific RNA of the ribosomes has a molecular weight which may be as high as 1,500,000, but this seems to be synthesized without the help of DNA. Thus, each molecule of DNA possesses a certain surplus capacity. It may be suggested that each molecule of DNA determines the synthesis of different forms of RNA which receive their information from different parts of the DNA. This, at least, seems true for phage DNA. It might, however, also be suggested that the extra "surplus" polynucleotide capacity of the molecule of DNA might contain the information as to the time and place at which the whole system is to be put into operation. If this were so, how might one imagine that such a system would work? Might one assume that the chromosomes contained some sort of genetic structures acting as "operators" or "regulators" arranged in a strictly determinate linear sequence, A, B, C, D, E, F, etc. so that they could only come into action in the correct order, but if one came into action the others would be switched off? Each such regulator gene would be associated with a series of genes linked to one another e.g. W, X, Y, Z which bring about increased specialization of the cell. When the cell divides, as shown in Fig. 62, into four

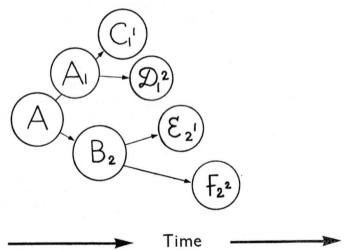

Time

Fig. 62. Appearance of differences with time in the course of differentiation (for explanation see text).

cells, different genes of the series are brought into action in their descendants according to the progress of their division, their location and other features. Each such regulator gene brings into operation a different group of interconnected genes W, X, Y, Z; $W^1$, $X^1$, $Y^1$, $Z^1$, etc. and, associated with the completion of the reactions promoted by each series of genes, there is a new regulator gene which will ensure a further deviation in differentiation, away from the original course and from that of other cells. The order in which these systems are brought into action seems to correspond with their evolutionary development. How could this "bringing into action" occur? There might be suppression of the synthesis of the molecules of histone which block some particular part of the stock of DNA. This might also be the mechanism of the "linkage" of the genes which are brought into action. Each type of DNA which is unblocked may carry the information which will ensure the unblocking of another strictly determinate part of the chromosomes.

Thus, in different parts of the body and at different stages of development different systems of genes are brought into operation in the cells.

The replacement of histone blockade by protamine blockade during spermatogenesis gives rise to total blockade of all the uncovered parts of

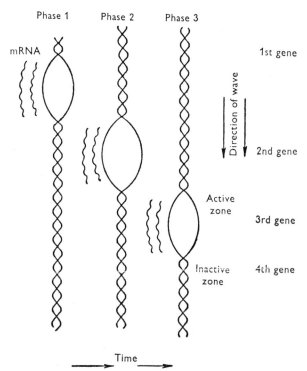

FIG. 63. Diagram of the successive activation of genetic loci by a wave principle. The unwound section is the activated locus of the DNA producing RNA.

the DNA, which is necessary if the new development is to begin at the original loci.

What could be a possible molecular mechanism for such a "switching on" of genes? The different genes have to work in a strictly consequential manner in time, one after the other, according to the principle that the beginning of the work of one gene signifies the ending of the work of others. In Fig. 63 we present a simple hypothetical scheme for this process. There exists evidence that the formation of messenger RNA takes place on a single polynucleotide chain of DNA and that unwinding of the double helix is a prerequisite for activation of the genetic locus, just as it is for the replication of the DNA itself. It is also known that the length of the polynucleotide chain of DNA is adequate for the encoding of the information for several genes, if by "gene" we mean that zone corresponding to a single protein. There have recently been preliminary publications describing DNA having molecular weight 90,000,000, which could be the material carrier for several hundred such genes. If such a linear aggregate of genes could undergo a wave-like unravelling and this wave were equal in length to one gene (Fig. 63), then this would bring about, as the wave moved along the aggregate, a successive activation of genes in a manner strictly consequential in time, so that the "switching on" of any gene would coincide with the "switching off" of the gene immediately preceding it. Of course, this model raises a number of problems—the marking of limits of the gene, the halting of the wave at these limits, and so forth. All the same, it does give an explanation in molecular terms for successive automatic changes in the activity of genes both in space and in time. It also permits variations, such as simultaneous unwinding over the length of several genes, or unwinding taking the form of two successive waves with an interval between them.

A second stage in creating a molecular model for the functional differentiation of chromosomes could be to assume the occurrence of a very wide heterogeneity in the histones and to postulate a special complementarity of their various fractions to particular loci in the chromosome, as in the reaction of antibodies with antigens. DNA, incidentally, possesses antigenic activity. And if the various forms of messenger RNA which determine synthesis of the various histones are produced from the same single helix of DNA, functioning on the wave principle described above, then successive replacement of the histone fractions could ensure successive blockade of different zones of the chromosomes. One could also assume that there exist molecules of DNA whose wave-like unravelling would promote not the synthesis of various histones but of substances capable of "blockading" histones, which would thus selectively remove particular individuals from the total array of repressors; in this way repression of particular loci of the genome could be selectively taken off. This type of DNA molecule could be described as "genetic clocks of development".

The above scheme is, of course, highly hypothetical, and we have introduced it merely to show that a "programme" for development with differentiation is theoretically possible: the true picture of the genetic regulation of the process of development will only be revealed by direct experiment.

Of course it is possible that the mechanism for successive "switching on" of the various genetic loci does not exist at the molecular level but at that of complicated chromosomal structures. From this point of view the recent work of Gall & Callan (66a) is of interest. They used auto-radiography to study the synthesis of protein and nucleic acid in giant ("lamp brush") chromosomes of *Triturus cristatus*. In one such chromo-some they observed a peculiar loop composed of two projecting portions. The incorporation of [³H]uridine into the RNA of this loop proceeded in a strict sequence, beginning at the narrow end of the loop and proceeding round it in the course of ten days to the other end. The speed of this movement was increased on administration of gonadotrophic hormone. On the other hand [³H]phenylalanine was simultaneously and rather quickly incorporated into all regions of the loop. According to the observations of these authors, the loop was not a static phenomenon, but was continually being unwound from the thin end and wound up at the thick end.

We should also mention the work of Khesin & Shemyakin (41a), who studied the synthesis of various forms of messenger-RNA on the DNA of bacteriophage at various stages of the reproduction of the phage in cells of *Escherichia coli*. Their results led them to the conclusion that the synthetic activity of the phage DNA, which is qualitatively different at different times, is controlled by the sequential unwinding of the DNA. The views of these authors are thus consistent with the scheme illustrated in Fig. 63.

## Conclusion

The study of the biochemical mechanisms of morphogenesis is a relatively new field in biochemistry and genetics in which not enough facts have yet been collected. This line of work in biochemistry and genetics must not be confused with what is called chemical embryology, which deals with a different aspect of the way in which development occurs and uses different methods. The difference between these two lines of work is that the former is directed towards the discovery of the primary mechanism of development connected with the hereditary apparatus of the cell, while the latter is concerned with the study of the chemical factors which determine particular embryonic processes such as gastrulation or the formation of the rudiments of the nervous system, which develop under the influence of causes and substances which have arisen under the control of hereditary factors.

In this chapter we have given a short account of the facts of the morphogenetic alteration of proteins and have tried to sketch, although only in general terms, a possible mechanism of genetic biochemical control of the processes of development. The theoretical approach to ontogenesis, in which it is regarded as a strictly controlled development predetermined by heredity, is sometimes referrred to as preformism, vitalism, fatalism, Weismannism or Morganism. This principle is seen in opposition to the principle of development as something which happens of its own accord, being brought about by its own inherent laws (contradictions). Such dialectical analysis from outside is, however, a completely unreasonable use of the dialectical method. The classical dialectical principle of development can only be applied to the development of living matter as a whole, that is to the evolution of life or to phylogenesis. Ontogenesis is a controlled development which is strictly regulated and determined by particular forms of biochemical information which record the evolutionary path of the living material and present it ready made.

At the same time we may be sure that the genetic system of the cells does not exert an absolute and comprehensive control over all the processes of morphogenesis. A large number of auxiliary regulators come into being successively in the course of ontogenesis. These determine systems at a higher level, such as the humoral and nervous connections between organs, etc. Nevertheless, some such strict primary determination prevails over complete chaos in bringing about these extremely complicated interactions.

## REFERENCES

1. BUTENKO, R. G., NICHIPOROVICH, A. A. & PROTASOVA, N. N. (1960). *Doklady Akad. Nauk S.S.S.R.* **135**, 210.

2. GULYĬ, M. F. (1953). *Ukrain. biokhim. Zhur.* **25**, 367.

3. — (1955). In collective work *Belki, ikh spetsificheskie svoĭstva* (ed. V. A. Belitser & M. F. Gulyĭ), p. 7. Moscow: Izd. AN SSSR.

4. GUPALO, P. I. (1954). *Uspekhi sovremennoĭ Biol.* **38**, No. *1* (4), p. 111.

5. — (1957). *Byull. Mosk. Obshchestva Ispytateleĭ Prirody, Otd. biol.* **62**, No. *5*, p. 77.

6. GURVICH, A. E. & KARSAEVSKAYA, N. G. (1956). *Biokhimiya* **21**, 746.

7. DENISOVA, A. A. & ZAĬDES, A. L. (1957). *Doklady Akad. Nauk S.S.S.R.* **114**, 1287.

8. DORFMAN, V. A. (1958). *Uspekhi sovremennoĭ Biol.* **45**, 313.

9. ZAĬDES, A. L. (1956). *Biofizika*, **1**, 279.

10. ZBARSKIĬ, I. B. & PEREVOSHCHIKOVA, K. A. (1954). *Byull. eksptl. Biol. i Med.* **38**, No. *10*, p. 61.

11. IVANOV, I. I., YUR'EV, V. A., KADYKOV, V. V., KRYMSKAYA, B. M., MOISEEVA, V. P. & TUKACHINSKIĬ, S. E. (1956). *Doklady Akad. Nauk S.S.S.R.* **111**, 649.

12. — (1956). *Biokhimiya*, **21**, 591.

13. KAZARYAN, V. O. (1952). *Stadiĭnost' razvitiya i stareniya odnoletnikh rastenii.* Erevan: Izd. AN Arm. SSR.

14. KASAVINA, B. S. (1954). *Trudy vses. Obshchestva Fiziologov, Biokhimikov i Farmakologov, Akad. Nauk S.S.S.R.* **2**, 151.

15. KASAVINA, B. S. & TORCHINSKIĬ, YU. M. (1956). *Biokhimiya*, **21**, 510.

16. KIKNADZE, I. I. (1960). *Tezisy dokladov. Konferentsiya po voprosam tsito- i gistokhimii*, p. 12. Moscow: Izd. AN SSSR.

17. KOK, I. P. (1960). *Doklady Akad. Nauk S.S.S.R.* **133**, 1216.

18. KONAREV, V. G. (1948). *Doklady Akad. Nauk S.S.S.R.* **59**, 773.

19. — (1959). *Nukleinovye kisloty i morfogenez rastenii.* Moscow: Gosudarst. Izd. "Vysshaya Shkola".

20. KRENKE, N. P. (1940). *Teoriya tsiklicheskogo stareniya i omolozheniya rastenii i prakticheskoe ee primenenie.* Moscow: Sel'khozgiz.

21. — (1950). *Regeneratsiya rastenii.* Moscow & Leningrad: Izd. AN SSSR.

22. LEĬKINA, E. M., TONGUR, V. S., LIOZNER, L. D., MARKELOVA, I. V., RYABININA, Z. A., SIDOROVA, V. F. & KHARLOVA, V. G. (1960). *Biokhimiya*, **25**, 96.

23. LUCHNIK, N. V. (1959). *Doklady Akad. Nauk S.S.S.R.* **126**, 417.

24. MEDVEDEV, ZH. A. (1960). *Izvest. Timiryazev. sel'skokhoz. Akad.* No. *1*, p. 103.

25. NICHIPOROVICH, A. A. (1958). *Trudy vsesoyuz. nauch.-tekh. Konf. Primenen. radioaktiv. i stabil. Izotopov i Izluchenii v. narod. Khoz. i Nauke, Fiziol. Rastenii, Agrokhim., Pochvoved., 1957* (ed. V. M. Klechkovskii *et al.*), p. 56. Moscow: Izd. AN SSSR.

25a. NEĬFAKH, A. A. (1962). *Problema vzaimootnoshenii yadra i tsitoplazmy v razvitii.* Moscow: Izd. Inst. Morfol. AN SSSR.

26. OPPEL', V. V. (1958). *Uspekhi sovremennoi Biol.* **46**, 281.

27. OREKHOVICH, V. N. (1952). *Prokollageny, ikh khimicheskii sostav, svoistva i biologicheskaya rol'.* Moscow: Izd. AMN SSSR.

28. OREKHOVICH, V. N. & PAVLIKHINA, L. V. (1957). *Voprosy med. Khim.* **3**, 195.

29. OREKHOVICH, V. N., ALEKSEENKO, L. P. & LEVDIKOVA, G. A. (1957). *Vestnik Akad. med. Nauk S.S.S.R.* No. *1*, p. 12.

30. ORLOVSKAYA, G. V., ZAĬDES, A. L. & TUSTANOVSKIĬ, A. A. (1956). *Doklady Akad. Nauk S.S.S.R.* **111**, 1396.

31. PASHEVICH, V. YU. (1940). *Uch. Zap. Mosk. gos. Univ.* No. *35*, p. 63.

32. REĬMERS, F. E. (1957). *Botan. Zhur.* **42**, 1465.

33. RON'ZHINA, O. A. (1957). *Doklady Akad. Nauk S.S.S.R.* **113**, 462.

34. RUSEV, G. K. (1960). *Zhur. obshch. Biol.* **21**, 130.

35. SISAKYAN, N. M. & MELIK-SARKISYAN, S. S. (1956). *Biokhimiya*, **21**, 329.

36. SISAKYAN, N. M., MELIK-SARKISYAN, S. S. & BEZINGER, E. N. (1952). *Biokhimiya*, **17**, 626.

37. SISAKYAN, N. M. & FILIPPOVICH, I. I. (1953). *Zhur. obshch. Biol.* **14**, 215.

38. SLEPCHENKO, N. V. (1959). In collective work *Biokhimiya nucleinovogo obmena rastenii* (ed. V. G. Konarev), p. 18. Ufa: Izd. AN SSSR.

39. SPITKOVSKIĬ, D. M., TONGUR, V. S. & DISKINA, B. S. (1958). *Biofizika*, **3**, 129.

40. TUSTANOVSKIĬ, A. A., ZAĬDES, A. L., ORLOVSKAYA, G. V. & MYAGKAYA, G. L. (1961). *Proc. V int. Congr. Biochem., Moscow*, **9**, 110.

41. FERDMAN, D. L. & NECHIPORENKO, Z. YU. (1948). *Ukrain. biokhim. Zhur.* **20**, 124.

41a. KHESIN, R. B. & SHEMYAKIN, M. F. (1962). *Biokhimiya*, **27**, 761.

42. CHAĬLAKHYAN, M. KH. (1959). *Byul. Moskov. Obshchestva Ispytatelei Prirody, Otd. biol.*, **64**, no. 1, p. 61.

43. SHMERLING, ZH. G. & USPENSKAYA, V. D. (1955). *Biokhimiya*, **20**, 31.

44. AGRELL, I. & PERSSON, H. (1956). *Nature (Lond.)*, **178**, 1398.

45. ALLFREY, V. G. & MIRSKY, A. E. (1959). *Trans. N. Y. Acad. Sci.*, Ser. II, 21, 3.

46. ANDERSON, N. G. (1953). *Nature (Lond.)*, 172, 807.

47. BANGA, I., BALÓ, J. & SZABÓ, D. (1956). *Acta physiol. Acad. Sci. Hung.* 9, 61.

48. BANGA, I., GUBA, F. & SZENT-GYÖRGYI, A. (1947). *Nature (Lond.)*, 159, 194.

49. BECKER, F. F. & GREEN, H. (1960). *Exptl. Cell Research*, 19, 361.

50. BENDICH, A., PAHL, H. B. & BEISER, S. M. (1956). *Cold Spring Harbor Symposia quant. Biol.* 21, 31.

51. BLOCH, D. P. & HEW, H. Y. C. (1960). *J. biophys. biochem. Cytol.* 7, 515.

52. — (1960). *J. biophys. biochem. Cytol.* 8, 69.

53. BRACHET, J. (1960). *The biochemistry of development*. New York: Pergamon Press.

53a. PAVAN, C. & BREUER, M. E. (1955). *Chromosoma*, 7, 371.

54. BURTON, D., HALL, D. A., KEECH, M. K., REED, R., SAXL, H., TUNBRIDGE, R. E. & WOOD, M. J. (1955). *Nature (Lond.)*, 176, 966.

54a. CHANTRENNE, H. (1961). *The biosynthesis of proteins*. Oxford: Pergamon.

55. CHARGAFF, E. (1955). In *The nucleic acids* (ed. E. Chargaff & J. N. Davidson), Vol. *1*, p. 307. New York: Academic Press.

55a. CLEVER, U. (1961). *Chromosoma*, 12, 607.

56. CROISILLE, Y. (1959). *Compt. rend. Acad. Sci.*, *Paris*, 249, 1712.

57. — (1960). *J. Embryol. exptl. Morphol.* 8, 216.

58. DANIELLI, J. F. (1953). *Cytochemistry: a critical approach*. New York: Wiley.

58a. DINGMAN, W. & SPORN, M. B. (1962). *Biochim. biophys. Acta*, 61, 164.

59. DOUNCE, A. L. (1959). *Ann. N. Y. Acad. Sci.* 81, 794.

60. EDDS, M. V. (1958). *Proc. natl. Acad. Sci. U.S.* 44, 296.

60a. EDSTRÖM, J. E. & BEERMANN, W. (1962). *J. Cell Biol.* 14, 371.

61. FELIX, K., FISCHER, H. & KREKELS, A. (1956). *Progr. Biophys. biophys. Chem.* 6, 1.

62. FEUGHELMAN, M., LANGRIDGE, R., SEEDS, W. E., STOKES, A. R., WILSON, H. R., HOOPER, C. W., WILKINS, M. H. F., BARCLAY, R. K. & HAMILTON, L. D. (1955). *Nature (Lond.)*, 175, 834.

63. FICQ, A. (1959). *Lab. Invest.* 8, 237.

64. FICQ, A., PAVAN, C. & BRACHET, J. (1959). *Exptl. Cell Research*, Suppl. 6, 105.

65. FINAMORE, F. J. (1955). *Exptl. Cell Research*, 8, 533.

65a. — (1961). *Quart. Rev. Biol.* 36, 117.

65b. FISCHBERG, M. & BLACKLER, A. W. (1961). *Scient. Amer.* 205, No. *3*, p. 124.

66. FRANKEL, F. R., KNAPP, J. & CRAMPTON, C. F. (1960). *Fed. Proc.* 19, 306.

66a. GALL, J. G. & CALLAN, H. G. (1962). *Proc. natl. Acad. Sci. U.S.* 48, 562.

67. GERGELY, J., GOUVEA, M. A. & KARIBIAN, D. (1955). *J. biol. Chem.* 212, 165.

68. CHIEFFI, G., SINISCALCO, M. & ADINOLFI, M. (1960). *Atti Accad. naz. Lincei, Rend., Classe Sci. fis. mat. nat.* 28, 233.

68a. GOLDSTEIN, A. & BROWN, B. J. (1961). *Biochim. biophys. Acta*, 53, 19.

69. HADORN, E. (1958). In *A symposium on the chemical basis of development* (ed. W. D. McElroy & B. Glass), p. 779. Baltimore, Md.: Johns Hopkins Press.

70. HALBRECHT, I. & KLIBANSKI, C. (1956). *Nature (Lond.)*, 178, 794.

70a. HNILICA, L., JOHNS, E. W. & BUTLER, J. A. V. (1962). *Biochem. J.* 82, 123.

70b. HOMMES, F. A., LEEUWEN, G. VAN. & ZILLIKEN, F. (1962). *Biochim. biophys. Acta*, **56**, 320.
70c. HUANG, R.-C. C. & BONNER, J. (1962). *Proc. natl. Acad. Sci. U.S.* **48**, 1216.
71. HERNER, A. E. & FRIEDEN, E. (1960). *J. biol. Chem.* **235**, 2845.
71a. — (1961). *Arch. Biochem. Biophys.* **95**, 25.
71b. HOWARD, A. & DEWEY, D. L. (1961). *Exptl. Cell Res.* **24**, 623.
72. ITANO, H. A. (1957). *Advanc. Protein Chem.* **12**, 215.
73. JACOB, F. & MONOD, J. (1959). *Compt. rend. Acad. Sci., Paris*, **249**, 1282.
74. — (1961). *Proc. V int. Congr. Biochem., Moscow*, **1**, 132.
74a. — (1961). *Cold Spring Harbor Symposia quant. Biol.* **26**, 193, 389.
74b. JOHNS, E. W. & BUTLER, J. A. V. (1962). *Biochem. J.* **82**, 15.
75. KEECH, M. K. & REED, R. (1957). *Ann. rheumatic Diseases*, **16**, 35.
76. KEECH, M. K., REED, R. & WOOD, M. J. (1956). *J. Path. Bact.* **71**, 477.
77. KENNEY, F. T., REEM, G. H. & KRETCHMER, N. (1958). *Science*, **127**, 86.
78. KIT, S. (1960). *Biochem. biophys. Research Commun.* **3**, 361.
79. — (1960). *J. biol. Chem.* **235**, 1756.
80. — (1960). *Arch. Biochem. Biophys.* **88**, 1.
81. KNOX, W. E. (1951). *Brit. J. exptl. Path.* **32**, 462.
82. KONDO, N. & OSAWA, S. (1959). *Nature (Lond.)*, **183**, 1602.
82a. KRETCHMER, N. & McNAMARA, H. (1956). *J. clin. Invest.* **35**, 1089.
83. LASH, J. W., HOMMES, F. A. & ZILLIKEN, F. (1962). *Biochim. biophys. Acta*, **56**, 313.
83a. LIMA-DE-FARIA, A., REITALU, J. & BERGMAN, S. (1961). *Abstracts of 1st Annual Meeting, American Society for Cell Biology*, p. 127.
84. LOENING, U. E. (1961). *Biochem. J.* **81**, 254.
84a. LOWTHER, D. A., GREEN, N. M. & CHAPMAN, J. A. (1961). *J. biophys. biochem. Cytol.* **10**, 373.
85. MADAN, C. L. (1956). *Planta*, **47**, 53.
86. MARKERT, C. L. (1958). In *A symposium on the chemical basis of development* (ed. W. D. McElroy & B. Glass), p. 3. Baltimore, Md.: Johns Hopkins Press.
87. MARRIAN, D. H., PHILLIPS, A. F. & WERBA, S. M. (1957). *Biochim. biophys. Acta*, **24**, 576.
87a. MAZIA, D. (1961). *Ann. Rev. Biochem.* **30**, 669.
88. MAURITZEN, C. M. & STEDMAN, E. (1960). *Proc. roy. Soc. Lond.* **153B**, 80.
89. McMASTER-KAYE, R. D. & TAYLOR, J. H. (1959). *J. biophys. biochem. Cytol.* **5**, 461.
90. METZNER, H. (1955). *Planta*, **45**, 493.
90a. MIRSKY, A. E. (1961). *Proc. V int. Congr. Biochem., Moscow*, **2**, 73.
91. MIRSKY, A. E. & ALLFREY, V. (G.) (1958). In *A symposium on the chemical basis of development* (ed. W. D. McElroy & B. Glass), p. 94. Baltimore, Md.: Johns Hopkins Press.
91a. MOORHEAD, P. & DEFENDI, V. (1961). *Abstracts of first annual Meeting of American Society for Cell Biology*, p. 145.
92. NANNEY, D. L. (1960). *Amer. Naturalist*, **94**, 167.
92a. NEELIN, J. M. & BUTLER, G. C. (1961). *Can. J. Biochem. Physiol.* **39**, 485.
92b. NASATIR, M., BRYAN, A. M. & RAKE, A. (1961). *Science*, **134**, 666.
93. NEMETH, A. M. (1954). *J. biol. Chem.* **208**, 773.
94. — (1959). *J. biol. Chem.* **234**, 2921.
95. NEMETH, A. M. & NACHMIAS, V. T. (1958). *Science*, **128**, 1085.
96. NICHIPOROVICH, A. A., ANDREEVA, T. F., VOSKRESENSKAYA, N. P., NEZGOVOROVA, L. A. & NOVITSKIĬ, YU. I. (1958). *Radioisotopes sci. Research, Proc. intern. Conf., Paris, 1957*, **4**, 411.

97. ONESON, I. & ZACHARIAS, J. (1960). *Arch. Biochem. Biophys.* **89**, 271.
98. PELLING, G. (1959). *Nature (Lond.)*, **184**, 655.
99. PELC, S. R. (1959). *Exptl. Cell Research*, Suppl. **6**, 97.
99a. POORT, C. (1961). *Biochim. biophys. Acta*, **51**, 236.
100. PELAUTSCH, M. (1960). *Embryologia*, **5**, 139.
100a. PHILLIPS, D. M. P. & SIMSON, P. (1962). *Biochem. J.* **82**, 236.
101. RACUSEN, D. & HOBSON, E. L. (1959). *Arch. Biochem. Biophys.* **82**, 234.
102. RALL, D. P., SCHWAB, P. & ZUBROD, C. G. (1961). *Science*, **133**, 279.
103. ROBINSON, D. S. (1952). *Biochem. J.* **52**, 621.
104. ROCHE, J., DERRIEN, Y. & ROQUES, M. (1953). *Bull. Soc. Chim. biol.* **35**, 933.
105. SADRON, C., POUYET, J. & VENDRELY, R. (1957). *Nature (Lond.)*, **179**, 263.
106. SAKAMOTO, M. (1957). *Tohoku J. exp. Med.* **65**, 269.
107. SAX, K. (1935). *Harvard University : Arnold Arboretum : Journal*, **16**, 301.
108. SCHACHMAN, H. K. (1957). *J. cell. comp. Physiol.* **49**, Suppl. *1*, 71.
109. SCHMITT, F. O. (1959). *Revs. modern Phys.* **31**, 5.
109a. SCHROEDER, W. A., JONES, R. T., SHELTON, J. R., SHELTON, J. B., CORMICK, J. & McCALLA, K. (1961). *Proc. natl. Acad. Sci. U.S.* **47**, 811.
109b. SCHOLTISSEK, C. (1962). *Nature (Lond.)*, **194**, 353.
110. SIAKOTOS, A. N. (1960). *J. gen. Physiol.* **43**, 999.
110a. SCHURIN, M. & MARMUR, J. (1961). *Abstracts of first annual Meeting of American Society for Cell Biology*, p. 193.
111. SIRLIN, J. L. (1960). *Exptl. Cell Research*, **19**, 177.
112. SIRLIN, J. L. & KNIGHT, G. R. (1960). *Exptl. Cell Research*, **19**, 210.
113. SPIRO, R. G. (1960). *J. biol. Chem.* **235**, 2861.
114. STEDMAN, E. & STEDMAN, E. (1950). *Nature (Lond.)*, **166**, 780.
115. STEDMAN, E., STEDMAN, E. & PETTIGREW, F. W. (1944). *Biochem. J.* **38**, xxxi.
116. SUSKIND, S. R. (1957). In *A symposium on the chemical basis of heredity* (ed. W. D. McElroy & B. Glass), p. 123. Baltimore, Md.: Johns Hopkins Press.
116a. TAYLOR, J. H. (1960). *J. biophys, biochem. Cytol.* **7**, 455.
117. TIMMER, R., HELM, H. J. VAN DER & HUISMAN, T. H. J. (1957). *Nature (Lond.)*, **180**, 240.
118. VENDRELY, R., KNOBLOCH, A. & VENDRELY, C. (1958). *Compt. rend. Acad. Sci., Paris*, **246**, 3128.
119. VOGEL, H. J. (1957). In *A symposium on the chemical basis of heredity* (ed. W. D. McElroy & B. Glass), p. 276. Baltimore, Md.: Johns Hopkins Press.
120. WANG, D. Y., SLATER, T. F. & GREENBAUM, A. L. (1960). *Nature (Lond.)*, **188**, 320.
121. WILKINS, M. H. F. (1956). *Cold Spring Harbor Symposia quant. Biol.* **21**, 75.
122. WILKINS, M. H. F. & ZUBAY, G. (1959). *J. biophys. biochem. Cytol.* **5**, 55.
123. WRIGHT, S. T. C. (1960). *Nature (Lond.)*, **185**, 82.
124. WRIGHT, B. E. (1960). *Proc. natl. Acad. Sci. U.S.* **46**, 798.
125. ZUBAY, G. & WATSON, M. R. (1959). *J. biophys. biochem. Cytol.* **5**, 51.

# CHAPTER XIX

# CHANGES IN PROTEINS AND NUCLEIC ACIDS OCCURRING WITH AGEING AND THE PROBLEM OF AGEING AT THE MOLECULAR LEVEL

## Introduction

The morphogenesis of individual organs and tissues is not continuous. The period of formation of the organ is usually followed by a period of functioning. This active period is, however, by no means everlasting; it is gradually superseded by a period of ageing and dying which, in most species of animals, coincides with the ageing and death of the individual, though in many plants the individual organs may grow old and die autonomously. Ageing differs in a qualitative way from morphogenesis in that during ageing no proteins with essentially new structures or functions arise, nor do new matrices or new organs. What already exists becomes changed and the changes are of a degenerative nature. Externally they resemble simple wearing out. The only exceptions to this state of affairs are members of some species in which death occurs as a result of ontogenesis and which die almost at once when they have carried out their reproductive functions.

Ageing is a biological phenomenon which guarantees a feature of living things which is very important from the point of view of evolution, namely the mortality of the individual. The death of individuals gives rise to a rapid succession of generations in some species of animals and plants and is a necessary condition for the evolution of living things as a whole. During the actual process of evolution there therefore arise various forms of death. It may be quick, just after fertilization (in many lower forms, certain annual plants and some insects) or it may be slow and result from the gradual ageing and incapacitation of the organism, which spontaneously loses its ability to resist unfavourable circumstances (diseases, environmental conditions, predators, etc.). Ageing is the commonest cause of the death of the individual and its purposive character in regard to the evolution of living things in general suggests that it may occur by a biological mechanism which is common to all forms of living matter.

Nevertheless there is a direct material connection between successive generations of living things which has not been broken since the moment

469

when life first came into being more than 1,000,000,000 years ago and which will obviously exist for ever.

From the point of view of biology as a whole ageing is, therefore, a purposive characteristic and has a common biological basis. There can be no doubt that there must be some mechanism of ageing which is common to all forms of living material but which yet does not affect the living link which connects the generations with one another.

Naturally, if ageing is a phenomenon having a common biological basis and occurring in all forms of living matter it must be connected primarily with some feature of the metabolism of proteins and nucleic acids, because the synthesis and metabolism of these compounds are the processes which are common to all living things.

Obviously when a complicated organism grows old, certain changes will occur in the course of time at all levels of specific organization and in all systems, molecular, subcellular, cellular, physiological, functional and morphological, and it is still hard to say which of many hundreds of variations of the living body occurring after maturity will be the limiting factor determining any particular length of life.

It is, however, clear that before we take up the question of the mechanisms by which changes in proteins and nucleic acids occur in adult life we must first convince ourselves that such changes really do occur. This preliminary analysis of the nature of ontogenetic changes in proteins is particularly important in understanding the processes of ageing which occur in most organisms in the absence of any obvious morphological changes and therefore sometimes give rise to an impression of simple physiological and functional incapacitation which has no molecular basis.

We have already written a specialized review and theoretical analysis (69) covering some of the evidence which will be brought forward in the present chapter.

## 1. Changes occurring in the proteins of the fully formed adult organism

We cannot yet say that the scientific literature concerned with the investigation of the nature of the changes occurring in proteins in adult life contains an extensive or satisfactory amount of material. The evidence in this field is very fragmentary and often contradictory.

In the first place we must note that in most investigations of this kind the authors have been mainly concerned with the total protein, which consists of individual proteins having different properties and compositions. Zbarskiĭ (27) has tried to show that, from a theoretical point of view, it is necessary to analyse the total protein, but many objections have been raised to this way of approaching the study of the changes occurring in adult life.

Without any doubt the analysis of individual proteins is more valuable from the point of view of establishing the nature of the possible changes in proteins than is the study of any preparations of the total proteins of an organ or an organism. There can, however, also be objections to adhering too strictly to this methodological desideratum. If we take as our criteria of the individuality of a protein those features of it by means of which we assess its structural homogeneity (behaviour during electrophoresis, ultracentrifugation, etc.), then any alteration in the structure with increasing age or due to other causes will change the constants in question and will thus remove the affected protein from the individual group under study and make it incomparable with its former self. If we admit that proteins can vary we cannot set about comparing them as absolute chemical individuals. By trying to find among objects of different ages some chemically homogeneous proteins for comparison we commit ourselves to discovering that proteins do not change. Furthermore, even if we could find such proteins it would be legitimate to suggest that some part of the protein being studied had been altered and had thus left the category being studied while the fraction analysed, which was the same at all ages, represented a group of molecules which had not yet undergone alteration owing to local inequalities in the process of ageing.

These facts give rise to a number of difficulties in the study of the changes occurring in proteins during ageing. It would therefore be most satisfactory if we could study families of proteins as a whole (e.g. a complex of albumins or a complex of haemoglobins), both in regard to the composition and the quantitative relationships between the fractions, and also in regard to their amino acid composition and (in the future, we hope) their amino acid sequence. Furthermore, very valuable material could be obtained by a similar study (on the fractional composition with an analysis of each fraction) of the variation of proteins which are associated morphologically or biologically into complexes which carry out particular functions and can reproduce themselves autonomously during the lifetime of the organism. The elementary intracellular structures such as the various types of microsomes, chondrosomes, plastids and so forth are all made up of such groups of proteins.

### a) Changes in the ratios between proteins and protein fractions occurring with increasing age

Our ideas as to the fractional composition of proteins change rapidly and we attribute this mainly to the development of modern techniques of fractionation. This means that many of the results which were obtained earlier in respect of "individual" proteins really represent the properties of the sum of several different fractions or so-called "families" of related proteins (96).

Obviously then, when studying changes in the proportions of proteins occurring with ageing we must divide the problem into two. On the one hand we must look at changes in the proportions of proteins (more exactly, families of proteins) with different functions (e.g. collagen and elastin, albumin and globulin, etc.) while, on the other hand, we must analyse the alterations in relative proportions of various subfractions of a protein all of which have the same function although they may be present in varying amounts (e.g. $\alpha$, $\beta$, $\gamma$, etc. chymotrypsins). It is also possible for intermediate cases to occur, e.g. the isolation from a single organ of several ribonucleases acting on different types of bonds in RNA (186).

At present it is very difficult to deal shortly with all the evidence concerning the "microheterogeneity" of proteins with a particular function, as a very extensive literature on this subject has already been built up. The chemical nature of the microheterogeneity of proteins is not the same in all cases. Sometimes it consists in differences in amino acid composition and in the terminal amino acids while in other cases the difference is only in the form of the molecule or the degree of polydispersity. Finally, in complex proteins, heterogeneity may be due to the properties and the ability of the protein to combine with the non-protein component. The sources of these subfractions may also be different (tissue and organ specificity, the heterogeneity of organelles, etc.). This question has been discussed in detail by Orekhovich and his colleagues (97).

The subject which is now being studied most intensively in this connection is the heterogeneity of haemoglobin (191, 138, 139, 164, 170, 195, 196, 197, 241, 242) and that of the serum proteins (18, 198, 180, 264 and others) because refined analysis of the composition of these proteins is beginning to be used for diagnostic purposes. Many of the proteins of plants and animals which had previously been thought to be chemical individuals and had been obtained in crystalline form are now known to be made up of several components. These include, among others, procollagen (99), collagen (19), insulin (178), phosphorylase (193), lactic dehydrogenase (283), chymotrypsin (149), myoglobin (247), histone (184, 277), glycinin (45) and myosin (240a).

It must be noted that the changes in the proportions of the components occurring during ageing have hardly been studied, haemoglobin alone being somewhat of an exception to this rule.

Changes in the proportions of proteins having different functions have, however, been studied rather more fully, but most of this material is concerned with the morphogenetic (embryonic and immature) stage of ontogenesis while the period with which we are dealing (from young adulthood to old age) has only been very inadequately studied. The most extensive evidence of this kind refers to the proteins of the blood in various animals and man (20, 25, 39, 104, 124, 234) and in this case the characteristic change is an increase in the proportion of globulin to albumin with advancing age.

Studies of the proportions of the cytoplasmic proteins are very limited and often do not include the necessary electrophoretic analysis of the fractions. Detailed analysis of a number of quantitative and qualitative indicators of the total cytoplasmic protein in different organs and of their variation with age have been carried out by Bulankin and his colleagues (9, 115). It was found that there were definite changes with age in the proportions of the different soluble proteins in the lens of the eye (128, 179). It was also found that there were characteristic changes in the electrophoretic pattern of the proteins of the cytoplasm and plastids in plants, to some of which we have already drawn attention (117, 284). We shall deal later with a number of results which have been obtained with regard to changes in enzymic activity associated with ageing and also, it would seem, connected to some extent with the changes during ageing in the quantities of particular enzymes present in the organism.

However, the most characteristic change in the protein composition of organs and tissues occurring during ageing is certainly the quantitative change in the relative amounts of cytoplasmic and metaplasmic proteins which Nagornyĭ (76-79) used as one of the foundations for the detailed working out of a theory of the ageing of organisms. The occurrence of characteristic quantitative changes in the metaplasmic proteins has also been confirmed by later researches. The most characteristic process of "metaplasmatization" of the organism is the regular relative and absolute increase in collagen in organs and tissues (30, 101, 159, 169, 172, 194, 207, 267, 265, 206). At the same time the amount of procollagen and soluble collagen decreases with age (100).

The further detailed study of these indicators of protein composition during the development and ageing of the individual will certainly play a very important part in working out a complete biochemical picture of ageing. Changes in the ratios of proteins tending towards a preponderance of metabolically inert forms will certainly also be one of the main indicators of this process. It is, however, important to find out the primary mechanism of these phenomena by means of analyses of the quantitative and qualitative changes in the synthetic systems, because it may be that the metaplasmatization of the organism is a secondary feature associated with change and degradation of the system for the synthesis of the active cytoplasmic proteins, the "gaps" being filled by other processes. The most direct indicator of qualitative changes in protein synthesis is the study of changes in the amino acid composition and internal structure of particular proteins occurring during ageing.

### b) Changes in the amino acid composition of proteins occurring during ageing

Researches dealing with the actual amino acid composition of individual biological objects already amount to several thousands, but only a few of them are concerned with the way it changes during ageing. Finally, there has been, literally, only one piece of research on the changes in the amino acid composition of proteins which are real individuals. The theoretically important question as to whether such changes really occur and, if so, in what direction they tend, is therefore still unsolved. Several studies of the changes in the amino acid composition of the total protein of whole organs, which occur as the organism gets older, show that there are certain changes which occur during the life of the individual. As yet, however, they do not prove the occurrence of changes in individual proteins as the organism gets older. They might just as well be simply due to changes in the quantitative relationships between the proteins and between intracellular structures, especially as the ratios between different protein fractions and different structures do, in fact, alter during the life of the individual. These studies are, of course, of decisive importance in establishing the characteristic biochemical features of ageing but they can hardly be used for working out the mechanisms whereby proteins actually change with increasing age.

We shall, therefore, first discuss the work done on the changes occurring in individual proteins. We shall, at this stage, call to mind several older studies of this sort, while remembering that the methods of purification of proteins and determination of their amino acid composition were not sufficiently accurate at the time when these studies were made. A review of this work may be found in a paper by Kizel' (34).

Among the numerous examples discussed by Kizel' only three (those of Abderhalden & Fuchs, Deseo et al. and Schenk) deal with changes in "individual" proteins with increasing age. In the first, the authors observed a decrease in the glutamic acid present in the keratin of horns and hooves as the animals grew older. In the second, variations were observed in the diamino acids in the albumin and globulin of the blood while the third also related to inconstancy in the composition of the globulin with age.

As to studies of variations with age in the amino acid composition of individual proteins, we only know of three, the work of Lansing and his colleagues (218) on the analysis of elastin, that of Korinenko (39, 41) on the changes with age in the composition of serum albumin and myosin and that which has been carried out in the laboratory of Sisakyan (118, 119) and deals with the changes with age in the composition of one of the protein components of the plastids and leucoplasts of the sugar beet.

Lansing and his colleagues studied the composition of elastin derived from the aorta of old and young people. The changes were found to affect only the amino acids listed in Table 14. At the same time they observed the accumulation of calcium in the aorta.

In looking at the figures given by Lansing we must not fail to notice that the change in the elastin consists in a considerable increase in those amino acids of which it normally contains less than do the cytoplasmic proteins and, on the other hand, a decrease in the amount of glycine in which many metaplasmic proteins are rich. If these results are not due

TABLE 14

Amino acid composition of old and young elastin (218)

| Amino acid | g./100 g. of protein | |
| --- | --- | --- |
| | young | old |
| Aspartic acid | 0·38 | 1·11 |
| Glutamic acid | 1·83 | 3·01 |
| Glycine | 26·1 | 21·3 |
| Lysine | 0·49 | 1·17 |
| Arginine | 1·78 | 4·35 |
| Histidine | 0·15 | 0·75 |
| Methionine | 0·06 | 0·35 |
| Tryptophan | 0·06 | 0·24 |
| Serine | 0·20 | 0·70 |
| Threonine | 0·65 | 1·13 |

to the chance contamination of the old elastin with some other protein they are of considerable interest as they show something which looks like an "approximation" of the composition of the protein to that of the metabolic pool of nitrogenous compounds.

We must mention that in analysing the total protein of the aorta, 40-50% of which is elastin, Gortner (182) did not find any marked differences in amino acid composition. However, the methods used for estimating the amino acids were very crude and had an error of the order of ± 100%.

In a recent research Korinenko (39) has studied the amino acid composition of the serum albumin and globulin of rats of different ages. He made quantitative determinations of 10 amino acids and did not find any noteworthy changes in the composition of these proteins as the animals got older. In another research he found that the composition of myosin remained constant with age in respect of 10 amino acids (41).

The work of Sisakyan and his colleagues (118, 119) was concerned with the amino acid composition of preparations of plant proteins derived from the leucoplasts of the roots and the plastids of the leaves of the sugar beet at different stages of its growth. In several cases the authors observed changes in the amounts of aspartic acid, alanine and lysine as well as an increase in the total sulphur content, indicating an increase in the sum of the sulphur-containing amino acids (cystine, cysteine and methionine).

We must also mention the detailed work of Muto & Araki (230) who studied the changes occurring in the amino acid composition of various organs of the rice plant as it grew older. The authors did not isolate individual proteins but fractionated the total protein into five fractions according to their solubility, and this made their work far more useful than it would have been had it been carried out on samples of the total

PB 21

protein. The authors observed considerable changes in the amount of several amino acids, but their results showed no consistency, as different fractions of proteins behaved differently in this respect, sometimes giving quite contradictory results.

The rest of the work on the amino acid composition of proteins, especially those from animal sources, is mainly concerned with the total proteins of organs and tissues and it is therefore difficult to use the results for theoretical analysis of the mechanisms of change (21, 26, 29, 42, 123, 125).

Several studies of this sort have been carried out on plant materials but the proteins have been analysed as a whole, owing to the poorness of the methods used for the separation of individual preparations. Furthermore, the leaves of plants do not contain typical intercellular metaplasmic proteins which have a markedly different composition from that of the cytoplasmic ones.

If we survey the available evidence in this connection we shall also find ourselves unable to discover any definite consistency in the changes occurring in the total proteins of plants. Kizel' (33) analysed the proteins of the leaves of the water melon at two stages of their development and found that as they grew older there was a decrease in the amounts of arginine and proline present while the amount of histidine increased. Lugg & Weller (222) observed that, as the leaves of clover grow older, the amount of methionine in them gets less while that of cystine and tyrosine gets greater. Waite *et al.* (280) determined the amounts of lysine, histidine and arginine in four species of grass at different stages of development but did not find any substantial variations in the amounts of these amino acids. Reber & MacVicar (243) determined the amounts of nine amino acids present in preparations of the whole plant of several members of the Gramineae. It was found that while the ears were forming and ripening the amount of glutamic acid and isoleucine in the leaves increased while that of lysine and methionine decreased.

In an interesting paper Steward *et al.* (270) reported that in the lupin they observed considerable changes in amino acid composition of the proteins as between growing and mature leaves and young and old roots. In a later series of experiments, workers in the same laboratory (238, 268, 269) found a very specific process of change of proline into hydroxyproline in some plant proteins. The authors believe that this change enhances the inertness of the proteins.

In the work of Kemble & Macpherson (211) comparing the proteins of young and old rye-grass plants in respect of their content of 10 monoaminomonocarboxylic acids, hardly any alterations were noticed (in this work the amounts of glutamic acid, lysine and methionine were not determined).

Very recently the changes with increasing age in the proteins of the leaves of two plants (narcissus and barley) have been the subject of detailed investigations in the laboratory of Fowden (113, 156, 237) but in this case no marked changes were found. The authors found that during ageing there was slight decrease in the amount of basic amino acids in the leaves of the narcissus and they believed this to be connected with the relatively small amounts of nuclear proteins in these cells.

We must pay special attention to those experiments in which the changes in the composition of the proteins, in particular intracellular fractions, were studied. These studies are more valuable from the point of view of the protein biochemistry of ageing than are analyses of tissues as a whole, because they deal with the composition of specialized intracellular structures, the biochemical activity of which is very closely connected with some particular physiological function and which are the simplest structures which can carry out the fundamental biological activities. Unfortunately this group of studies is still extremely meagre.

In one such investigation, carried out by Osipova & Timofeeva (102), a comparison was made between young (June) and old (September) leaves of beans. Their results showed definite differences (cystine, dicarboxylic amino acids, tryptophan and histidine).

Using a microbiological method Yemm & Folkes (286) determined the amounts of 18 amino acids in the proteins of the plastids and cytoplasm of mature and young leaves of barley. The amounts of all the amino acids remained at the same level (variations within the error of the method) and only lysine showed a statistically significant difference (decreased with age from 7·9-6·5%).

In 1955-6 we ourselves (68) studied the changes with age in the amino acid composition of individual intracellular fractions of the proteins of bean and sunflower leaves. We only determined the amounts of the sulphur-containing amino acids cystine and methionine in the proteins, using a radiochromatographic method on hydrolysates of proteins obtained from plants which had been grown on a medium with a constant ratio of radioactive to stable isotopes of sulphur. This made the accuracy of the determination of the sulphur-containing amino acids greater than it would have been had the ordinary method of paper chromatography been used and made it possible to use the two-dimensional rather than the one-dimensional method. The proteins were analysed after they had been divided into four fractions according to their intracellular localization (proteins of the vacuolar juice and vascular bundles, proteins of the plasma juice, proteins of the plastids and proteins of the cytoplasmic structures).

As a result of these analyses, which were carried out during the whole period of growth of the plant, it was found that certain changes occur in the amounts of these amino acids, mainly in the form of increases which were not always the same but depended on location in plant, nature of the fraction and the conditions of cultivation. Bearing in mind that an increase in the amount of these amino acids is the most easily noticable change in the proteins of plants (102, 119, 222, 226, 230) we suggested that the amount of sulphur-containing amino acids in the proteins of a plant is a dynamic biochemical indicator associated with the occurrence of various ontogenetic phases and with the age of the plant. The actual mechanism of this connection requires further study. It must, however, be reckoned that the sulphydryl groups of proteins play an extremely important part in the regulation of enzymic processes and in the formation of the specific internal structures of proteins. In this connection we must also mention the interesting results obtained by Vinter (279a) who, having found an increase in the cystine and cysteine content of the spores of *Bacillus megaterium*, concluded that the increase in the number of sulphide bridges leads to an increase in the stability of the proteins. In their recent papers Parhon & Oeriu (109, 110, 235) attach great importance to the substitution of disulphide and methyl groups in proteins as part of the general picture of ageing. We must, however, bear in mind the results of Nikitin, Silin & Moroz (93) who did not find any change associated with ageing in the cystine and methionine content of the proteins of the liver and muscles of white rats. Korinenko (39) also did not find changes in the sulphydryl groups in proteins in the blood of rats of various ages.

To summarize the evidence given here on the changes in the amino acid composition of proteins with increasing age, it must be said that it is still hard to draw any conclusion. Judging from most of the evidence, a small change occurs in the amino acid composition of some proteins during ageing while other proteins differ in having a constant composition. It is not possible to notice any definite tendency in these changes. It is clear that further systematic studies in this direction are required, with correlation of the analyses and improvement in their accuracy. The study of changes occurring in proteins during ageing requires methods of analysis which can show up very small changes. We must also take account of the fact that, if there is random variation in the composition of proteins resulting, for example, from errors of biosynthesis (which is, in fact, extremely probable), other changes which may occur will be very hard to identify by present-day methods. Individual molecules may alter in different ways and in the analysis even of so-called individual proteins this variety of changes may be averaged out so that it does not appear in the analysis.

## c) Changes in the physical and physico-chemical properties of proteins with ageing

Although the available evidence concerning the changes in the amino acid composition of proteins occurring with increasing age is contradictory and fragmentary, that concerning their physico-chemical alteration presents a more definite picture. However, these researches too are, for the most part, not concerned with chemically individual proteins of the protoplasm, but with preparations of all the proteins of a tissue, including such extracellular fibrillar proteins as collagen.    The evolution of the physico-chemical properties of the total protein of tissues and organs with age is quite clearly demonstrated.    The most characteristic alteration in the protoplasm of the cell as it grows older is a lowering of the hydrophilia and water-retaining power of the protein colloids.    The literature contains so many series of studies of this problem that it is hardly worth while to discuss any particular one here, the phenomenon is so generally accepted and of such regular occurrence.

The dehydration of the colloids of protoplasm consists primarily in the dehydration of its proteins.    Physiologically reversible dehydration of colloids may be associated with a number of factors such, for example, as entering on a period of rest.    The irreversible dehydration of proteins with ageing is primarily associated with changes in those properties and peculiarities of the protein which are responsible for the extreme hydrophilia of living material, that is to say, essentially with changes in the specific native structure of the protein, with the appearance of protein particles in the cell and with a decrease in the number of free hydrophilic groups.

Ageing of plant and animal cells is, as a rule, characterized by an increase in the viscosity of their protoplasm (48, 49).    It has also been observed that ageing is associated with a decrease in the adsorptive powers of proteins.    Saïchuk (cf. 28) has drawn attention to the fact that the erythrocytes of young horses sorb rather more amino acids and peptides than those of old animals.

Salganik (114) has recently found that the same is true of the plasma proteins in in vitro experiments.    He made such studies of the binding of [35S]methionine by adsorption or chemical linkage to the plasma proteins of adults and old rats.    The experiments which he carried out showed that the age of the animals is of considerable importance in connection with the intensiveness of this process.    The amount of [35S]methionine bound by the plasma proteins of young animals in vitro was almost 40% higher than that bound by the plasma proteins of old animals.    The plasma protein of middle-aged animals occupied an intermediate position in this respect.

Salganik believes that the changes in the amount of bound [35S]methionine which he observed are due to changes in the ability of the functional groups of the proteins to react.    The results obtained by Pasynskiĭ and his colleagues (103, 112) suggest that SH groups play an important part in this process.

Working with plant proteins, we ourselves (67) have also found changes in the adsorptive properties of proteins as they affect methionine.    The difference between young and old leaves was again between 30 and 40%.

The physico-chemical structure of proteins also undergoes characteristic changes with ageing, the most detailed studies of this question having been made on the metaplasmic proteins, in particular collagen (53, 57, 74, 75, 129, 160, 202, 231, 143, 258, 213, 279, 223). The collagen fibre becomes thicker, more rigid and more resistant to the effects of temperature as it grows older. Its elasticity and ability to shorten also diminish sharply. The ratio of gel to fibrils in the collagen of the connective tissues also changes considerably (267). Collagen also becomes more resistant to hydrolysis (261). It may also become chemically linked with elastin (153a).

In the scientific literature, especially in older papers (those of Ruzicka, Marinesco, Minot, Shade and others) but also in some modern ones, the suggestion has been made that the changes occurring in the physico-chemical properties of the colloids of the cells may account for the ageing of the organism as a biological phenomenon. Present-day evidence as to the metabolism and self-renewal of the proteins of the organism, however, indicates that the conditions within the living cell are not such as to cause the typical, spontaneous, physico-chemical processes of hysteresis of the cytoplasmic colloids. Any process of change in the colloid-chemical properties occurring in the organism does so as a result of its own vital activities and of the processes of self-renewal characteristic of any protein. However, for many proteins such as elastin and collagen, self-renewal takes place very slowly and this provides the necessary conditions for the occurrence of spontaneous changes. The theory of ageing worked out in 1935-40 by Nagornyĭ (76) lays great stress on the importance of these changes for the process of ageing of the organism as a whole. It has now been found that in the organism of the rat, in the tail for example, there are fractions of collagen which are renewed so slowly that the "lifetime" of the individual molecules may be up to a year or more. Such stability of certain proteins also provides the basis for their spontaneous physical and physico-chemical variation so that, according to the ideas of Nagornyĭ, each altered, "stabilized" molecule loses its functional significance, is only dissimilated with great difficulty and therefore accumulates as "ballast" which only changes in the direction of further inactivation. Kao and his colleagues (205) have recently shown that this possibility may really occur. They found that the insoluble collagen of the skin and tails of rats aged from 8 to 24 months is not renewed at all. The rate of renewal of the cytoplasmic proteins also decreases with age and this increases the tendency towards the occurrence of spontaneous physico-chemical variation as a secondary process.

One must agree with the view put forward by Lansing (217) that the old organism does not contain old colloids; it contains newly formed colloids which have the characteristic properties of old age. In this connection we cannot avoid noting that even collagen, the physico-chemical evolution of which is most clearly observable, becomes metabolically inert during the very process of "pure" ageing in the post-embryonic stage of development (47, 208).

*d*) *Changes in the digestibility of the proteins of the organism
and their susceptibility to degradation associated with ageing*

In his works on ageing the Czech biochemist Ruzicka showed, several decades ago, that the older the tadpole, the harder it is to digest it with trypsin *in vitro*. Similar results are reported in other old papers by different authors. Results concerning autodigestion or the digestion of the organism as a whole are not, however, very helpful, as they may be affected, not only by inactivation of proteins, but also by changing the proportions in which the different proteins are present, as well as by a decrease in the activity of the enzymes. Studies of the action of proteinases on individual proteins and the proteins of individual tissues and organs are much more valuable.

The work of Gol'denberg & Kondrashina (17) on the rate of tryptic proteolysis of the serum proteins of rabbits of different ages is very interesting in this connection. They found a marked decrease in the ease with which the serum proteins as a whole (albumin + globulin) could be split as the animals got older.

However, by similar experiments Korinenko, Lobanov & Denisov (40) have shown that the changes in digestibility by pepsin of the serum proteins of rats and horses as they grow older is mainly due to an increase in the proportion of the more stable globulins. Another paper from this laboratory reported the finding of changes in the effect of pepsin on myosin with increasing age (41).

A particularly striking change in the digestibility of mixed proteins has been demonstrated by Bulankin & Blyumina (8) who studied this aspect of the proteins of the liver, brain and heart muscle and of collagen derived from animals of different ages.

As a result of their studies it was shown that the susceptibility of the proteins of the organs as a whole (liver, heart, brain) to enzymic proteolysis decreases with age. In the experiments on collagen mentioned above, the collagen of the tails of old rats and of young ones showed the same behaviour in relation to trypsin. However, collagen is, in general, very resistant to trypsin and is mainly hydrolysed by a specific proteinase called collagenase (168). In this connection we must draw attention to two papers by Keech (209, 210) which showed that collagen derived from the skins of people of different ages was not digested in the same way by collagenase. The collagen of old people was considerably more resistant to hydrolysis by this enzyme.

Keech suggests two possible causes for this phenomenon: the inertness of "old" collagen may be due to metabolic (chemical) or to purely physical causes. In his opinion the coarse, thick and compact collagen fibres of old people are, for purely physical reasons, less accessible to the action of collagenase than are the finer and more delicate fibres of young ones. Such an explanation is very probable but the second of these reasons, the physical structure of the fibre, must surely reflect, to some extent, the internal structure of the protein molecules.

Kohn & Rollerson (213) have also shown that there is a marked decrease in the susceptibility of collagen to hydrolysis by collagenase with increasing age. The collagen used in these experiments was obtained from the tendons and the diaphragms of people of different ages. The ability of the collagen to swell also decreases with age.

Among other studies of the changes in the digestibility of proteins with age we must mention that of Orekhovich & Sokolova (98) in which they showed that various newly regenerated tissues (in amphibians) were always more susceptible to hydrolysis by cathepsin than were the rest of the old tissues. The authors attribute this difference to the structure of the proteins, as the amounts of activators of the enzyme (glutathione, cystine) were not found to be significantly different.

Nikitin & Prokopenko (92) have recently observed characteristic changes associated with ageing in the susceptibility of proteins to autolysis in the ontogenesis of the oak silkworm. A decrease in the digestibility of proteins with increasing age has also been observed in preparations of vegetable origin (63, 65, 165).

The susceptibility of proteins to hydrolysis by enzymes is an important indicator because it is indirect evidence for the occurrence of changes in the internal structures of proteins. At the same time it is a constant which is undoubtedly connected with the rate of intracellular metabolism, for the breakdown and synthesis of proteins, which form the basis of self-renewal, are interdependent. It is, however, difficult to assess, purely on the basis of the rate of breakdown of a particular protein, the nature of those changes in its structure which determine the alteration in its behaviour in regard to the enzymes. Changes may affect the sequence and proportions of the amino acid residues but they may be confined to purely physico-chemical characteristics such as the closeness of the packing of the polypeptide chains, ability to swell, aggregation of molecules, and the masking of active groups which will also be reflected in the interaction between the enzyme and substrate. All the same, the absence of any change in the response of a protein to the digestive enzymes cannot be assumed to extend to its response to the intracellular proteases and peptidases, which have a weaker action on proteins. In order to get more accurate evidence on this problem we shall need to take account, not only of the rate of breakdown under the influence of the intracellular proteases, but also of the nature of the breakdown products derived from proteins of different ages.

The author (65) has made a few observations along these lines. These experiments were designed on the basis of the hypothesis that if the changes occurring in proteins with ageing are accompanied by an accumulation in them of bonds which are not easily broken by proteases, then it might be possible to find, among the products of proteolysis of such proteins, a relatively larger proportion of unsplit peptides and poly-peptides than one would find in a hydrolysate of young proteins.

There is considerable difficulty in choosing a suitable method for such an investigation. The proteins are being changed within systems of intracellular metabolism and primarily by the enzymes of intracellular metabolism. From a methodological point of view one ought, therefore, to study the relationship between the proteins and enzymes of those microstructures in which the proteins undergo their ontogenetic evolution. The need for this will be shown in a later theoretical analysis of the mechanism whereby proteins are inactivated during ageing. However, our relatively slight knowledge of the proteases of the intracellular organelles and the methods for isolating and purifying them make this approach unsuitable. The study of the products of the breakdown of proteins in living tissues when synthesis is inhibited approach this ideal but this is only possible in objects of plant origin (such as leaves which have been cut and placed in the dark) because in animals the breakdown products are quickly moved from organ to organ and become dispersed. However, if we use this approach we become involved, not only with changes in the proteins forming the substrate, but also with changes in the proteins constituting the enzymes, and these may affect both their specificity and their activity. It has been shown in several papers that the quality of the proteolytic enzymes, by which we mean their ability to lower the energy of activation, decreases with age (7, 8). This factor may also be important if, instead of using living tissues, we make our observations on the composition of the products of autolysis of crushed tissues. Furthermore, the autolysis of crushed tissues brings into contact enzymes and substrates which would, perhaps, normally not react with one another and in this way it still further confuses the picture of the physiological breakdown of proteins. Analyses of the hydrolysis of proteins derived from sources of different ages by the enzymes of the digestive tract are also artificial on account of the greater "power" of these enzymes in comparison with the intracellular proteases. Finally, we may form an opinion as to the structure of proteins of various ages by studying their acid and alkaline hydrolysis.

So far we have only carried out one experiment along these lines, using the perennial, rhizome-bearing leguminous plant *Galega orientalis*. The objects studied were young leaves gathered in June and old leaves gathered in August from the same level of the plants. Immediately after they had been gathered from the shoots the leaves were placed in a moist chamber in the dark at a temperature of 21°C. If the leaves were supplied with nutrient substances in the absence of light the proteins in them were broken down with the liberation of amino acids and peptides, the relative amounts of which were calculated by the usual methods. From the results of this experiment given in Table 15, it is clear that during the physiological breakdown of proteins a considerable quantity of peptides is formed and in the older leaves a considerably higher proportion of the breakdown products is in the form of peptides. This indicates that a greater number of the peptide bonds in the proteins of these leaves were resistant to the proteolytic enzymes present there.

Although the rate of degradation was faster in the young leaves, this feature was not characteristic as it does not depend entirely on the stability of the proteins but also on the rate of respiration. The overall loss of dry weight in 6 days resulting from respiration was $14 \cdot 3\%$ for young leaves and only $7 \cdot 7\%$ for old ones.

The differences between the amino acid compositions of breakdown products obtained from young and old leaves in these experiments were also very interesting. The method of "dark" degradation (during starvation) makes it possible to observe certain differences in the course and end-products of the breakdown of proteins from young and old organs.

It is tempting to try to study the breakdown of proteins of different ages under the influence of protoplasmic preparations derived, for example, from young tissues of the same species of plants and animals. A serious obstacle to this is the autolysis of the proteins of the preparation so that it would be difficult to set up a control for such an experiment because the autolysis of enzymes occurs in different ways according to whether the substrate is present or absent (132, 133).

The method of "starvation breakdown" of proteins has also been used for the study of changes occurring in the tissues of animals as they grow older (15, 80). In these experiments it was found that the endogenous breakdown of proteins during starvation proceeded considerably more slowly in old animals than in young ones. Typically the endogenous breakdown hardly affected the so-called structural proteins like the liponucleoproteins in old animals while in young ones the breakdown extended rapidly to all kinds of proteins.

A very original approach to the determination of the connection between the susceptibility of the proteins and tissues to decomposition and the "age" of the proteins has been adopted recently by Fleischer & Haurowitz (176). These authors decided to test whether any difference whatever could be observed in the dissimilation of blood albumin under experimental conditions in which albumin from embryonic and adult rabbits was introduced into the blood of one and the same adult rabbit. The authors did not, however, find any significant difference in the rate of dissimilation of albumin derived from animals of different ages. In another series of experiments Walter & Haurowitz (281) showed that the rate of elimination of molecules of "young" albumin formed shortly after the introduction into the organism of [$^{14}$C]amino acids did not differ from the rate of elimination of "old" molecules formed 0-4 days after the introduction of [$^{14}$C]amino acids.

We also studied the same problem somewhat earlier in experiments on the total proteins of plants and yeast (63). In these experiments the

TABLE 15

Characteristics of the products of "dark" degradation of proteins in young and old leaves of *Galega*

| | Decrease in protein % | Amino N in mg./30 g. wet wt. | | % of amino N combined as peptides |
|---|---|---|---|---|
| | | Free before hydrolysis | Free after hydrolysis | |
| *Young leaves* | | | | |
| Control | — | 21·0 ± 1·2 | 30·6 ± 2·8 | 31·4 |
| "Dark" degradation for 24 hours | 3·08 | 27·4 ± 3·4 | 37·6 ± 3·4 | 27·1 |
| ,, ,, 72 ,, | 13·9 | 39·8 ± 1·3 | 59·2 ± 3·4 | 32·8 |
| ,, ,, 144 ,, | 33·7 | 61·6 ± 4·6 | 77·6 ± 5·4 | 20·6 |
| *Old leaves* | | | | |
| Control | — | 9·8 ± 0·42 | 15·05 ± 1·1 | 35·0 |
| "Dark" degradation for 24 hours | 5·09 | 11·9 ± 0·45 | 22·4 ± 1·2 | 46·9 |
| ,, ,, 72 ,, | 8·03 | 20·3 ± 2·0 | 40·2 ± 1·9 | 49·6 |
| ,, ,, 144 ,, | 23·0 | 22·8 ± 1·4 | 34·6 ± 2·1 | 34·9 |

materials (leaves of plants or suspensions of yeast) were autolysed for various lengths of time after the introduction of radioactive label ($^{35}$S) *in vivo* and from the changes in the specific activity of the remaining protein during the process of autolysis it was possible to determine the relative rates of breakdown of the "fresh" molecules which had only just been formed and the older fractions of completely labelled proteins. In general, when the label had only been introduced for short periods the specific activity of the proteins increased during autolysis, suggesting that the "fresh" "young" molecules were rather more stable.

We drew a similar conclusion from the results of a series of experiments on the autolysis of nucleic acids formed at various intervals after the introduction of $^{33}$P (62). These experiments were carried out in 1953-4. In our experiments with plants, however, we only dealt with complete preparations, not with individual proteins.

*e) Changes with age in certain other biochemical
properties of proteins*

The serological analysis of proteins at different stages of ontogenesis is of great interest in connection with the study of the changes occurring in proteins with increasing age.

The results of a number of experiments along these lines show that the immunological, antigenic properties of proteins isolated from tissues of different ages are in fact different (1, 37, 154).

## 2. Changes in the processes of protein synthesis with age

The results described in the previous section concerning the changes in composition and certain other properties of proteins occurring with increasing age do not in themselves enable us to make any assessment of the way in which ontogenetic evolution takes place. It is quite obvious that they must be supplemented by studies of the changes in the metabolism of protein associated with ageing. It is necessary to work out the way in which the internal structure of the proteins and the changes in the biochemical processes of protein metabolism are related to the continual breakdown and synthesis of protein which forms the basis of life.

At present, however, the only aspect which has been much studied is the change with ageing of the quantitative indicators of protein synthesis. Evidence concerning qualitative changes in the synthetic systems is still very full of gaps, but, nevertheless, analysis of the material obtained in this field, even although it is very meagre, has a certain interest.

It is now generally accepted that during ageing there is a falling off of the process of protein synthesis with a decrease in the effectiveness of self-renewal. This fact by itself does not, however, tell us anything about the nature of the changes occurring with increasing age or their internal mechanisms. If we are to solve these problems we must look at the changes with age in the synthesis and breakdown of proteins in conjunction with the changes in their composition and the evolution of the metabolism of the organism as a whole.

There is a fairly large amount of literature on the question of the changes in protein metabolism with increasing age. There is no need to carry out a complete review of this work here because a considerable proportion of the results have already been published in several papers and monographs (76-79, 81, 82) but we must deal shortly with certain aspects of the subject.

In the first place we must mention that, in themselves, the decrease of synthetic activity with increasing age, the gradual equalization of the synthesis and breakdown of living material, the establishment of a so-called equilibrium and the cessation of growth are, in the main, not the result of ageing in its pure form but are consequences of the morphogenetic variation, differentiation and specialization of tissues. Continuous growth and a constant positive balance occur in many members of the living world (some perennial plants and some fish) and this sometimes serves (especially in plants) as the basis for a phenomenal prolongation of life. In most cases the relationship between synthesis and breakdown follows a definite curve. This has been demonstrated specially clearly in the work of Nagornyï (77), who showed that in the ontogenesis of animals an equilibrium is gradually established between the synthesis and breakdown of proteins.

It must, however, be noted that it is not only the new synthesis of protein, calculated on the basis of the excess of synthesis over breakdown, which decreases during ageing, but there is also a diminution in the intensity of the self-renewal of proteins, i.e. their complementary breakdown and resynthesis. The lowering of the rate of self-renewal continues throughout the whole of adult life, including the period of stable equilibrium when the overall rates of synthesis and breakdown balance one another. This phenomenon is also shown in experiments using labelled amino acids, in which it is not just the "incorporation" of amino acids into proteins which is calculated but the ratio between incorporation involving an increase in the amount of protein present and incorporation as part of a process in which breakdown is balanced against the resynthesis of protein molecules.

Even the earliest observations of Schoenheimer and his colleagues showed that the "inclusion" of labelled amino acids into proteins and their elimination from proteins took place considerably more rapidly in the young organism than in the old one. This conclusion has since been confirmed repeatedly.

The more intensive renewal of the proteins of young tissues was also demonstrated in work with minced tissues (homogenates) in which there is no significant increase in protein under the experimental conditions in use (153, 183, 228, 285).

Detailed investigations of the rate of the correlated breakdown and synthesis of proteins in various organs and tissues of rats, studied by means of [$^{35}$S]methionine, have been carried out by Toropova (125a). She found that the interval between synthesis and breakdown, i.e. the lifetime, of the protein molecule is considerably shorter in young animals.

It would be possible to adduce many other results which have been obtained in the last few years and which confirm these findings in respect of other animals and by the study of other proteins (10, 14, 105-107, 116, 205, 208).

Particularly detailed researches devoted to the study of the changes with age in the synthesis of proteins in various organs have been carried out recently by Bulankin & Parina (11-13). In this work the authors did not confine themselves to any one method but made a complicated study of the changes associated with ageing in different forms of protein synthesis. In 1954 Nikitin (82) proposed the following classification of the different forms of protein synthesis:

1. Growth synthesis, i.e. synthesis providing for the growth of the organ or of the organism as a whole.
2. Regenerative synthesis, which occurs when the animal is fed after complete starvation, protein starvation, loss of blood, etc.
3. "Stabilizing" synthesis, associated with the renewal of protein which has been lost by dissimilatory processes (the self-renewal of tissue protein).

Bulankin & Parina (11) consider it necessary to add one more to this list of three forms of protein synthesis and that is functional synthesis. There can be no doubt that functional synthesis is qualitatively different from the other forms in that, as a rule, it is accomplished in organelles which are specialized for the purpose and is associated with secretion out of the cell of the protein which has been synthesized.

The functional synthesis of protein is a characteristic feature of many organs. Among these are, in the first place, the liver and after it the glands which produce protein secretions, namely the pancreas, the mucous glands of stomach, the intestinal glands, the salivary glands, the mammary glands (which function periodically), some of the glands of internal secretion and so on.

It is very important to separate functional synthesis from other forms of protein synthesis when studying the changes associated with ageing because, during the formation of some particular organ, functional synthesis may increase even in spite of a falling-off of growth synthesis. The considerable part played by the element of functional synthesis in the general protein balance of any particular organ is clearly visible from the following table produced by Bulankin & Parina:

TABLE 16

The synthesis of protein in the livers of middle-aged rats
(6-12 months) (11)

mg./24 hr.

| | | |
|---|---|---|
| Growth synthesis | 2·3 | (0·24%) |
| Self-renewal synthesis | 190 | (20·0%) |
| Functional synthesis | 760 | (79·76%) |
| Total of all newly formed protein | 952·3 | (100%) |
| Reparative synthesis | | |
| (a) Stimulated synthesis (after starvation) | 107 | |
| (b) Regenerative synthesis | 162 | |

According to the authors the ways in which these various forms of synthesis changed with age were very different. Growth synthesis decreases and almost ceases in adult animals.

If we compare growth synthesis with another form, namely reparative synthesis, on the basis of the results obtained by Bulankin & Parina we shall find that during the regeneration of internal organs the following picture will be obtained.

The excision of two lobes of the liver (comprising about 65% of its weight) will lead to the rapid restoration of the weight of the liver by an increase in the size of the remaining lobes, not only in 45-60 day-old rats but also in old rats of 730 days old. In the old animals the rate of new formation of protein was slower than in the young ones but it was still quite considerable and amounted to 40% of that in the 60 day-old animals. Similar results were obtained in experiments on hypertrophy of the kidney. Comparison of the increase in the weight of the remaining kidney after unilateral nephrectomy showed that in rats aged 360 days the rate of increase amounted to 14% of that in 30 day-old animals.

Analysis of these results leads Bulankin & Parina to the conclusion that, although both of the forms of protein synthesis tend to decrease with age, the amount of the decrease is different for each. While the growth synthesis diminishes sharply as early as the first half of ontogenesis and almost ceases by the time the individual is fully formed, the decrease in reparatory synthesis associated with ageing takes place more gradually and in the old organism it is still at a relatively high level.

Therefore, in spite of the marked decrease during ontogenesis in the new formation of protein associated with growth, the possibilities of synthesizing proteins are still fairly considerable in old age. They are brought into play under unfavourable conditions when, if the organism is to survive at all, it must either quickly replace its normal protein content or compensate for the lost tissue.

According to these authors, the synthesis associated with self-renewal also decreases with increasing age.

A clear-cut decrease in the intensity of the synthesis of various forms of protein has also been observed in plants. This process has been shown particularly clearly as a result of the use of radioactive and stable isotopes for determining the rate of protein synthesis (68, 72, 73, 126, 127, 130, 131, 239). One of the most important features of the synthesis of proteins in leaves is that it must not only provide for the growth and self-renewal of the cells of the leaf but it is, at the same time, a functional process which ensures the supply of plastic substances in all the other growing organs. In addition to the structural proteins of the protoplasm, the leaves also produce the so-called reserve proteins, the periodic breakdown and synthesis of which ensures the outflow of amino acids and peptides from photosynthesizing leaves into other organs, especially growing points. The maximal rate of synthesis of these proteins is usually to be found in the leaf which is already fully formed, that is to say, it does not coincide with the maximal rate of growth synthesis of the proteins of the intracellular structures.

A typical example of the differences in the rate of synthesis of proteins in the leaves of plants with increasing age is given in Table 17, which is taken from the author's own work and demonstrates the rate of incorporation of labelled sulphur into proteins when it is applied to the zone of the root system of plants (72).

One reason for the diminution of the rate of protein synthesis during ageing may be suggested by the results of Racusen & Aronoff (240) who studied the protein metabolism of soya leaves. In their experiments they investigated the utilization of $^{14}CO_2$ by exposing the plants to an atmosphere containing labelled carbon dioxide. In the young leaves 27% of all the radioactive material assimilated by the leaf was in the protein, while in the middle-aged leaves the maximal radioactivity of the proteins under the same conditions amounted to 11%. The distribution of the radioactivity between the amino acids and proteins shows that young leaves synthesize considerably more of the "essential" amino acids than do middle-aged ones (phenylalanine, leucine, tyrosine, arginine and lysine). The authors believe that a deficiency of certain amino acids at the time when the leaves become mature may limit the length of their life. Naturally this mechanism for limiting synthesis can only occur in autotrophs.

The changes in the intensity of protein synthesis with increasing age do not, however, mean that there is any degeneration or any qualitative change in the working of the biochemical systems of protein synthesis and breakdown. As experiments on regeneration and "stimulated" synthesis show, old organs retain their general synthetic potentiality to a very considerable extent (14, 81, 94, 84, 85, 89). A far more interesting set of facts than those concerned merely with the intensiveness of synthesis are those concerned with the changes in the energetics of protein synthesis with increasing age. Most of the evidence in this respect has also been

## TABLE 17

The radioactivity of different groups of proteins in the leaves of beans of different ages 26 hours after immersion in a nutritive mixture of [35S]sulphates

| | Radioactivity of the protein (counts/min./2 mg. protein) derived from: | | |
| | Plasma juice | Plastids | Microsomes and mitochondria |
| --- | --- | --- | --- |
| 1st tier (old leaves) | 171 | 123* | |
| 2nd-3rd tier (middle-aged leaves) | 469 | 328 | 343 |
| 4th-5th tier (young leaves) | 2949 | 2404 | 2011 |

* Proteins isolated from the structures as a whole.

obtained by the Kharkov school of ontophysiologists (10, 82, 83, 85-90, 54, 108).

The energetic processes in cells are associated with particular complicated biochemical mechanisms and it is obvious that we must look to some internal changes in these systems for an explanation of the changes in the energetics and renewal of proteins occurring with increasing age.

In recent experiments Nikitin & Golubitskaya (87, 88) have studied the changes with age in the stimulation of protein synthesis in homogenates of liver by means of ATP. These authors consider that, as concerns the study of changes with age in the synthesis of proteins, a special interest attaches to the question of the extent to which the decrease of protein synthesis with age is determined by central and inter-organ regulators and how far it is determined by the changes in the biochemical processes and internal organization of individual cells, which occur with increasing age. One may imagine that one of these latter factors which would play a certain part in this process would be the deterioration of the ability of the tissues to use the energy of ATP for protein synthesis.

The results of the experiments which have been carried out show that the stimulating effect of ATP on the synthesis of proteins by liver homogenates gradually decreases with increasing age. Thus, if month-old rats are used the addition of ATP to the homogenate increases the synthesis of protein by 14.25% while the increase is only 12.69% if 3 month-old animals are used, 10.30% for year-old rats and 9.92% for 2 year-old ones.

Recently Weinbach & Garbus (282) have made a detailed study of the successive changes in oxidative phosphorylation during the ageing of rats. It was found that, in addition to the purely morphological differences between the mitochondria of old and young animals (the mitochondria of the brain and liver were studied), the ability of the mitochondria to carry out the oxidative phosphorylation of a number of substrates also decreased with age.

A short review of the finds in connection with the morphological changes in the mitochondria during ageing is given in an article by Rouiller (248).

It should be noted that the speed of renewal of proteins in liver mitochondria of mature and of aged rats was, according to Fletcher & Sanadi (176a), much the same. These workers reached this conclusion by studying the rate of loss of radioactivity from mitochondria which had been previously labelled with $^{14}$C. Their measurements related not only to the total, but also to individual fractions of the mitochondria (cytochrome $c$, soluble protein and lipids). It is also interesting to note changes reported in the ratios of various kinds of ribosomes on passing from larvae to adult forms of *Drosophila virilis* (252a).

## 3. Changes in the nucleic acids as matrices for protein synthesis associated with increasing age

If the nucleic acids really do undergo some qualitative change during ontogenesis, then this process must affect their participation in protein synthesis as the matrices which ensure the reproduction of the specificity of this synthesis.

The facts concerning the quantitative kinetics of the nucleic acids with increasing age are very numerous and well-known so it is hardly worthwhile to illustrate the position in detail. It seems to us that it is enough to give several characteristic examples.

The kinetics of nucleic acids are particularly clearly demonstrable in the organs of plants. It must be noted that it has been found that there is a particularly sharp fall in the RNA content of leaves as they grow older and that this affects the soluble fraction which does not form part of the organelles (66, 71, 221). In this connection we may mention the earlier work of Belozerskiĭ (4, 5) on this subject. He showed clearly that, in cultures of plants and bacteria, a particularly sharp decrease in RNA is observed in determinations of the free RNA which is not in the form of a complex with a protein. The RNA content of the cellular organelles also falls off with increasing age and Sisakyan and his colleagues (95, 120, 121) have made a specially detailed study of this process by investigating the changes occurring in the chemical composition of the plastids with increasing age. The authors also note that in the plastids of young leaves almost all the RNA is readily extractable (0·14 M-NaCl and 1·7 M-NaCl) while, when the plastids of old leaves are submitted to a similar extraction procedure, a considerable amount of RNA remains behind and can only be extracted with a weak alkali. Sisakyan & Odintsova therefore suggest that, as the organism develops and grows old, there is a "fixation" of the RNA on the structures and that it is this which brings about the "exclusion" of the RNA from the kinetics of metabolism.

In many earlier researches it has, in fact, been shown that nucleic acids may be more or less closely bound to proteins in different ways (4, 5, 50), the metabolism of the conjoined proteins and nucleic acids depending on the nature of the bonds. On the basis of these ideas, Nikitin (82) has suggested that in the course of cytomorphosis and during ontogenesis "the degree of association of nucleic acids and proteins into nucleoproteins gradually becomes closer and more stable and it would seem that, at first, this leads to an increase in the synthetic powers of these compounds but later, as cytomorphosis continues (beyond the limits of its optimal level), the association 'hardens' and leads to the inactivation of the nucleoproteins as leading factors in protein synthesis". The authors bring forward a considerable amount of biochemical evidence for the occurrence of this process.

It is interesting that in the course of studying the changes with age in the DNA of bacteria and plants it was noticed that the opposite phenomenon occurs, there is a decrease in the proportion of the DNA which is closely bound to protein (36, 38). On the basis of his cytochemical studies of the nuclei of ageing plant cells Konarev has also suggested that there may be degradative depolymerization of the nuclear DNA associated with ageing.

As to the changes occurring in the qualitative composition of DNA and RNA as age increases, we have not yet gathered more than the very slightest information and what we have is only concerned with the proportion in which the nucleotides are present and not with their sequence in the polynucleotide chains, although this would, in fact, have been more valuable. As polynucleotides are made up of only four components it is clear that very substantial, though random, changes in their order might not be revealed by determination of the nucleotide composition, though what is important for the reproduction of the specificity of the proteins is the actual nature of the "message" impressed on the RNA.

As early as 1953 we looked at several materials with regard to qualitative and quantitative changes in their RNA and DNA (59). Since then there has been no worthwhile progress in this field. In several papers it has been shown that the nucleotide composition of RNA and DNA remains the same during the development of bacterial cultures in spite of very considerable changes in their quantitative composition (6, 24). It has, however, been established that small changes occur in the nucleotide composition of RNA and DNA during the life cycle of the unicellular alga *Chlorella* (201).

Chargaff and his colleagues (173) did not find any significant change in the nucleotide composition of the RNA of the developing sea urchin. On the other hand Gold & Sturgis (160a) found very marked changes in the nucleotides in the RNA during the development of the endothelium of man. De Lamirande *et al.* (216) found some changes in the composition of the RNA of the livers of rats as they regenerated and during starvation and this indicates the theoretical possibility that qualitative alterations in RNA may take place.

Olmsted & Villee (233) have observed changes in the ratios between the purine and pyrimidine bases of the nucleic acids of the human liver during ontogenesis.

Bulankin and colleagues (14) have obtained some interesting results on the changes occurring in the RNA of the livers of rats as they grow older. They found that the ratio of phosphorus to pentose was much higher in the new-born rat than in the adult. They suggest that this decrease may indicate a change in the degree of phosphorylation of the RNA as the animals get older and, at the same time, a decrease in their ability to pass on energy for protein synthesis.

Makhin'ko & Blok (55) found something similar in their studies of the livers of domestic birds during embryogenesis. On the basis of this work the authors came to the conclusion that the embryonic liver contains over-phosphorylated RNA.

In our own studies of the nucleotide composition of the RNA of bean leaves (70) by autoradiography of the products of hydrolysis of RNA isolated from plants generally labelled with $^{32}$P a comparison was made between the ratios of the nucleotides in the RNA of the cytoplasm of old and young leaves. In this work we noticed very small percentage differences (cytidylic acid 32·8 and 29·6, guanylic acid 27·5 and 30·2) which were almost within the limits of error of the determination.

Having studied both the changes in the amount of RNA in the plastids of leaves with increasing age and the nucleotide composition of this RNA, Sisakyan & Odintsova (120, 121) concluded that although it is sometimes possible to detect changes in this composition they do not occur with any regularity. Cooper & Loring (163) reached the same conclusion after studying the nucleotide composition of the RNA of tobacco leaves of various ages.

Nikitin & Shereshevskaya found no notable changes with age of the total RNA and DNA of rat liver (93a).

Detwiler & Draper (167) have recently studied the nucleotide composition of RNA from the different organelles of the cells of the livers and muscles of middle-aged (11 months) and old (30 months) rats and the rate of metabolism of the individual nucleotides (specific activity 1 hour after injection of $^{32}PO_4^{3-}$). Only in the nuclear RNA were statistically significant changes found. In addition to an increase with age in the RNA in the nuclei there was an increase in the amount of cytidine in this RNA but its specific activity was less. The RNA of the mitochondria, microsomes and hyaloplasm did not change substantially.

We have indirect evidence for the occurrence of changes with increasing age occurring in the biochemical systems associated with heredity in facts which suggest that the age of the parents has a fairly definite effect on the development and biochemical characteristics of their offspring, in the sense that the offspring of very old parents are also, from the start, "older" in their biochemical and morphogenetic potentialities (3, 16, 134, 250).

An interesting hypothesis concerning the role of DNA in the ageing of the organism has recently been advanced by Sinex (258). He believes that, as DNA is only synthesized in connection with mitosis and is renewed to only a very slight extent in cells which have finished growing (nerve cells, muscle cells, etc.), there may be physico-chemical inactivation of the DNA of these cells, like that which occurs in collagen. This inactivation of the genetic material may also bring about the cessation of the reproductive properties of the old cells and the synthesis of RNA and proteins in them.

We must mention that the metabolic stability of DNA in mitotically inactive, mature tissues has recently been studied by Bennett *et al.* (147). They concluded that $^{14}$C incorporated in the DNA of the brain and liver of the embryo remains fixed there for the whole lifetime of the animal.

There is a certain amount of evidence concerning changes in the fractional composition of nucleic acids and nucleotides with increasing age (4, 236) and during regeneration (46). It seems that we must assume that among the fractions of RNA and DNA there are some which are inactive as matrices and as purveyors of information. Dunaevskiĭ (23) has obtained some very suggestive evidence in this connection. He found, in nervous tissue, so-called "packed" nerve cells. These cells were specially numerous in persons who were suffering from psychological disorders and in very old people. They were distinguished by having a very weak function and by being very rich in RNA. The RNA with which these cells were filled was, however, different from the ordinary RNA of nervous tissue in its unusual resistance to the effect of ribonuclease (it was 3-6 times as stable), and it seems to be functionally inert.

In view of the importance of DNA for the synthesis of RNA, which has been commented upon in many papers (cf. Chaps. XIII and XIV), it would be interesting to find out how the relation of these polymers to one another in these cells differs from that in ordinary cells.

Manukyan (51, 52) has shown clearly that during the ontogenesis of the rabbit the DNA content of the nerve cells (per unit weight of tissue) decreased considerably faster than the RNA content. The ratio RNA/ DNA therefore increased considerably. In 20-day embryos it was 0·69 while in 30-day rabbits it had risen to 2·8. In recent experiments Krasil'nikova (43, 44) observed a similar occurrence by cytochemical methods.

Having made a detailed study of the RNA/DNA ratio in a number of organs, Nikitin and his colleagues (91) found that this ratio increases in the heart and, while the RNA content of the tissues falls to a half, their DNA content falls to a third of its previous value. A similar picture is found in the kidneys, though the changes occur more slowly. In the small intestine, however, the composition of the nucleic acids remains very stable during ageing. The authors associate this with the rapid division of the cells of the intestinal epithelium, which enables it to function. It is important to note that the process of renewal of RNA also falls off with age (12).

According to the results of Barrows, Shock and their colleagues (145, 146, 175), however, the DNA content per nucleus of the liver cells of rats is the same at the age of 12-14 months as it is at the age of 24-27 months. Apparently, therefore, the change in the DNA content calculated in terms of the dry weight of the organ, which has been found by other authors, is only a relative change. It reflects an increase in the relative content of the tissues of intercellular and other substances of low activity. At the same time, the number of polyploid cells in the liver increased markedly with age. A number of interesting studies relating to changes in nucleic acids with age have been published in a recent collection of papers from the Kharkov school of "ontophysiologists" (113a).

## 4. Changes in enzymes concerned with the metabolism of proteins and nucleic acids

About 250-300 papers have already been published on the subject of the activity of particular enzymes during the ontogenesis of micro-organisms, plants and animals. However, most of the work dealt with the morphogenetic stage of ontogenesis or with oxidative enzymes and those affecting the metabolism of carbohydrates and fats, etc. and only comparatively few of the authors studied the successive changes in the enzymes of protein and nucleic acid metabolism during typical ageing processes. The results obtained by various authors agreed as to the nature of these changes. It was found, for example, that during ageing there was a definite decrease in the activity of proteolytic enzymes in various objects (7, 31, 151, 246).

The nature of the curve for the activity of ribonuclease is rather peculiar. During the period of formation of the organs and gradual slowing of growth, the activity of the ribonuclease diminishes markedly both in various tissues of rats (122) and in the human placenta (155). In the final stages of the individual development of rats, however, the activity of this enzyme increases again, according to the results of Stavitskaya (122), and she thinks that this is connected with the possibility of breaking down RNA at this time of life without complementary synthesis. Similar results have been obtained by Nikitin and his colleagues (90).

In summarizing this short review of the evidence as to the changes occurring with increasing age in the direct process of protein synthesis and in the properties of the biochemical system undertaking this synthesis we may conclude that although this evidence is still very inadequate it indicates that certain ontogenetic changes do occur in this system. It is only natural that there should not yet be any direct evidence for changes in the sequence of nucleotides in RNA and DNA associated with ageing and we cannot hope that any will soon be forthcoming, but all the indirect evidence of the activity of matrices and systems of activation of amino acids indicates that the ability of RNA to bring about the qualitatively specific synthesis of proteins becomes impaired during ageing. These changes occurring in the matrices and ancillary systems of protein synthesis during ageing are not of a morphogenetic nature like those characteristic of embryogenesis but are a clear mark of gradual degradation.

But are these changes in proteins and the systems synthesizing them primary and spontaneous, or do they develop only as a result of the degradation during ageing of some more complicated biological systems? The answer to this question is of very great theoretical and practical importance.

## 5. Reasons for the occurrence of random changes in the structure of molecules with increasing age and some theories of ageing

Although the results which we have just discussed are contradictory and fragmentary they do still, in the main, suggest that, as the organism gets older, changes occur in the physico-chemical, chemical and biological properties of proteins and in the biochemical systems which synthesize them, among which are the nucleic acids. It is not impossible that these may be the primary changes of ageing and that the study of how they happen and the mechanisms which produce them may create the basis required for building up a biochemical theory of ageing which will not only reveal the reasons for certain anomalies of ageing but will also elucidate the biological nature of ageing as a process which occurs in all individual members of the living world against the background of the pre-eminence and immortality of living matter as a whole. The study of the nature and mechanism of these changes is naturally, therefore, of very considerable interest from all points of view.

In Chapter XVI we have already discussed the evidence concerning "errors" in the synthesis of proteins, RNA and DNA which occur spontaneously and inevitably from the action of internal causes (inaccuracy of the reactions whereby the polymers and their substrates are synthesized, damage to the structures by free radicals, endogenous formation of analogues and toxins, etc.) and by external factors (the action of radiation, pathological processes, etc.). The inevitable nature of these errors, especially those of endogenous origin, naturally brings to mind the idea that it is just the accumulation of them which leads to the random and scattered alteration of those essential substrates of living matter, nucleic acids and proteins. We must point out that the random nature of these changes makes it very difficult to identify them, as different molecules change in different ways and individually, not all at the same time. Because of this the functional inadequacy of a particular protein may manifest itself sooner than the biochemist would be in a position to expect on the basis of chemical or physico-chemical indications.

It is important to note that it is just these principles of molecular variation which lie at the root of many modern hypotheses about ageing.

### a) Hypotheses of molecular ageing of the organism

At the beginning we must remark that hypotheses as to the causes of biological ageing cannot be based purely on evidence about molecular occurrences, they can only be based on a study of all aspects of the subject, chemical, physical, biochemical, physiological, histological, cytochemical, etc. as very many of the degenerative phenomena of ageing can be more or less autonomous. We do not intend to try in this book to undertake any study even of all sides of the biochemical aspect of the problem of ageing. We do, however, consider it worthwhile to indicate a few theoretical approaches to the study of the biochemical problems of ageing which have been developed actively over the past few years and which supplement the excellent review of theories of ageing made by Nagornyĭ (76, 78, 79) between 1940 and 1950.

Since biology first began there have been at least a hundred different attempts to assign a cause to ageing. In the case of most of the hypotheses, the authors started by assuming that there was some deficiency in some of the physiological and functional or even morphological or morphologo-biochemical processes and, as a rule, this deficiency actually does appear during ontogenesis and actually causes the accumulation of pathological changes which may take the form of endocrinological changes, auto-intoxication, oxygen deficiency, trophic changes, the accumulation of metaplasmic proteins, etc.

All these hypotheses were essentially based on real phenomena but none of them could give more than a partial explanation of the following features of ageing: (1) its universality, affecting every individual at every stage of development; (2) the differences in the rate of ageing from species to species; (3) the hereditary nature of the characteristics which determine the duration of life; (4) the randomness of ageing; and (5) the immortality of living material as a whole.

The biological purposiveness of the mortality of the individual upon which we remarked in the introduction to this chapter also generally falls outside the scope of most of the hypotheses about ageing.

Even the simple enumeration of all these phenomena which are not explained by most of the hypotheses concerned with ageing shows clearly that our search for the cause of ageing as a universal biological phenomenon must start at the molecular level and only then proceed successively from one degree of the organization of living material to the next. Considerable progress in molecular biology has very effectively made it possible to make such an approach to analysing the causes of ageing of the organism.

The idea that it is necessary to look for the primary mechanisms of ageing at the molecular level is also being developed by many authors. We may, for instance, point to the hypothesis of the hysteresis of colloids developed by Ruzicka, the hypothesis of the hysteresis of metaplasmic proteins advanced by Nagornyĭ (76) and the recent hypothesis of spontaneous irreversible changes in the so-called irreplaceable macromolecules (259-262), the hypothesis of the transformation of protein molecules to an energetically more stable state (111), the suggestion that bonds of a diketopiperazine type accumulate in proteins (8, 32), the hypothesis of the non-equivalent metabolism of proteins and nucleic acids, increasing the number of bonds in their molecules which are of types resistant to the action of enzymes advanced by us in 1952-3 (56-58, 61), the concept, worked out in detail by Nikitin, concerning the alteration with age in the nature of the bonds joining proteins to nucleic acids in nucleoproteins (82), the suggestion that as age increases there is an accumulation of optical isomers in proteins (2, 214) or crossed connections between proteins and protein complexes (152), the hypothesis of the accumulation of damaged structures (203, 204), the hypothesis of the accumulation of macromolecules which have been damaged by free radicals formed during metabolism (187, 188), the idea that it is impossible to maintain the ideal conditions for the resynthesis of specific macromolecules (249), the hypothesis of ageing based on recognition of the importance of the accumulation of —S—S— groupings in the organism as it grows older and their methylation (110, 235), the hypothesis of the inactivation of proteins by transformation of proline into hydroxyproline (268, 269), the hypothesis of incoordination between successive stages in complicated reactions (35), the hypothesis of the accumulation of the consequences of autoimmune reactions (150, 225) and a number of others.

Recently the well-known American physicist Szilard (275) has entered the field with a very detailed hypothesis of molecular ageing. He suggested that ageing of the organism is brought about by accumulation of changes in the genetic stock of the organism which control biochemical processes. Szilard's hypothesis assumes that the fundamental unit of the process of ageing is the "ageing hit", which puts out of action some chromosomes or part of a chromosome in the sense that the genes in the chromosome become inactive. The author further assumes that the "ageing hits" occur at random and that their frequency may be constant throughout life. The author assumes still further that the rate at which the somatic cells are damaged by the action of these "hits" is characteristic for a given species. According to the author the accumulation of such "hits", which are essentially somatic mutations, leads to gradual ageing. On the basis of these assumptions Szilard has carried out a detailed mathematical study of the possible ways in which this process could occur. We must mention that the mathematical side of this theory has been criticized (244, 263).

An interesting attempt at elucidating the molecular basis of ageing has been made in the comprehensive and intensive work of Strehler (271). In this work Strehler put forward a very systematic analysis of ageing at all levels, starting at the molecular level. He points out that ageing, as a biological phenomenon, consists of a gradual diminution of the adaptation of the organism even to normal external circumstances following on sexual maturity. This decrease in adaptation manifests itself in a decrease in the ability of the individual organism to carry out its various specialized functions. In the last analysis this functional weakness presents itself as an increase in the probability of the occurrence of pathological processes and of death with the passage of time.

According to Strehler the essential criteria of ageing are its universality, its causation by internal factors, its progressiveness or cumu-lativeness (in the sense that the number of changes increases) and the destructive nature of the changes which accumulate.

In his work Strehler gives a very detailed and logical classification of the changes occurring with increasing age and the connections between them. According to him there are several processes which act as the primary factors determining the occurrence of ageing; they are, in the first place, physical or chemical molecular changes (molecular accidents) and changes of a genetic nature. He gives a detailed classification of the possible causes of these accidents, singling out from among them the following: (1) deficiencies in the actual organization of the living organism as a result of which the system is incapable of repairing the results of large-scale or small-scale injuries. This imperfection of the organization also leads to the constant occurrence within the cell of weak but harmful side-reactions of both an enzymic and a non-enzymic nature; (2) continual small-scale damage to molecules arising from the action of a number of causes (mutation, denaturation, disruption of molecular structure as a result of high concentrations of energy occurring locally). The author attaches great importance to the localized liberation of large quantities of energy. The sources of this energy may be statistically localized fluctua-tions in the emission of heat, absorption of actinic energy or particles of high energy or local emission of heat from exothermic chemical reactions.

According to Strehler all these factors can give rise to three main types of changes of the systems which are responsible for the synthesis and replication of specific macromolecules: (1) loss of the code; (2) loss of the ability to pass on the code during synthesis; (3) loss of the conditions required for carrying out the process of reproduction accurately.

This whole collection of primary changes, along with the larger-scale changes brought about by illnesses, autointoxication and external factors, leads to consequent changes in all stages of the physiological, morpho-logical and biochemical organization of the living body and these, in their turn, have secondary effects on the determining factors. The author divides his subject into ageing at the cellular level, ageing at the level of tissues, etc.

Among the intracellular changes which develop as a result of the primary factors determining ageing Strehler distinguishes the following: loss of the ability to replace subcellular structures, loss of the power to differentiate in cells which replaced others, loss of plasticity, arrest of growth, increase in "errors" in catalytic and non-enzymic reactions, changes in the nuclei, membranes, organelles and so on. As a result of this, changes occur in transport and permeability, in tissues and functions, etc. in such a way that all these destructive changes enhance the primary variability, thus increasing the probability of death, that is to say, they bring about typical ageing, affecting all the systems and functions of the organism.

This is a summary of the main points in Strehler's ideas which are put forward in his own work in a considerably fuller and more systematic way and are illustrated by a number of well-conceived schemes of the interaction of the various factors.  In this work the author does not develop any new, simple hypothesis as to the causes of ageing.  He makes a thorough analysis of the phenomenon, devises a detailed system of relationships between the various factors and shows their action at various levels of physiological and biochemical organization of the living system.

Such a systematization of the extensive factual and theoretical material concerned with the processes of ageing is, at the present moment, considerably more valuable than the erection of some partial hypothesis because it reveals the weakest links in biological systems, control of which may prevent the development of further cycles of changes.  In a new series of papers Strehler (229, 272-274) has carried out a further systematization and specification of his interesting ideas.

### (b) Internal and external causes of random changes in molecular structures with ageing

A considerable number of the hypotheses which we have just surveyed concentrate on some particular forms of changes which may occur in proteins.  In general, all these hypotheses can be divided into two groups. One group includes those hypotheses which focus attention on the results of changes (such as non-equivalent exchange and increase in sulphide bonds) which are caused by many factors.  The second group contains those hypotheses which concentrate mainly on the reasons for the changes (such as the action of free radicals, the fluctuation of thermal energy or autoimmune reactions) which may have various consequences.  If we look at the hypotheses in this way we can understand why each is so one-sided when taken by itself, for it is surely necessary to take account both of the nature of the changes and the nature of the causes which bring the changes about.  The nature of the changes in proteins and nucleic acids does not, as we have seen in earlier sections, show any definite regularity. It is also very likely that there could not be any such regularity on account of the random (chaotic) nature of the process and because if, among 1000 molecules of protein or RNA in an old organism, there are 100 or 200 which have some defect, all the defects may be different from one another and this makes it very difficult to discover them.  At the same time, the majority of these 1000 molecules would undoubtedly be unchanged in their specificity because, obviously, death occurs before 100% of the molecules have become altered.  Therefore, when we are isolating individual proteins with a view to determining their state we shall naturally, in the first place, isolate those molecules which have not undergone change and we may therefore fail to detect the nature of the changes which have actually occurred.

It is, however, possible to obtain indirect demonstration of the significance of molecular variation and one typical demonstration of this sort is the increase in the rate of general ageing of the organism when it is exposed to ionizing radiation (219, 232, 266, 278). It has also been found that the action of heat on DNA may lead to a disruption of its molecule (analogous to "ageing hits") (245).

Having recognized the existence of spontaneous variations at the molecular level which lead, in the first place, to an accumulation of changes in the ageing organism, we must naturally look for the causes of this variation. We may classify these as primary and secondary causes. It is clear, for example, that a change in the way RNA is bound to a protein is a secondary change which is accounted for by some alteration in the composition of the protein or the RNA. There may be quite a large number of primary, "independent", causes of molecular variation acting in parallel and it would be valuable to give a short systematic survey of the available evidence about some factors which could bring about the occurrence of random changes at the molecular level and alter the "information" of specific macromolecules.

1. *The action of free radicals.* We have already mentioned the hypothesis of Harman (187-190) which lays special stress on the importance of the effect of free radicals in producing molecular damage. Such damage occurs in the course of biochemical reactions, it is a perfectly real phenomenon (22, 181) and is the very factor responsible for the mutagenic effect of free radicals.

Free radicals are either fragments of molecules or electrically charged whole molecules. They have at least one free valency and, as a rule, a high degree of chemical activity, as a result of which they easily react chemically with ordinary saturated molecules. This is the explanation of both their active role in chemical reactions and the extensiveness and variety of the reactions in which the radicals take part.

The action of $^-OH$ radicals on amino acids leads to the formation of a wide range of amino acid analogues (162). The formation of free radicals within cells leads to all kinds of side-reactions with a local disturbance of the orderliness of the biochemical cycles and damage to the macromolecules. The occurrence of free radicals in biological systems was established long ago and now the problem of the biological consequences of their action is being studied more intensively as part of the study of many other biological phenomena (carcinogenesis, radiation damage, etc.) (cf. the Proceedings of the Symposium on the action of free radicals in biological systems at Stanford University (177)). Free radicals are constantly arising in biological systems for a variety of reasons (chemical reactions, photochemical processes, the effects of radiation, thermal dissociation, etc.).

2. *Damage from radiation.* Radiation exerts an overall destructive influence on orderly systems and in living things the system which is

most sensitive to it is that of DNA.   Irradiation from without accelerates
the process of ageing (219, 182a, 199, 237, 229a, 266, 278) but the ageing
induced by radiation is not quite the same as normal ageing (273).   This
question has been treated in a special symposium (274a).   The organism
is continually subject to the action of radiation from without (cosmic) and
from within (from the naturally occurring radioactive elements such as
potassium).   Furthermore, other forms of radiation such as exogenous
and endogenous ultraviolet radiation have the same effect on biological
structures.   The mechanism by which radiations have their harmful
effects is complicated and involves ionization, the formation of free
radicals, inhibition of enzymic reactions, disruption of polymers, etc. and
it is important to emphasize that this factor is not accidental, it is, as it
were, "built into" the biological system.   At present the effect of radiation
is being studied with special intensity, particularly as it affects large
molecules (199, 200, 200a, 142, 215, 252).

3. *The incorporation of analogues of normal metabolites.*   We have
already seen (in Chaps. XVI and XVII) that when analogues of nucleotides are incorporated during the synthesis of nucleic acids this leads to
a localized change in the information contained in them and therefore
has a mutagenic effect.   In experiments on artificial mutagenesis synthetic
analogues are generally used but small amounts of analogues do, in fact,
arise endogenously.   We have also dealt (in Chaps. IV and XVII) with
the evidence indicating the presence in RNA and DNA of small amounts,
sometimes only traces, of unusual nucleotides which are analogous to the
ordinary nucleotides.   The biological significance of these compounds is
still obscure and we suggest that they are natural endogenous mutagenic
factors built into the genetic system.   During ontogenesis somatic mutations of a similar sort lead to the accumulation of random alterations in
the information in the RNA and DNA.   In addition to this it is to be pre-
sumed that even during the normal processes of synthesis of amino acids
and proteins some one in hundreds of thousands or hundreds of millions
of the compounds formed will be changed to form an analogue of a normal
metabolite.   It is practically impossible to achieve a yield of exactly 100%
of the required product either *in vitro* or *in vivo* and under biological conditions the accuracy of the result is due to a progression from inaccuracy
to accuracy which *will only cease when selection ceases.   However, selection
will cease when the lifetime reaches the most appropriate length from an
evolutionary point of view and therefore the development of adaptations
which would ensure the constancy of living organization at all levels of complexity will never achieve ideal completion under natural conditions.*

4. *Side reactions and non-enzymic unplanned reactions.*   In such a
multifarious biochemical system as protoplasm it is always possible that
a certain number of side-reactions and non-enzymic reactions will occur,
leading to the formation of analogues or accidental compounds, the precipitation of insoluble complexes or salts and so on.   Some results and

hypotheses suggest that, for example, lipofuscin, the pigment characteristic of ageing, which is usually found in nerve cells and others which have
a long lifetime, is the product of unplanned processes at the polymeric
level (273). People have begun to study some non-enzymic side-reactions
during the past few years (136, 137, 212).

5. *Statistical errors in enzymic reactions.* Errors of this sort also seem
to have a general significance. The formation of analogues is, in part, an
error of this sort. The range of theoretically possible errors must be
very wide and a start is only just being made on the study of their biological consequences. In a recent and interesting paper Almond &
Niemann (140) have made a theoretical survey of the possible consequences of systematic errors in enzymic kinetics.

The reaction of haemoglobin with oxygen is an interesting example
of an error in a biochemical process which has been thoroughly investigated. In this reaction haemoglobin and oxygen form $99.8\%$ of ordinary
oxyhaemoglobin ($HbO_2$) and $0.2\%$ of an oxyhaemoglobin which does not
dissociate quickly and reversibly ($HbO_3$). In view of the very rapid
succession of cycles of oxygenation of haemoglobin it would soon all be
converted into the stable oxyhaemoglobin if the blood did not contain a
special system for the "correction" of these errors by turning the stable
oxyhaemoglobin back into free haemoglobin (148). According to Gitel'zon
and his colleagues (16a) there is a difference between the relative proba-
bilities of formation of $HbO_2$ and $HbO_3$ in old and young erythrocytes.
This is obviously connected with a lowering of methaemoglobin reductase
activity (148).

6. *Thermal emissions at the molecular level.* Strehler attaches great
importance to the harmful effects of thermal emissions on biochemical
reactions (271, 273).

7. *The binding of active groups of macromolecules.* Reactions binding
the active groups may be a common phenomenon in long-lived proteins.
In a number of recent papers it has been shown that there may be a
change in the properties of proteins which are associated with ions or
other small molecules (157, 158, 200).

8. *Autoimmune reactions.* Proteins which have become changed may
behave as foreign proteins and react with the systems producing immunity.
In Chapter XI, which was concerned with the synthesis of antibodies,
we have already seen that the antibody-forming system can distinguish
between what is its own and what is foreign and this ability develops
during the final stages of morphogenesis. It is thus clear that, if changed
molecules appear anywhere in the circulation, whether it is in the blood,
lymph or tissue fluids, the antibodies will react with them as if they were
foreign and antibodies which penetrate into the tissues will, of course,
react with such molecules even inside the cells, where they will form insoluble complexes. The range of antibodies will, of course, hardly be
likely to coincide with the range of all the molecular changes. However,

if such processes really do take place in the organism then, on the one hand they will bring about the elimination of the changed molecules, while on the other hand it would appear that in some cases they would lead to the formation and accumulation of insoluble complexes.

9. *Molecular changes in the system for regulating "structural" genes.* Under the influence of factors described above, there may occur changes in the regulation of the complicated system of "structural" genes which control the synthesis of proteins typical for each type of differentiated tissue. In this way, without any change of genotype the metabolic system of the cell may cease to be completely in tune with its particular kind of differentiation and there may appear in it a certain number of chaotic and unregulated processes. At present considerable significance is attached to this kind of change to explain the appearance of malignant cells (133a); it must be remembered that this tendency is closely correlated with the processes of ageing. This kind of molecular change could be described as "chaotic (random) morphogenesis" and it may well be that there exists a genetically built-in mechanism which intensifies these changes at a particular stage in development of the organism ("genetic clock of ageing"). In this case it may be that with increasing age there occur changes in the state of the repressors (histones) and of the messenger RNA.

In addition to these factors, alteration of the information in proteins and nucleic acids may be caused by many other factors such as changes in the proportions of substrates, the direct transformation of amino acids when they are already forming part of a protein (251, 192), small-scale changes occurring during denaturation, the formation of optical isomers and several others. One may compare these factors to "noises" which develop during the transmission of particular pieces of information and gradually change that information.

The spontaneous variation which results from the action of all these factors is enhanced by many other external and internal causes which bring about ordinary and premature ageing.

The organism can undoubtedly oppose these changes and natural selection may give rise to counteracting factors which will increase its span of life. If a longer and longer life span were an evolutionarily favourable characteristic then undoubtedly man would also have a lifetime of thousands of years, for there are examples in nature of individual trees which live for 10,000-12,000 years. Species specificity of the length of the lifetime exists just because there are factors in nature which oppose ageing and variation, and selection gives each species its particular length of life, just as it gives it the particular shape of its wing or its circulatory system.

It is also quite natural that biologists should also be able to control these factors in the course of time; but the important point here is, of course, to discover and study their action.

It seems to us that the main line of work in this connection must be the study of the biological mechanisms of accumulation and elimination (removal) of molecular and subcellular damage.

## 6. Biological mechanisms of accumulation and elimination of molecular changes in proteins and nucleic acids as factors determining the life span

One may encounter the phenomena of accumulation and elimination of changes at all levels of the morphological, physiological and biochemical profile of living systems and each level makes its contribution to the determination of a particular life span. Here we shall only deal with the molecular level.

### a) Some theoretical considerations in regard to the biological mechanisms of accumulation and elimination of molecular changes occurring during ageing

Living material can, on the one hand, reproduce almost exact copies from generation to generation over thousands and millions of years while, on the other hand, the organisms which are formed cannot maintain themselves as they are in the best state for carrying out their vital activities. They regress and die. This double-sidedness of living material shows clearly that there must exist in nature two different mechanisms for reproducing the specificity of polymers, one of which can reproduce the specificity of polymers with almost perfect accuracy during successive reproductions, while the other cannot do this and therefore cannot provide for the eternal life of complicated biological systems, the normal activity of which requires an accurate correlation of the composition and functional role of the compounds.

The material, which we have discussed in earlier chapters, concerning the reproduction of the specificity of proteins, RNA and DNA makes it easy for us to guess the nature of these mechanisms. The most accurate reproduction of specificity, ensured within the extremely narrow stereochemical limits of the interaction of complementary forms of nucleotides takes place by the autoreproduction of DNA. The current syntheses of protein and RNA can also provide for the reproduction of specificity but the mechanism for this reproduction does not have the degree of accuracy which would enable it to maintain the eternal replication of the same cycle of breakdown and synthesis.

The factors which we have already mentioned as causing molecular variation do not all act in the same way on proteins, RNA and DNA. The nucleus is not the centre of multifarious biochemical reactions and therefore damage to its structure by free radicals, the giving off of heat, side-reactions and non-enzymic reactions is minimal. The sluggish metabolism of DNA protects it from statistical "errors" and the strictly complementary synthesis protects it from variations in the proportions of the substrates. Irradiation has a strong effect on DNA but under natural conditions some of the changes caused in this way are so small that they only manifest themselves in the course of evolution. Also the errors in the synthesis of DNA which are deleterious are eliminated by evolution. A molecule of protein which has undergone alteration may vanish without trace and be replaced by an intact molecule; a molecule of DNA which has undergone alteration will retain the error which has arisen and so will its descendants. Selection roots out and eliminates these changes and in this way it stabilizes the stock of DNA, selecting only intact or improved examples for further reproduction. Owing to this evolutionary elimination it is possible to reproduce types of DNA which existed hundreds of millions of years ago. Rubbish is not preserved and this is why the intact structure is reproduced. Rubbish is, however, conserved during ontogenetic syntheses and it accumulates; disturbances of the systems of synthesis of proteins and RNA undoubtedly are the main sources of this molecular rubbish.

The synthesis of RNA is partly localized in the chromosomes and is controlled by DNA. Any errors which arise in the DNA system will therefore be handed on to the RNA. But, in addition to this hereditary change in the "code" of DNA, RNA is subject to further possibilities of alteration, both under the influence of those factors which also affect DNA (accidental alterations, the incorporation of analogues, the action of free radicals and thermal damage) and by the effect of several other factors, especially when it is being synthesized autonomously in the nucleolus and cytoplasm (changes in the proportions of the substrates, possibilities of linear elongation). Furthermore, it must be reckoned that the turnover of RNA is about 50-100 times as fast as that of DNA. RNA functions in the cell in the form of a single-stranded polynucleotide and occupies a central position in the biochemistry of the cytoplasm; this greatly increases the number of errors which can occur during the synthesis of the polymer and may even affect the synthetic activities of the molecules, including their function as a matrix.

However, although there is a considerably greater opportunity for the occurrence of errors in the synthesis of RNA than in that of DNA, it is considerably more difficult to detect the errors in the RNA.

Not having subjected the synthesis of RNA to such strict limitations as that of DNA nature "took care" that alterations in RNA should not be hereditary. Errors in the synthesis of RNA seem to be more frequent but they are more episodic, their effects are confined to one individual and each new generation begins its development from a zygote which contains a stock of information which has not been substantially affected by variations occurring in the RNA of the previous generation. In other words, the number of errors occurring in the synthesis of RNA in the nucleus under the direct control of DNA is always equal to the number of errors which have become established in the DNA or, more correctly, somewhat greater than this, while the main mass of the errors occurring in RNA as a result of the action of factors which have already been mentioned is mainly localized in the cytoplasm.

Changes in DNA (during the reproductive cycle) can easily be detected because, as a rule, they manifest themselves as hereditary characteristics. It is also theoretically possible, especially by indirect signs, to detect changes in proteins for these change, upset or put out of action some particular biochemical function. The changes occurring in RNA are, however, especially haphazard, they are hidden and it is only possible to find out that they exist by means of changes in the proteins which are formed under the influence of the RNA.

Thus, in regard to the perfection of the mechanism of reproduction of specificity of structure, DNA is the most constant polymer in living matter and for this reason it plays the main part in maintaining the material connection between one generation and the next and in ensuring the occurrence of inheritance. RNA is very different from DNA in this respect and in determining the "current" synthesis of individual forms it cannot preserve the identity of intracellular tissues and structures; by its variations it provides the molecular prerequisites for ageing. Finally, the proteins are the most easily altered components of the living body because, while they reflect many (but not all) of the changes in RNA, they also have many additional opportunities for non-equivalent exchange. While they are extremely specific compounds, the structure of which is directly connected with the carrying out of their functions, it is proteins, which cannot be reproduced with perfect accuracy in the system of RNA during the course of the innumerable cycles of reproduction, which act as the limiting factor of ontogenesis. It is, in the last analysis, these changes in the proteins which serve as the molecular basis for the structural alteration of living material at the level of the organelles and above, disturbing the unity of structure and function and ensuring the ultimate mortality of the individual.

*It may, however, be assumed that if autonomous synthesis and reproduction of RNA did not occur and the whole stock of RNA was synthesized under the strict determining influence of DNA, then this would decrease the rate of change of RNA many times over and accordingly it would decrease the rate of accumulation of molecular changes in proteins with increasing age.*

DNA controls the synthesis of RNA, but only part of the RNA which is synthesized in the cell is subject to this control. In connection with this we have recently made a hypothesis according to which the essential condition for the accumulation of harmful changes during ageing is the gradual weakening of the hereditary control over cytoplasmic syntheses after morphogenesis has been completed, especially in those highly-specialized cells which cannot divide (227, 69).

We believe that this weakening of control takes place by the gradual loosening of the connections between DNA and RNA. RNA (both nuclear and cytoplasmic) begins to be formed in a more autonomous way which cannot continue for ever because of its imperfections. Its duration depends on the individual and specific peculiarities of the particular organism.

Further analysis of this possible way in which the road to molecular variation may be opened up shows that this autonomous synthesis of RNA does occur in every cell, though to a varying extent. It must be mentioned that some geneticists, such as Hadorn (185), suggest that on passing through the functional to the morphogenetic period genetic loci then become inactive.

The autonomous synthesis of RNA would seem to be inevitable, as those parts of the DNA which control the current synthesis of RNA in the cells only represent part of the general pool of DNA because much of this pool is not directly concerned with current syntheses but is the basis of the whole of ontogenesis. The small proportion of the DNA which is concerned with current syntheses in the cell is simply not in a position to carry out the whole system of protein synthesis by matrices and adaptors in the cell, it can only provide the model for this purpose. A large amount of the RNA synthesized in the nucleolus is certainly also not formed under the direct control of DNA but autonomously on models synthesized in the chromosomes (cf. Chap. XIII).

The absence of complete control by DNA over the synthesis of all the RNA is not the cause but the condition for the accumulation of changes with increasing age. It is, of course, possible to remove causes but it is also possible not to create the conditions under which these causes can exercise their harmful influence. It would seem that, in its regulation of the length of life, nature acts in both of these ways and man must also look for all possible ways of directing this important process.

It is also possible that the "disconnection" of the system of DNA and RNA is under genetic control and that the beginning of this "disconnection" is determined.

The classical example of the complete abdication by DNA of control over cytoplasmic processes when differentiation has been completed is that of the reticulocytes of mammals, which usually lose their nucleus and a great part of their DNA. However, in many other highly specialized cells as well, the function of the nucleus becomes weaker and the ratio of DNA to RNA declines sharply. We have already described the evidence for this in connection with the cells of the nervous system, the muscles and the kidneys.

### b) *Concerning the possibility of the autonomous accumulation of changes in proteins with increasing age independently of nucleic acids*

The living organism consists of structures which are constantly being renewed. As this takes place any changes, including those caused by increasing age, are preserved and become more widespread, not only because they have happened but because they are constantly reproduced during self-renewal and do not vanish without a trace like tiredness after a good sleep.

When increasing age is accompanied by changes in polymers which are undergoing continual cycles of breakdown and resynthesis, the conservation of errors is important but so is the conservation of the ability to make errors. Episodic errors in protein synthesis (independent of DNA) may not be preserved, for example, but the possibility or likelihood of their occurrence increases with age.

When we were discussing the possible causes of the accumulation of changes as the organism gets older we started from the assumption that protein molecules do not really bequeath their structure to those synthesized later but that new molecules always arise *de novo*. In other words, we took the view that the information which determines the sequence of amino acids cannot be handed on from one protein to another but only from RNA to a protein. This view, though accepted by almost everyone, is not, however quite accurate in relation to the accumulation of changes.

The transmission of specific information and the transmission of random variations are two different phenomena and must not be regarded as the same. We may, for example, consider the possible occurrence of sequences of events such as change in matrix → change in enzymes → decrease in the precision of reactions and increasing production of analogues → incorporation of analogues in matrices → increase in the variability of matrices.

Long-lived proteins may also accumulate changes like closed systems. This refers to collagen, elastin, keratin and a number of complicated protein complexes occurring in nervous tissue (166) and also to haemoglobin (244).

By their very nature the changes which occur in collagen fibres with increasing age cannot be attributed to changes in matrices because, for the most part, they concern the secondary and tertiary structure of the molecules (223).

Davison (166) has recently determined the "lifetimes" of certain stable complexes (of lipoproteins, collagen, etc.) in the nervous system and other tissues of rats. In many cases it amounted to 250 or more days.

If we take the view that the inactivation of parts of molecules decreases the likelihood of their being broken down, then it strikes us that

the occurrence in several organs of fractions of collagen which are renewed quickly and others which are only renewed slowly may be due to
dissociation into changed and unchanged molecules.

It may, however, be held theoretically, that the autonomous accumulation of changes does not only take place in fibrillar and inert proteins
but also occurs in proteins which are being actively renewed in the cells
and tissues. We put forward a hypothetical scheme of such inactivation
in 1952-3 (56-61) and although many of the assumptions made then are
now out of date the factors, which we suggested might operate, may turn
out to be real as auxiliary mediators of variation with age even if not as
the main ones.

The essential requirement of our scheme was that the products of incomplete breakdown of a protein could be used locally for new synthesis
of the same protein. It has now been shown that this does happen (cf.
Chap. VI).

Peptide fragments of proteins are, however, specific compounds which
exhibit particular features of protein structures and the interdependence
between proteins may extend to the level of peptides. *The direct use of
peptides in the synthesis of proteins is one form of the transfer of information
from those proteins which form the original substrate for the liberation of the
peptides; it is the direct transfer of their specificity to the proteins which are
newly formed on the matrices.*

The conservation of changes by the formation of new proteins from
the products of the partial breakdown of old ones is very probable and it
was the assumption of changes of just this sort which served as the basis
of the theory of ageing which we put forward in several earlier papers (56-
58, 60, 61). We also suggested the possibility of accumulation of changes
in cycles of successive synthesis of DNA and RNA in accordance with the
scheme RNA → oligonucleotides → new RNA (59) but the complementary
theory provides a simpler solution to this problem. It is, however, possible that the variant which we suggested may play a supplementary part.

The main assumption underlying this hypothesis of the inactivation
of proteins during ageing was that three factors interacted. These were:
non-equivalence of metabolism; the possibility of reutilization of peptides
formed from the breakdown of proteins for the resynthesis of new proteins; and differences in the resistance of the various bonds in proteins
to proteolytic enzymes which determine the differences in the rates of
breakdown and resynthesis of these bonds and create inequalities in their
variability.

If we take no notice of any possible outside effects we may theoretically
postulate some polypeptide structure which undergoes breakdown and
resynthesis in a system of enzymes which affect the bonds of the polypeptide at different rates and assume that in the course of this there is a
certain likelihood that some cases of non-equivalent substitution will
occur. As time goes on the accumulation of changes in this polypeptide

can only occur in respect of amino acids joined by stable bonds. Non-equivalent exchange of amino acids joined by easily split bonds is episodic and such changes are lost during breakdown and not inherited by the new generation of molecules. It is clear that if we assume that the molecule contains different numbers of bonds which are broken quickly and slowly and that the percentage of transfers of these bonds which can occur as a result of non-equivalent substitution is proportional to the number of cycles then, in the course of self-renewal, the number of slowly broken bonds will increase as the number of transfers per unit of time from the labile to the stable state will, in absolute terms, be greater than the number in the opposite direction. This state of affairs is brought about by the fact that the more frequent the metabolic cycles the more transfers from one type of bond to another will be possible in a given time. In the case of the most labile bonds it is quite reasonable that they can only be converted into more stable ones. Of course, slow cycles of breakdown and resynthesis of more stable bonds may also be converted into faster ones but in the overall total of transformations the gradual tendency away from quick cycles of exchange of labile bonds and towards slow cycles must become quite clear just because the transformations occur more often in quick cycles than in slow ones.

The possibility of the accumulation of changes by this type of direct transfer of information between the generations of proteins seems to us to be real but the relative importance of changes of this type in the general mass of changes is, of course, only small.

We have already mentioned two of the main factors in this variation: the ability of the synthesizing system to use peptides and the non-equivalence of protein metabolism. The third and, perhaps the most important factor in this sort of evolution of proteins is the real occurrence of interlacing of the processes of breakdown and synthesis in their intermediate stages. The possibility of such interlacing is quite clear from a survey of the biochemical mechanisms for the intracellular breakdown of proteins both during autolysis and *in vitro*, showing that these processes do, in fact, take place by stages.

This hypothesis of autonomous variability of proteins during ageing is based on the recognition of the very wide differences in the resistance of the internal peptide bonds of proteins to the action of the intracellular proteases and peptidases so that the breakdown of these proteins takes place by stages. During this breakdown intermediate polypeptide and peptide products appear in the cell and it seems that these can be incorporated into the metabolic cycle again without passing through all the stages of breakdown to free amino acids. This idea of the mechanism of the breakdown of proteins inside the organism is based mainly on results concerning the specificity of proteolytic enzymes and the actual presence in proteins of bonds which are sensitive and insensitive to the effects of particular proteases or peptidases.

## 7. The "quantity" of molecular changes and the stages of life

The question of the "quantity" of molecular changes corresponding to the concepts of "youth", "maturity", "advancing years" and "old age" is very complicated and cannot be answered in the same way for different proteins. The "quantity" of change in the structural proteins like collagen corresponding with old age seems to be very great and covers the whole mass of the protein. The "quantity" of changes in enzymic proteins, for example, may be quite insignificant; but the presence of even 0·01% of improper enzymes may lead to considerable disharmony in the metabolism and hasten ageing. Thus we cannot, as yet, make conclusions about the absence of changes with age in any particular proteins on the basis, for example, of analysis of their amino acid composition or of results obtained by other methods which have an experimental error of the order of 3-5%.

What distinguishes the concept which we have developed here from earlier hypotheses about molecular variation is *that the changes occurring in proteins and nucleic acids with the passage of time are not directed in any orderly way but are chaotic.* The direction of attention towards the fact that even random and chaotic changes can accumulate should serve as a stimulus towards the working out of suitable methods for detecting them.

We also believe that, even in spite of their insignificant extent of the order, perhaps, of 0·01%, these changes diminish the accuracy of enzymic reactions and increase the occurrence of "anarchic" or "chaotic" biochemical activity and thus augment the accumulation of changes at higher levels of organization of the living body. It therefore seems to us that the direct analysis of proteins does not provide the most suitable approach to the detection of their random variation but that the study of indicators of the biochemical activity of particular groups of proteins would be more fruitful.

## 8. Genetic aspects of ageing

The genetic aspects of ageing comprise a large subject which we cannot discuss in detail here. We shall only touch very briefly on a few problems. Gowen (182a) has recently organized a special symposium on the genetic aspects of radiation damage and of normal ageing which dealt with the very topical question of the connection between genetic phenomena, in particular, somatic mutations, and the phenomenon of ageing. In his own paper Gowen produced many facts which indicated that as age increases there really is an increase in the genetic mosaic character of the cells in the tissues. It is well known that the rate at which somatic mutations take place is several orders higher than the rate of mutation of the sex cells and that the mutations are of a mosaic (chaotic) nature. The occurrence of these mutations undoubtedly introduces elements of disorganization into the whole sum of all the processes in the organism and an increase in these changes with the passage of time is certainly one of the universal manifestations of ageing in general.

In this connection we are interested in the function of the recessive lethal characteristics and mutations which are to be found in virtually all crossings of different species and become apparent on inbreeding. It is not yet known whether or not recessive lethal genes exert a biochemical activity which is compensated for by the parallel functional dominant genes of the other chromosome of the pair. If these lethal genes are biochemically active this is yet one more avenue leading to the accumulation, as time goes on, of "noises" and disharmony in the biochemical complex of the living organism.

It would be interesting to find out whether the range of lethal genes increases with increasing age. For this purpose it would be necessary to study the results of inbreeding animals of different ages.

The genetic approach to ageing leads to the assumption that the number of possible mutants in the progeny of old parents ought to be increased relative to that in the progeny of young parents. There are in fact a number of pieces of evidence in support of this. Thus in humans, the incidence of the genetically determined disease of mongolism increases with increasing age of the mother (235a, 195a). Similar observations have been made for a number of hereditary pathological conditions (175a, 270a).

There is, of course, no contradiction between the mutational and molecular concepts of ageing; they are two facets of the same theory. In this connection it is interesting to take note of the fact that, behind the evolutionary variation which makes it possible for life to maintain itself and develop on our planet and the variation with age which limits each ontogenesis, we find the same property of living matter.

## 9. Is it possible to control changes associated with ageing at the molecular level?

The existence of many levels of ageing and many means of accumulation of variations makes it certain that very many agents may be used to regulate the process of variation with increasing age (they may be hormonal, nervous, pharmacological, trophic, nutritional, etc.). At present we shall not put forward any assessment of these lines of work but shall only refer shortly to one question of principle concerning the possibility of direct actions on the systems of synthesis of proteins and nucleic acids of the organism as a whole, although they may be limited to the regulation of the processes of accumulation of molecular variation.

There can be no doubt that during ageing changes occur in all the molecular, supramolecular, subcellular, cellular, physiological and functional systems and it is possible that, if there were a possibility of eliminating all the destructive changes in proteins and nucleic acids, then, maybe, the process of ageing would become slower though it would continue to take place on account of the accumulation of changes in more complicated and specific structures. It seems that there are a great many independent ways in which living systems can change in the course of time and that there are many types of irreversible reaction and that these are only connected with one another through the general interdependence of all metabolic processes.

The gradual discovery of the fundamental features of protein synthesis enables us also to find ways of intervening actively in this process with the help of particular products which are involved in the synthetic side of protein and nucleic acid metabolism. This line of work has been specially extensively developed by the study of the possibility of suppressing various forms of pathological syntheses by means of a tremendous assortment of antimetabolites affecting the metabolism of amino acids and nucleotides. However, a similar approach could certainly be made to the stimulation of synthesis and also to its direction, that is its direction towards the correction of any defects which have arisen in the synthesis of proteins and nucleic acids with increasing age.

The material which we have discussed in this book shows clearly that biochemistry, genetics and gerontology are progressively approaching the possibility of intervening in the vital processes which may enable them in the near future to exert effective control over the length of life.

Analysis of the factors affecting molecular variation and the factors which determine the accumulation of the variations leads logically towards finding the line which is likely to produce the best results in regard to the experimental control of these processes.

Three parallel approaches may be adopted to intervention in the process of ageing at the molecular level: (1) the suppression or inhibition of factors which cause molecular variation; (2) strengthening control at the level DNA → RNA; and (3) repair and replacement of altered structures.

One example of action on the factors which change macromolecules is that of controlling the formation of free radicals which appear spontaneously in the cells during biochemical and biophysical reactions (187-190).

It is interesting to note that in Harman's experiments the systematic use of inhibitors of free-radical reactions increased the mean lifetime of rats by 15-20% (189). A similar increase in the mean length of the life of rats has also been observed as a result of the systematic use of so-called radiation protectors (188). The mechanism whereby these substances act would seem to be common to all of them, as the inhibitors of free-radical reactions also decreased the amount of radiation damage (135).

Another way of affecting the changes occurring with increasing age is the use of immune reactions. If it is really possible to show that reactions between antibodies and altered molecules (autoimmune reactions) increase the disharmony of growing older by the formation of insoluble complexes then it should surely be possible to prevent the occurrence of these phenomena by using the phenomenon of tolerance (cf. Chap. XI), that is to say, by introduction into the embryo of substances characteristic of the old organism. This eliminates the ability of the immunological "memory" to "recognize" subsequent changes of this kind. If it is found that auto-immune reactions of this kind are valuable in promoting the precipitation and selective elimination of "deformed" molecules the formation of antibodies of this kind can be increased by immunization of the organism while still young.

The question of strengthening the control of DNA over the synthesis of matrices can still only be treated abstractly, although such an approach is by no means unprofitable. The tissues certainly contain endogenous factors which affect this control and if these could be discovered it might also show the way to intervention in this most intimate biological process.

The problem of replacement and repair of altered structures can be subjected to intensive investigation straight away. For instance, it might be quite possible to "renew" the stock of soluble RNA or to introduce into the cell a messenger RNA which also had a low molecular weight. In such a case the transfer of the ability to synthesize "young" proteins is similar in principle to the transfer by means of RNA of the ability to synthesize adaptive enzymes, antibodies, viral proteins, etc. If this were done it is possible that a single injection of easily assimilated, soluble, low-molecular, "young" RNA could be effective as, according to the principle of complementarity, it would find unchanged matrices among many which had been changed and would activate their participation in the synthesis of proteins.

If it is found that it is possible to use larger oligonucleotides as well as nucleotides for the synthesis of nucleic acids, then it will also be possible to begin experimental work on the "correction" of the forms of RNA which have a higher molecular weight, especially if the evidence for the selective incorporation of oligonucleotides into homologous organs is confirmed.

We still have no direct evidence of the possibility of using oligonucleotides in the synthesis of nucleic acids without their first being broken down to nucleotides. We have indirect evidence of this possibility in the works of Lu & Winnick (220, 221) and especially in the work of Ardry et al. (141). The authors of the latter work injected solutions of labelled RNA and DNA isolated from various sources (liver, pancreas and yeast) intramuscularly into rats and observed the way in which they were distributed and assimilated within the organism. Four days after the beginning of daily injections of labelled nucleic acids it was noticed that there was a well-defined tropism in the products of the disintegration of RNA and

DNA derived from the liver and pancreas, which were incorporated into the nucleic acids of the corresponding organs of the experimental rats. Such an effect might result from the reutilization of the products of incomplete depolymerization which retained some of their tissue specificity.

It would seem that a similar approach could be made to the "repair" of proteins by direct incorporation into them during resynthesis of the products of the incomplete hydrolysis of proteins characteristic of youth and obtained, for example, from tissue cultures.

In many laboratories experiments are now going on which have shown that when the products of the incomplete hydrolysis of proteins of particular organs are injected into an organism they can be used selectively for the synthesis of the proteins of homologous organs (171, 224). These facts indicate the possibility of the selective reconstruction of individual organs and tissues which have become altered, by the use of the products of the incomplete hydrolysis of homologous young organs and tissues. When proteins are partly disintegrated there comes a time when their antigenicity is lost and the lysate obtained is easily assimilated, but the peptide fragments which are obtained still have many of their specific properties (e.g. weak enzymic activity) and a tissue and species specificity which is accounted for by a particular sequence of amino acid residues. It would seem that the same is true of nucleic acids. Mixtures and various combinations of these products certainly have unlimited possibilities for biological stimulation and inhibition and it is no accident that in living nature a large proportion of antibiotics and a large assortment of hormones are specific peptides. The study of the way in which proteins and nucleic acids of different organs and tissues are synthesized and change with age and the control of these processes through the substitution of particular products of an intermediate nature seems to us to be a completely realistic though difficult line of approach to the control of the ontogenesis of the organism.

These ideas have been put forward partly on the basis of a small experiment which we carried out on yeast cultures in an attempt to see whether it was possible to change some of the indicators of age in yeast cells by their assimilation of the products of the breakdown of proteins from yeast cultures of different ages (64). In carrying out this work we managed to show that when young yeast cultures assimilate peptic hydrolysates of proteins from old yeast cultures (which are distinguished by their great resistance to autolysis and by their amino acid composition) the proteins of the young culture "grow old" and their resistance to autolysis and amino acid composition become similar to those of the proteins of old cultures of yeast.

In experiments on unicellular organisms and lower animals it might be possible to study the analogous activity of peptides "taken away" from RNA. It is quite possible that the most radical way of replacing and reconstructing "old" matrices to get "young" ones will form the basis of the final victory of mankind over the processes of ageing. This will not require the isolation of "young" matrices from living things, it will be possible to prepare them synthetically by means of "samples" or to isolate them from tissue cultures.

It would appear that the very randomness of the changes occurring in proteins and amino acids with increasing age may be used as a means for putting an end to their accumulation. Owing to this randomness the systems for the synthesis of specific proteins always contain, in addition to "deformed" matrices a definite percentage (which is clearly the greater part) of unchanged structures and unchanging cycles. Thus it is necessary to find some way of blocking and isolating the altered matrices and increasing the load and intensity of the self-reproduction of unaltered matrices. It is therefore our task to find some way of selectively accelerating the elimination of altered macromolecules and matrices.

Attempts have already been made to change the physico-chemical state of old metaplasmic proteins which have been inactivated by the use of reducing agents which can break hydrogen, sulphide and cross-linking bonds (144, 160).

In this present book we do not intend to go into any detailed analysis of the problem of rejuvenation, because we have regarded ageing only from its molecular aspect. The study of the various biological mechanisms which compensate for ageing is the main task which the gerontologists have before them. Many valuable ideas in this field have been put forward and worked out in many gerontological laboratories (82, 84, 109, 161, 174, 253-257 and others). The only real way of making any perceptible progress in this field is by the painstaking and systematic study of all types and forms of variation which occur with increasing age at all levels of complexity and the working out of a wide and varied range of stabilizing, compensatory and reparative agents. It seems that no individual agency will be able to counteract the all-pervading changes of this infinitely complicated system which is bound up with the evolution of the organic world. Although the problem is extremely complicated, the rapid development of biology and especially biochemistry, genetics and gerontology shows that it is not insoluble.

## Conclusion

The process of ageing is an extremely complicated phenomenon of living nature. This does not, however, prevent us from looking for relatively simple processes and mechanisms which determine the onset of the evolution of the ageing of that extremely well-adjusted and complicated system, the organism of higher animals, which has been created by morphogenetic processes out of the relatively simple biological systems which existed during the first stages of the development of the biological world. The functions of the essential specific biological polymers, DNA, RNA and protein, are very closely correlated with their composition and it is changes in this structure resulting from the spontaneous non-equivalence of metabolism and from external and internal causes which forms the basis of all forms of biological variation as well as of the changes occurring in the individual with increasing age. These changes occur in many ways and it is not yet known which is the limiting one but this is a question relevant to the problem of the length of life. The hypotheses which we have discussed in relation to the mechanisms of variation of protein and the various ways in which ontogenesis is controlled are certainly very schematic. A great deal of work is still required before we can approximate them to what actually happens in life and discuss intracellular metabolism in terms of real facts.

In Fig. 64 we have tried to produce a diagrammatic summary of the reactions and interconnections between the factors responsible for molecular variation and its consequences.

A considerable number of these factors fall into the category of "imperfections". In Strehler's interesting papers (271, 273) these "imperfections" were divided into four groups: inadequacy of design, errors of design, omissions of design and contradictions in design. However, these imperfections of biological systems are only defects from the point of view of people who subjectively desire to prolong life by increasing the ability of the organism to maintain its constancy. From the evolutionary point of view these "imperfections" are valuable, ensuring the succession of generations of the species. If nature has found the ways and means of controlling the length of life of biological species, then biologists must set the same goal before themselves and endeavour to attain it in the shortest possible time.

# REFERENCES

1. AVREKH, V. V. & GERONIMUS, E. S. (1937). *Byull. eksp. Biol. Med.* **4**, 505.
2. ALPATOV, V. V. & NASTYUKOVA, O. K. (1948). *Doklady Akad. Nauk S.S.S.R.* **59**, 1365.
3. ANOROVA, N. S. (1956). *Doklady Akad. Nauk S.S.S.R.* **110**, 494.
4. BELOZERSKIĬ, A. N. (1941). *Mikrobiologiya*, **10**, 185.
5. — (1949). *Vestn. Mosk. Univ.* **2**, 125.
6. — (1956). *Uspekhi sovremennoĭ Biol.* **41**, 144.
7. BLAGOVESHCHENSKIĬ, A. V. (1950). *Biokhimicheskie osnovy evolyutsionnogo protsessa u rastenii.* Moscow: Izd. AN SSSR.
8. BULANKIN, I. N. & BLYUMINA, M. A. (1947). *Uchenye Zupiski Khar'kov. Univ.* **25**, 61.
9. BULANKIN, I. N., LANTODUB, I. Yu., NOVIKOVA, N. M., PAPAKINA, I. K. & FRENKEL', L. A. (1954). *Uchenye Zapiski Khar'kov. Univ. 53*, Trudy nauch.-issledovatel. Inst. Biol. **21**, 87.
10. BULANKIN, I. N. & PARINA, E. V. (1954). *ibid.* **21**, 135.

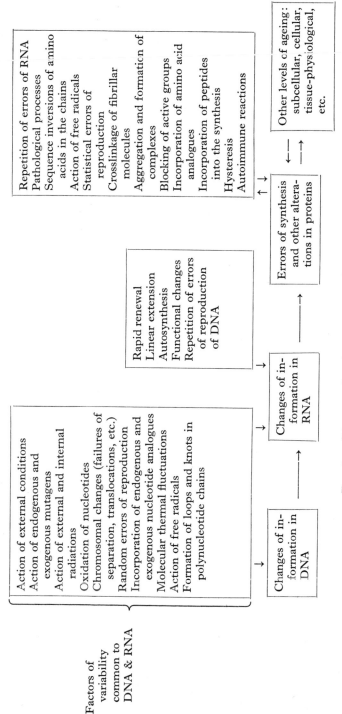

FIG. 64. Diagram of ageing at the molecular level.

11. — (1959). In collective work *Aktual'nye problemy sovremmenoĭ biokhimii.* *1. Biokhimiya belkov* (ed. V. N. Orekhovich), p. 205. Moscow: Izd. AMN SSSR.

12. — (1960). *Uchenye Zapiski Khar'kov. Univ.: Trudy nauch.-issledovatel. Inst. Biol.* **29**, 23.

13. BULANKIN, I. N., PARINA, E. V. & GOLOVKO, N. I. (1960). *Doklady Akad. Nauk S.S.S.R.* **134**, 1461.

14. BULANKIN, I. N., PARINA, E. V. & KURILENKO, R. P. (1956). *Trudy nauch.-issledovatel. Inst. Biol. i biol. Fak. Khar'kov. gos. Univ.* **24**, 5.

15. BULANKIN, I. N., PARINA, E. V. & SERGIENKO, E. F. (1951). *Trudy konferentsii po vozrastnym izmeneniyam obmena veshchestv i reaktivnosti organizma* (ed. N. N. Sirotinin), p. 27. Kiev: Izd. AN Ukr. SSR.

16. VITT, V. O. (1953). In *Trudy nauchnoĭ konferentsii po voprosam znacheniya vozrasta pri razvedenii sel'skokhozyaĭstvennykh zhivotnykh* (ed. V. O. Vitt), p. 9. Moscow: Izd. Timiryazev. sel'skokhoz. Akad.

16a. GITEL'ZON, I. I. & TERSKOV, I. A. (1960). In collective work *Voprosy biofiziki, biokhimii i patologii eritrotsitov* (ed. I. A. Terskov & I. I. Gitel'zon), p. 148. Krasnoyarsk: Izd. AN SSSR.

17. GOL'DENBERG, E. E. & KONDRASHINA, M. P. (1940). *Byull. eksp. Biol. Med.* **9**, 226; [in French] *Bull. Biol. Med. exptl. U.R.S.S.* **9**, 288.

18. GRABAR, P. (1957). *Biokhimiya*, **22**, 49.

19. DENISOVA, A. A. & ZAĬDES, A. L. (1957). *Doklady Akad. Nauk S.S.S.R.* **114**, 1287.

20. DERVIZ, G. V. (1939). In collective work *Starost'* (ed. A. A. Bogomolets), p. 213. Kiev: Izd. AN Ukr. SSR.

21. DOBRYNINA, V. (1940). *Fiziol. Zhur. S.S.S.R.* **29**, 225.

22. DUBININ, N. P., SIDOROV, B. N. & SOKOLOV, N. N. (1959). *Doklady Akad. Nauk S.S.S.R.* **128**, 172.

23. DUNAEVSKIĬ, F. R. (1960). *Doklady Akad. Nauk. S.S.S.R.* **133**, 954.

24. ZAĬTSEVA, G. N. & BELOZERSKIĬ, A. N. (1957). *Mikrobiologiya*, **26**, 533.

25. ZAMARIN, L. G. (1953). *Trudy Saratov. zoovet. Inst.* No. 4, p. 31.

26. ZAKHAROVA, A. V. (1954). *Byull. eksp. Biol. Med.* **17**, 53.

27. ZBARSKIĬ, B. I. (1949). *Problemy Sovetskoĭ fiziologii, biokhimii i farmakologii. Trudy 7-go s"ezda fiziologov*, Vol. 2, p. 608. Moscow: Izd. AN SSSR.

28. ZBARSKIĬ, I. B. & DEMIN, N. N. (1949). *Rol' eritrotsitov v obmene belkov.* Moscow: Izd. AMN SSSR.

29. ZBARSKIĬ, I. B. (1949). In collective work *Voprosy meditsinskoĭ khimii* (ed. V. N. Orekhovich *et al.*), Vol. *1*, Nos. 1 & 2. Moscow: Izd. AMN SSSR.

30. IOFFE, K. G. & SOROKIN, V. M. (1954). *Biokhimiya*, **19**, 652.

31. KANTSEL'SON, R. S. (1955). *Izvestiya estestvennogo nauchnogo Instituta imeni P. F. Lesgafta*, **27**, 98.

32. KIZEL', A. R. (1934). In collective work *Problema belka* (ed. A. R. Kizel' *et al.*), p. 27. Moscow: Izd. AN SSSR.

33. — (1935). *Uchenye Zapiski Moskov. gos. Univ.* **4**, 194.

34. — (1938). *Uspekhi sovremennoĭ Biol.* **8**, 151.

35. KOMAROV, L. V. (1960). *Tezisy Dokladov Vtorogo Soveshchaniya po Gerontologii i Geriatrii pri Moskovskom Obshchestve Ispytateleĭ Prirody*, p. 68. Moscow: Izd. Mosk. Univ.

36. KONAREV, V. G. (1958). *Izvest. Akad. Nauk S.S.S.R., Ser. biol.* No. 4, p. 395.

37. KONYUKHOV, B. V. (1958). *Uspekhi sovremennoĭ Biol.* **45**, 97.

38. KORNEEVA, A. M. (1957). *Vestnik Moskov. Univ.* 12, Ser. biol., pochvoved., geol. i geograf. No. 4, p. 45.

39. KORINENKO, V. M. (1960). *Trudy nauch.-issledovatel. Inst. Biol. Khar'kov. gos. Univ.* **29**, 29.
40. KORINENKO, V. M., LOBANOV, A. V. & DENISOV, V. M. (1960). *Ibid.* **29**, 41.
41. KORINENKO, V. M. & MARTYNENKO, A. A. (1960). *Ibid.* **29**, 44.
42. KOSYAKOV, K. S. (1940). *Uchenye Zapiski Khar'kov. gos. Univ.*, No. *34*; *Antropologiya,* p. 189.
43. KRASIL'NIKOVA, V. I. (1960). *Tezisy Dokladov Pervoĭ Konferentsii po Voprosam Tsito- i Gistokhimii*, p. 98. Moscow: Izd. AN SSSR.
44. — (1960). *Tsitologiya*, **2**, 29.
45. KRETOVICH, V. L., SMIRNOVA, T. I. & FRENKEL', S. YA. (1958). *Biokhimiya*, **23**, 135.
46. LEĬKINA, E. M., TONGUR, V. S., LIOZNER, L. D., MARKELOVA, I. V., RYABININA, Z. A., SIDOROVA, V. F. & KHARLOVA, V. G. (1960). *Biokhimiya*, **25**, 96.
47. MAZUROV, V. I. & OREKHOVICH, V. N. (1960). *Biokhimiya*, **25**, 814.
48. MAKAROV, P. V. (1948). *Fiziko-khimicheskie svoĭstva kletok i metody ikh izucheniya.* Leningrad: Izd. Leningradsk. gos. Univ.
49. MAKSIMOV, N. A. & MOZNAEVA, L. V. (1944). *Doklady Akad. Nauk S.S.S.R.* **42**, 229, 277.
50. MANOĬLOV, S. E. & ORLOV, A. S. (1953). *Biokhimiya*, **18**, 456.
51. MANUKYAN, K. G. (1955). *Doklady Akad. Nauk S.S.S.R.* **101**, 1085.
52. — (1955). *Doklady Akad. Nauk S.S.S.R.* **102**, 567.
53. MARTSINKOVICH, L. D. (1956). *Doklady Akad. Nauk S.S.S.R.* **111**, 1105.
54. MAKHIN'KO, V. I. (1951). In collective work *Vozrastnye izmeneniya obmena veshchestv i reaktivnosti organizma* (ed. N. N. Sirotinin), p. 36. Kiev: Izd. Akad. Nauk Ukr. SSR.
55. MAKHIN'KO, V. I. & BLOK, L. N. (1960). *Uchenye Zapiski Khar'kov. Univ. Trudy Inst. Biol. i biol. Fak.* **29**, 337.
56. MEDVEDEV, ZH. A. (1952). *Uspekhi sovremennoĭ Biol.* **33**, 202.
57. — (1952). *Referaty Dokladov nauchnoĭ Konferentsii Timiryazev. sel'skokhoz. Akad.* No. *16*, p. 249 (ed. G. M. Loza).
58. — (1953). *Uspekhi sovremennoĭ Biol.* **35**, 338.
59. — (1953). *Uspekhi sovremennoĭ Biol.* **36**, 161.
60. — (1953). *Priroda*, No. *3*, 101.
61. — (1953). In *Sbornik materialov konferentsii po voprosam znacheniya vozrasta pri razvedenii sel'skokhozyaĭstvennykh zhivotnykh* (ed. V. O. Vitt), p. 40. Moscow: Izd. Timiryazev. sel'skokhoz. Akad.
62. — (1955). In collective work *Mechenye atomy v issledovaniyakh pitaniya rastenіĭ i primeneniya udobrenіĭ* (ed. V. N. Klechkovskiĭ et al.), p. 61. Moscow: Izd. AN SSSR.
63. — (1956). *Doklady Timiryazev. sel'skokhoz. Akad.* **22**, 345.
64. — (1956). *Trudy nauch.-issledovatel. Inst. Biol. Khar'kov. gos. Univ.* **24**, 65.
65. — (1956). *Doklady Timiryazev. sel'skokhoz. Akad.* No. *23*, p. 214.
66. — (1957). *Doklady Timiryazev. sel'skokhoz. Akad.* No. *29*, p. 55.
67. — (1957). *Doklady Akad. Nauk S.S.S.R.* **117**, 860.
68. — (1957). *Biokhimiya*, **22**, 855.
69. — (1961). *Uspekhi sovremennoĭ Biol.* **51**, 299.
70. MEDVEDEV, ZH. A. & ZABOLOTSKIĬ, N. N. (1958). *Izvest. Timiryazev. sel'skokhoz. Akad.* No. *3*, p. 207.
71. MEDVEDEV, ZH. A., ZABOLOTSKIĬ, N. N., SHEN', Ts.-S., MO, S.-M., DAVIDOVA, E. G. & DAVIDOV, E. R. (1960). *Biokhimiya*, **25**, 1001.
72. MEDVEDEV, ZH. A. & FEDOROV, E. A. (1956). *Fiziologiya Rastenіĭ*, **3**, 547.
73. MEDVEDEV, ZH. A. & U TSZYUN' (1956). *Doklady Timiryazev. sel'skokhoz. Akad.*, No. *26*, pt. 1, p. 273.

74. MIKHAĬLOVA, V. N. (1958). *Izvest. Akad. Nauk S.S.S.R., Ser. biol.* No. *1*, p. 89.

75. MOLCHANOVA, V. V. (1957). *Doklady Akad. Nauk S.S.S.R.* **112**, 1119.

76. NAGORNYĬ, A. V. (1940). *Problema stareniya i dolgoletiya.* Kharkov: Izd. Khar'kov. gos. Univ.

77. — (1947). *Uchenye Zapiski Khar'kov. Univ. im Gor'kogo*, **25**, 18.

78. — (1950). *Starenie i prodlenie zhizni.* Moscow: Izd. "Sovetskaya Nauka".

79. — (1951). *Trudy Konferentsii po vozrastnym izmeneniyam obmena veshchestv i reaktivnosti organizma* (ed. N. N. Sirotinin), p. 5. Kiev: Izd. AN Ukr. SSR.

80. NIKITIN, V. N. (1947). *Uchenye Zapiski Khar'kov. gos. Univ.* **25**, 95.

81. — (1948). *Zhur. obshcheĭ Biol.* **9**, 113.

82. — (1954). *Trudy nauch.-issledovatel. Inst. Biol. Khar'kov. gos. Univ.* **21**, 29.

83. — (1957). *Trudy Inst. Morfol. Zhivotnykh im. Severtseva*, No. *22*, p. 26.

84. — (1958). *Priroda*, No. *2*, p. 39.

85. NIKITIN, V. N. & GOLUBITSKAYA, R. I. (1954). *Trudy nauch.-issledovatel. Inst. Biol. Khar'kov. gos. Univ.* **21**, 143.

86. — (1954). *Trudy nauch.-issledovatel. Inst. Biol. Khar'kov. gos. Univ.* **21**, 113.

87. — (1959). *Biokhimiya*, **24**, 1023.

88. — (1960). *Trudy nauch.-issledovatel. Inst. Biol. Khar'kov. gos Univ.* **29**, 95.

89. NIKITIN, V. N., GOLUBITSKAYA, R. I., DRYUCHINA, L. A., SEMENOVA, Z. L. (1951). *Trudy Konferentsii po vozrastnym izmeneniyam obmena veshchestv i reaktivnosti organizma* (ed. N. N. Sirotinin), p. 17. Kiev: Izd. AN Ukr. SSR.

90. NIKITIN, V. N., GOLUBITSKAYA, R. I., SILIN, O. P. & STAVITSKAYA, L. P. (1956). *Trudy nauch.-issledovatel. Inst. Biol. Khar'kov. gos. Univ.* **24**, 102.

91. NIKITIN, V. N., NOVIKOVA, A. I. & TSIKALO, A. P. (1960). *Trudy nauchno.-issledovatel. Inst. Biol. Khar'kov. gos. Univ.* **29**, 91.

92. NIKITIN, V. N. & PROKOPENKO, R. K. (1957). *Ukr. biokhim. Zhur.* **29**, 329.

93. NIKITIN, V. N., SILIN, O. P. & MOROZ, YU. A. (1959). *Dopovidi Akad. Nauk. Ukr. R.S.R.* No. *11*, p. 1280.

93a. NIKITIN, V. N. & SHERESHEVSKAYA, TS. M. (1961). *Biokhimiya*, **26**, 1062.

94. NOVIKOVA, N. M. (1960). *Trudy nauch.-issledovatel. Inst. Biol. Khar'kov. gos. Univ.* **29**, 161.

95. ODINTSOVA, M. S. (1959). In collective work *Problemy fotosinteza* (ed. I. I. Krasnovskiĭ *et al.*), p. 242. Moscow: Izd. AN SSSR.

96. OREKHOVICH, V. N. (1956). *Priroda*, No. *5*, p. 35.

97. OREKHOVICH, V. N., ALEKSEENKO, L. P. & LEVDIKOVA, G. A. (1957). *Vestnik Akad. med. Nauk S.S.S.R.* **1**, 12.

98. OREKHOVICH, V. N. & SOKOLOVA, T. P. (1940). *Doklady Akad. Nauk S.S.S.R.* **28**, 748.

99. OREKHOVICH, V. N. & SHPIKITER, V. O. (1958). *Biokhimiya*, **23**, 285.

100. OREKHOVICH, V. N. (1950). *Doklady Akad. Nauk S.S.S.R.* **71**, 521.

101. ORLOVSKAYA, G. V., ZAĬDES, A. L. & TUSTANOVSKIĬ, A. A. (1956). *Doklady Akad. Nauk S.S.S.R.* **111**, 1396.

102. OSIPOVA, O. P. & TIMOFEEVA, I. V. (1949). *Doklady Akad. Nauk S.S.S.R.* **67**, 105.

103. PAVLOVSKAYA, T. E., VOLKOVA, M. S. & PASYNSKIĬ, A. G. (1955). *ibid.* **101**, 723.

104. PADUCHEVA, A. L. (1954). *Byull. eksptl. Biol. i Med.* **38**, No. 7, 33.

105. PALLADIN, A. V., BELIK, YA. V. & KRACHKO, L. I. (1957). *Biokhimiya*, **22**, 359.

106. PANCHENKO, L. F. (1957). *Uchenye Zapiski 2-go Moskov. med. Inst.* **6**, 41.
107. — (1958). *Fiziol. Zhur. S.S.S.R.* **44**, 243.
108. PARINA, E. V. & SOPONINSKAYA, E. B. (1956). *Trudy nauch.-issledovatel. Inst. Biol. i biol. Fak. Khar'kov. gos. Univ.* **24**, 43.
109. PARKHON, K. I. [PARHON, C. I.] (1960). *Vozrastnaya biologiya.* Bucharest: Ed. Acad. R.P.R.
110. PARKHON, K. I. [PARHON, C. I.] & OERIU, S. (1960). *Biokhimiya,* **25**, 61.
111. PASYNSKIĬ, A. G. (1951). *Doklady Akad. Nauk S.S.S.R.* **77**, 863.
112. PASYNSKIĬ, A. G., VOLKOVA, M. S. & BLOKHINA, V. P. (1955). *Doklady Akad. Nauk S.S.S.R.* **101**, 317.
113. PLESHKOV, B. P. & FOUDEN [FOWDEN], L. (1959). *Izvest. Timiryazev. sels'kokhoz. Akad.,* No. *5,* p. 95. Cf. *Nature (Lond.)* **183**, 1445.
113a. NIKITIN, V. N. (ed.) (1962). *Problemy vozrastnoĭ fiziologii i biokhimii.* Khar'kov: Izd. Khar'kov. gos. Univ.
114. SALGANIK, R. I. (1956). *Voprosy meditsinskoĭ Khimii,* **2**, 424.
115. SERGIENKO, E. F. & TIMKOVITSKAYA, A. M. (1954). *Trudy nauch.-issledovatel. Inst. Biol. Khar'kov. gos. Univ.* **21**, 81.
116. SILIN, O. P. (1960). *Trudy Inst. Biol. Khar'kov. gos. Univ.* **29**, 53.
117. SISAKYAN, N. M. (1953). *Uspekhi sovremennoĭ Biol.* **36**, 332.
118. SISAKYAN, N. M., BEZINGER, E. N. & GUMILEVSKAYA, N. A. (1953). *Doklady Akad. Nauk S.S.S.R.* **91**, 907.
119. SISAKYAN, N. M., BEZINGER, E. N., GUMILEVSKAYA, N. A. & LUK'YANOVA, N. F. (1955). *Biokhimiya,* **20**, 368.
120. SISAKYAN, N. M. & ODINTSOVA, M. S. (1954). *Doklady Akad. Nauk S.S.S.R.* **97**, 119.
121. — (1956). *Biokhimiya,* **21**, 577.
122. STAVITSKAYA, L. I. (1956). *Trudy nauch.-issledovatel. Inst. Biol. i biol. Fak. Khar'kov. Univ.* **24**, 59.
123. STEPANENKO, A. S. (1939). *Sbornik Rabot po Fiziologii Moskovskogo meditsinskogo Instituta,* **1**, 156.
124. TARANOV, M. T. (1953). *Konevodstvo,* **4**, 28.
125. TUSTANOVSKIĬ, A. A. (1938). *Biokhimiya,* **3**, 218.
125a. TOROPOVA, G. P. (1955). *Voprosy Pitaniya,* **14**, No. *4,* p. 12.
126. TURCHIN, F. V., GUMINSKAYA, M. A. & PLYSHEVSKAYA, E. G. (1953). *Izvest. Akad. Nauk S.S.S.R.,* Ser. biol. No. *6,* p. 66.
127. — (1955). *Fiziol. Rastenii* **2**, No. *1,* p. 3.
128. FIRFAROVA, K. F. (1956). *Voprosy med. Khim.* **2**, No. *1,* p. 69.
129. FUKS, B. B. & KOLAEVA, S. G. (1960). *Tezisy Dokladov Pervoĭ Konferentsii po Voprosam Tsito- i Gistokhimii,* p. 107. Moscow: Izd. AN SSSR.
130. TSERLING, V. V., SHCHEGLOVA, G. M., PLYSHEVSKAYA, E. G. & ZERTSALOV, V. V. (1957). *Fiziol. Rastenii.* **4**, 3.
131. — (1958). In collective work *Fiziologiya Rastenii, Agrokhimiya, Pochvovedenie (Trudy Vsesoyuznoĭ Konferentsii po Primeneniyu Izotopov)* (ed. V. M. Klechkovskiĭ et al.), p. 158. Moscow: Izd. AN SSSR.
132. CHERNIKOV, M. P. (1955). *Biokhimiya,* **20**, 71.
133. — (1955). *Biokhimiya,* **20**, 657.
133a. SHAPOT, V. S. (1962). *Priroda,* No. *12,* p. 19.
134. EĬDRIEVICH, E. V. & POLYAKOV, E. V. (1953). *Zhur. obshcheĭ Biol.* **6**, 635.
135. EMANUEL', N. M., KRUGLYAKOVA, K. E., ZAKHAROVA, N. A. & SAPEZHINSKIĬ, I. I. (1960). *Doklady Akad. Nauk S.S.S.R.* **131**, 1451.
136. ALIVASATOS, S. G. A., MOURKIDES, G. A. & JIBRIL, A. (1960). *Nature (Lond.),* **186**, 718.
137. ALIVASATOS, S. G. A., UNGAR, F., JIBRIL, A. & MOURKIDES, G. A. (1961). *Biochim. biophys. Acta.* **51**, 361.

138. ALLEN, D. W., SCHROEDER, W. A. & BALOG, J. (1958). *J. Amer. chem. Soc.* **80**, 1628.

139. ALLISON, A. C. & TOMBS, M. P. (1957). *Biochem. J.* **67**, 256.

140. ALMOND, H. R., jr. & NIEMANN, C. (1960). *Biochim. biophys. Acta,* **44**, 143.

141. ARDRY, R., KIRSCH, F. & BRUX, J. DE (1956). *Thérapie,* **11**, 658.

142. AUGENSTINE, L. G., CARTER, J. G., NELSON, D. R. & YOCKEY, H. P. (1960). *Radiation Res. Suppl.* **2**, 19.

143. BANFIELD, W. G. (1960). *Proc. V int. Congr. Gerontol, San Francisco,* Abstr. p. 29.

144. BANGA, I., BALÓ, J. & SZABÓ, D. (1956). *Experientia, Suppl.* No. *4*, 28.

145. BARROWS, C. H., jr. (1960). In *The biology of aging* (ed. B. L. Strehler) p. 116. Washington D.C.: American Institute of Biological Sciences (Publication No. 6).

146. BARROWS, C. H., jr., YIENGST, M. J. & SHOCK, N. W. (1958). *J. Gerontol.* **13**, 351.

147. BENNETT, L. L., jr., SIMPSON, L. & SKIPPER, H. E. (1960). *Biochim. biophys. Acta,* **42**, 237.

148. BERGER, H., ZUBER, C. & MIESCHER, P. (1960). *Gerontologia,* **4**, 220.

149. BETTELHEIM, F. R. & NEURATH, H. (1955). *J. biol. Chem.* **212**, 241.

150. BILLINGHAM, R. E. (1958). In *A symposium on the chemical basis of development* (ed. W. D. McElroy & B. Glass), p. 575. Baltimore, Md.: Johns Hopkins Press.

151. BIRMINGHAM, M. K. & GRAD, B. (1954). *Cancer Res.* **14**, 352.

152. BJORKSTEN, J. (1958). *J. Amer. geriatr. Soc.* **6**, 740.

153. BORSOOK, H., DEASY, C. L., HAAGEN-SMIT, A. J., KEIGHLEY, G. & LOWY, P. H. (1949). *J. biol. Chem.* **179**, 705.

153a. BOUCEK, R. J., NOBLE, N. L. & MARKS, A. (1961). *Gerontologia,* **5**, 150.

154. BRENNER, L. O., WAIFE, S. O. & WOHL, M. G. (1951). *J. Gerontol.* **6**, 229.

155. BRODY, S. (1957). *Biochim. biophys. Acta,* **24**, 502.

156. BRYANT, M. & FOWDEN, L. (1959). *Ann. Botany* [N.S.], **23**, 65.

157. CANN, J. R. (1960). *J. biol. Chem.* **235**, 2810.

158. — (1960). *J. chem. Phys.* **33**, 1410.

159. CHVAPIL, M. (1955). *Physiol. Bohemoslov.* **4**, 303.

160. CHVAPIL, M. & HRUZA, Z. (1959). *Gerontologia,* **3**, 241.

160a. GOLD, N. I. & STURGIS, S. H. (1954). *J. biol. Chem.* **206**, 51.

161. COMFORT, A. (1956). *The biology of senescence.* London: Routledge & Paul.

162. NOFRE, C., CIER, A., MICHOU-SAUCET, C. & PARNET, J. (1960). *Compt. rend. Acad. Sci., Paris,* **251**, 811.

163. COOPER, W. D. & LORING, H. S. (1957). *J. biol. Chem.* **228**, 813.

164. DALY, M. M., ALLFREY, V. G. & MIRSKY, A. E. (1952). *J. gen. Physiol.* **36**, 173.

165. DAVIES, M., EVANS, W. C. & PARR, W. H. (1952). *Biochem. J.* **52**, xxiii.

166. DAVISON, A. N. (1961). *Biochem. J.* **78**, 272.

167. DETWILER, T. C. & DRAPER, H. H. (1962). *J. Gerontol.* **17**, 138.

168. DRESNER, E. & SCHUBERT, M. (1955). *J. Histochem. Cytochem.* **3**, 360.

169. DREYFUS, J. C., SCHAPIRA, G. & BOURLIÈRE, F. (1954). *Compt. rend. Soc. Biol.* **148**, 1065.

170. DUSTIN, J. P., SCHAPIRA, G., DREYFUS, J. C. & HESTERMANS-MEDARD, O. (1954). *Compt. rend. Soc. Biol.* **148**, 1207.

171. EBERT, J. D. (1954). *Proc. natl. Acad. Sci. U.S.* **40**, 337.

172. EHRENBERG, R., WINNECKEN, H. G. & BIEBRICHER, H. (1954). *Z. Naturforsch.* **9B**, 492.

173. ELSON, D., GUSTAFSON, T. & CHARGAFF, E. (1954). *J. biol. Chem.* **209**, 285.

174. EVERITT, A. V., MUGGLETON, A., BINGLEY, M. & DANIELLI, S. F. (1962). In *Biological aspects of aging* (ed. N. W. Shock), p. 231. New York: Columbia University Press.

175. FALZONE, J. A., jr., BARROWS, C. H., jr. & SHOCK, N. W. (1959). *J. Gerontol. Moscow*, **9**, 115.

175a. FERGUSON-SMITH, M. A. (1960). *Arch. internal Med.* **105**, 627.

176. FLEISCHER, S. & HAUROWITZ, F. (1960). *Arzneimittel-Forsch.* **10**, 362.

176a. FLETCHER, M. J. & SANADI, D. R. (1961). *J. Gerontol.* **16**, 255.

177. BLOIS, M. S., BROWN, H. W., LEMMON, R. M., LINDBLOM, R. O. & WEISS-BLUTH, M. (eds.) (1961). *Free radicals in biological systems.* New York: Academic Press.

178. FREDERICQ, E. (1953). *Arch. intern. Physiol.* **61**, 424.

179. FUCHS, R. & KLEIFELD, O. (1956). *Albrecht von Graefe's Arch. Ophthalmol.* **158**, 29.

180. GIRI, K. V. (1956). *Naturwiss.* **43**, 448.

181. GORDY, W. (1958). In *A symposium on information theory in biology* (ed. H. P. Yockey, R. L. Platzman & H. Quastler), p. 353. New York: Pergamon Press.

182. GORTNER, W. A. (1954). *J. Gerontol.* **9**, 251.

182a. GOWEN, J. W. (1961). *Fed. Proc.* **20**, Suppl. *8*, p. 35.

183. GREENBERG, D. M., FRIEDBERG, F., SCHULMAN, M. P. & WINNICK, T. (1948). *Cold Spring Harbor Symposia quant. Biol.* **13**, 113.

184. GRÉGOIRE, J. & LIMOZIN, M. (1954). *Bull. Soc. Chim. biol.* **36**, 15.

185. HADORN, E. (1958). In *A symposium on the chemical basis of development* (ed. W. D. McElroy & B. Glass), p. 779. Baltimore, Md.: Johns Hopkins Press.

186. HAKIM, A. A. (1957). *J. biol. Chem.* **225**, 689.

187. HARMAN, D. (1956). *J. Gerontol.* **11**, 298.

188. — (1957). *J. Gerontol.* **12**, 257.

189. — (1962). In *Biological aspects of aging* (ed. N. W. Shock), p. 267. New York: Columbia University Press.

190. — (1960). *J. Gerontol.* **15**, 38.

191. HAUROWITZ, F. & HARDIN, R. L. (1954). In *The Proteins* (ed. H. Neurath & K. Bailey), Vol. *2*, Part A, p. 279.

192. HAUSMANN, E. & NEUMAN, W. F. (1961). *J. biol. Chem.* **236**, 149.

193. HENION, W. F. & SUTHERLAND, E. W. (1957). *J. biol. Chem.* **224**, 477.

194. HERRMANN, H. & BARRY, S. R. (1955). *Arch. Biochem. Biophys.* **55**, 526.

195. HILL, R. L. & SWENSON, R. T. (1961). *Proc. V int. Congr. Biochem., Moscow*, **9**, 115.

195a. HSIA, D. Y. Y. (1961). *Ann. N. Y. Acad. Sci.* **91**, 674.

196. HUISMAN, T. H. J. (1954). *Arch. int. Physiol.* **62**, 564.

197. HUISMAN, T. H. J. & DRINKWAARD, J. (1955). *Biochim. biophys. Acta,* **18**, 588.

198. HUGHES, W. L. (1954). In *The proteins* (ed. H. Neurath & K. Bailey), p. 663. New York: Academic Press.

199. HUTCHINSON, F. (1960). *Am. Naturalist,* **94**, 59.

200. — (1960). *Radiation Research Suppl.* **2**, 49.

200a. — (1961). *Science*, **134**, 533.

201. IWAMURA, T. & MYERS, J. (1959). *Arch. Biochem. Biophys.* **84**, 267.

202. JOLLES, B., GREENING, S. G. & DUN, G. B. (1956). *Nature (Lond.)*, **178**, 148.

203. JONES, H. B. (1956). *Advances biol. med. Phys.* **4**, 281.

204. — (1958). In *A symposium on information theory in biology* (ed. H. P. Yockey, R. L. Platzman & H. Quastler), p. 341. New York: Pergamon Press.

205. KAO, K.-Y. T., HILKER, D. M. & McGAVACK, T. H. (1961). *Proc. Soc. exp. Biol. Med.* **106**, 335.

206. KAO, K.-Y. T., BOUCEK, R. J. & NOBLE, N. L. (1957). *J. Gerontol.* **12**, 153.

207. KAO, K.-Y. T. & McGAVACK, T. H. (1959). *Proc. Soc. exp. Biol. Med.* **101**, 153.

208. KAO, K.-Y. T., TREADWELL, C. R., HILKER, D. M. & McGAVACK, T. H. (1962). In *Biological aspects of aging* (ed. N. W. Shock), p. 343. New York: Columbia University Press.

209. KEECH, M. K. (1954). *Yale J. Biol. and Med.* **26**, 295.

210. —(1955). *Ann. rheumatic Diseases*, **14**, 19.

211. KEMBLE, A. R. & MACPHERSON, H. T. (1954). *Biochem. J.* **58**, 44.

212. KING, M. E. & CARTER, C. E. (1960). *Biochim. biophys. Acta*, **44**, 232.

213. KOHN, R. R. & ROLLERSON, E. J. (1960). *J. Gerontol.* **15**, 10.

214. KUHN, W. (1955). *Experientia*, **11**, 429.

215. LAJTHA, L. G. (1960). In *Nucleic acids* (ed. E. Chargaff & J. N. Davidson), Vol. **3**, p. 527. New York: Academic Press.

216. LAMIRANDE, G. DE, ALLARD, C. & CANTERO, A. (1958). *Biochim. biophys. Acta*, **27**, 395.

217. LANSING, A. I. (ed.) (1952). *Cowdry's Problems of aging: biological and medical aspects* (3rd edn.). Baltimore, Md.: Williams & Wilkins Co.

218. LANSING, A. I., ROBERTS, E., RAMASARMA, G. B., ROSENTHAL, T. B. & ALEX, M. (1951). *Proc. Soc. exp. Biol. Med.* **76**, 714.

219. LINDOP, P. J. & ROTBLAT, J. (1962). In *Biological aspects of aging* (ed. N. W. Shock), p. 216. New York: Columbia University Press.

220. LU, K.-H. & WINNICK, T. (1954). *Exptl. Cell Research*, **6**, 345.

221. —(1955). *Exptl. Cell Research*, **9**, 502.

222. LUGG, J. W. H. & WELLER, R. A. (1948). *Biochem. J.* **42**, 412.

223. LUPIEN, P. J. & McCAY, C. M. (1960). *Gerontologia*, **4**, 90.

224. MAHLER, H. R., WALTER, H., BULBENKO, A. & ALLMANN, D. W. (1958). *Symposium Inform. Theory Biol., Gatlinburg Tenn., 1956* (ed. H. P. Yockey, R. L. Platzman & H. Quastler), p. 124. New York: Pergamon Press.

225. MAKARI, J. G. (1960). *J. Am. geriat. Soc.* **8**, 604.

226. McCOY, T. A., SUBLETT, T. H. & DOBBS, V. W. (1953). *Plant Physiol.* **28**, 89.

227. MEDVEDEV, ZH. A. (1962). In *Biological aspects of aging* (ed. N. W. Shock), p. 255. New York: Columbia University Press.

228. MELCHIOR, J. B. & HALIKIS, M. N. (1952). *J. biol. Chem.* **199**, 773.

229. MILDVAN, A. S. & STREHLER, B. L. (1960). In *The biology of aging* (ed. B. L. Strehler), p. 216. Washington, D.C.: American Institute of Biological Sciences (Publication No. 6).

229a. MOLE, R. H. & THOMAS, A. M. (1961). *Int. J. Radiation Biol.* **3**, 493.

230. MUTO, T. & ARAKI, T. (1955). *J. agric. Chem. Soc. Japan*, **29**, 577.

231. NAYLOR, A., HAPPEY, F. & MACRAE, T. (1954). *Brit. med. J.* pt. **2**, 570.

232. NEARY, G. J. (1960). *Nature (Lond.)*, **187**, 10.

233. OLMSTED, P. S. & VILLEE, C. A. (1955). *J. biol. Chem.* **212**, 179.

234. PARFENTJEV, I. A. & JOHNSON, M. L. (1955). *Geriatrics*, **10**, 232.

235. PARHON, C. I. & OERIU, S. (1962). In *Biological aspects of aging* (ed. N. W. Shock), p. 268. New York: Columbia University Press.

235a. PENROSE, L. S. (1934). *Proc. roy. Soc.* **B115**, 431.

236. PETERMANN, M. L. & HAMILTON, M. G. (1958). *J. biophys. biochem. Cytol.* **4**, 771.

237. PLESHKOV, B. P. & FOWDEN, L. (1959). *Nature (Lond.)*, **183**, 1445.

238. POLLARD, J. K. & STEWARD, F. C. (1959). *J. exptl. Botany*, **10**, 17.
239. RACUSEN, D. W. (1955). *Iowa State Coll. J. Sci.* **29**, 481.
240. RACUSEN, D. W. & ARONOFF, S. (1954). *Arch. Biochem. Biophys.* **51**, 68.
240a. RICE, R. V. (1961). *Biochim. biophys. Acta*, **52**, 602.
241. ROCHE, J., DERRIEN, Y. & ROQUES, M. (1953). *Bull. Soc. Chim. biol.* **35**, 933.
242. ROCHE, J., DERRIEN, Y., REYNAUD, J., LAURENT, G. & ROQUES, M. (1954). *Bull. Soc. Chim. biol.* **36**, 51.
243. REBER, E. & MACVICAR, R. (1953). *Agron. J.* **45**, 17.
244. ROSA, J., SCHAPIRA, G. & DREYFUS, J.-C. (1961). *Proc. V int. Congr. Biochem., Moscow*, **9**, 105.
245. ROSENKRANZ, H. S. & BENDICH, A. (1959). *J. Amer. chem. Soc.* **81**, 6255.
246. ROSS, M. H. & ELY, J. O. (1954). *J. Franklin Inst.* **258**, 63.
247. ROSSI-FANELLI, A. & ANTONINI, E. (1956). *Arch. Biochem. Biophys.* **65**, 587.
248. ROUILLER, C. (1960). *Intern. Rev. Cytol.* **9**, 227.
249. SACHER, G. A. (1958). In *A symposium on information theory in biology* (ed. H. P. Yockey, R. L. Platzman & H. Quastler), p. 317. New York: Pergamon Press.
250. SAWIN, P. B. (1954). *Ann. N.Y. Acad. Sci.* **57**, 564.
251. SCHAPIRA, G., DREYFUS, J.-C., KRUH, J. & LABIE, D. (1961). *Proc. V int. Congr. Biochem., Moscow*, **9**, 116.
252. SCHOLES, G. & WEISS, J. (1960). *Mezhdunarodnyĭ Simpozium po makromolekulyarnoĭ Khimii, Referaty*, Section **3**, p. 360. Moscow: Izd. AN SSSR.
252a. SCHURIN, M. (1961). *Abstr. Papers First Ann. Meeting, Amer. Soc. Cell Biol.*, p. 192.
253. SHOCK, N. W. (1956). *Fed. Proc.* **15**, 938.
254. — (1956). *Bull. N.Y. Acad. Med.* **32**, 268.
255. — (1957). *Geriatrics*, **12**, 40.
256. — (1951). *Trends in gerontology*. Stanford, Calif.: Stanford University Press.
257. — (1960). In *Aging: some social and biological aspects* (ed. N. W. Shock), p. 241. American Association for Advancement of Science, Monograph No. *65*.
258. SINEX, F. M. (1957). *J. Gerontol.* **12**, 190.
259. — (1959). *J. Gerontol.* **14**, 496.
260. — (1961). *Science*, **134**, 1402.
261. — (1960). *J. Gerontol.* **15**, 15.
262. — (1962). In *Biological aspects of aging* (ed. N. W. Shock), p. 307. New York: Columbia University Press.
263. SMITH, J. M. (1959). *Nature (Lond.)*, **184**, 956.
264. SMITHIES, O. (1957). *Nature (Lond.)*, **180**, 1482.
265. SMITS, G. (1957). *Biochim. biophys. Acta*, **25**, 542.
266. SOBEL, H. H. (1959). *J. Gerontol.* **14**, 496.
267. SOBEL, H. [H.] & MARMORSTON, J. (1956). *J. Gerontol.* **11**, 2.
268. STEWARD, F. C. & POLLARD, J. K. (1958). *Nature (Lond.)*, **182**, 828.
269. — (1958). *Proc. IV int. Congr. Biochem., Vienna*, **6**, 193.
270. STEWARD, F. C., WETMORE, R. H., THOMPSON, J. F. & NITSCH, J. P. (1954). *Am. J. Botany*, **41**, 123.
270a. STEWART, J. S. S. (1960). *Nature (Lond.)*, **187**, 804.
271. STREHLER, B. L. (1959). *Quart. Rev. Biol.* **34**, 117.
272. — (1960). In *The biology of aging* (ed. B. L. Strehler), p. 309. Washington, D.C.: American Institute of Biological Sciences (Publication No. *6*).

273. — (1960). In *Aging : some social and biological aspects* (ed. N. W. Shock), p. 273. American Association for Advancement of Science, Publication No. *65*.

274. STREHLER, B. L. & MILDVAN, A. S. (1960). *Science*, **132**, 14.

274a. HANDLER, P. (Ed.) (1961). *Radiation and aging. Fed. Proc.* **20**, Suppl. *8*.

275. SZILARD, L. (1959). *Proc. natl. Acad. Sci. U.S.*, **45**, 30.

277. UI, N. (1954). *Bull. chem. Soc. Japan*, **27**, 392.

278. UPTON, A. C. (1960). *Gerontologia*, **4**, 162.

279. VERZÁR, F. (1962). In *Biological aspects of aging* (ed. N. W. Shock), p. 319. New York: Columbia University Press.

279a. VINTER, V. (1959). *Nature (Lond.)*, **183**, 998.

280. WAITE, R., FENSOM, A. & LOVETT, S. (1953). *J. Sci. Food Agric.* **4**, 28.

281. WALTER, H. & HAUROWITZ, F. (1958). *Science*, **128**, 140.

282. WEINBACH, E. C. & GARBUS, J. (1959). *J. biol. Chem.* **234**, 412.

283. WIELAND, T. & PFLEIDERER, G. (1957). *Biochem. Z.* **329**, 112.

284. WILDMAN, S. & JAGENDORF, A. (1952). *Ann Rev. Plant Physiol.* **3**, 131.

285. WU, H. & SNYDERMAN, S. E. (1951). *J. gen. Physiol.* **34**, 339.

286. YEMM, E. W. & FOLKES, B. F. (1953). *Biochem. J.* **55**, 700.

# GENERAL CONCLUSION

In concluding this monograph we do not propose to make any general assessment of all the questions and problems which we have discussed in detail because these assessments have already been made in the conclusions to the chapters and parts of the book. Throughout the book we have tried to analyse the general scheme of protein synthesis from the point of view of the way in which it affects the development of the organism while analysing the scheme of ontogenesis from the point of view of the way in which it is affected by the mechanisms of synthesis of proteins and nucleic acids.

Proteins first came into being as a result of chemical processes, but highly-developed, specific proteins result from prolonged biological evolution. The specific sequence of amino acid residues and the unique structure of proteins are together associated with their functions and constitute the original foundation of all biological phenomena, which is concealed from our view by the complexity of the many interlocking functional and biochemical processes. The biosynthesis of proteins is not merely a chemical process which may be reproduced *in vitro*. It is not merely a synthesis but a process of development, a molecular ontogenesis which reflects the molecular phylogenesis of the molecule in question, frozen in the structure of self-reproducing matrices.

It is now quite clear that the system of protein synthesis is, at the same time, the system of realization of the genetic information which is concentrated in the nucleus. In order to realize the information of some individual gene the desoxyribonucleic acid must determine the formation of a particular protein or group of proteins and this is accomplished by the formation by the nucleus of desoxyribonucleic acids which are similar to the matrices of protein synthesis. In this way DNA is responsible for its own reproduction while creating a system of synthesis of nucleoside triphosphates and enzymes which will only combine these nucleoside triphosphates into new polynucleotides on the surface of an already existing polynucleotide.

Thus this system of biochemical connections (Fig. 65) lies behind the reproducibility of the organization of vital processes and the successive appearance of vital structures over a series of generations is ensured, in the first place, by DNA.

The organization of the chromosomes is, in essence, devoted to ensuring the conditions for the parallel activity, co-ordinated in space and time, of hundreds or thousands of such unitary systems and the organization of the cytoplasm to ensure the co-ordinated and interlocking function

of the proteins which are formed in these systems. In any real, complicated organism there is a colossal multitude of the most various of biochemical cycles, controlling systems, systems of interaction, co-ordination,
adaptation etc. However, all these are the superstructure; it is the genetic
system which is fundamental, the system which accumulates within itself
information about the phylogenesis of living material. This system is

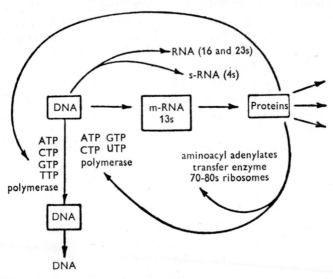

FIG. 65

mutable, it can change; but selection plays a stabilizing role. While
preserving the small proportion of useful variations and thus ensuring
the evolution of living material, selection maintains the constancy of most
of the links in the genetic system by eliminating most of the changes
which arise.

In ontogenesis, however, the stabilizing effect of selection does not
act on the thousands and millions of unitary systems (DNA → RNA → protein) or else it is insufficient and this limits the duration of life to some
particular length. If we are trying to control this length of life we must
find some way of artificially bringing about or increasing selection with
its stabilizing effect in the biochemical and cellular systems of the living
body.

SUPPLEMENT FOR THE ENGLISH EDITION

# CHAPTER XX

# SOME OF THE MORE IMPORTANT ADVANCES IN THE STUDY OF THE BIOSYNTHESIS OF PROTEINS AND THE MOLECULAR ASPECTS OF ONTOGENESIS MADE IN 1963 AND 1964

## *Introduction*

The preceding chapters are an English translation of a Russian book which was finished and sent to the publisher in September 1962. Some supplementary passages added in proof included the results of work published up to the end of 1962. During the preparation of the English edition for the press it has been possible, owing to the kindness of the translator and her husband and their continuous contact with the author, to include some references to work published in 1963 where this was necessary. However, it was thought not to be suitable or convenient to undertake any radical revision of the chapters but better to write a supplementary chapter for the English edition in which the more important experimental results and theoretical considerations relevant to our problem published in 1963 and early 1964 might be discussed. The author was very happy to undertake this task and the present chapter is the outcome. The subject of each section of the chapter is connected with that of some particular chapter of the book and forms, as it were, a supplement to that chapter. The order of the sections is, however, different from that of the chapters as the new chapter naturally deals first with those problems which have developed most rapidly during the period. We have also tried to ensure that this chapter shall not be just a collection of supplements but should constitute a logical concluding chapter to the book. With this in mind we have only discussed the most important problems.

### 1. The study of information-transferring "messenger" RNA and its function in protein synthesis. New models of protein synthesis

In 1963 and 1964 information-transferring (messenger) RNA was very extensively studied. Chapter XV contains material on the early stages of the development of this problem during which "messenger"

534

RNA was discovered and it was shown, for several systems, that this form of RNA is the actual matrix for protein synthesis. In 1962 the mechanism of synthesis of this form of RNA was established and it was shown that it receives its code message from DNA through the agency of DNA-dependent RNA-polymerase. Methods for isolating messenger RNA were suggested and some features of its structure and composition were established. As a result of this work the study of messenger RNA fanned out over a very wide front towards the end of 1962 and during 1963 with the most varied experimental material as its object. This quickly became the main approach to the study of protein synthesis. In this supplementary chapter we naturally cannot give any comprehensive account of this work but will only set out the results of the work which seems to us most important.

### a) Further studies on the synthesis of RNA controlled by DNA (the synthesis of DNA-like RNA)

The occurrence of synthesis of RNA under the control of DNA was established in principle in 1960-1 and the relevant work is discussed in Chapters XIV and XV. Since then, this synthesis has been studied intensively from extremely different points of view (purification of the enzyme concerned, experiments with synthetic polynucleotides, the kinetics of the reaction, the demonstration of the reaction of synthesis of RNA under the control of DNA in different species of bacteria, plants and animals and in tumours and cultures of cells, the selective inhibition by actinomycin, etc.).

The most important advance in this field in 1963-4 was, however, the discovery of the mechanism of the synthesis and the use of the system for the elucidation of some of the basic problems of molecular genetics, concerned with the nature of the coding of information in the system DNA-RNA-protein.

For many years one of the puzzles of molecular genetics was what information did each of the complementary chains of the double helix of DNA contain and whether one of them was not "nonsense" from the point of view of the synthesis of polypeptide chains. It appears that this problem has recently been solved through the study of the synthesis of RNA under the control of DNA.

As we have already shown on pages 288-298, each of the chains of the double helix of DNA can act as a primer for the synthesis of the complementary strand whereas it has been shown recently that in vivo only one of the polynucleotides of the DNA acts as the source of genetic information. Hayashi, Hayashi & Spiegelman (76) studied the question by using the very interesting system for the synthesis of RNA found in the phage $\phi$X 174, which is known to have single-stranded DNA in its mature form. In Chapter XIV (p. 296) we have already pointed out that when this

DNA is introduced into the cell the process of reproduction of the phage precedes the formation of the double helix (the so-called replicative form or RF). Experiments carried out by Spiegelman and colleagues on hybridization of each of the polynucleotides of this replicative form of DNA with RNA from phage-infected cells (the formation of a DNA-RNA helix) showed that this RNA is only complementary to one of the poly-nucleotides of the replicative form of the DNA, namely the "vegetative" strand, while this, in turn, is complementary to the single strand found in the mature phage. In later work (77) these authors made a remarkable observation, which explained the difference between the priming effect of the DNA of this phage *in vivo* and *in vitro*. They found by electron microscopy that *in vivo* the molecules of DNA were in the form of rings, while the rings were opened out *in vitro*. In the ringed form only one of the polynucleotides of the double helix could act as a primer. When the ring was broken and two ends appeared during the procedure of separation of the DNA each of the polynucleotide helices of the DNA acquired the ability to act as a matrix.

The conclusion that information is only handed on by one of the strands of the double helix of DNA has recently been confirmed for more complicated cellular systems by the work of McCarthy & Bolton (106) who used the ingenious method of passing RNA through DNA-agar columns (38). The interaction between the DNA and RNA in the agar showed that the population of the cellular RNA contained molecules which were homologous only with half the possible nucleotide sequences of the cellular DNA. If the DNA of the cell is separated mechanically into two portions and one of these is deposited on agar in the form of single-stranded polynucleotides, then, when the other half of the DNA is passed through the agar in the denatured form, the DNA which was in the agar goes over completely to the two-stranded form by combining with that which was added. If cellular RNA is passed through such a column instead of DNA only half the molecules of DNA in the agar will form DNA/RNA helices. Thus the molecules of messenger RNA formed in the chromosomes are only complementary to one of the polynucleotide strands of each two-stranded molecule of DNA and are homologous with the other. According to McCarthy & Bolton this is the reason why this form of RNA does not spontaneously form double helices in the cells.

Nevertheless, when complementary RNA is synthesized *in vitro* it forms secondary double helices (60).

It is interesting to note that the DNA/DNA double helix is more stable than the DNA/RNA helix (34).

It has been found that in the presence of a DNA primer RNA poly-merase can catalyse reactions of two types, first the synthesis of comple-mentary RNA and secondly the synthesis of homopolymers of RNA (polyadenylic acid, polyuridylic acid, etc.) (41, 58, 153). Reactions of the first type require the presence of the four ribonucleoside triphosphates,

$Mn^{++}$, $Co^{++}$ and $Mg^{++}$. The formation of homopolymers such as poly A or poly U requires the presence of the appropriate polynucleotide phosphate and $Mn^{++}$. For this purpose denatured DNA is more active than the native material and the presence of a second nucleotide in the solution inhibits the process. The mechanism by which homopolymers are formed in such a system is still unclear. It is suggested that the matrices, in this case, may be provided by small uniform sections of the DNA chain, e.g. the polythymidine sequence in such a chain as ATTTTTTTTAU. However this may be, the rate of synthesis is related to the nucleotide composition of the DNA. The formation of long homopolymers of RNA in this way is explained at present on the basis of a preliminary hypothesis concerning sliding reactions.

The first stage in the formation of complementary messenger RNA is the "stripping" of some part of one of the polynucleotides of the double helix of DNA and the formation of a mixed DNA/RNA hybrid. Free RNA does not appear until the ratio of DNA to RNA in the hybrid exceeds a certain figure (46, 145, 164). When single-stranded DNA such as that from phage $\phi$X 174 is used as a matrix for the synthesis of RNA, a DNA-RNA hybrid is first formed and this becomes the active matrix for the synthesis of free informational RNA (45a). The biological activity of RNA synthesized *in vitro* as a matrix for protein synthesis has been demonstrated by the work of Bonner and colleagues (35a). Having formed a single system for carrying out DNA-dependent synthesis of RNA, using DNA from pea-shoots and RNA polymerase from *E. coli*, the authors observed the synthesis of typical pea-shoot protein. It has also been established that DNA-dependent synthesis of RNA forms the basis of the cytoplasmic genetic systems which determine, for example, the autonomous reproduction of mitochondria and chloroplasts (61a, 142a).

### b) The physical and chemical characteristics of messenger RNA

The earliest evidence suggesting that the molecular weight of messenger RNA was about 300,000 and its sedimentation constant 12-14 Svedberg units has been largely corrected by recent work. It has been found that this form of RNA varies in these respects within rather wide limits. Chemical fractionation (on columns of methylated albumin for example) in normal colonies of *Escherichia coli* and on colonies infected with $T_2$ phage has been used to obtain four undegraded components of messenger RNA with sedimentation constants of 8 S, 10-12 S, 19-21 S and 23-30 S (93, 127). The last of these forms of messenger RNA formed complexes with ribosomes. In another study (137), when messenger RNA was isolated from cells infected with $T_2$ phage by using its complementary reaction with the DNA of the phage while preventing enzymic degradation of the RNA, the distribution of the size of the molecules of the messenger RNA was also very wide (from 4 S to 25 S).

Studies of the informational or "quickly labelled" RNA of animal cells, liver cells of rats (79) and HeLa cells (128), also showed that they varied over a wide range (from 6 S to 39 S), the molecules of RNA in the latter case being apparently linked with a protein. Such complexes of messenger RNA with a protein have recently been given the name "informosomes" by Spirin (13).

The molecular heterogeneity of messenger RNA would seem to be correlated with its metabolic heterogeneity (4). *In vitro* informational or messenger RNA acquires a secondary structure in the form of a double helix giving a characteristic X-ray diffraction picture similar to that of DNA (156).

### c) Messenger RNA in protein synthesis

A very extensive range of studies of the functional processes associated with the participation of messenger RNA in protein synthesis in ribosomes has been undertaken in 1963 and 1964. The results of these studies have already been discussed in reviews (104, 148) and here we shall only deal with a few aspects of the problem in a very condensed form.

The study of messenger RNA, which was begun very intensively on bacteria and phages, has been extended to materials of animal and plant origin and it has been shown to have essentially the same action on protein synthesis in these systems as in bacteria or phages (105, 61, 36, 84, 141). However, in the cells of plants and animals, this RNA is more stable and functions as a matrix over a longer period. This RNA manifests extreme variation in its composition, metabolism and other properties. Owing to its greater stability the method of pulse-labelling is not always suitable for the identification of this fraction and the isolation of DNA-like RNA from animal tissues is usually carried out by preparative methods. The demonstration of the complementarity of DNA and messenger RNA is usually carried out by the formation of DNA/RNA hybrids using DNA-agar columns (38, 35, 85). In this way the messenger RNA is not only isolated by its reaction with the DNA but the extent of the homology or "relatedness" of the nucleic acids of animals and bacteria of different species is demonstrated.

In studies on the function of messenger RNA carried out on bacteria and tissues work has also proceeded by the isolation, identification and purification of the individual fractions of RNA which are concerned with the synthesis of a particular protein. In this field successes have been achieved in the isolation of the messenger RNA determining the synthesis of tryptophan synthetase (161), catalase (78), enzymes for the utilization of galactose (23), serum albumin, tryptophan pyrrolase and glucose-6-phosphatase (25), thyroglobulin (143) and other proteins.

Other very interesting experiments have been done on the synthesis of particular fractions of messenger RNA which are responsible for the appearance of some of the specific functions of specialized cells, especially nerve cells (86, 87, 45).

## 2. Stimulation of the synthesis of proteins and polypeptides by synthetic polyribonucleotides and further advances towards the cracking of the nucleotide code

As we showed in chapter XV, the use of synthetic polynucleotides as model matrices made it possible, as early as the middle of 1962, to calculate the composition of about 22 coding triplets of nucleotides (codons) for 19 amino acids. As we showed, the nucleotide code is a "degenerate" one, that is to say, each amino acid may be coded by several combinations of nucleotides. Altogether the four nucleotides can form 64 triplets, and work on particular sequences of nucleotides has shown that all of these can exist in RNA. Later work in this field has been directed in the first place towards finding out the "meaning" of the remaining 42 possible triplets of nucleotides and in the second place towards determining the sequence of nucleotides in each triplet. At the same time certain other properties and peculiarities of the nucleotide-amino acid code have been cleared up. It should be pointed out that the optimistic forecast, made at the end of 1961, that the problem of the genetic code would be completely solved within a year has not been fulfilled. The task was found to be more difficult and its accomplishment will certainly be delayed for some years.

The main part in the working out of the nucleotide-amino acid code by the building of polynucleotide model matrices in 1963 has certainly been that of the work done in the laboratories of Nirenberg and Ochoa, although several other laboratories have recently become associated with this work.

In a new series of publications Ochoa, Wahba and their colleagues (59, 158, 159, 160) have described how they established the "meaning" of more than 20 further nucleotide triplets, thus increasing the number of codons to 49 (out of 64). They made certain improvements in their previously described experimental methods in order to achieve these results. As certain types of polypeptides (e.g. polylysine) were found to be soluble in trichloroacetic acid, they used tungstic acid for the precipitation of their proteins. Furthermore, as well as using random copolymers of ribonucleotides as model matrices they also used non-random copolymers by using oligomers of known composition (such as AAU, AU, AUU ... U GUU .... U and so on) as primers during the operation of the polynucleotide phosphorylase. The use of such polynucleotides then makes it possible to establish the sequence of the nucleotides within the triplets. At present other supplementary theoretical and experimental methods are also used to determine the nucleotide sequence.

PB 2 E

In recent work from Nirenberg's laboratory (89, 123) improved methods have also been used and these, together with calculations on the composition of the triplets, have made it possible to establish some of the general properties of the nucleotide-amino acid code. The authors experimented with copolymers, not only of one, two and three nucleotides, but also with those of four nucleotides, using which it was found that there was stimulation of the incorporation of several amino acids. This fact confirms the view that there is some degree of degeneracy of the code so that most of the nucleotide sequences can be read. In another investigation Nirenberg and colleagues (144) tried to determine the significance of the secondary structure of RNA in relation to its function as a matrix. Before these experiments were carried out, the polynucleotides used as model matrices had a predominance of one particular kind of base, for example uracil, such that the ratio of it to the other, for example guanine, was of the order of 5:1. In such cases the formation of secondary bonds was weak. The authors showed that the ability of such copolymers to act as matrices fell sharply when the ratio U/G fell below 1 (between 0·6 and 1). This falling off (nearly to $\frac{1}{60}$ of the former value) is understandable when phenylalanine (coding triplet UUU) is being incorporated into the protein but it is not understandable when the case concerns leucine, valine or tryptophan. The authors found that copolymers of this sort show a marked tendency to secondary structures in the form of double helices formed by the interaction of guanine with uracil and they suggested that the parts involved were inactive as matrices for protein synthesis. In this connection, they put forward the interesting suggestion that the places at which the interaction occurs may determine the beginning and end when polypeptide chains are being formed (like knots on a string), as is shown in the following diagram (Fig. 66).

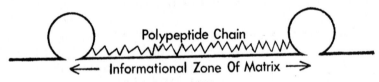

FIG. 66

Recently Nirenberg and colleagues (103) have succeeded in making a synthetic model of the system DNA → RNA → protein. Having synthesized chains of polydesoxythymidylate 6-14 nucleotides in length they showed that, in the presence of ATP and RNA polymerase, these could serve as "primers" for the synthesis of polyadenylic acid and that this latter stimulated the synthesis within the system of polylysine. Falaschi, Adler & Khorana (53) used chemically synthesized desoxyribonucleotides of known nucleotide sequence as primers for the synthesis of RNA matrices. By using this method they hope to give an unequivocal answer to the question of the sequence of nucleotides in the codons.

Among other approaches to the study of polynucleotide model matrices for protein synthesis we must mention in the first place the work of Martin & Ames (112), Marcus et al. (109) and Jones et al. (90) who determined the correlation between the rate of incorporation of amino acids and the size of the molecules of polyuridylic and polyadenylic acids used as matrices. It was found that, for the chain of polyuridylic acid, the length which gave the best results was 450-700 nucleotide residues. However, shorter polynucleotides also stimulated incorporation. Polynucleotides (polyuridylic acid) of about 12-42 residues had only a slight activity while even shorter ones exerted an inhibitory effect. On studying polyadenylic acid (90) it was found that polynucleotides of about 11-16 residues possessed a marked activity in the synthesis of polylysine.

In the second place there is the work of Bretscher & Grunberg-Manago (37) in which they used cell-free systems derived from Escherichia coli and used Nirenberg's method to determine the composition of some triplets. They showed that the ratios of the diphosphates used for the synthesis of the polyribonucleotides might differ markedly from the ratio of bases in the polymer finally obtained and that conclusions as to the composition of triplets in cases in which analysis of the polymers had not been undertaken might be false. In the third place we have the work of Grossman (69, 70) who found that treatment of polyuridylic acid with ultraviolet radiation gives rise to a "model mutation" as a result of which the incorporation of [14C]phenylalanine into acid-precipitated material is markedly decreased while incorporation of [14C]serine is increased, leading to the synthesis of polyphenylalanylserine. One may also be sure that the studies which have been made of the way in which synthetic polynucleotides, made partly from analogues of normal nucleotides, function as matrices will prove fruitful (160, 70a, 130b).

Stimulation of the incorporation of amino acids into proteins by synthetic polynucleotides has been observed in other preparations besides the ribosomal system of E. coli; these include ribosomes from reticulocytes (21, 22, 165, 167), extracts of tumour cells and liver cells (54, 114, 166), ribosomes from sea-urchin embryos (119, 171), and ribosomes of algae (136). In these experiments the effects of particular triplets on the incorporation of particular amino acids were analogous to those found in the E. coli preparations (UUU affects phenylalanine, 2U 1G affects leucine 1U 2G affects glycine and tryptophan, etc.). This suggests that there is a universal code.

A start has also been made recently on the detailed study of the nature of the peptide products which are formed in the reactions controlled by synthetic polynucleotides (138, 98, 147). Different methods of fractionation and identification were used for this purpose so that it was possible to make sure whether any particular amino acid was really being incorporated into the peptide chain.

Weisblum, Benzer & Holley (168) have adopted a very interesting

TABLE 18

Summary of views on the composition of codons (early 1964)

| Amino acids | Based on work done in the laboratories of Ochoa and Nirenberg | | Based on theoretical considerations and evidence from the substitution of amino acids in mutants | | |
|---|---|---|---|---|---|
| | Adequately verified | Doubtful or inadequately supported | Luchnik (6) | Eck (51) | Ratner (10) |
| Ala | $2C_1G$ | AGC, GCU | $2G_1C$ ($2G_1C$) | $2G_1C$ | — |
| Arg | $2C_1G$, $2A_1G$ | GCU | $2G_1C$, AGU | $2G_1A$ | AGC |
| Asp | AGC, AGU | — | — | — | — |
| Asp.NH₂ | $2A_1C$, $2A_1U$ | ACU | $2G_1U$, AGU, GCU (AGC) | GCU | GCU |
| Val | $2U_1G$ | — | | GCU | — |
| His | $2C_1A$, ACU | — | | — | — |
| Gly | $2G_1U$, $2G_1A$ | $2G_1C$ | 3G | 3G | — |
| Glu | $2A_1G$ | AGU | $2G_1A$ | $2G_1A$, $2G_1U$ | AGC |
| Glu.NH₂ | $2A_1C$ | $2G_1A$ ACU | (AGC) | AGC | — |
| Ileu | $2U_1A$, $2A_1U$, ACU | 3U | | AGU | — |
| Leu | $2U_1G$, $2U_1C$, $2C_1U$, $2U_1A$ | $2A_1U$, $2A_1C$ | | — | AGC |
| Lys | 3A, $2A_1G$ | — | | AGU | — |
| Met | AGU | — | | AGC | — |
| Pro | 3C, $2C_1A$, $2C_1U$ | $2C_1G$ | $2G_1A$ (AGU) | $2G_1C$, ACU | — |
| Ser | $2U_1C$, GCU, $2C_1U$, AGC | — | AGC | — | — |
| Tyr | $2U_1A$, ACU | — | | AGC | — |
| Thr | $2A_1C$, $2U_1A$, $2C_1G$, ACU | $2C_1U$ | | AGU | — |
| Try | $2G_1U$ | — | | — | — |
| Phe | 3U, $2U_1C$ | — | AGU | GCU | — |
| Cys | $2U_1G$ | $2G_1U$ | | — | — |

approach to the experimental solution of the problem of the code. Having divided the lysine-accepting soluble RNA (S-RNA) of *E. coli* into two fractions by countercurrent distribution these authors found that the incorporation of leucine into the ribosomes from one fraction is mainly stimulated by synthetic polynucleotides of the type of poly-UC, while the other fraction is more strongly stimulated by polynucleotides of the type of poly-UG.

In later work another group of authors (52) found that it is possible to separate three different fractions of acceptor of [¹⁴C]leucine from the S-RNA of *E. coli*. They all had different coding properties, one was stimulated by poly-UC, another by poly-U and copolymers with a high U content, while the third fraction was activated by poly-UG.

The work of Davies, Gilbert & Gorini (49a) has recently attracted a lot of attention. These authors found that the coding properties of polyuridylic acid (poly-U) and the copolymer of guanylic and adenylic acid (poly-GA; 2:1) altered when the ribosome (of *E. coli*) was disrupted by the action of streptomycin. The incorporation, for example, of phenylalanine into polypeptide material, which is typical of poly-U, was inhibited by the effect of streptomycin while the incorporation of other amino acids which is normally not coded by poly-U was stimulated. The same was observed in the case of other nucleotides. Deviations from "normal" coding have also been observed to occur when the concentration of magnesium is altered and under the influence of several organic solvents (147a) and changes of temperature (155a). These facts indicate that the correct "reading" of the information of the matrix depends, not only on the linear sequence of the nucleotides, but also on the secondary structure of the RNA.

Table 18, which is a new version of the table of coding triplets given on page 347 of this book, shows the composition of 50 coding triplets (codons) as established by early 1964. This table was compiled by Luchnik (6) and extended by him specially for use in this chapter by the inclusion of material published in the latter half of 1963 and early 1964.

## 3. Localization and mechanism of synthesis of acceptor S-RNA and the high-polymeric RNA of the ribosomes

As we saw in an earlier chapter, there were many problems relating to the study of S-RNA which were still unsolved at the end of 1962. The secondary structure of the molecule, the site of the coding nucleotide group (anticodon) which interacts with the matrix RNA, the part played by unusual bases in the composition of S-RNA, how many forms of S-RNA exist for each amino acid and whether this heterogeneity corresponds with the degree of degeneracy of the code, the species-specificity of S-RNA, the localization of its synthesis, etc., were all uncertain. Developments in this field in 1963 and early 1964 have been of the greatest interest.

Studies of the secondary structure of S-RNA (40, 107) have given good grounds for suggesting that this secondary structure is of the nature of a double helix in which the unpaired nucleotides also form peculiar loops in which the anticodons are situated. It may be that the unusual, "minor", bases mark the limits of these zones (113, 122). The question of the existence of several forms of S-RNA for each amino acid (confirmation of the degeneracy of the code) has been answered in the affirmative, but so far this only applies to a few amino acids, especially leucine (30, 31, 154). In these last three papers it was actually shown that each of the leucine-RNA complexes isolated from *E. coli* or yeast reacted selectively with a different type of codon in the matrix of the messenger RNA.

When artificial synthetic systems are used and one synthetic poly-nucleotide is substituted for another we find that one form of leucine-S-RNA reacts with a matrix of poly-UC while another reacts with poly-UG. As we have seen, studies of this kind give us yet another possibility for working out the genetic code. The problem may also be solved by experimental substitution of one nucleotide in the S-RNA. Yu & Zamecnik (177) found that when S-RNA is brominated the bromine reacts first with the uracil and cytosine in it. When the level of bromination was such that there was one atom of bromine to every 40 nucleotide residues, phenylalanyl S-RNA lost 67% of its activity while the activity of lysyl S-RNA fell to 32% and that of valyl S-RNA to 16%. It is thus very probable that the anticodon for phenylalanine contains at least one bromine-sensitive residue and thus, in part at least, differs from the suggested AAA sequence.

A similar method of revealing the degree of inactivation of different forms of soluble RNA under the influence of other factors (treatment with hydroxylamine which inactivates uracil and with *O*-methyl-hydroxy-lamine which inactivates cytosine) was used in the work of Kiselev & Frolova (5a) to establish the presence of particular nucleotides in "anti-codons". Owing to the work of Schweet and colleagues (22a) it has been shown that the reaction of transfer by transfer-RNA associated with an amino acid on a ribosome can be subdivided into two reactions catalysed by different enzymes; first, combination with the ribosome associated with the action of the coding triplets and stimulated by the binding enzyme, and second, the formation of peptide bonds on the matrix with neighbouring amino acids which are already combined with it, which is stimulated by another enzyme known as peptide synthetase.

It must also be mentioned that, as was to be expected, the sequence of nucleotides in S-RNA is determined by DNA. Using DNA/RNA hybrids it has been shown that in the total DNA of the cells there are quite extensive zones which are complementary to the S-RNA fraction of the cells (65). Synthesis of this fraction in the chromosomes has also been demonstrated by other methods (130).

It is interesting to note that the genetic (DNA) system of bacteria has been found to contain parts (cistrons) which are complementary to the high-molecular-weight RNA of the ribosomes (173, 174). Other forms of RNA did not compete with the ribosomal RNA in reacting with single-stranded DNA. It was found that 23 S and 16 S RNA each reacted with a different DNA. This provides indirect evidence that ribosomal RNA may be synthesized in the nucleus. It has also been found that non-nucleolated variants of *Xenopus laevis* do not synthesize ribosomal RNA (39). In studying the problem of the localization within the nucleus of the synthesis of ribosomal RNA, evidence has been obtained which suggests the possibility that the nucleolus may be pre-eminent in this synthesis (114a). The DNA which is complementary to ribosomal RNA is concentrated in the chromatin associated with the nucleolus.

In addition to this, material evidence has been obtained for the occurrence of RNA-controlled synthesis of RNA, not only during the reproduction of viruses, but also in the course of normal RNA metabolism (42). It may be that the chromosomes, which have a very limited number of cistrons for the synthesis of RNA, do not bring about the synthesis of all of this sort of RNA, which accounts for about 80% of the total RNA of the cell. If this is so, it may be that they simply synthesize the prototypes from which the functional molecules of ribosomal RNA are synthesized. When viral RNA is introduced into the cell it induces the formation there of a special enzyme called RNA-replicase (73).

A special high-molecular 45 S RNA has been found in HeLa cells. Kinetic studies suggest that it serves as a precursor of 28 S and 16 S ribosomal RNA (141).

Two recent papers by Spirin (11, 14) describe the detailed analysis of different aspects of the macromolecular structure and synthesis of ribosomal RNA.

## 4. The discovery of polyribosomes—native complexes of messenger RNA and ribosomes (polysomes, ergosomes) and the setting up of new models of the terminal stages of protein synthesis

The discovery of polyribosomes, which are complicated, temporary, dynamic complexes of ribosomes with messenger RNA, was undoubtedly the most important achievement of 1963 in the field of the study of the mechanism of protein synthesis.

Until the end of 1962 it was generally considered that the RNA matrix combined with individual two-subunit 70 S ribosomes and, being coiled around its surface, as it were, served as a site for the formation of polypeptide chains. With the ordinary methods of preparation of ribosomes, aggregates with sedimentation rates about 100 S were not formed. However, another picture gradually emerged when the method of centrifugation in a sucrose gradient was introduced into biochemistry. (In this method the cytoplasmic suspensions are diluted with a 15-30% solution of sucrose and then centrifuged in bucket rotors and a series of fractions from the sucrose gradient is collected.) With this method it was found, even in 1962, that there were certain "heavy" active ribosomes which occurred when the messenger RNA of a $T_2$ phage reacted with ribosomes of *E. coli* (134). It was also found that when polyuridylic acid reacts with ribosomes a rapidly precipitated complex is also formed (26, 149). At that time the earlier scheme of reaction of messenger RNA and ribosomes had already come under question because the length of the molecule of RNA, which is of the order of 1500-3000 Å, and the diameter of the ribosomes, which is about 230 Å, gave rise to difficulties in explaining their interaction in terms of a stable coiling of the matrix around the ribosome. However, the questions which arose were quickly answered in the beginning of 1963 when independent publications from several laboratories appeared almost simultaneously describing a new class of structures, the polyribosomes. These are dynamic, linear complexes of messenger RNA and several ribosomes and they bring about the synthesis of polypeptide structures (62, 63, 129, 146, 163, 169, 178). The work was carried out on different material (reticulocytes, liver, *E. coli*, HeLa cells, etc.) but the results were essentially the same in all cases.

In this connection we shall discuss in more detail the results obtained with reticulocytes and HeLa cells. From the methodological point of view the experiments with reticulocytes are most convincing, as the ribosomes and messenger RNA in the non-nucleated reticulocytes are uniform because, broadly speaking, only one protein, namely haemoglobin, is synthesized by them. The structure and function of reticulocytes and HeLa cells were studied in great detail by Rich and colleagues (66, 163). Using the sucrose-gradient method the authors found that the particles which were most active in incorporating labelled amino acids in the reticulocytes had a sedimentation constant of 170 S. Electron microscopy showed that they are aggregates of 5-6 ribosomes strung together by a single-stranded polynucleotide which is broken by ribonuclease. It was naturally suggested that these structures might be dynamic and that the ribosomes, as they move along the matrices, in some way "work out" the information contained in them. According to the current scheme the ribosomes combine with one end of the linear polyribosomal aggregate and then, as a reult of their movement along the polynucleotide matrix, they cause the polypeptide chain to grow. When they reach the end the ribosomes are liberated and separate off the finished polypeptide chain. Goodman & Rich (66) have represented this process by means of the following model (Fig. 67). Similar models have also been suggested in other papers.

Experimental testing of this model (163) by the introduction of labelled ribosomes into the mixture and by the addition of unlabelled ribosomes to the labelled polyribosomes actually confirmed that both the process of combination and the process of liberation of the ribosomes from the polyribosomal structures do really take place.

The formation of polyribosome aggregates may be brought about by the addition of messenger RNA to the ribosomes of reticulocytes. This process also starts off the synthesis of protein (62).

The beginning of the synthesis of the polypeptide chain is related to the combination of the 80 S ribosomes with the polysomes, while the ending of the growth of the polypeptide chain is associated with the setting free of these ribosomes (72, 111). A similar phenomenon is also observed when the synthesis of messenger RNA is inhibited by actinomycin (151).

The formation of polyribosomes is also observed during the synthesis of viral particles (by the viruses of animal cells). In this case the RNA becomes a component of the polyribosomes (129, 140). However, in the case of TMV and the yellow mosaic virus of turnips the synthetic activity was exhibited even by monosomal complexes (RNA-ribosome) (74).

This evidence concerning the existence of polysomes and the theoretical considerations as to their functions has been confirmed by other studies using bacterial systems (63, 99a, 139, 142, 150), plants (47) and fungi (130a). The only things which vary are the findings as to the sizes and numbers of ribosomes on any matrix and this would seem to depend on the size of the programme required for the protein being synthesized.

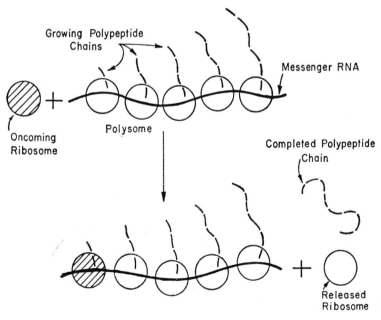

FIG. 67. A schematic model of polysome function.

The largest polyribosomes are formed in cells which have been infected with the polio-virus. They contain up to 50 individual ribosomes (132, 133). The polysomes of HeLa cells may contain up to 40 ribosomes (172). A start has now also been made on the study of the mechanism of inter-action of acceptor S-RNA with the polyribosomes in the process of protein synthesis (95, 44). According to some of the evidence part of the poly-ribosomes can be formed directly in the nucleus (80).

An interesting observation has been made recently by Monroy & Tyler (116). They found that in unfertilized sea-urchins' eggs a large proportion of the ribosomes is present in the non-aggregated form, as monosomes. A considerable number of polysomes is found after fertiliza-tion and the proportion of polysomes increases as embryological develop-ment proceeds. In the development cycle of *Neurospora crassa* poly-ribosomes are absent from the cytoplasm in those stages in which active synthesis of proteins is not going on (resting conidia, ascospores) but appear in large numbers in the germinating spores and growing hyphae (78a).

## 5. Study of the general properties of the genetic code, its universality and the determination of the sequences of the nucleotides in the coding triplets

In the period from the autumn of 1962 to April 1964 the study of the general properties of the genetic code followed the same lines as it had in the years 1961 and 1962 (cf. Chap. XVII). There were no essential changes in the treatment of the problem and the fundamental ideas as to the general properties and particular features of the genetic code which had been formulated in the earlier period were fully confirmed by the researches of the later one.

One important advance towards the establishment of the details of the code is the method we have already discussed of studying the substi-tution of one amino acid for another in a protein in the presence of in-duced mutations associated with changes in the RNA of viruses and also, more generally, the theoretical analysis of the evidence as to the substitu-tion of amino acids arising both as a result of spontaneous genetic muta-tions (for example, the pathological genetic anomalies of haemoglobin) and also in the more general case of the evolutionary variability of proteins. On p. 401 of this book we present a table from Ochoa's review in which he surveys the evidence from such substitutions (16 types of substitution) from the point of view of the coincidence of the triplets in their com-position as revealed by experiments on their stimulation of the synthesis of synthetic polynucleotides. At the end of 1963 the possibilities for such comparisons were largely extended and Ratner (10) has recently surveyed from this angle 53 types of substitution in the protein of TMV, in haemo-

globin and tryptophan synthetase in mutants discovered in various laboratories.   In 45 of these cases the substitutions agreed theoretically with the
composition of the codons postulated by Nirenberg and Ochoa but in 8
they did not.   If we take into account the fact that the composition of a
number of possible triplets has not yet been established these calculations
may be taken as a very serious pointer to the universality of the genetic
code.

Recently the working out of the problem of the amino acid-nucleotide
code has attracted a great deal of attention among scientists.   This sort
of theoretical working out must be based on factual material obtained by
the use of synthetic polynucleotides, by analysis of the results of substitution of amino acids in mutant phages and from the molecular variations
of specific proteins (haemoglobins, cytochromes, tryptophan synthetases,
etc.) of a mutational and evolutionary nature and from many other molecular-genetic studies.   The main aim of the theoretical study of the
problem of the code has been to find out what nucleotide sequences can
act as coding groups (codons) in RNA and DNA, the size of the codons
(two, three, four, . . . etc. nucleotides), the mode of decipherment of the
code (whether it is overlapping or not, whether or not it contains "commas"), the nature of the marginal parts which determine the length of
the polypeptide chains, the regularity of the composition of the coding
groups, the universality of the code and many of its other features.   Although the theoretical solution of these problems has been slower than
was expected at first it has made considerable progress and many solutions
proposed in different laboratories and based on different methods of
analysis of the facts have coincided in many respects.

Many very comprehensive theoretical discussions of this subject have
been published in late 1962 and during 1963 and early 1964 (2, 6, 10,
49, 91, 92, 51, 16, 83, 152, 175, 100, 155, 126, 122a, 176).   Most of these
papers mainly confirm the theoretical conclusions as to the nature of
the code which we set out in Chapters XV and XVII and which are
based on the original work in this field (i.e. that the codons are triplets,
that the code is degenerate and universal and that there are no "commas",
etc.).   However, as concerns the sequences of nucleotides within the
codons and the regularity of their composition, there are quite a number
of models to choose from, depending on the assumptions on which each
author based his calculations.   The deepest and most comprehensive
analysis of this problem with a comparison of all the main codes has been
recently published (in May 1964) in a special monograph by Luchnik (6).

In determining the sequence of nucleotides all authors proceed from
the composition of triplets determined in cell-free systems and evidence
concerning the substitution of amino acids in natural proteins when
mutations occur, especially those caused by nitrous acid.   In principle
these facts could give unequivocal information about the sequence of
nucleotides concerned; that at present available is not unequivocal owing

to the shortage of factual material. As to the establishment of the absolute sequence of nucleotides, the only information we have is derived from the results of the work to which we have already referred (159). However, this work is only of a preliminary nature and does not provide many facts so the conclusions as to the absolute sequence of the nucleotides in most triplets cannot yet be considered fully reliable. Several authors, on the basis of different assumptions and arguments, have independently reached the conclusion that the degeneracy of the code is not of a random nature (16, 51, 6, 89, 123, 160). They all consider that most of the triplets coding any one amino acid have two nucleotides in common, which will greatly facilitate the elucidation of the sequences of the nucleotides in the triplets. On the basis of rather different hypotheses some of these authors have suggested a sequence of nucleotides for most of the triplets. In a considerable number of cases the suggestions of these authors agree with one another, which argues for their truth.

The direct study of the composition of triplets in two organisms which differ widely from one another in the overall composition of their DNA (*Alkaligenes faecalis*: $G + C = 66\%$ and *Escherichia coli*: $G + C = 52\%$) has shown that the composition of the 18 coding triplets studied in these experiments were the same in both cases and there seemed to be no deviation from the universality of the code (131).

An original method for working out the size of the codons and the sequence of the nucleotides in the coding triplets has recently been devised by Nirenberg and colleagues (102b, 122a). As a matrix for the synthesis of polypeptides these authors used short oligonucleotides combined with ribosomes and they studied only the first phase of the synthesis, that is the specific association of the complex S-RNA-amino acid with the ribosome. If, for example, [$^{14}$C]phenylalanyl-S-RNA is incubated with the ribosomes the process of binding of the S-RNA is stimulated by oligonucleotides of uridylic acid. Under these conditions the dinucleotide pUpU had a weak activity while the trinucleotide pUpUpU markedly stimulated the binding of [$^{14}$C]phenylalanyl-S-RNA to the same extent as did the tetra-, penta- and hexanucleotides. This confirmed earlier views as to the trinucleotide nature of the codons. By varying the type of trinucleotide matrix and the type of S-RNA-amino acid complex the authors could begin a direct study of the sequence of nucleotides in the triplets of codons.

## 6. The molecular-genetic mechanism of development

Questions as to the molecular-genetic mechanisms of development have recently been attracting more and more attention and the number of investigations in this field has increased sharply. It is quite natural that, in connection with the solution of the fundamental problems of the biosynthesis of proteins and nucleic acids—i.e. problems of the reproduction and replication of specificity—as well as the working out of the various details, the interests of many laboratories should be directed towards the solution of the next problem, that of the explanation and clarification of the molecular mechanisms of development and differentiation.

During 1963 and 1964, however, no essentially new lines of work in this field developed. The main work followed the course of those problems which were briefly surveyed in Chapter XVIII. Nevertheless, several interesting results and new facts which are bringing us nearer to the solution of this problem must certainly be mentioned. Some of this material has been considered in several reviews published recently (18, 67, 68, 88, 89, 157, 56, 115, 50, 102).

### a) Changes in proteins and nucleic acids during the processes of development

The number of proteins, especially enzymic proteins, which have been studied from the point of view of their synthesis and manifestation in connection with embryonic development has increased markedly. Besides the genesis of the proteins of muscle (actin, myosin, myoglobin), haemoglobin and the serum proteins, collagen, elastin, tryptophan pyrrolase and certain others which have been discussed in Chapter XVIII, the study of the morphogenetic shift in the enzymic pattern has been extended to cover a series of new enzymes, i.e. those concerned with the metabolism of glycogen and other carbohydrates (101, 24, 43), those concerned with phosphorylation of glucose and fructose (25, 162), those activating amino acids (108), glutamotransferase (118) and others.

Special interest has been aroused by the discovery of tissue specificity in the collection of isoenzymes, above all iso forms of lactic dehydrogenase, in respect of the origin of this heterogeneity during embryonic development (64, 110, 55, 170). In mammals almost every tissue contains five isoenzymes of lactic dehydrogenase, the relative proportions of each being different but constant for each tissue in the adult organism. However, these proportions change markedly during embryonic and post-embryonic development and this process provides very interesting material for the study of the genetic factors affecting development. An investigation of this sort has also been started in relation to other enzymes which also typically occur in multiple forms (117).

As regards the changes in nucleic acids associated with morphogenesis, the purely quantitative relationships between DNA and RNA at different stages of development and in different objects have been studied but, as well as this, there has recently been an attempt to determine the qualitative changes in the composition of RNA which would reveal differences in information supplied by the nucleus to the cytoplasm in different stages of development and types of specialization. The work of Nemer (120) has shown that in the unfertilized and newly fertilized eggs of the sea-urchin the RNA belongs mainly to fractions 17 S and 22 S. Before the beginning of gastrulation the nature of the RNA changes and RNA with a sedimentation rate of about 10 S begins to predominate. The authors suggest that this change may be associated with intensive synthesis of "messenger" RNA. As development continued the RNA became more polydisperse.

A study of the fractional composition of RNA at different stages of embryogenesis of the loach (*Misgurnus fossilis*) has recently been made by Spirin and colleagues (1, 1a, 12). They found that in the stages of morula and blastula formation the embryo did not synthesize new ribosomes; the cytoplasm of its cells obtained ready-made ribosomes from the yolk where a stock had been laid down during oogenesis. At the same time they observed intensive synthesis of certain fractions of "messenger" RNA. However, the changes in nucleotide composition of the RNA as a whole during the embryogenesis of the loach were extremely slight (15).

The work of Neïfakh (9, 121) is very interesting. He found that complete inactivation of the nuclear structures of loach embryos does not put an end to their development up to a certain stage. The development comes to a stop at a determinate moment regardless of whether the embryo has received a lethal dose of radiation 1, 2, 3 or 4 hours beforehand. The author thinks that this phenomenon is associated with the periodic functioning of the nucleus which, having transferred the information required for a particular stage of morphogenesis in the form of "messenger" RNA, then undergoes reprogramming so that even when the nucleus is destroyed the process of development continues to the end of its phase on account of the presence of the RNA which was already synthesized.

The connection between the induction of particular proteins and enzymes in animals and the activity of strictly specific forms of RNA has been established in several researches (7, 124, 125).

### b) Molecular mechanisms of differential (selective) activity of the genome during morphogenesis

The cytomorphological indicator of differential or selective activity of the genome which gives rise to specialization is, as we have already pointed out, a peculiar activity of certain zones of the giant chromosomes in the shape of the formation of bulges, the so-called puffs, which mark the place where the synthesis of RNA and proteins is going on most actively. In Chapter XVIII we brought forward evidence from several authors to show that the pattern of the formation of puffs on the chromosomes of the salivary glands of insects underwent regular changes associated with morphogenesis and with the action of several factors affecting morphogenesis.

In 1963-4 the study of the chromosomal puffs began to follow a definite direction, towards the elucidation of the connection between this process and the molecular-genetic aspects of development and differentiation.

The real connection between the puffs and the morphogenesis and differentiation of insects has been established more clearly (5, 48, 29, 102a). For example, it has been found that there is a definite connection between the development and extinction of puffs and particular phases of the development of the pupa. In the pupae of *Drosophila* and of butterflies, puffs are formed in the first and third chromosomes while in the larvae the second chromosomes are the most active (5). The order in which the puffs work is changed by mutation. When inbred lines of *Drosophila* are used a number of puffs are extinguished while, on the other hand, in interspecific hybrids new puffs are formed (5). On the basis of these and other observations the author of the work in question came to the conclusion that the gene material brought together in particular groups of the chain has a tendency to be incorporated in an appropriate place.

A biochemical approach to the nature of the puffs shows that their formation is associated, above all, with the DNA-dependent synthesis of "messenger" RNA (127a). The DNA content of the puffs is constant (99, 135). Inhibition of protein synthesis by puromycin does not prevent the formation of puffs but actinomycins C and D, which inhibit the DNA-dependent synthesis of RNA, completely inhibit the formation of puffs on the chromosomes of the salivary glands (135) and that of loops on the lampbrush chromosomes of the oocytes of *Triton* (94).

Suppression of synthesis of "messenger" RNA in the chromosomes and an associated inhibition of the formation of puffs has also been observed when histones, which are rich in arginine, act on the chromosomes (20, 94). The possibility that histones may act as genetic repressors in differentiated cells has already been discussed in Chapter XVIII, as the material published before the end of 1962 gave sufficient grounds for this. In the succeeding year and a half the part played by histones as genetic repressors has been confirmed by a number of new studies (19, 28, 82, 3, 110, 116a, 27, 32, 33), in the course of which it has been found that differences in the amino acid composition of the histone fractions are correlated with differences in their ability to inhibit the synthesis of RNA in the chromosomes (132). It has also been shown that the nature of the heterogeneity of the histones changes during embryonic development (103a).

Interesting details have also been obtained by a more precise study of the action of actinomycin D, which inhibits the synthesis of RNA, on particular morphogenetic processes. For example, if the gastrula of the frog *Rana pipiens* is placed in a solution of actinomycin for two days and then returned to normal conditions there will be an arrest in the development of certain systems only, most of the tissues and organs, which become specialized earlier or later, will be formed normally (57). This demonstrates that different types of "messenger" RNA are synthesized in a chronological sequence.

Experiments on the use of tissue-specific RNA to induce the beginning of the formation of the appropriate organ in embryos have also been begun (81). Interesting evidence bearing on the general analysis of the problem of differentiation has also been obtained in experiments in which nuclei have been transplanted in the early stages of embryogenesis. A review of this work has recently appeared in a paper by Gurdon (71). From the point of view of the molecular mechanisms of development, however, these researches have not yielded any information which is radically new.

### c) On the nature of the molecular clocks which regulate the chronological sequence and localization of morphogenetic processes

In Chapter XVIII, which dealt with questions of morphogenesis, we have already touched shortly on the problem of morphogenetic molecular "clocks" and have put forward several tentative suggestions in this connection. In the simplest case, the morphogenesis of a phage, we believe that the appropriate chronological arrangement of the syntheses could be connected with the linear uncurling of the DNA, different cistrons being brought into play successively (Fig. 63). In 1962 the only indirect experimental confirmation of the possibility of this sort of regulation was that in the work of Khesin & Shemyakin on phages and that of Gall & Callan (p. 463) who observed the shifting of loops in the chromosomes of the oocyte of *Triton*.

The factual material concerning the nature of the autonomous genetic mechanism regulating morphogenesis was extended somewhat during 1963 and early 1964 and this increased the possibilities for construction by theoretical workers of hypotheses explaining the repetition in each generation of the exact sequence and spatial localization of the morphogenetic processes.

The starting-point for the analysis of these phenomena is the more general problem of the organization of the maintenance and inheritance of the so-called biological rhythms. As we know, rhythms are characteristic of all biological processes and they are based on some sort of molecular mechanisms or genetic "clocks". There are microrhythms or pulsations, there are the more fully studied diurnal sort, seasonal rhythms with alteration of the activities of systems measured over periods of months and so forth. From the point of view of the continuity of the genome (the transfer of genes from generation to generation) the morphogenetic process is also an individual act in the rhythmic process of succession of generations. The nature of the genetic clock and the genetic calendar is unknown. Its biochemical nature may, however, be inferred from the fact that many inhibitors of metabolism influence this rhythm (75). One very interesting observation (97) is that actinomycin D, which is a specific inhibitor of DNA-dependent synthesis of "messenger" RNA, inhibits the manifestation of diurnal rhythm in the alga *Gonyaulax polyedra*. This observation is direct proof of the part played by DNA in the mechanism of the biological clock.

Further experimental confirmation has recently been obtained for the hypothesis that in phages the successive synthesis of certain so-called early proteins (enzymes) followed by late ones which are constituents of the mature viral particles is associated with the successive calling into play in the synthesis of RNA of different linear sections of the enormous molecule of the phage DNA (17, 96).

It would seem that something similar also takes place in bacteria. For example, recent research (50a) has shown that "messenger" RNA from *Bacillus subtilis* isolated at different stages of the life-cycle of the bacterium (sporulation, germination, step-down or transition) reacts with different loci when "hybridized" with DNA. This shows that in bacteria, as in phages, differential transcription of the genome occurs during morphogenesis.

Thus, the possibility of setting up a molecular-genetic model of the "automatic" regulation and appropriate organization of morphogenetic processes in time and space has become a bit greater. In 1963 we put forward the following model for this regulation (8, 115) (Fig. 68) which seems to us to account for many of the facts described above.

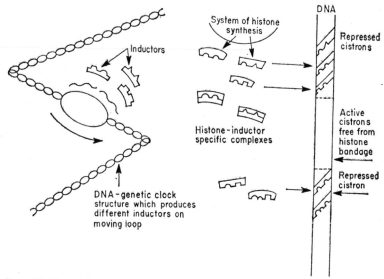

Fig. 68. Possible mechanism for control by DNA of cell differentiation in time.

## 7. Ageing as a prolonged chaotic morphogenesis

In the year and a half which have elapsed since sending for publication the Russian edition of this book, no important changes have taken place in the experimental study of the problem of ageing. Nevertheless we should like to supplement the book in this respect with some purely theoretical considerations.

It is quite obvious that to the question of which of the factors of molecular variability makes the greatest contribution to the general sum of molecular and subcellular damage we must add the no less important one of which *molecular-genetic or functional system is most sensitive to the action of these factors.*

It would seem to us that *the material concerned with morphogenesis which we have already discussed gives us reason to suppose that, statistically, it is a disturbance of the genetic control of the maintenance of some particular type of specialization which is the most frequent result of the action of any factor causing intracellular damage.* In any specialized cell only a very small part of the general stock of genes is "active". The majority of the genes (more than 99%) are specifically repressed. If we assume, for example, that free radicals do not have a selective effect in damaging just those genes which are "active", then the probability of damage to repressed genes, systems of synthesis of repressors and inductors will, statistically, be far greater than the "incidence" in active genes. For this reason the conversion of some repressed gene to the active state as a result of damage to the repressor system or to the acceptors of repressors must, statistically, occur more often than damage to already active genes which would lead, for example, to production by them of anomalous proteins of some sort. As a result of these occurrences there must develop permanent, chaotic, senseless, "loud" microdisturbances of specialized cells which are the equivalent of somatic mutations. Morphogenesis, which had stopped when sexual maturity was reached, is, as it were, prolonged but occurs slowly and chaotically and is directed towards the creation of disharmony.

Such slow and chaotic morphogenesis is not, however, always the basis of ageing. There is an extensive group of species in which the death of the individual is determined, not by the slow process of wearing out, but by some rapid morphogenetic process following on sexual multiplication. Such types of ontogenesis are specially frequent in the plant world (monocarpal plants). In monocarpal plants the sexual process and the ripening of the seeds lead to the process of complete withdrawal of plastic material from the leaves into the reproductive organs. The accession of nutrients from the soil ceases and the vegetative part of the plant quickly succumbs, though artificial removal of the flowers or buds can prolong the function of the vegetative organs for many months or years.

Such a relationship between the sexual process and death also exists in many insects, lower animals, etc. Even in higher animals many authors recognize a correlation between the rate of sexual maturing and the length of life. In any case, there is no doubt of the connection between sexual reproduction and the death of the individual in that the onset of sexual maturity and the sexual process or the process of multiplication (the formation of seeds, eggs, etc.) leads to the "incorporation" of the supplementary morphogenetic process which leads to death.

In both the animal and the plant kingdoms increasing complexity and size of the individual has led to an increase in the complexity of the reproductive process. In this case conservation of form has meant the replacement of the ability to reproduce a large number of offspring in a single act of reproduction by the ability to carry out several reproductive acts one after another. In many animals there has arisen the need to feed, protect and train the offspring until they become sexually mature. A new and prolonged phase of maturity has arisen in ontogenesis; it separates the attainment of sexual maturity from death by a long period of time which is very different in different species. Such species-specific ageing also suggests the coming into action of an active genetic control of ageing and that there is a particular centre in the genetic system of the cells giving rise to this involutionary process. It is still not clear what is the nature of this control. In Chapter XIX we suggested, in the first place, that the "incorporation" of ageing might be connected with a weakening of nuclear control over the synthesis of RNA such that the autonomous system of RNA became unable to carry out such accurate reproduction of the matrices as that carried out by DNA, so that various forms of error and damage accumulated.

Furthermore, we suggested that there might exist within cells a special generator of errors, that is to say, of genes which govern the synthesis of very small amounts of the so-called "unusual" nucleotides which are found as contaminants to the main nucleotides of RNA and DNA. We consider it possible that there may be genetically controlled endogenous production of such compounds acting on RNA and DNA, acting as a peculiar slow mutagenic brake, bringing disharmony into the perfect coordination of all the hundreds of thousands of individual biochemical processes.

Finally, it is very probable that, when the morphogenetic programme has been completed, when, in the course of ageing, a chaotic or random "incorporation" of certain genetic loci occurs, changing the character of the biochemical specialization of the cells in a direction which is away from the normal and is not harmonious, the process of this "incorporation" may also be part of the programme of the genetic system. In other words we may suppose that so-called somatic mutations leading to the accumulation of disharmony with increasing age may be a functional process in its own right, different from mutation of the reproductive cells which has an evolutionary significance. We know, for instance, that there is no

direct proportional relationship between the frequency of genetic mutations manifesting themselves in the offspring and shortening of the
duration of life brought about by irradiation and that an increase in the
daily dose of radiation which will multiply the number of mutations
produced several times over will only shorten life by a small percentage.
This and many other facts show that there is no direct parallelism between
the frequency of mutation and the rate of ageing and that there are many
somatic factors involved in ageing. It is not because they are not subject
to the genetic control of the nucleus that they are not genetic but because
they are not associated with hereditary, mutational alterations of the genes
in the reproductive cells.

## REFERENCES

1. Aĭtkhozhin, M. A., Belitsina, N. V. & Spirin, A. S. (1964). Biokhimiya, 29, 169.
1a. Belitsina, N. V., Aĭtkhozhin, M. A., Gavrilova, L. P. & Spirin, A. S. (1964). Biokhimiya, 29, 363.
2. Vol'kenshteĭn, M. V. (1963). Biofizika, 8, 394.
3. Vorob'ev, V. I. (1964). Tezisy Dokladov pervogo vsesoyuz. biokhim. S''ezda, (ed. E. M. Kreps), Vol. 1, p. 40. Leningrad: Izd. AN SSSR.
4. Georgiev, G. P. & Lerman, M. I. (1963). Voprosy med. Khimii, 9, 218.
5. Kiknadze, I. I. & Filatova, I. T. (1963). Doklady Akad. Nauk S.S.S.R. 152, 450.
5a. Kiselev, L. L. & Frolova, L. Yu. (1964). Biokhimiya, 29, 1177.
6. Luchnik, N. V. (1964). Statisticheskiĭ analiz problemy aminokislotnogo koda. Sverdlovsk: Izd. Ural. Fil. AN SSSR.
7. Mazurov, V. I. & Orekhovich, V. N. (1963). Voprosy med. Khimii, 9, 436.
8. Medvedev, Zh. A. (1963). Zhur. vsesoyuz. Khim. Obshchestva im. D. I. Mendeleeva, 8, 384.
9. Neĭfakh, A. A. (1963). Zhur. vsesoyuz. Khim. Obshchestva im D. I. Mendeleeva, 8, 403.
10. Ratner, V. A. (1964). In collective work Problemy kibernetiki (ed. A. A. Lyapunov), No. 12, p. 181. Moscow: Izd. fiz.-mat. Lit.
11. Spirin, A. S. (1963). Nekotorye problemy makromolekulyarnoĭ struktury ribonukleinovykh kislot. Moscow: Izd. AN SSSR. (The macromolecular structure of RNA. New York: Reinhold (1964)).
12. —(1964). Tezisy Dokladov pervogo vsesoyuz. biokhim. S''ezda (ed. E. M. Kreps), Vol. 1, p. 63. Leningrad: Izd. AN SSSR.
13. —(1964). Tezisy Dokladov pervogo vsesoyuz. biokhim. S''ezda (ed. E. M. Kreps), Vol. 1, p. 41. Leningrad: Izd. AN SSSR.
14. —(1964). Ribonukleinovye kisloty: sostav, stroeniye i biologicheskaya rol' (19-oe Bakhovskoe Chteniye). Moscow: Izd. "Nauka".
15. Timofeeva, M. Ya. & Kafiani, K. A. (1964). Biokhimiya, 29, 110.
16. Frank-Kamenetskiĭ, M. D. (1963). Biokhimiya, 28, 361.
17. Khesin, R. B., Gorlenko, Zh. M., Shemyakin, M. F., Bass, I. A. & Prozorov, A. A. (1963). Biokhimiya, 28, 1070.
18. Shmal'gauzen, I. I. (1964). Regulyatsiya formoobrazovaniya v individual'nom razvitii. Moscow: Izd. "Nauka".

19. ALLFREY, V. G. & MIRSKY, A. E. (1962). *Proc. natl. Acad. Sci. U.S.* **48**, 1590.
20. ALLFREY, V. G., LITTAU, V. C. & MIRSKY, A. E. (1963). *Proc. natl. Acad. Sci. U.S.* **49**, 414.
21. ARLINGHAUS, R. & SCHWEET, R. (1962). *Biochem. biophys. Res. Commun.* **9**, 482.
22. ARLINGHAUS, R., FAVELUKES, G. & SCHWEET, R. (1963). *Biochem. biophys. Res. Commun.* **11**, 92.
22a. ARLINGHAUS, R., SHAEFFER, J., & SCHWEET, R. (1964). *Proc. natl. Acad. Sci. U.S.* **51**, 1291.
23. ATTARDI, G., NAONO, S., GROS, F., BRENNER, S. & JACOB, F. (1962). *Compt. rend. Acad. Sci., Paris,* **225**, 2303.
24. BALLARD, F. J. & OLIVER, I. T. (1963). *Biochim. biophys. Acta,* **71**, 578.
25. BALLARD, F. J. & OLIVER, I. T. (1962). *Nature (Lond.),* **195**, 498.
26. BARONDES, S. H. & NIRENBERG, M. W. (1962). *Science,* **138**, 813.
27. BARR, G. C. & BUTLER, J. A. V. (1963). *Nature (Lond.),* **199**, 1170.
28. BAZILL, G. W. & PHILPOT, J. ST. L. (1963). *Biochim. biophys. Acta.* **76**, 223.
29. BEERMANN, W. & CLEVER, U. (1964). *Scient. Amer.* **210**, No. 4, 50.
30. BENNETT, T. P., GOLDSTEIN, J. & LIPMANN, F. (1963). *Proc. natl. Acad. Sci. U.S.* **49**, 850.
31. BERG, P., LAGERKVIST, U. & DIECKMANN, M. (1962). *J. mol. Biol.* **5**, 159.
32. BILLEN, D. & HNILICA, L. S. (1963). *J. Cell. Biol.* **19**, No. 2, 7A.
33. BLOCH, D. P. (1963). *J. cell. comparat. Physiol. Suppl. 1* to Vol. **62**, p. 87.
34. BOLTON, E. T. & MCCARTHY, B. J. (1964). *J. mol. Biol.* **8**, 201.
35. BOLTON, E. T., BRITTEN, R. J., BYERS, T. J., COWIE, D. B., HOYER, B., MCCARTHY, B. J., MCQUILLEN, K. R. & ROBERTS, R. B. (1963). In *Carnegie Institution of Washington Year Book* **62**, for the year July 1 1962-June 30 1963, p. 303.
35a. BONNER, J., HUANG, R.-C. & GILDEN, R. V. (1963). *Proc. natl. Acad. Sci. U.S.* **50**, 893.
36. BRAWERMAN, G., GOLD, L. & EISENSTADT, J. (1963). *Proc. natl. Acad. Sci. U.S.* **50**, 630.
37. BRETSCHER, M. S. & GRUNBERG-MANAGO, M. (1962). *Nature (Lond.),* **195**, 283.
38. BRITTEN, R. J. (1963). *Science,* **142**, 963.
39. BROWN, D. D. & GURDON, J. B. (1964). *Proc. natl. Acad. Sci. U.S.* **51**, 139.
40. BROWN, G. L., KOSINSKI, Z. & CARR, C. (1962). *Colloq. Intern. Centre nat. Rech. scient. No. 106,* 183.
41. BURDON, R. H. (1963). *Biochem. biophys. Res. Commun.* **13**, 37.
42. BURDON, R. H. & SMELLIE, R. M. S. (1962). *Biochim. biophys. Acta,* **61**, 633.
43. BURCH, H. B., LOWRY, O. H., KUHLMAN, A. M., SKERJANCE, J., DIAMANT, E. J., LOWRY, S. R. & VON DIPPE, P. (1963). *J. biol. Chem.* **238**, 2267.
44. CAMMARANO, P., GIUDICE, G. & NOVELLI, D. (1963). *Biochem. biophys. Res. Commun.* **12**, 498.
45. CHAMBERLAIN, T. J., ROTHSCHILD, G. H. & GERARD, R. W. (1963). *Proc. natl. Acad. Sci. U.S.* **49**, 918.
45a. CHAMBERLIN, M. & BERG, P. (1963). *Cold Spring Harb. Symp. quant. Biol.* **28**, 67.
46. — (1964). *J. mol. Biol.* **8**, 297.
47. CLARK, M. F., MATTHEWS, R. E. F. & RALPH, R. K. (1963). *Biochem. biophys. Res. Commun.* **13**, 505.
48. CLEVER, U. (1963). *Chromosoma,* **14**, 651.

49. CRICK, F. H. C. (1963). *Science,* **139**, 461.

49a. DAVIES, J., GILBERT, W. & GORINI, L. (1964). *Proc. natl. Acad. Sci. U.S.* **51**, 883.

50. DEAN, A. C. R. & HINSHELWOOD, C. (1964). *Nature (Lond.),* **201**, 232.

50a. DOI, R. H. & IGARASHI, R. T. (1964). *Proc. natl. Acad. Sci. U.S.* **52**, 755.

51. ECK, R. V. (1963). *Science,* **140**, 477.

52. EHRENSTEIN, G. VON, & DAIS, D. (1963). *Proc. natl. Acad. Sci. U.S.* **50**, 81.

53. FALASCHI, A., ADLER, J. & KHORANA, H. G. (1963). *J. biol. Chem.* **238**, 3080.

54. FESSENDEN, J. M. & MOLDAVE, K. (1963). *Nature (Lond.),* **199**, 1172.

55. FINE, I. H., KAPLAN, N. O. & KUFTINEC, D. (1963). *Biochemistry,* **2**, 116.

56. FLICKINGER, R. A. (1963). *Science,* **141**, 608.

57. — (1963). *Science,* **141**, 1063.

58. FOX, C. F. & WEISS, S. B. (1964). *J. biol. Chem.* **239**, 175.

59. GARDNER, R. S., WAHBA, A. J., BASILIO, C., MILLER, R. S., LENGYEL, P. & SPEYER, J. F. (1962). *Proc. natl. Acad. Sci. U.S.* **48**, 2087.

60. GEIDUSCHEK, E. P., MOOHR, J. W. & WEISS, S. B. (1962). *Proc. natl. Acad. Sci. U.S.* **48**, 1078.

61. GEORGIEV, G. P., SAMARINA, O. P., LERMAN, M. I. & SMIRNOV, M. N. (1963). *Nature (Lond.),* **200**, 1291.

61a. GIBOR, A. & GRANICK, S. (1964). *Science,* **145**, 890.

62. GIERER, A. (1963). *J. mol. Biol.* **6**, 148.

63. GILBERT, W. (1963). *J. mol. Biol.* **6**, 374.

64. GOLDBERG, E. & CATHER, J. N. (1963). *J. cell. comp. Physiol.* **61**, 31.

65. GOODMAN, H. M. & RICH, A. (1962). *Proc. natl. Acad. Sci. U.S.* **48**, 2101.

66. — (1963). *Nature (Lond.),* **199**, 318.

67. GROBSTEIN, C. (1964). *Science,* **143**, 643.

68. — (1963). In Symp. *Cytodifferentiation and macromolecular synthesis* (ed. M. Locke) p. 1. London: Acad. Press.

69. GROSSMAN, L. (1962). *Proc. natl. Acad. Sci. U.S.* **48**, 1609.

70. — (1963). *Proc. natl. Acad. Sci. U.S.* **50**, 657.

70a. GRUNBERG-MANAGO, M. & MICHELSON, A. M. (1964). *Biochim. biophys. Acta,* **80**, 431.

71. GURDON, J. B. (1963). *Quarterly Rev. Biol.* **38**, 54.

72. HARDESTY, B., HUTTON, J. J., ARLINGHAUS, R. & SCHWEET, R. (1963). *Proc. natl. Acad. Sci. U.S.* **50**, 1078.

73. HARUNA, I., NOZU, K., OHTAKA, Y. & SPIEGELMAN, S. (1963). *Proc. natl. Acad. Sci. U.S.* **50**, 905.

74. HASELKORN, R. & FRIED, V. A. (1964). *Proc. natl. Acad. Sci. U.S.* **51**, 308.

75. HASTINGS, J. W. & BODE, V. C. (1962). *Ann. N.Y. Acad. Sci.* **98**, 876.

76. HAYASHI, M., HAYASHI, M. N. & SPIEGELMAN, S. (1963). *Proc. natl. Acad. Sci. U.S.* **50**, 664.

77. — (1964). *Proc. natl. Acad. Sci. U.S.* **51**, 351.

78. HAYWOOD, A. M., GRAY, E. D. & CHARGAFF, E. (1962). *Biochim. biophys. Acta,* **61**, 155.

78a. HENNEY, H. R. & STORCK, R. (1964). *Proc. natl. Acad. Sci. U.S.* **51**, 1177.

79. HIATT, H. H. (1962). *J. mol. Biol.* **5**, 217.

80. HILL, M., MILLER-FAURÈS, A. & ERRERA, M. (1964). *Biochim. biophys. Acta,* **80**, 39.

81. HILLMAN, N. W. & NIU, M. C. (1963). *Proc. natl. Acad. Sci. U.S.* **50**, 486.

82. HINDLEY, J. (1963). *Biochem. biophys. Res. Commun.* **12**, 175.

83. HINEGARDNER, R. T. & ENGELBERG, J. (1963). *Science,* **142**, 1083.

84. HOAGLAND, M. B. & ASKONAS, B. A. (1963). *Proc. natl. Acad. Sci. U.S.* **49**, 130.

85. HOYER, B. H., McCARTHY, B. J. & BOLTON, E. T. (1963). *Science*, **140**, 1408.
86. HYDEN, H. & EGYHAZI, E. (1962). *Proc. natl. Acad. Sci. U.S.* **48**, 1366.
87. — (1962). *J. Cell Biol.* **15**, 37.
88. JACOB, F. & MONOD, J. (1963). In Symp. *Cytodifferentiation and macromolecular Synthesis* (ed. M. Locke). London: Acad. Press,
89. JONES, O. W. & NIRENBERG, M. W. (1962). *Proc. natl. Acad. Sci. U.S.* **48**, 2115.
90. JONES, O. W., TOWNSEND, E. E., SOBER, H. A., HEPPEL, L. A. (1964). *Biochemistry*, **3**, 238.
91. JUKES, T. H. (1963). *Amer. Scientist*, **51**, 227.
92. — (1963). *Biochem. biophys. Res. Commun.* **10**, 155.
93. ISHIHAMA, A., MIZUNO, N., TAKAI, M., OTAKA, E. & OSAWA, S. (1962). *J. mol. Biol.* **5**, 251.
94. IZAWA, M., ALLFREY, V. G. & MIRSKY, A. E. (1963). *Proc. natl. Acad. Sci. U.S.* **49**, 544.
95. KAJI, A. & KAJI, H. (1963). *Biochem. biophys. Res. Commun.* **13**, 186.
96. KANO-SUEOKA, T. & SPIEGELMAN, S. (1962). *Proc. natl. Acad. Sci. U.S.* **48**, 1942.
97. KARAKASHIAN, M. W. & HASTINGS, J. W. (1962). *Proc. natl. Acad. Sci. U.S.* **48**, 2130.
98. KAJIRO, Y., GROSSMAN, A. & OCHOA, S. (1963). *Proc. natl. Acad. Sci. U.S.* **50**, 54.
99. KEYL, H. G. (1963). *Exptl. Cell. Res.* **30**, 245.
99a. KIHO, Y. & RICH, A. (1964). *Proc. natl. Acad. Sci. U.S.* **51**, 111.
100. KING, J. C. (1963). In *Biological organization at the cellular and supercellular level* (ed. R. J. C. Harris), p. 129. New York: Academic Press.
101. KORNFELD, R. & BROWN, D. H. (1962). *J. biol. Chem.* **238**, 1604.
102. KRETCHMER, N., GREENBERG, R. E. & SERENI, F. (1963). *Ann. Rev. of Medicine*, **14**, 407.
102a. KROEGER, H. (1964). *Chromosoma (Berl.)*, **15**, 36.
102b. LEDER, P. & NIRENBERG, M. (1964). *Proc. natl. Acad. Sci. U.S.* **52**, 420.
103. LEDER, P., CLARK, B. F. C., SLY, W. S., PESTKA, S. & NIRENBERG, M. W. (1963). *Proc. natl. Acad. Sci. U.S.* **50**, 1135.
103a. LINDSAY, D. T. (1964). *Science*, **144**, 420.
104. LIPMANN, F. (1963). *Progress in Nucleic Acid Research*, **1**, 135.
105. LOENING, U. E. (1962). *Nature*, **195**, 467.
106. McCARTHY, B. J. & BOLTON, E. T. (1964). *J. mol. Biol.* **8**, 184.
107. McCULLY, K. S. & CANTONI, G. L. (1962). *J. mol. Biol.* **5**, 497.
108. MAGGIO, R. & CATALANO, C. (1963). *Arch. Biochem. Biophys.* **103**, 164.
109. MARCUS, L., BRETTHAUER, R. K., BOCK, R. M. & HALVORSON, H. O. (1963). *Proc. natl. Acad. Sci. U.S.* **50**, 782.
110. MARKERT, C. L. (1963). In *Cytodifferentiation and macromolecular synthesis* (ed. M. Locke), p. 65. New York: Academic Press.
111. MARKS, P. A., RIFKIND, R. A. & DANON, D. (1963). *Proc. natl. Acad. Sci. U.S.* **50**, 336.
112. MARTIN, R. G. & AMES, B. N. (1962). *Proc. natl. Acad. Sci. U.S.* **48**, 2171.
113. MATTHEWS, R. E. F. (1963). *Nature (Lond.)*, **197**, 796.
114. MAXWELL, E. S. (1962). *Proc. natl. Acad. Sci. U.S.* **48**, 1639.
114a. McCONKEY, E. H. & HOPKINS, J. W. (1964). *Proc. natl. Acad. Sci. U.S.* **51**, 1197.
115. MEDVEDEV, ZH. A. (1964). *Adv. gerontological Research*, **1**, 181.
116. MONROY, A. & TYLER, A. (1963). *Arch. Biochem. Biophys.* **103**, 431.

116a. MOORE, B. C. (1963). *Proc. natl. Acad. Sci. U.S.* **50**, 1018.

117. MOORE, R. O. & VILLEE, C. A. (1963). *Science,* **142**, 389.

118. MOSCONA, A. A. & HUBBY, J. L. (1963). *Develop. Biol.* **7**, 192.

119. NEMER, M. (1962). *Biochem. biophys. Res. Commun.* **8**, 511.

120. — (1963). *Proc. natl. Acad. Sci. U.S.* **50**, 230.

121. NEYFAKH, A. A. (1964). *Nature (Lond.),* **201**, 880.

122. NIHEI, T. & CANTONI, G. L. (1963). *J. biol. Chem.* **238**, 3991.

122a. NIRENBERG, M. & LEDER, P. (1964). *Science,* **145**, 1399.

123. NIRENBERG, M. W., MATTHAEI, J. H., JONES, O. W., MARTIN, R. G. & BARONDES, S. H. (1963). *Fed. Proc.* **22**, 55.

124. NIU, M. C. (1963). *Develop. Biol.* **7**, 379.

125. NIU, M. C., CORDOVA, C. C., NIU, L. C. & RADBILL, C. L. (1962). *Proc. natl. Acad. Sci. U.S.* **48**, 1964.

126. OCHOA, S. (1963). *Federat. Proc.* **22**, 62.

127. OTAKA, E., MITSUI, H. & OSAWA, S. (1962). *Proc. natl. Acad. Sci. U.S.* **48**, 425.

127a. PELLING, C. (1964). *Chromosoma (Berl.),* **15**, 71.

128. PENMAN, S., SCHERRER, K., BECKER, Y. & DARNELL, J. E. (1963). *Proc. natl. Acad. Sci. U.S.* **49**, 654.

129. PERRY, R. P. (1962). *Proc. natl. Acad. Sci. U.S.* **48**, 2179.

130. — (1963). *Exptl. Cell. Res.* **29**, 400.

130a. PHILLIPS, W. D., RICH, A. & SUSSMAN, R. R. (1964). *Biochim. biophys. Acta,* **80**, 508.

130b. POCHON, F., MICHELSON, A. M., GRUNBERG-MANAGO, M., COHN, W. E. & DONDON, L. (1964). *Biochim. biophys. Acta,* **80**, 441.

131. PROTASS, J. J., SPEYER, J. F. & LENGYEL, P. (1964). *Science,* **143**, 1174.

132. RICH, A. (1963). *Scient. Amer.* **209**, No. 6, 44.

133. RICH, A., PENMAN, S., BECKER, Y., DARNELL, J. & HALL, C. (1963). *Science,* **142**, 1658.

134. RISEBROUGH, R. W., TISSIÈRES, A. & WATSON, J. D. (1962). *Proc. natl. Acad. Sci. U.S.* **48**, 430.

135. RITOSSA, F. M. & PULITZER, J. F. (1963). *J. Cell Biol.* **19**, No. 2, 60A.

136. SAGER, R., WEINSTEIN, I. B. & ASHKENAZI, Y. (1963). *Science,* **140**, 304.

137. SAGIK, B. P., GREEN, M. H., HAYASHI, M. & SPIEGELMAN, S. (1962). *Biophysical J.* **2**, 409.

138. SANGER, F., BRETSCHER, M. S. & HOCQUARD, E. J. (1964). *J. mol. Biol.* **8**, 38.

139. SCHAECHTER, M. (1963). *J. mol. Biol.* **7**, 561.

140. SCHARFF, M. D., SHATKIN, A. J. & LEVINTOW, L. (1963). *Proc. natl. Acad. Sci. U.S.* **50**, 686.

141. SCHERRER, K., LATHAM, H., DARNELL, J. E. (1963). *Proc. natl. Acad. Sci. U.S.* **49**, 240.

142. SCHLESSINGER, D. (1963). *J. mol. Biol.* **7**, 569.

142a. SCHWEIGER, H. G. & BERGER, S. (1964). *Biochim. biophys. Acta,* **87**, 533.

143. SEED, R. W. & GOLDBERG, I. H. (1963). *Proc. natl. Acad. Sci. U.S.* **50**, 275.

144. SINGER, M. F., JONES, O. W. & NIRENBERG, M. W. (1963). *Proc. natl. Acad. Sci. U.S.* **49**, 392.

145. SINSHEIMER, R. L. & LAWRENCE, M. (1964). *J. mol. Biol.* **8**, 289.

146. SLAYTER, H. S., WARNER, J. R., RICH, A. & HALL, C. E. (1963). *J. mol. Biol.* **7**, 652.

147. SMITH, M. A. & STAHMANN, M. A. (1963). *Biochem. biophys. Res. Commun.* **13**, 251.

147a. SO, A. G. & DAVIE, E. W. (1964). *Biochemistry,* **3**, 1165.

148. SPIEGELMAN, S. (1963). *Federat. Proc.* **22**, 36.

149. SPYRIDES, G. J. & LIPMANN, F. (1962). *Proc. natl. Acad. Sci. U.S.* **48**, 1977.
150. STAEHELIN, T., BRINTON, C. C., WETTSTEIN, F. O. & NOLL, H. (1963). *Nature (Lond.)*, **199**, 865.
151. STAEHELIN, T., WETTSTEIN, F. O. & NOLL, H. (1963). *Science*, **140**, 180.
152. STAEHELIN, T., WETTSTEIN, F. O., OURA, H. & NOLL, H. (1964). *Nature (Lond.)*, **201**, 264.
153. STEVENS, A. (1964). *J. biol. Chem.* **239**, 204.
154. SUEOKA, N. & YAMANE, T. (1962). *Proc. natl. Acad. Sci. U.S.* **48**, 1454.
155. TAVLITZKI, J. (1962). *Bull. Soc. Chim. biol.* **44**, 697.
155a. SZER, W. & OCHOA, S. (1964). *J. mol. Biol.* **8**, 823.
156. TOMITA, K.-I. & RICH, A. (1964). *Nature (Lond.)*, **201**, 1160.
157. WADDINGTON, C. H. (1962). *New patterns in genetics and development.* New York: Columbia Univ. Press.
158. WAHBA, A. J., BASILIO, C., SPEYER, J. F., LENGYEL, P., MILLER, R. S. & OCHOA, S. (1962). *Proc. natl. Acad. Sci. U.S.* **48**, 1683.
159. WAHBA, A. J., GARDNER, R. S., BASILIO, C., MILLER, R. S., SPEYER, J. F. & LENGYEL, P. (1963). *Proc. natl. Acad. Sci. U.S.* **49**, 116.
160. WAHBA, A. J., MILLER, R. S., BASILIO, C., GARDNER, R. S., LENGYEL, P. & SPEYER, J. F. (1963). *Proc. natl. Acad. Sci. U.S.* **49**, 880.
161. WAINWRIGHT, S. D. & McFARLANE, E. S. (1962). *Biochem. biophys. Res. Commun.* **9**, 529.
162. WALKER, D. G. (1963). *Biochim. biophys. Acta*, **77**, 209.
163. WARNER, J. R., KNOPF, P. M. & RICH, A. (1963). *Proc. natl. Acad. Sci. U.S.* **49**, 122.
164. WARNER, R. C., SAMUELS, H. H., ABBOTT, M. T. & KRAKOW, J. S. (1963). *Proc. natl. Acad. Sci. U.S.* **49**, 533.
165. WEINSTEIN, I. B., SAGER, R., OSSERMAN, E. F. & FRESCO, J. R. (1963). In *Symp. on informational macromolecules* (ed. H. J. Vogel, B. Bryson, J. O. Lampen), p. 246. New York: Academic Press.
166. WEINSTEIN, I. B. & SCHECHTER, A. N. (1962). *Proc. natl. Acad. Sci. U.S.* **48**, 1686.
167. WEINSTEIN, I. B., SCHECHTER, A. N., BURKA, E. R. & MARKS, P. A. (1963). *Science*, **140**, 314.
168. WEISBLUM, B., BENZER, S. & HOLLEY, R. W. (1962). *Proc. natl. Acad. Sci. U.S.* **48**, 1449.
169. WETTSTEIN, F. O., STAEHELIN, T. & NOLL, H. (1963). *Nature (Lond.)*, **197**, 430.
170. WIGGERT, B. O. & VILLEE, C. A. (1964). *J. biol. Chem.* **239**, 444.
171. WILT, F. H. & HULTIN, T. (1962). *Biochem. biophys. Res. Commun.* **9**, 313.
172. WINCKELMANS, D., HILL, M. & ERRERA, M. (1964). *Biochim. biophys. Acta*, **80**, 52.
173. YANKOFSKY, S. A. & SPIEGELMAN, S. (1962). *Proc. natl. Acad. Sci. U.S.* **48**, 1069.
174. — (1963). *Proc. natl. Acad. Sci. U.S.* **49**, 538.
175. YANOFSKY, C. (1963). In *Symp. on informational macromolecules* (ed. H. J. Vogel, B. Bryson & J. O. Lampen), p. 195. New York: Academic Press.
176. YČAS, M. (1962). *Internat. Rev. Cytol.* **13**, 1.
177. YU, C. T. & ZAMECNIK, P. C. (1963). *Biochim. biophys. Acta*, **76**, 209.
178. ZIMMERMAN, E. F. (1963). *Biochem. biophys. Res. Commun.* **11**, 301.

# INDEX OF AUTHORS

The numbers in the three-column list refer to the Chapters (Roman small caps) and the Signals in the Text and Lists of References (Arabic figures). For convenience the two-column list below gives the opening page-numbers of the Chapter References Lists.